Biomedical Engineering for Global Health

Rebecca Richards-Kortum's pioneering work and teaching are captured in this motivating text written for non-science majors and biomedical engineers, inspiring them to engage in solving the global health issues that face us all.

Studying with *Biomedical Engineering for Global Health*, students will:

- begin to understand the medical, regulatory, economic, social, and ethical challenges facing the development of health systems;
- see how these constraints affect the design of new devices and therapies;
- learn through case studies, including cancer screening, imaging technologies, implantable devices, and vaccines; and
- read profiles of undergraduate students who have participated in international technology development internships in Africa.

Rebecca Richards-Kortum is the Stanley C. Moore Professor of Bioengineering at Rice University, where her research group is currently developing miniature microscopes and low-cost imaging systems to enable early detection of pre-cancerous changes in living tissue. In collaboration with colleagues at Rice University and the Texas Medical Center, she has developed a four-year multi-disciplinary education and training program that promotes engineering and engineering technology on a global scale. Prior to working at Rice University, Dr. Richards-Kortum was a Professor of Biomedical Engineering at the University of Texas at Austin, where she was elected to the Academy of Distinguished Teachers and received the Chancellor's Council Outstanding Teaching Award. She has also received many other awards, including being named a Piper Professor for excellence in teaching by the Minnie Stevens Piper Foundation in 2004, and receiving the Sharon Keillor Award for Women in Engineering (2004) and the Chester F. Carlson Award (2007) from the American Society for Engineering Education. In 2008, she was elected to the National Academy of Engineering.

"This beautifully written volume by Rebecca Richards-Kortum will inspire and empower the next generation of engineers to make global health their calling."
Thomas Kalil, UC Berkeley and Clinton Global Initiative

"This book will become the most influential biomedical text of our generation... No other book has made such a transformation of young scientists in such a short time."
Nicholas Peppas, The University of Texas at Austin

"This book is an excellent first step in educating engineers about medical problems in the developing worlds and ways in which bioengineers can make a difference."
Paul Yager, University of Washington, Seattle

Cambridge Texts In Biomedical Engineering

Series Editors

W. Mark Saltzman, *Yale University*

Shu Chien, *University of California, San Diego*

Series Advisors

William Hendee, *Medical College of Wisconsin*

Roger Kamm, *Massachusetts Institute of Technology*

Robert Malkin, *Duke University*

Alison Noble, *Oxford University*

Bernhard Palsson, *University of California, San Diego*

Nicholas Peppas, *University of Texas at Austin*

Michael Sefton, *University of Toronto*

George Truskey, *Duke University*

Cheng Zhu, *Georgia Institute of Technology*

Cambridge Texts in Biomedical Engineering provides a forum for high-quality accessible textbooks targeted at undergraduate and graduate courses in biomedical engineering. It covers a broad range of biomedical engineering topics from introductory texts to advanced topics including, but not limited to, biomechanics, physiology, biomedical instrumentation, imaging, signals and systems, cell engineering, and bioinformatics. The series blends theory and practice, aimed primarily at biomedical engineering students, it is also suitable for broader courses in engineering, the life sciences and medicine.

Biomedical Engineering for Global Health

Rebecca Richards-Kortum

Rice University, Houston, Texas

CAMBRIDGE
UNIVERSITY PRESS

CAMBRIDGE UNIVERSITY PRESS
Cambridge, New York, Melbourne, Madrid, Cape Town, Singapore, São Paulo, Delhi

Cambridge University Press
The Edinburgh Building, Cambridge CB2 8RU, UK

Published in the United States of America by Cambridge University Press, New York

www.cambridge.org
Information on this title: www.cambridge.org/9780521877978

First published 2010

Printed in the United Kingdom at the University Press, Cambridge

A catalog record for this publication is available from the British Library

Library of Congress Cataloguing in Publication data
Richards-Kortum, Rebecca, 1964–
Biomedical engineering for global health / Rebecca Richards-Kortum.
    p. ; cm. – (Cambridge texts in biomedical engineering)
Includes bibliographical references and index.
ISBN 978-0-521-87797-8 (hardback)
1. Biomedical engineering.   2. World health.   I. Title.
II. Series: Cambridge texts in biomedical engineering.
[DNLM: 1. Biomedical Technology.   2. Biomedical Engineering.
3. World Health. W 82 R516b 2009]
R856.R53   2009
610.28 – dc22   2009022583

ISBN 978-0-521-87797-8 hardback

Additional resources for this publication at www.cambridge.org/9780521877978

Contents

Dedication

To my mostly patient children, Alex, Max, Zach and Kate, and to their endlessly patient father, Phil.

Acknowledgments

I have been blessed to work with a wonderful team of creative, dedicated, and inspiring colleagues. Without their tireless efforts, this project would never have reached completion. Their contributions are most gratefully acknowledged. My thanks to Deanna Buckley, Christine Edwards, Dan Erchick, Jessica Gerber, Allison Lipper, Yvette Mirabal, Guadalupe Rodriguez, Richard Schwarz, Lauren Vestewig, and Rachel Wergin. Finally, my thanks to the many BME301, BIOE 301 and BTB students who provided thoughtful feedback and inspiration along the way.

Preface

Biomedical Engineering for Global Health gives students a cohesive overview of how biomedical technologies are developed and translated into clinical practice. The text integrates the major diseases facing developed and developing countries with the recent technological advances and the economic, social, ethical and regulatory constraints which impact the development of new technologies.

Biomedical Engineering for Global Health is accessible to students from all disciplines. The text responds to student interest in the fields of bioengineering and global health. As the world becomes more interconnected, students seek more opportunities to learn about disease and health, and how science and engineering can be used to solve global health challenges. In a global context, the text introduces students to bioengineering, epidemiology, health disparities, and the development of medical drugs and devices. For introductory courses in bioengineering, global health, epidemiology or related fields, this text serves as a comprehensive overview of global health challenges and the methods to improve health and prevent disease. The text answers four primary questions.

1. What are the major health problems worldwide and how do these differ throughout the world?

2. Who pays to solve problems in healthcare and how does this vary throughout the world?

3. How can we use technology to solve world health problems?

4. How do new technologies move from the laboratory to the bedside?

Throughout the text, three major case studies are used to illustrate the development, assessment and global diffusion of new medical technologies, including development of new vaccines to prevent infectious disease, development of imaging technologies to improve early cancer screening and development of implantable devices to treat heart disease. The case studies and other examples help students understand the economic challenges associated with developing health systems. Frequent examples are used to contrast health systems in both developed and developing countries.

The text includes profiles of leaders in translational research to expose students to the variety of career paths taken by individuals with MD, PhD or MD/PhD degrees. Also included are profiles of undergraduate students who have participated in international technology development internships in four countries in Africa. Students can directly relate what they are learning in the text to the experiences of their peers. These profiles will help young bioengineering or global health students understand the important aspects of their discipline in the context in which it is practiced.

Homework problems engage a broad audience in mathematical and graphical analysis of real biomedical data, as well as in writing about the social implications of technology development. In addition, a project assignment spans the text, guiding students through the design of a clinical trial to test a new technology. The project provides an opportunity for students to develop, expand and test their knowledge of subject matter in a global context. The project asks students to select a disease of global health significance that is of interest to them. Students research current medical technologies to diagnose or treat the disease, and the limitations of those technologies in a resource-constrained setting. Design constraints are outlined for a new technology to operate in a resource-limited setting. Finally students propose a new medical technology to diagnose or treat the disease which meets these constraints and design a clinical trial to test the technology.

To further engage students in real world problems, a series of interactive classroom activities have been developed to accompany the lecture materials for the course. These activities contextualize real global health problems so that students can better understand and begin to view problems and solutions simultaneously. Multi-media materials and connections to new accounts of scientific developments increase student engagement.

Instead of simply focusing on the study of science and technology, this text takes an engaging student-centered, contextual approach to the study of bioengineering and biotechnology. Unlike other similar texts, *Biomedical Engineering for Global Health* is designed for students from all disciplines. It places a strong emphasis on the need for new health technologies, the process of technology development and the impact of technology development in a personalized, global perspective. Understanding these processes is vitally important to students throughout their lives as they make decisions about their own medical care and contribute to discussion of public policy issues affecting healthcare throughout the world.

Emerging medical technologies: high stakes science and the need for technology assessment

In the past century, advances in medical technology have yielded enormous improvements in human health. For example, our scientific understanding of the immune response and the resulting development of vaccines has vastly reduced the incidence of many infectious diseases. Smallpox has killed more people throughout history than perhaps any other infectious disease. Yet, in 1980, the World Health Organization announced that smallpox had been eradicated worldwide through a program of vaccination (Figure 1.1). Despite these advances, many medical technologies are available to only a small segment of the world's population that can afford them.

Today, emerging technologies have the potential to transform the future of healthcare, offering the potential to diagnose and prevent disease before it strikes, to treat disease in a targeted manner, and to utilize cells and genes for patient-specific therapies. For example, gene therapy offers the promise to cure fatal genetic diseases such as cystic fibrosis and to reprogram a patient's immune system to more effectively fight HIV/AIDS, the leading cause of death in sub-Saharan Africa. Sequencing the genome of *M. tuberculosis* has pointed to new molecular targets for more effective drugs to treat tuberculosis. Small silicon chips containing every gene in the human genome may soon be used to detect cancer at the earliest and most curable stages and to

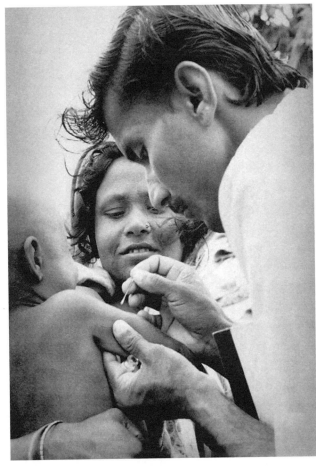

Figure 1.1. *The development of the smallpox vaccine and the subsequent eradication of the disease is an example of a powerful medical technology. CDC/ World Health Organization; Stanley O. Foster.*

individually tailor therapeutic agents for each patient. Tissue engineering holds the promise to create artificial organs, overcoming problems with the limited supply of donor organs. Novel, biologically active materials may be used to coat blood vessels within the heart to prevent heart attacks, one of the leading causes of death in the United States.

Medical technology
The use of novel technologies to develop new drugs, biologics, or medical devices designed to diagnose, treat or prevent disease.

Bioengineering
The application of engineering design to develop new medical technologies.

Biotechnology
The use of living systems to make or improve new products, frequently targeted toward improving human health.

What is needed to bring these new technologies from the research laboratory to your physician's office in a safe and affordable way? As a society, how should we invest our limited financial and human resources to develop new medical technologies? Can new technologies reduce global disparities in health or will they simply widen the gap in health status between developing and developed countries? In this textbook, we examine how bioengineers integrate advances in the physical, information and life sciences to develop new **medical technologies**. To be effective, new healthcare technologies must provide a better means of preventing, detecting or treating disease. At the same time, technologies must also be affordable to those who need them. The goal of bioengineering is to harness science to solve health problems in the face of such constraints. Our study of bioengineering for world health is organized to first understand both global health needs and resource limitations – as we will see, the healthcare problems and economic constraints vary dramatically throughout the world. With this beginning, we profile new technologies emerging from **biotechnology** and **bioengineering**

which can significantly impact world health. Throughout the book, we present and apply tools to systematically evaluate these new medical technologies. The book is organized to address **four central questions**.

Four central questions addressed
(1) What are the major human health problems worldwide and how do these differ throughout the world?
(2) Who pays to solve problems in healthcare and how does this vary throughout the world?
(3) How can we use technology to solve world health problems?
(4) How do new technologies move from the laboratory to the bedside?

---

(1) What are the major health problems worldwide?

Global mortality data show a significant gap in health status between developed and **developing countries**. Leading causes of death in the developed world include cancer, ischemic heart disease, and stroke. In the developing world, infectious diseases like tuberculosis and malaria are far more prevalent owing to widespread poverty, poor infrastructure, and a lack of healthcare resources. A child born today in one of the least developed countries is more than 1000 times more likely to die of measles, an easily preventable and curable disease, than one born in an industrialized country. Worldwide, more than 31 million adults and 2.0 million children are living with HIV/AIDS, most in developing countries. Over the next decade, noncommunicable diseases such as diabetes and heart disease are expected to overtake infectious diseases and malnutrition as leading causes of death in developing countries. The fraction of the global burden of disease linked to lifestyle and behavior choices, currently 20–25%, is expected to increase throughout the world – for example, by 2020 tobacco is expected to kill more people than any single disease, even HIV/AIDS [2]. Understanding how health needs differ throughout the world and how these needs are projected to change in the coming years is the first

UN Millenium Development Goals

Some 80% of the world's population live in **developing countries**. In 2000, 189 countries committed to a broad set of goals to meet the needs of the world's poorest citizens. The goals include the following.

Eradicate extreme poverty and hunger

- Halve the proportion of people whose income is less than one dollar a day by 2015.
- Halve the proportion of people who suffer from hunger by 2015.

Achieve universal primary education

- Eliminate gender disparity in primary and secondary education in all levels of education by 2015.

Reduce child mortality

- Reduce the under-five mortality rate by two thirds by 2015.

Improve maternal health

- Reduce the maternal mortality ratio by 75% by 2015.

Combat HIV/AIDS, malaria and other diseases

- Halt and begin to reverse the spread of HIV/AIDS by 2015.
- Halt and begin to reverse the incidence of malaria and other major diseases by 2015.

Ensure environmental sustainability

- Halve the proportion of people without sustainable access to safe drinking water and sanitation by 2015.

Develop a global partnership for development

The Millenium Country Profiles (http://unstats.un.org/unsd/mi/mi.asp) provide a source of data to compare economic and health status of countries and to monitor progress toward these goals [1].

Table 1.1. *Average health care expenditures per capita of selected WHO nations [3].*

Country	Avg. Health Care Expenditure per capita, 2001 (US$)
Liberia	$1
India	$24
China	$49
Colombia	$105
Mexico	$370
Portugal	$982
Israel	$1641
Switzerland	$3779
United States	$4887

step to enable the development of new technologies to address these needs.

(2) Who pays to solve problems in healthcare?

Despite recent advances, many medical technologies are available only to a small segment of the world's population. As a result, standards of medical care differ radically between the developed and developing world. Average annual healthcare expenditures in high income countries are more than $1800 per person, compared to only $16 per person in the world's least developed countries (Table 1.1). Even in high income countries, the cost of new medical technologies is of great concern. Over the past two decades, healthcare spending has risen dramatically in the United States and throughout the industrialized world, and this rise is expected to continue through the next decade. In the USA, healthcare costs now account for one seventh of the nation's expenditures. The increasing use of new, expensive technologies, an aging population, and increased administrative costs all contribute to the overall rise in healthcare spending. As we will see later, increasing health expenditures does not always improve health status. As health spending grows beyond a minimum value, there is a decreasing rate of return on investment, with fewer years of life gained per dollar invested [4]. In order to achieve the promise of new technologies worldwide,

Major areas of bioengineering

Tissue engineering and regenerative medicine

The use of engineering design principles to regenerate natural tissues and create new tissues using biological cells and three dimensional scaffolds of biomaterials.

Molecular and cellular engineering

Engineering approaches to modify properties of molecules and cells to solve biotechnological and medical problems.

Computational bioengineering

Use of computational tools to analyze large biological data sets such as in genomics or proteomics; computational models to predict structure and behavior of large biological molecules and to guide design of new drugs.

Biomedical imaging

Design of imaging systems (e.g. ultrasound), image analysis tools, and contrast agents to record anatomic structure or physiologic function.

Biomaterials

The engineering design of materials compatible with biological organisms that can be used to make implants, prostheses, and surgical instruments that do not provoke immune rejection.

Drug delivery

Design of materials and systems to achieve controlled release of drugs in physiologic systems.

Biomechanics

The study of mechanical forces in living systems and the use of engineering design to create prosthetic devices and tools for rehabilitation.

Biosensors

Engineering design of systems to identify and quantify biological substances. Advances in microelectronics have aided in developing miniature, implantable biosensors.

Biosystems engineering

Modeling complex, interacting networks of biological systems within cells and organisms to understand physiology and disease and suggest therapeutic strategies to modify behavior.

---

our society must develop and evaluate technologies in a cost-conscious manner.

(3) How can bioengineering solve global health problems?

Technology development begins with scientific knowledge; in health issues this often means an understanding of a disease and its effects on the body. **Bioengineers** build on this scientific knowledge to create new technologies that solve healthcare problems. Magnetic resonance imaging, radiation therapy, and vaccines are all examples of health-related technologies that have become widespread within the past century. The heart–lung bypass machine, pacemakers and other technologies have revolutionized the treatment of heart disease, reducing cardiovascular mortality by half over the past 50 years. In this book, we will consider how new technologies can be used to diagnose, treat, and ultimately prevent the three leading causes of death throughout the world: infectious disease, cancer, and heart disease.

As we will see later, the development of new healthcare technologies must take into account the societal and economic context in which they will be used and their potential status as a priority or a luxury at a given time. For example, development of a totally implantable artificial heart may provide a solution to the problem of end-stage heart failure in developed countries, but owing to differences in infrastructure and resources is unlikely to be a practical solution in many developing countries. To help illustrate these challenges, throughout this book, we will profile the experiences of several undergraduate students who carried out internships in sub-Saharan Africa as a part of a course in Bioengineering and World Health. Their experiences highlight both the opportunities and challenges of developing new technologies to improve world health.

(4) How do new technologies move from the laboratory bench to the patient's bedside?

New medical technologies developed in research laboratories must be subjected to a rigorous testing procedure to ensure that they are both safe and effective. In many cases, this involves carrying out experiments with human subjects. How can we ensure that these experiments are carried out in an ethical way? How can we balance the desire to bring promising new treatments to patients who need them as soon as possible against the risk of harming patients by allowing them access to therapies that haven't been sufficiently tested? As healthcare consumers we are often faced with conflicting media reports of the safety of new medical technologies. In order to make choices about our own healthcare, it is necessary to understand how medical research is funded and how new drugs and medical devices are regulated.

Learn more about breast cancer

Breast Cancer Facts and Figures 2005–2006. (Atlanta, GA: American Cancer Society, Inc.; 2005). [5]

http://www.cancer.org/downloads/STT/CAFF2005BrF.pdf

Answers to these four questions are complex and interrelated. We begin our journey to understand how bioengineering can be used to improve world health by examining a case study of the development of a new technology – the use of high dose chemotherapy and bone marrow transplant to treat advanced breast cancer. This case study illustrates the difficult personal and social issues that can arise as new technologies are developed and tested, and will introduce many of the issues that we will examine in more detail throughout the text. We conclude our case study with a look at how the process of healthcare technology assessment can be systematically used to address these complex and sensitive issues in a scientifically sound manner.

Case study: breast cancer and bone marrow transplant

Breast cancer is both a devastating and a common disease. If you are female and live in the United States, you have a one-in-eight (12.5%) chance of developing breast cancer sometime in your life [5]. When detected early, there are many effective treatments for breast cancer. However, few effective treatments exist for the disease in its later stages. Less than 20% of women are alive five years after the detection of Stage IV metastatic breast cancer, the most advanced form of the disease. In the 1980s a promising new therapy was developed for women with metastatic breast cancer: high dose chemotherapy followed by bone marrow transplant (HDCT+BMT).

Small, early clinical trials of this technique were very promising. The effectiveness of a new cancer treatment is initially measured by the fraction of patients who experience a complete or total response following treatment. In the 1980s, a number of small studies showed a substantial increase in the number of patients with metastatic breast cancer who responded to this new therapy compared to historical experience for patients treated with standard chemotherapy. Although these results were exciting, they were viewed with caution until the patients could be followed for a longer period of time. Many patients who initially respond to therapy may relapse; thus long term

Beyond Traditional Borders: Reports from Student Interns

Kim Bennett accompanied Dr. Ellie Click across Malawi conducting intensive training at hospitals as part of a pilot project for the use of bloodspot PCR for infant HIV diagnosis.

Dave Dallas and **Tessa Elliott** assisted in the design and implementation of World Food Program food distribution system at a pediatric AIDS clinic in Mbabane.

Lindsay Zwiener and **Rachel Solnick** pilot-tested software that generates pictorial medication guides, which were developed as their Bioengineering & World Health course projects. They assessed whether these guides help caregivers in Botswana in the proper dosing and timing of anti-retroviral (ARV) medications, promoting adherence to ARV therapy.

Christina Lagos and **Sophie Kim** rolled out their Bioengineering & World Health course project in the SOS Village in Maseru. The project was an after-school activities club to promote interest in science and health education with a focus on HIV/AIDS. They also implemented a Reach Out and Read program at a pediatric AIDS clinic.

The course in Bioengineering & World Health was developed and offered at The University of Texas at Austin and at Rice University. Through a new initiative called Beyond Traditional Borders, made possible by a grant to Rice University from the Howard Hughes Medical Institute through the Undergraduate Science Education Program, students at Rice University can travel to Africa for a summer and implement the projects they developed as part of this course. The inaugural class of interns kept a blog describing their experiences. Throughout the book, we include excerpts from the blog to provide a student's view of how bioengineering can improve world health.

You can find more student blogs at:
www.owlsbeyondborders.rice.edu

Departure: June 8th, 2007
Christina Lesotho

Coming from a close family, I have been doing a lot of explaining about my goals and purpose for this trip and doing my best to calm the fears of my family. I know that they simply want me to be safe and are concerned about me while I am gone, and I am used to the ways of overprotective Greek relatives. In the end, I think I have convinced them that this will be the experience of a lifetime and that I have been looking forward to something like this since I began college.

I was getting ready to record something in my personal journal last night and found that the last sentence I wrote the last time I made an entry had to do with Africa. From my last weeks in Washington, D.C., working on health policy in Africa, I expressed a desire to go and experience the challenges and situations first hand. "I want to go to Africa . . . why not me?", that is what I had written as I wondered why it always seemed so far-fetched or impossible that I would one day be able to visit. And now it's quickly approaching, and I feel so fortunate and excited for this opportunity.

I am prepared for some of the best and worst emotions I have ever experienced and am ready to fully immerse myself in the work I am about to do in Lesotho. I feel almost guilty for having somehow cheated during this pre-departure period . . . I have been looking at tons of Google images of Maseru, Lesotho, and the surrounding area, and I feel like I have some sort of unfair advantage as I travel. When I was younger and did not use or have access to the Internet as much, traveling to a new place was always so much more of a mystery and I always envisioned my destination so differently than it turned out to be. I know that a bunch of Google images and travel sites will not do Lesotho justice, but I still feel like I have done away with at least a bit of the mystery of travel. Maybe I won't do that next time.

I am looking forward to spending the next few days in Johannesburg with a family-friend who grew up there. I will be there until the 12th when I will be meeting up with Sophie at the airport to head to Maseru.

It will be nice to leave the hot and humid start of summer here in Florida and find the cold beginnings of winter in southern Africa!

survival rates are often used as a better metric to determine the effectiveness of a new cancer therapy. The three year survival rate measures the number of patients still alive three years after beginning cancer therapy. In the early 1990s, a small study indicated that women with high risk breast cancer treated with HDCT+BMT had a 72% three year survival rate, dramatically higher than the historical experience for women treated with standard dose chemotherapy, which was only 38–52% [6].

These studies offered new hope to women who faced high risk or metastatic breast cancer. HDCT+BMT is a grueling treatment that has been described by Dr. Jerome Groopman as "an experience beyond our ordinary imaginings – the ordeal of chemotherapy taken to a near-lethal extreme [7]." In desperation, more than 41 000 American women with advanced breast cancer endured HDCT+BMT in the 1990s, even though there

was little clinical evidence to show that it was superior to standard therapy [8]. The story of what happened as this technology was developed and tested illustrates how political pressures can overwhelm science, leading to substantially increased medical costs and dramatically reduced quality of life for patients.

Breast cancer in the USA

After skin cancer, breast cancer is the most common cancer among women, and accounts for almost one of every three cancers diagnosed in women in the United States [5]. In 2005, more than 40,000 American women are expected to die of breast cancer; only lung cancer causes more cancer deaths in women. An estimated 211,240 new cases of breast cancer occurred in the USA in 2005, and there are over 2.3 million women living in the USA who have been diagnosed with breast cancer.

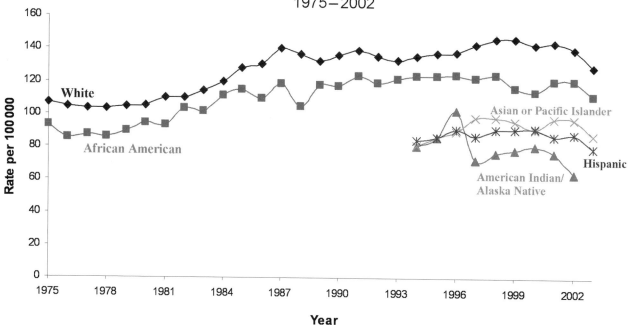

Figure 1.2. *Female breast cancer incidence rates by race and ethnicity in the United States as reported by SEER [9]. The rates are age adjusted to the 2000 USA standard population.*

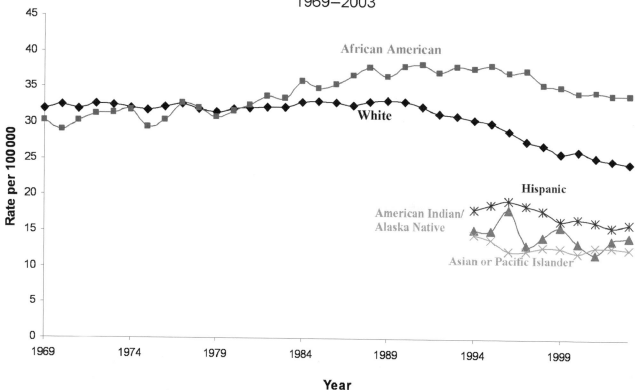

Figure 1.3. *Female breast cancer death rates by race and ethnicity in the United States as reported by SEER [10]. The rates are adjusted to the 2000 USA standard population.*

Female breast cancer incidence rates have risen in the USA from 1973 to 1998, as reported by the NCI Surveillance, Epidemiology and End Results (SEER) Program (Figure 1.2). Incidence rates have increased owing to a combination of changes in reproductive patterns (delayed childbearing, having fewer children) and better early detection with mammography. Female breast cancer death rates in the USA during the same period have decreased (Figure 1.3), primarily owing to better early detection of more treatable cancers and to improvements in breast cancer treatments.

Figure 1.4 shows an illustration of the female breast. After childbirth, milk is produced in glandular tissue in the breast, leading to milk ducts [11]. This glandular tissue is where most breast cancers develop. When cancer cells are confined to these ducts, and have not spread to surrounding fatty tissue, the disease is called Stage 0, and is completely curable with surgical excision. Lesions which have spread to the surrounding fatty tissue but

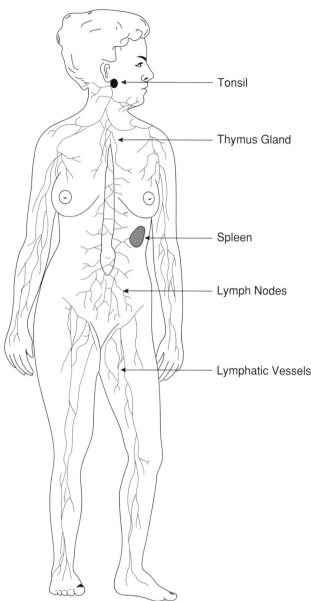

Figure 1.5. *Lymphatic system. Source: SEER Training Modules, Lymphoma. US National Cancer Institute. 2009. http://training.seer.cancer.gov/*

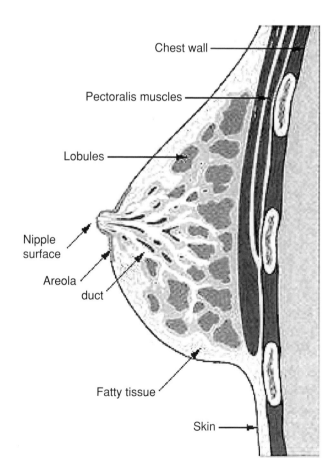

Figure 1.4. *The human female breast. Source: SEER Training Modules, Breast Cancer. US National Cancer Institute. 2009. http://training.seer.cancer.gov/*

are less than 2 cm in diameter are referred to as Stage I lesions, and also have excellent prognosis, with a 100% five year survival rate [12]. A series of lymphatic vessels, leading to lymph nodes under the armpit (axillary lymph nodes), drain breast tissue (Figure 1.5) [11]. Breast cancer cells can migrate from the initial lesion and enter these lymphatic vessels, providing a way for breast cancer cells to spread to other distant organ sites (metastasize). If the cancer has spread to one–three lymph nodes close to the breast but not to distant sites, it is referred to as a Stage II lesion, and the five year survival rate is in

Table 1.2. *Breast cancer staging [12].*

Stage	Definition	5 yr survival
Stage 0	Cancer cells are located within a duct and have not invaded the surrounding fatty breast tissue	100%
Stage I	The tumor is 2 cm or less in diameter and has not spread to lymph nodes or distant sites.	100%
Stage II	The cancer has spread to 1–3 lymph nodes close to the breast but not to distant sites	81–92%
Stage III (High risk)	The cancer has spread to 4–9 lymph nodes close to the breast but not to distant sites	54–67%
Stage IV (Metastatic)	Cancer has spread to distant organs such as bone, liver or lung or to lymph nodes far from the breast.	20%

the range 81–92%. Stage III breast cancers involve more than four nodes, and because the five year survival rates are so low (54–67%) are referred to as "high-risk breast cancers." In metastatic breast cancer (Stage IV), the disease has spread from the lymphatics to other organ sites far from the breast, such as the brain. The five year survival rate for metastatic breast cancer is only 20%. The stages of breast cancer and the prognosis for each stage are summarized in Table 1.2 [12].

Treatments for breast cancer

There are many treatments for breast cancer. Treatment for most early cancers involves some form of surgery to remove the cancer cells. If the lesion is small, only a portion of tissue may be removed (lumpectomy), or the entire breast may be removed (mastectomy). Larger tumors may be treated using chemotherapy. In some cases, chemotherapy may be used to shrink larger tumors so that they can be removed surgically; in others it may be used following surgery to reduce risk of recurrence. In chemotherapy, drugs which are toxic to cancer cells are given intravenously or by mouth. These drugs travel through the bloodstream, reaching cancer cells throughout the body. Chemotherapeutic drugs interfere with ability of cells to divide; many cancer

cells cannot repair damage caused by chemotherapy drugs so they die.

Rapidly dividing normal cells may also be affected by chemotherapy drugs, but they can repair this damage. Because chemotherapy drugs affect rapidly dividing normal cells, they give rise to many undesirable side effects. The cells which line the gastrointestinal tract divide rapidly; thus chemotherapy can lead to nausea, vomiting, mouth sores and loss of appetite. Cells in the hair follicles divide rapidly and chemotherapy can lead to hair loss. Rapidly dividing cells in the bone marrow which produce oxygen carrying red blood cells, infection fighting white blood cells, and platelets important in blood clotting are also affected by chemotherapy drugs. Chemotherapy patients are thus at high risk for infection, bleeding and fatigue. While these side effects are temporary, chemotherapy can also produce permanent side effects such as premature menopause and infertility.

High dose chemotherapy

Because chemotherapy can damage both cancer cells and rapidly dividing, but crucial, normal cells, cancer treatment must strike a balance between completely destroying all cancer cells while causing minimal damage to normal cells. In the 1980s a number of dose comparison studies of chemotherapy to treat metastatic breast cancer showed that a higher dosage of chemotherapy was associated with a higher response rate. Scientists and clinicians hypothesized that metastatic breast cancer could be treated more effectively with higher doses of chemotherapy. Unfortunately, such high doses completely destroy the bone marrow, leaving patients with no way to continue to produce the cells of the blood system and the immune system, which are necessary for life.

Our blood consists of four components: plasma, red blood cells, white blood cells, and platelets. Plasma carries nutrients and hormones throughout the body. Red blood cells deliver oxygen throughout the body, while white blood cells are necessary to fight infections. Platelets are necessary for blood clotting following injury. Throughout our lives, our blood cells are continually renewed within the bone marrow. The source

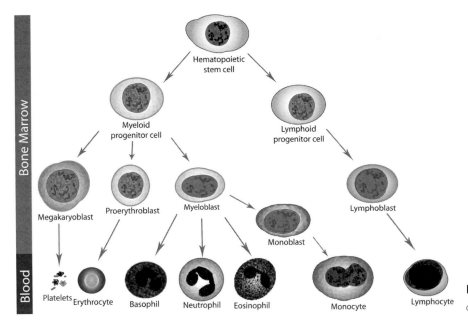

Bone Marrow

Blood

Hematopoietic stem cell

Myeloid progenitor cell

Lymphoid progenitor cell

Megakaryoblast

Proerythroblast

Myeloblast

Monoblast

Lymphoblast

Platelets Erythrocyte

Basophil

Neutrophil

Eosinophil

Monocyte

Lymphocyte

Figure 1.6. *Red and white blood cells are derived from cells of the bone marrow.*

of all these cells is the pluripotent hematopoeitic stem cell which can give rise to all the types of blood cells (Figure 1.6). Laboratory experiments in mice show that a single stem cell can yield the half-trillion blood cells of an entire mouse. Clinicians theorized that if the bone marrow was completely destroyed in high dose chemotherapy, a bone marrow transplant could be done to restore these hematopoeitic stem cells; in fact such bone marrow transplants had proven very successful in the treatment of cancers of the bone marrow.

Learn more about bone marrow transplants

In his article, "Bone marrow transplant: a healing hell," Dr. Jerome Groopman describes the experience of two patients who undergo bone marrow transplants [7].

J. Groopman. Bone marrow transplant: a healing hell. *The New Yorker* (19 October 1998), pp. 34–39.

Bone marrow transplants

Stem cells are found in high concentration in the bone marrow, and can be harvested for transplantation in a painful procedure. More recently, stem cells transplants have been carried out using peripheral blood stem cells

(PBSCs) which are found in the blood. In a transplant, these stem cells are isolated from the blood in a process known as **apheresis**. The patient is given medication to increase the number of stem cells in the bloodstream. Next, blood is removed from the body through a central venous catheter and passes through a machine that removes the stem cells (Figure 1.7). The blood is then returned to the patient and the collected stem cells are stored for future transplantation. The entire process takes 10–12 hours, and yields enough stem cells to fill one syringe.

The initial attempts to transplant bone marrow took place in Cooperstown, NY during the 1950s [7]. The effects of the atom bomb used at the end of World War II sparked a tremendous interest in identifying ways to restore bone marrow. One reason that the bomb's radiation was so deadly was because it destroyed the bone marrow cells of its victims, leading to hemorrhage (uncontrolled bleeding) and the inability to fight off infection. At the time physicians could successfully transfuse oxygen carrying red blood cells from compatible donor to needy recipient.

However, bone marrow cells could not be transfused. Invariably, the recipient's body identified them as foreign invaders and destroyed them [7].

One researcher who was especially interested in the bone marrow transplant problem was Don Thomas.

Apheresis technology

Most stem cells are found in the bone marrow, but some, called peripheral blood stem cells, can be found in the blood. It is typically much more difficult to harvest bone marrow than cells in peripheral blood – harvesting bone marrow requires hospitalization and general anesthesia. Typically, the concentration of stem cells in the peripheral blood is very low, so patients are given growth factors to increase the concentration of peripheral blood stem cells for several days prior to harvesting stem cells.

During apheresis, blood is removed from a large vein in the arm and sent to a machine which contains a centrifuge to separate white blood cells. Anticoagulants must be added to the blood to prevent it from clotting. The centrifuge spins the entering blood, and the resulting centrifugal force separates the various components of blood – plasma, red blood cells and white blood cells – based on differences in their density. The red blood cells are pushed to the outside of the centrifuge, while plasma remains near the center of the rotor. A layer of white blood cells called buffy coat separates the plasma and red blood cells. This layer contains the peripheral blood stem cells and is separated. The remaining blood is returned through a tube to the patient's other arm.

A successful transplant requires collection of a large number of peripheral blood stem cells – approximately five million stem cells per kilogram of body weight are required. Thus, we must quantify the number of peripheral blood stem cells harvested during apheresis to determine whether a sufficient number have been collected. When viewed through a standard microscope the stem cells can't be differentiated from other white blood cells. However, stem cells express a protein called CD34 on their membrane. The fraction of CD34 positive cells can be quantified by labeling the cells with a fluorescent dye linked to a molecule that binds to CD34 and using a special machine called a flow cytometer to count the number of CD34 positive cells. Over 20 liters of blood must be processed (the entire blood volume must be treated four times) to collect sufficient cells for later transplant, and apheresis is typically performed over several days. These cells are then treated with cryopreservatives and frozen to be injected into the patient following the high dose chemotherapy procedure [13, 14].

Thomas treated patients with cancer of the bone marrow (leukemia) with chemotherapy. He believed that providing new, healthy bone marrow cells was essential to curing leukemia. He tested various transplant techniques in dogs initially, and then in patients with late stage leukemia. In early trials, every patient who underwent transplantation died. "Things were pretty grim", Thomas later remarked [7]. After four years of unsuccessful transplantations attempts, he stopped human trials.

Eight years later, Thomas identified protein markers on the surface of white blood cells [7]. These histocompatibility markers are unique to each individual and are found on the surface of nearly every cell in the body, but are particularly numerous on the surface of white blood cells. Histocompatibility markers enable a patient's immune system to differentiate between for-eign invaders and the patient's own cells. The histocompatibility markers explained the failure of previous transplant attempts and held the key to future success. When not properly matched, the patient's immune system would reject transplanted cells. Proper matching of histocompatibility markers between donor and recipient led to successful results in dogs. With this advance, Thomas resumed human trials, which led to successful treatment for leukemia. Thomas (Figure 1.8) received the Nobel Prize in 1990 for his important work in this area.

Today bone marrow transplantation is a successful treatment for leukemia. In the past 40 years the five year survival rate for leukemia has more than tripled, from 14% in 1960–63 to 49% in 1995–2002 [15]. However, it is still a gruelingly difficult treatment. Dr. Jerome Groopman describes the experiences of two patients

Figure 1.7. *An apheresis machine. Copyright Caridian BCT, Inc. 2009. Used with permission.*

who received bone marrow transplants in his article, "Bone marrow transplant: a healing hell [7]." Courtney Stevens was a high school sophomore when she was diagnosed with leukemia. She received a bone marrow transplant, and recounts her experience in Groopman's article. "It was a complete nightmare. For days, I'd be on all fours and just retch and retch. I looked like a lobster, and thought I had bugs crawling on me. I'd hit myself and scream. I was in that sterile bubble, and forgot what skin against skin felt like. That was lost. I just wanted to hold on to my mom or dad, like a two-year-old, and I couldn't. I had terrible diarrhea, a blistering rash all over my body, and jaundice. I was the color of an egg yolk [7]."

A new technology for advanced breast cancer, HDCT+BMT

With the success of bone marrow transplant for leukemia, clinicians hypothesized that extremely high dose chemotherapy could be used to treat metastatic breast cancer if followed by a bone marrow transplant. In this case, the patient's own stem cells could be harvested prior to the chemotherapy and then reinfused following treatment, thus insuring a perfect histocompatibility match. Compared to standard chemotherapy, this procedure was initially very expensive (>$140,000) and initial trials had very high treatment associated mortality (death) rates, in the range 7–22% [16, 17]. Despite the extreme expense and side effects, the combination

Figure 1.8. *Don Thomas and his wife and partner in research, Dottie, with childhood leukemia survivors. Used with permission from © Jim Linna.*

of HDCT+BMT offered some of the only promise for the treatment of metastatic breast cancer. An early study showed that the three year survival rates of women with high risk breast cancer treated with HDCT+BMT were 40% higher than those of women who had not participated in the trial and had received standard chemotherapy [8]. While this study offered hope for the new treatment, it was criticized for several reasons. It was a small study, involving only 85 patients, and did not randomly assign women to receive either the new therapy or the standard therapy. It also only included women whose disease initially responded to standard chemotherapy and who therefore might be expected to do better than those whose disease was not responsive to standard treatment.

> ### Learn more about HDCT+BMT
> Drs. Michelle Mello and Troyen Brennan provide a more complete account of the early controversy surrounding the use of HDCT+BMT to treat breast cancer [8].
>
> M. M. Mello and T. A. Brennan. The controversy over high-dose chemotherapy with autologous bone marrow transplant for breast cancer. *Health Affairs*, **20**:5 (2001), 101–17.

In order to gain more evidence, several larger clinical trials were initiated in which women with advanced breast cancer were to be randomly selected to receive either standard chemotherapy or HDCT+BMT. Clinicians planned to compare the percentage of patients who were still alive (survival rates) three and five years following therapy in both arms of the trial as well as the percentage of patients whose cancer had not recurred (disease-free survival rates).

Such randomized clinical trials are considered to be the most important kind of clinical evidence to indicate whether a new therapy is better, the same, or worse than a standard therapy. Typically, in the absence of such evidence, a therapy is considered to be experimental and most insurance companies in the USA will not pay for it. Because there are so few effective treatments available for advanced breast cancer however,

there was a strong public demand for HDCT+BMT, even in the absence of good clinical evidence to indicate that it worked.

Public reaction to new hope

In 1991, the television show 60 Minutes aired a piece decrying the company Aetna's decision to deny **insurance coverage for HDCT+BMT** to treat breast cancer [8]. At the same time, Nelene Fox, a 38 year old mother of three who was diagnosed with advanced breast cancer, sued her insurance company [8]. The company, HealthNet, refused to pay for HDCT+BMT for Fox, even though it had recently paid for a relative of its CEO to receive the same treatment. Mrs. Fox and her family sued HealthNet for failure to provide coverage. In the meantime the family raised more than $210 000 so she could receive HDCT+BMT. Mrs. Fox died of breast cancer before a verdict was reached; her family argued that the delay in receiving the treatment contributed to her death. The family was awarded $89M, then the largest jury verdict ever against an HMO. The case received widespread publicity, and in 1993 the Massachusetts legislature mandated that insurers provide coverage for HDCT+BMT for advanced breast cancer. In 1994, insurers approved coverage for 77% of breast cancer patient requests for HDCT+BMT as part of clinical trial participation [8]. However, approval was highly arbitrary, even for similar patients covered by the same insurer. Nine of 12 large insurers surveyed indicated that the threat of litigation was a major factor in their decision to provide coverage.

In 1995, the results of a small, short randomized trial of 90 patients in South Africa was reported by the lead physician, Dr. Werner Bezwoda [20]. Dr. Bezwoda's study showed that, on average, women who received HDCT+BMT for metastatic breast cancer survived twice as long without a relapse than women who received standard chemotherapy. By this time, more than 80% of American physicians believed that women with metastatic breast cancer should be treated with HDCT+BMT, and these results seemingly supported that conclusion [8]. During the 1990s, more than 41,000 patients underwent HDCT+BMT for breast cancer despite a paucity of clinical evidence regarding

Breast cancer in developing countries

More than 1.2 million people worldwide were diagnosed with breast cancer in 2005. Women in developed countries have access to imaging technologies such as mammography and ultrasound to aid in early detection and to advances in hormonal treatments and chemotherapy. However, women in developing countries frequently do not have access to these lifesaving technologies.

Maria Saloniki is a 60 year old mother of ten living in the United Republic of Tanzania. This image of her (below) is used courtesy of WHO/ Chris de Bode. When she was 57 she experienced fever, a swollen armpit and pain. Over three years, she visited local healers, various clinic doctors, and even traveled to Nairobi, Kenya to seek treatment. She was prescribed herbal ointments, antibiotics, and told that nothing could be done for her condition. Finally, three years after her initial symptoms she traveled to Dar es Salaam, where a biopsy showed that she had breast cancer and she began chemotherapy.

Her husband has had to borrow a large sum to finance her care, and can't afford both the cost of the treatment and bus fare to come and visit her [33].

Learn more about insurance coverage for HDCT+BMT

The Aetna insurance company will only pay for HDCT+BMT as part of a controlled clinical trial sponsored by the Food and Drug Administration or the National Cancer Institute. This article explains their rationale [19].

Breast cancer: high dose chemotherapy with autologous sterm cell support. *Clinical Policy Bulletins* (Aetna Inc, 7 October 2005).

http://www.aetna.com/cpb/data/CPBA0507.html.

effectiveness. In fact, it was so difficult to recruit patients to randomized phase III clinical trials (because women were afraid they would be randomly selected to receive the standard therapy) that the trials took more than twice as long to complete than planned.

In 1999, at the meeting of the American Society of Clinical Oncology, the results of five randomized clinical trials were reported. Sadly and surprisingly, four of the studies showed no survival benefit with HDCT+BMT; some showed it took a little longer for cancer to return.

Figures 1.9a and 1.9b compare the survival and disease free survival rates over time in women receiving either HDCT+BMT or standard therapy in one of the trials; no meaningful differences were noted in either case. Only one South African study, again from Dr. Bezwoda, showed a survival benefit [22]. In his study, women with high-risk breast cancer had an 83% chance of five year survival if they received HDCT+BMT, compared to only a 65% chance of five year survival with standard chemotherapy. The average disease free survival time was 100 months for women receiving HDCT+BMT, versus only 47.5 months average disease free survival for those receiving standard chemotherapy. The poor results of the four negative HDCT+BMT trials were widely reported in the media.

Public reaction was again strong. Prior to the negative trial results, in 1996–98, Anthem Insurance saw the number of women requesting HDCT+BMT for breast cancer increase [25]. In 1999, prior to the trial results, the company expanded indications for which

(a)

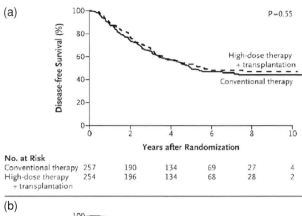

No. at Risk

Conventional therapy	257	190	134	69	27	4
High-dose therapy + transplantation	254	196	134	68	28	2

(b)

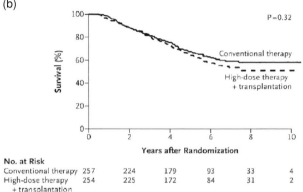

No. at Risk

Conventional therapy	257	224	179	93	33	4
High-dose therapy + transplantation	254	225	172	84	31	2

Figure 1.9. *Disease free survival (a) and survival (b) rates over time in women receiving either HDCT+BMT or standard therapy from a randomized clinical trial [21]. Copyright © 2003 Massachusetts Medical Society. All rights reserved.*

Learn more about public reaction

Denise Grady covered the announcement of these results for the New York Times [23].

D. Grady. Doubts raised on a breast cancer procedure. *New York Times* (16 April 1999).

Joanne Silberner covered the announcement of these results for National Public Radio [24].

J. Silberner. Breast cancer. Radio program, *National Public Radio* (16 April 1999).

http://www.npr.org/templates/story/story.php?storyId=1049404.

---

they would approve HDCT+BMT. After the trial results were reported in 1999, they received only four requests for such coverage, despite the expanded coverage. Most insurance companies now cover HDCT+BMT for breast cancer only as part of an FDA or NCI sponsored clinical trial.

Scientific misconduct

Scientists could not understand why one trial showed improved survival with HDCT+BMT, while four other trials showed no benefit. A team of scientists was sent to audit the results of the South African trial (Figure 1.10). Unfortunately, the audit team could not find records for many of the patients supposedly enrolled in the study. They found that the study showed little evidence of randomization, and that many patients whose records could be found did not meet the eligibility criteria for the trial [26]. They also found that the trial had not been properly approved by the Institutional Review Board at Dr. Bezwoda's university, which is required to approve all research involving human subjects in advance. The university conducted a formal ethics inquiry, and Dr. Bezwoda admitted to a "serious breach of scientific honesty and integrity [27]." The university fired Dr. Bezwoda, and many of his publications were formally retracted from the journal in which they had been published.

Where are we now?

Scientists continued to follow the patients enrolled in the five randomized clinical trials originally reported in 1999 (Table 1.3). Even with longer follow up, it appears that there is no survival benefit to HDCT+BMT at either three years or five years following treatment as compared to standard chemotherapy. There is a small but significant increase in disease free survival at three years with HDCT+BMT, but this advantage disappears at five years. Serious side effects are more common with HDCT+BMT compared to standard therapy, but most are reversible. Patients report that quality of life is lower at six months following treatment with HDCT+BMT, but similar to that of standard chemotherapy one year following treatment. The costs of HDCT+BMT have been reduced to about $60,000, which is still nearly two times that of standard chemotherapy.

Most physicians and insurance companies now agree that HDCT+BMT should not be used to treat high risk breast cancer outside of a randomized clinical trial. Research in this area continues, to identify if longer follow up (7–10 years) will show advantages of high dose therapy, or to determine if there are sub-groups of

Table 1.3. *Results of five randomized clinical trials of HDCT+BMT for breast cancer [21, 28–30].*

Study	# Randomized patients	% survival	Disease free survival
Stadtmauer Metastatic	184	32% 3 year BMT 38% 3 year control	9.6 months BMT 9.0 months control
Lotz Metastatic	61	29.8% 5 year BMT 18.5% 5 year control	9% disease free at 5 yrs BMT 9% disease free at 5 yrs control
Peters High Risk	783	79% 3 year BMT 79% 3 year control	71% disease free at 3 yrs BMT 64% disease free at 3 yrs control
Rodenhuis High Risk	885	75% 5 year BMT 73% 5 year control	65% disease free at 5 yrs BMT 59% disease free at 5 yrs control $p = 0.09^*$
Tallman High Risk	511	58% 6 year BMT 62% 6 year control	49% disease free at 6 yrs BMT 47% disease free at 6 yrs control

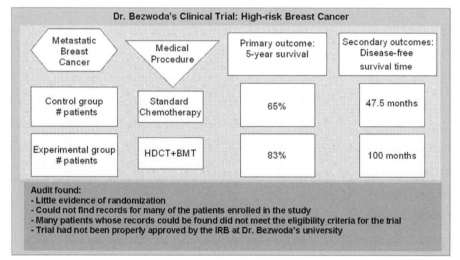

Figure 1.10. *The results of Dr. Bezwoda's controversial trial [26].*

women who benefit from high dose therapy (for example those whose tumors are negative for certain genetic markers or who have 10 or more axillary lymph nodes which show cancer cells). New technologies to completely rid the transplanted stem cells of any rogue cancer cells may also reduce recurrence rates in women treated with HDCT+BMT. However, all of these theories must be subject to rigorous testing if they ever are to become methods of standard treatment.

Lessons learned

The example of HDCT+BMT to treat breast cancer illustrates the dangers of allowing political pressures to overwhelm scientific evidence. What is the proper forum to resolve such controversies? Should it be the media, the courtroom or the laboratory? In an age where high-technology treatments are one of the most powerful drivers of healthcare costs, these are crucial questions.

Healthcare technology assessment

Professors Frazier and Mosteller, experts in health policy and management, have stated, "If we are to have good medical care, we need to know what works, and this cannot be known without systematic technology assessment. The intuitions of physicians and the guesses of biologists are not adequate guides to the best treatments [31]." How then do we assess new technologies objectively, avoiding political pressures that can lead us to waste precious healthcare resources and subject thousands of patients to punishing, but ineffective, treatments?

Healthcare technology assessment

The systematic process of evaluating the safety, short term and long term efficacy, acceptability and cost effectiveness of a new medical technology.

Answering these questions is increasingly important in a world where early studies of new medical advances can receive substantial publicity in the popular press before randomized clinical trials are completed. A recent study published in the *Journal of the American Medical Association* compared conclusions presented in highly cited articles in major general clinical journals to those of subsequent studies with larger sample size or better controlled design. Results showed that nearly 1/3 of highly cited studies were later contradicted and that this was most likely for nonrandomized studies [32]. As we examine these important issues in this book, we will build a toolkit to help us answer politically sensitive questions about how to use limited resources in a deliberate and unbiased manner. Technology assessment will be an important part of our toolkit, and it is the subject of Chapter 2.

Bioengineering and Global Health Project
Project overview

Design a new technology to solve a health problem, present a mock prototype of the new technology to a design review committee, and design a clinical trial to test the new technology.

Throughout this text, you will use the engineering design method to design a new solution to an important health problem. You will identify an important health problem, and carry out research to understand the scope of the problem and limitations of current health technologies. You will follow the engineering method to design a new solution which meets the constraints you identify. You will create a physical prototype of your design and will present it to the class as part of a design review exercise.

Homework

1. Advanced breast cancer has a high mortality. Initial clinical trials indicated that high dose chemotherapy followed by a bone marrow transplant could reduce the mortality rate by as much as 40%.

   a. Why did physicians and scientists believe that higher doses of chemotherapy would be more effective than standard therapy for advanced breast cancer?

   b. Why is it necessary to give patients a bone marrow transplant following high dose chemotherapy? What will happen if they do not receive a bone marrow transplant?

   c. In the context of this example, discuss how political pressures overwhelmed scientific evidence. How could this be avoided in the future?

   d. Find a news report describing a new health technology published in the last year. In your opinion, does this news report provide balanced discussion of the potential promise and the potential limitations of this technology?

2. The Pew Global Attitudes Project is a worldwide survey of public opinion. In 2002, more than 38,000 people in 44 countries were asked to assess the quality of their own lives, their level of optimism about their lives in the next five years, and to rank problems faced by themselves and their countries. In this exercise, you are asked to review the results of this survey and to prepare several graphs summarizing the results.

   Pew World Attitudes Website: http://people-press. org/reports/display.php3?ReportID=165

   Pew World Attitudes Report: http://people-press.org/reports/pdf/165.pdf

   You will examine results in countries profiled in Unit 2: the United States, Canada, China, India and Angola. For parts a–e, please construct graphs, for part f provide a discussion which supports your findings.

   a. What fraction of people surveyed in each country expressed satisfaction with their own lives?

   b. What fraction of people surveyed in each country report that they are unable to afford food?

c. What fraction of people in each country cite the following as a very big problem in their country?
Poor drinking water
Crime
AIDS and disease

d. What fraction of people in each country believe that the following is the greatest danger facing the world today?
Nuclear weapons
AIDS and other infectious diseases

e. What fraction of people surveyed in each country are optimistic that their lives will improve in the next five years?

f. Compare general agreement on questions 4 and 5 throughout countries in Africa and Europe.

References

[1] *Health in the Millenium Development Goals: Millenium Development Goals, Targets and Indicators Related to Health.* World Health Organization; 2004.

[2] *Glossary of Globalization, Trade and Health Terms*: Health Transition. World Health Organization; 2007.

[3] Beaglehole R, Irwin A, Prentice T. *The World Health Report 2004: Changing History.* Geneva, Switzerland: The World Health Organization; 2004.

[4] Coulter SL, Cecil B. *Assessing the Value of Health Care in Tennessee.* Chattanooga: Blue Cross Blue Shield of Tennessee; 2003.

[5] *Breast Cancer Facts and Figures 2005–2006.* Atlanta, GA: American Cancer Society, Inc.; 2005. http://www.cancer.org/downloads/STT/CAFF2005BrF.pdf

[6] Peters WP, Ross M, Vredenburgh JJ, Meisenberg B, Marks L, Winer E, *et al.* High-dose chemotherapy and autologous bone marrow support as consolidation after standard-dose adjuvant therapy for high-risk primary breast cancer. *Journal Of Clinical Oncology: Official Journal Of The American Society Of Clinical Oncology.* 1993 June; **11**(6): 1132–43.

[7] Groopman J. Bone marrow transplant: a healing hell. *The New Yorker.* 1998 October 19: 34–9.

[8] Mello MM, Brennan TA. The controversy over high-dose chemotherapy with autologous bone marrow transplant for breast cancer. *Health Affairs (Project Hope).* 2001 Sep–Oct; **20**(5): 101–17.

[9] *Surveillance, Epidemiology, and End Results (SEER) Program* (www.seer.cancer.gov) SEER*Stat Database: Incidence – SEER 13 Regs Public-Use, Nov 2005 Sub (1992–2003) and SEER 9 Regs Public-Use, Nov 2005 Sub (1973–2003), National Cancer Institute, DCCPS, Surveillance Research Program, Cancer Statistics Branch, released April 2006, based on the November 2005 submission.

[10] *Surveillance, Epidemiology, and End Results (SEER) Program* (www.seer.cancer.gov) SEER*Stat Database: Mortality – All COD, Public-Use With State, Total U.S. for Expanded Races/Hispanics (1990–2003) and All COD, Public-Use With State, Total U.S. (1969–2003), National Cancer Institute, DCCPS, Surveillance Research Program, Cancer Statistics Branch, released April 2006. Underlying mortality data provided by NCHS (www.cdc.gov/nchs).

[11] Silverthorn DU. *Human Physiology: An Integrated Approach.* 2nd edn. Upper Saddle River, NJ: Prentice Hall; 2001.

[12] American Cancer Society. *Detailed Guide: Breast Cancer: How is Breast Cancer Staged?* Sep 18 2006 [cited 2007]. Available from: http://www.cancer.org/docroot/CRI/content/CRI_2_4_3X_How_is_breast_cancer_staged_5.asp

[13] Walker F, Roethke SK, Martin G. An overview of the rationale, process, and nursing implications of peripheral blood stem cell transplantation. *Cancer Nursing.* 1994 Oct 10; **17**(2): 141–8.

[14] Hansson M, Svensson A, Engervall P. Autologous peripheral blood stem cells: collection and processing. *Medical Oncology.* 1996 Jun; **13**(2): 71–9.

[15] *Leukemia, Lymphoma, Myeloma, Facts and Statistics 2006–2007.* White Plains, NY: Leukemia and Lymphoma Society; 2007.

[16] White K. Notebook: bone marrow transplant for breast cancer is questioned on basis of incomplete data. *Journal of Women's Health and Gender-Based Medicine.* 1999; **8**(5): 577–82.

[17] Brockstein BE, Williams SF. High-dose chemotherapy with autologous stem cell rescue for breast cancer: yesterday, today, and tomorrow. *Stem Cells.* 1996; **14**: 79–89.

[18] Abegunde D, Beaglehole R, Durivage S, Epping-Jordan J, Mathers C, Shengelia B, *et al. Preventing Chronic Diseases: A Vital Investment.* Geneva, Switzerland: World Health Organization; 2005.

[19] Aetna. Breast cancer: high-dose chemotherapy with autologous stem cell support. *Clinical Policy Bulletins* 2005 October 5 [available from: http://www.aetna.com/cpb/data/CPBA0507.html].

[20] Bezwoda WR, Seymour L, Dansey RD. High-dose chemotherapy with hematopoietic rescue as primary

treatment for metastatic breast cancer: a randomized trial. *Journal Of Clinical Oncology: Official Journal Of The American Society Of Clinical Oncology*. 1995 Oct; **13**(10): 2483–9.

[21] Tallman MS, Gray R, Robert NJ, LeMaistre CF, Osborne CK, Vaughan WP, *et al.* Conventional adjuvant chemotherapy with or without high-dose chemotherapy and autologous stem-cell transplantation in high-risk breast cancer. *The New England Journal Of Medicine*. 2003 Jul 3; **349**(1): 17–26.

[22] Bezwoda WR. Randomised, controlled trial of high dose chemotherapy versus standard dose chemotherapy for high risk, surgically treated, primary breast cancer. *Proceedings of the American Society of Clinical Oncology*. 1999; **18**: 2a.

[23] Grady D. Doubts raised on a breast cancer procedure. *New York Times*. 1999 April 16; Sect. A1.

[24] Silberner J. Morning Show: breast cancer. National Public Radio. 1999 April 16. http://www.npr.org/templates/story/story.php?storyId=1049404.

[25] Anthem. Insurance payments for bone marrow transplantation in metastatic breast cancer. *The New England Journal of Medicine*. 2000 April 13; **342**(15): 1138–9.

[26] New audit uncovers scientific misconduct in 1995 South African study on metastatic breast cancer. *American Society of Clinical Oncology*. 2001 April 26.

[27] Horton R. After Bezwoda. *Lancet*. 2000 Mar 18; **355**(9208): 942–3.

[28] Rodenhuis S, Bontenbal M, Beex LV, Wagstaff J, Richel DJ, Nooij MA, *et al.* High-dose chemotherapy with hematopoietic stem-cell rescue for high-risk breast cancer. *The New England Journal Of Medicine*. 2003 Jul 3; **349**(1): 7–16.

[29] Stadtmauer EA, O'Neill A, Goldstein LJ, Crilley PA, Mangan KF, Ingle JN, *et al.* Conventional-dose chemotherapy compared with high-dose chemotherapy plus autologous hematopoietic stem-cell transplantation for metastatic breast cancer. Philadelphia Bone Marrow Transplant Group. *The New England Journal of Medicine*. 2000 Apr 13; **342**(15): 1069–76.

[30] Antman KH. Randomized trials of high dose chemotherapy for breast cancer. *Biochimica Et Biophysica Acta*. 2001 Mar 21; **1471**(3): M89–98.

[31] Frazier HS, Mosteller F. *Medicine Worth Paying For: Assessing Medical Innovations*. Cambridge, Mass.: Harvard University Press; 1995.

[32] Ioannidis JP. Contradicted and initially stronger effects in highly cited clinical research. *JAMA: The Journal Of The American Medical Association*. 2005 Jul 13; **294**(2): 218–28.

[33] *Face to Face with Chronic Desease: Maria's Story: Fighting Cancer*. World Health Organization, 2005.

Bioengineering and technology assessment

In Chapter 1, we examined the development and introduction of a new technology that initially appeared as if it could provide new hope to women with advanced breast cancer. Small clinical trials showed that women with high risk or metastatic breast cancer treated with high dose chemotherapy and bone marrow transplant (HDCT+BMT) had substantially better response rates and survival compared to historical experience with standard chemotherapy. These early promising results were widely publicized, and even though the therapy had serious side effects, it was used to treat thousands of women. Usually, before a new technology is adopted, randomized clinical trials are conducted to compare the performance of the new technology to that of existing technologies. In such randomized clinical trials, patients are randomly selected to receive either the current standard therapy or the new therapy; outcomes such as response rate, survival and side effects are then compared for the two groups of patients. However, because patient demand was so high for HDCT+BMT, randomized clinical trials took much longer to complete than planned. Ultimately, randomized clinical trials showed that HDCT+BMT did not improve survival for most patients. The promising results of early trials were misleading due to a combination of factors, including their small size, selection bias and scientific misconduct.

The case study of HDCT+BMT for advanced breast cancer underscores the need for a systematic method to guide the development and introduction of new technologies. In Chapter 1, we saw how the interplay of desperate patients seeking the best treatment, early media publicity and a scientist who falsified data all combined to slow the progress of medical science. In the end, many patients unnecessarily underwent an expensive and highly toxic therapy. How can we prevent this from happening with future technologies? In this chapter, we will consider the methodology of technology assessment, which provides a systematic set of tools to determine the performance of a new technology and to assess the impact of using the technology both for individual patients and for society as a whole. When used properly, technology assessment can help ensure that new medical technologies are introduced on the basis of sound scientific evidence and not simply on the opinions of physicians and scientists, or the hopes of patients.

As a prelude to technology assessment, we consider the steps involved in bringing a new technology from the laboratory bench to the patient's bedside. Figure 2.1 shows a roadmap of this process. Bioengineers build on the scientific understanding of a disease to design new healthcare technologies. New technologies must be rigorously tested to determine whether they are safe and effective. This testing process can include preclinical

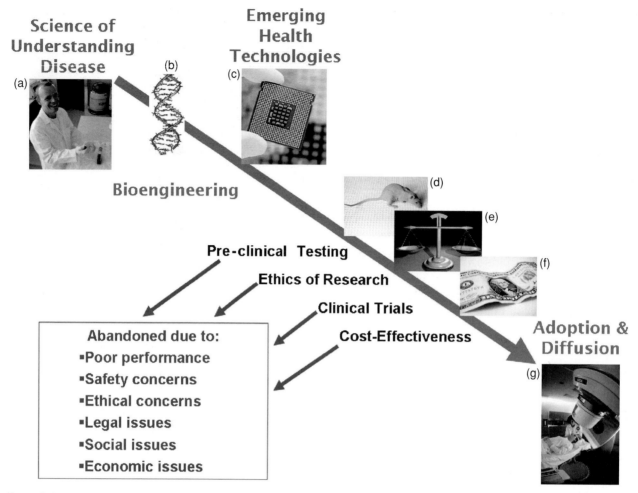

Figure 2.1. *A roadmap of the healthcare technology development process. Technology assessment spans the entire range of development activities. Parts (b), (e) and (f), source: Wikimedia. Part (c), source: Jupiter Images. Part (d), source: NCI/Lindia Bartlett. Part (g), source: NCI/Michael Anderson.*

testing in cell or animal models, as well as testing in human subjects. These tests must be carried out in an ethical manner. In addition, an important consideration in the adoption of new technologies is whether they are cost effective. The process of health technology assessment spans all the steps in the healthcare technology development process, from lab to patient.

Learn more about the Littenberg method

The Littenberg method of technology assessment is defined in this article and used to analyze screening tests for hypercholesterolemia [1].

Littenberg, B. Technology assessment in medicine. *Academic Medicine*, **67**(7), 424–428.

The Littenberg method of technology assessment

Benjamin **Littenberg** proposed a model of technology assessment that is particularly useful for new technologies [1]. The Littenberg method asks five questions regarding a new medical technology.

- *Biologic plausibility*: does our current understanding of the biology of the disease in question support the use of the technology?
- *Technical feasibility*: can we safely and reliably deliver the new technology to the target patients?
- *Clinical Trials*: do the results of randomized clinical trials comparing the new technology to current standards of care show a benefit?

- *Patient outcomes*: are patients better off for having used the new technology?
- *Societal outcomes*: what are the costs and ethical implications of the technology?

It is useful to consider our case study of HDCT+BMT in the context of the Littenberg model to see whether the technology was assessed appropriately at each level and whether that assessment supports the use of the technology.

Biologic plausibility: many scientific studies supported the promise of HDCT+BMT. In particular, as the dose of chemotherapeutic agent was increased to treat women with breast cancer, response rates increased. Based on these data, physicians believed patients with advanced breast cancer would benefit from doses of chemotherapy so high that it would destroy bone marrow.

Technical feasibility: mortality rates were initially quite high for breast cancer patients treated with HDCT+BMT, despite the advances in leukemia treatments showing that bone marrow transplantation could be safely performed. However, as more women were treated and regimens were refined, mortality rates dropped substantially, improving technical feasibility. Thus initially HDCT+BMT was supported by both biologic plausibility and technological feasibility, the first two criteria of Littenberg's method.

Clinical trials: there were many small, clinical trials carried out to assess the effectiveness of HDCT+BMT; however, these trials were not randomized clinical trials. As we will see later, a randomized clinical trial is the strongest source of scientific evidence to assess whether a new technology is effective compared to current standards of care (Figure 2.2). The tragedy of HDCT+BMT was the delay of completing randomized clinical trials, due to political and media pressures and scientific misconduct. By the time clinical trial results seriously questioning the benefit of HDCT+BMT compared to standard chemotherapy were available, many women had already undergone the treatment.

Patient outcomes: were women better off in the long term for having been treated with HDCT+BMT? The early clinical trials assessed only response rates to ther-

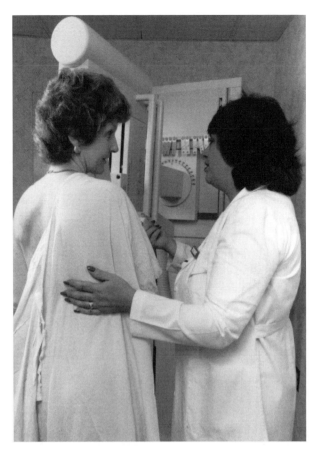

Figure 2.2. *Multiple clinical trials have proven mammography to be effective for screening for breast cancer in women. CDC.*

apy; later trials examined survival. In assessing patient outcomes it is important to consider both short term outcomes (e.g. response rates) and long term outcomes (e.g. survival) as well as quality of life issues. Clearly, patients treated with HDCT+BMT experienced a quality of life that was initially lower because of the side effects of the treatment. In the end, randomized trials showed that survival rates were not substantially higher than those for standard treatment.

Societal outcomes: was society better off for having used HDCT+BMT? The new technology was substantially more expensive than standard chemotherapy while adding no additional survival benefit. If HDCT+BMT had showed clinical benefit compared to standard therapy, then society would have to consider the difficult question of whether the increased benefit is worth the additional cost.

Thus, HDCT+BMT was not supported by the final three criteria of the Littenberg method, a conclusion

As we will see in Chapter 4, patients who are HIV positive are treated with drugs called anti-retrovirals or ARVs. Often, patients must follow complex regimes of drugs. The situation is further complicated for HIV positive children, because drug companies currently do not make ARVs in pediatric doses. So often parents must split pills in half to ensure their children receive the correct dosage. Patients must adhere closely to the schedule of taking their ARVs or HIV can develop resistance to the drugs. This is a major challenge in treating HIV/AIDS today.

I was thinking of different ways to help AIDS patients and I thought that a seven-day AM/PM pillbox would be useful. (I have used them, and I remember my grandmother used them for her multitude of pills.) Then at the beginning or end of every week, the patient or the caretaker can put all the pills (and half pills) in the proper compartments. Although this unfortunately

excludes all the people who need to take syrups, I was talking with one of the patients who came in for a guide yesterday about adherence and she said it would help. She has a son in his teenage years and, while she gives him the pills most of the days, sometimes he goes out to play football (soccer) with his friends and forgets to take the pills with him. When I asked about how he would feel carrying around a pill box, and if he was comfortable taking it around his friends, she said that he didn't mind and that some of his friends were on ARVs as well. I am getting a mixed view on exactly how prevalent or entrenched the stigma for AIDS really is. Driving down from Choebe this weekend, even in the smaller towns there were billboards about AIDS encouraging testing, a large ad for condoms, and a free condom box at the passport immigration. I also saw the first abstinence billboard, which seemed almost to contradict the "get tested" billboard on the opposite side of it.

that puts the effectiveness of the technology in serious doubt. Clearly, HDCT+BMT technology was not adequately assessed before entering widespread use, a failure that led to many women receiving a painful treatment that offered fewer benefits than initially believed.

Important vocabulary of technology assessment

In the rest of this chapter, we will outline in more detail the methods of technology assessment. They will guide our thinking as we examine new technologies throughout the rest of the text. We begin with definitions of some important terms. Technology does not necessarily involve sophisticated or expensive James Bond like gadgetry or devices. Healthcare technology can be any intervention to promote health, including specific tests

or treatments as well as systems that aid in the delivery of healthcare. The technology can address any component along the healthcare continuum: **prevention**, **screening**, **diagnosis**, **treatment**, or **rehabilitation**. In this text you will see examples of technologies ranging from simple childhood vaccinations to prevent disease to complex total artificial hearts to treat end stage heart disease. While the complexity of the technology can differ dramatically, the process of healthcare technology assessment is the same.

The ultimate goal of health technology assessment is to inform decision making, whether it is done from the perspective of an individual patient or from the larger perspective of society. The underlying questions which health technology assessment needs to address include the following [2].

Prevention: health interventions designed to prevent a patient from developing disease.

Screening: a test given to members of a defined population, not necessarily at risk for a disease, to identify those individuals who are most likely to be helped by further tests to diagnose the disease.

Diagnosis: the identification of disease through signs, symptoms, imaging, bloodwork, cultures, cytologic sampling, or biopsy.

Treatment: a health intervention to cure disease or to reduce symptoms of disease.

Rehabilitation: the process of restoring skills lost to illness or injury.

- What is the clinical impact of the intervention?
- What is the cost of the intervention?
- What is the clinical impact of the intervention weighed against its cost?

In order to answer these questions, it is necessary to evaluate the safety, effectiveness, cost effectiveness, and the social, ethical, and legal impacts of a technology [3]. In this evaluation, we must consider both the direct and indirect consequences of using a health technology.

Direct consequences of a health technology are the **intended benefits and costs** [3]. For example, if we develop a new, more accurate cancer screening test, the intended benefits include factors such as the accuracy of the test and the number of late stage cancers that could be prevented through early detection if the new test is widely adopted. The intended costs include the cost of the test, as well as the savings that would result from being able to treat patients for early stage cancer, which is less expensive than treatment for late stage cancer. In rare cases, the savings associated with a technology can actually be greater than the costs of using the technology. We will later see that this is the case with some childhood immunizations.

The indirect consequences of a technology are the unintended economic, social, or other technology effects

Examples of benefits and costs of prostate cancer screening

Monitoring the level of prostate specific antigen (PSA) is frequently used to screen older men to determine if they are likely to have prostate cancer. Invasive diagnostic testing is performed to determine whether men with elevated PSA levels have prostate cancer requiring therapy.

Intended benefits: screening can identify men with prostate cancer at an early stage, when it is still curable.

Intended costs: the cost of a PSA exam is less than $100.

Unintended costs: because prostate cancer grows slowly, the new test can identify men with prostate cancer that will never cause any symptoms. These men undergo further invasive, painful testing and treatment which may be unnecessary [4].

[3]. Let's imagine that our screening test is somewhat invasive and patients perceive it to be much more uncomfortable than the previous screening test. As a result, some patients who would have been screened with the old test now avoid cancer screening altogether, because of fear or embarrassment. In this case, the introduction of a new and more accurate test can actually decrease screening effectiveness because patient adherence to physician recommendations decreases. Health technology assessment must account for such unintended consequences.

As such, health technology assessment is a bridge between the basic research and development of a technology and its real-life application. In the ideal world, technology assessment provides an opportunity to assess the technology's effects *before* its widespread introduction, but in many cases it is used to analyze mature technologies already in routine clinical use to suggest strategies to use limited healthcare resources more effectively.

When do we assess technology?

A key question in health technology assessment regards the timing of when to perform a technology assessment. When health technology is assessed early, there are increased benefits, including potentially protecting public safety and identifying which populations should use the technology. However, assessment in the earlier stages also has risks. A health technology may not yet be perfected, and the populations for whom the technology should be used might not be appropriately identified. In addition, in an early assessment the data available about the performance of the technology are more likely to come from clinical trials rather than the settings where it will be routinely used. This can be a problem because clinical trials are often carried out by experts under well controlled conditions and may overestimate the performance of a technology in the community setting.

We have a special vocabulary to describe this change in performance. Efficacy refers to the performance of a technology under ideal, controlled conditions. Efficacy is studied in homogeneous populations by using standardized procedures under ideal testing conditions by expert practitioners. Effectiveness is the measure of performance in a normal clinical setting. Effectiveness is studied in heterogeneous populations, and the technology is implemented by ordinary practitioners under conditions of routine clinical care [5]. Often the efficacy of a technology is much higher than its effectiveness, because the same experts that developed a test are using it under the best circumstances. If a health technology assessment is carried out using efficacy data, then the assessment lacks critical information about how the technology will truly perform in the real world. The true effectiveness, or impact, of the technology cannot be known until the technology has widespread use.

The argument for later evaluation also has advantages and disadvantages. A later evaluation of technology will have more data, particularly regarding effectiveness, but the data may be biased. For example, if the technology has already been dispersed widely, then a randomized clinical trial may not be ethically feasible. If the initial effectiveness of the technology is favorable, researchers face an ethical dilemma by intentionally withholding the treatment to trial participants in the control group. Potential participants are also less likely to accept randomization in such a trial if significant benefits have already been demonstrated. Additionally, by the time a health technology assessment is completed the technology may already be outdated, either in how it is specifically used, or because superior alternatives for the given clinical problem have been identified.

Technology assessment in developing countries

Technology assessment is particularly important in settings where healthcare resources are extremely limited, such as developing countries. Unfortunately, few developing countries have health technology assessment programs.

The results of technology assessment applied in different settings cannot simply be used in developing countries – many factors, such as whether a disease is common or rare, the social acceptability of a technology, the efficacy of a technology and the cost – vary dramatically throughout the world [6].

Thus, the timing of healthcare technology assessment is a "moving target" problem with no obvious answer. It is for this reason that health technology assessment should be an iterative process, done on an ongoing basis to achieve evaluations throughout the life cycle of the technology development process.

Metrics of health technology assessment

Many times, a ratio of benefit to cost will be calculated as a quantitative metric of technology assessment. It is important to point out that the decision regarding the usefulness of a technology cannot be made in a vacuum. In other words, it must be recognized that, when health technology is assessed, one must think about the relative benefits and costs involved in the clinical situation. There is always an alternative to a new health technology being assessed. Alternatives include currently used treatments or technologies, termed the standard of care. Both the standard of care and the new technology have economic costs and clinical benefits that must be taken into account. Ranking strategies according to their benefit to cost ratio can often be a helpful way to compare the effectiveness of different approaches, including new technologies, the standard of care and "do nothing" strategies [7].

Collecting data for health technology assessment

Where do we obtain data about the costs and benefits of healthcare technologies? Generally data come from clinical trials – well controlled experiments designed to compare the performance of two technologies. In carrying out health technology assessment, we can obtain secondary data about a technology from published literature describing clinical trials, or we can collect primary data by carrying out our own clinical trial. Primary data can be analyzed for efficacy, effectiveness, safety, reproducibility, patient satisfaction, and cost effectiveness. Secondary data can also be used to assess the same outcomes – either by using data from one published study or by using meta-analysis, a statistical method that combines data from different studies to estimate the overall effect of an intervention on a specific outcome [5]. As we will see throughout this text, carrying out a clinical trial can be an expensive task; if we want to examine long term outcomes, we may have to wait decades for results. Many times we want to carry out a health technology assessment without waiting this long. As an alternative, we can write a computer program to simulate the clinical trial. This process is called decision analysis. In decision analysis, we simulate the likely outcomes from a group of hypothetical patients, using probabilistic methods [5]. Data from primary and secondary sources are used to estimate the efficiency of a technology as well as likely patient outcomes. The computer program follows each group of patients over time, using these data to "roll the dice" for each patient at important time points. Decision analysis can provide a very quick and inexpensive way to estimate what might happen in a clinical trial without having to spend millions of dollars and wait decades. However, it is not a substitute for actually carrying out a clinical trial.

There are many types of clinical trials that can provide useful data for health technology assessment. Nonrandomized clinical series often provide data on the efficacy of a technology. Randomized clinical trials, where participants are randomly divided into an experimental group receiving the new technology and a control group receiving existing technology, should be used to compare the performance of the new technology to the existing technology. In the hierarchy of information, randomized clinical trials are considered the strongest study design [8]. Randomized clinical trials may not be necessary if the patient benefit from the test is so dramatic as to leave little room for doubt that the new technology is as good or better and less expensive than existing options. However, few technologies meet those criteria.

Policy decisions and HTA

How is a health technology identified as a candidate for assessment? How do the results of health technology assessment affect the use of a technology? Who monitors health technology assessments? These are all crucial policy questions that are just as important as the scientific methodology of health technology assessment. As mentioned earlier, the purpose of health technology assessment is to aid and inform decision making regarding the use of a technology. It is important then to understand not only how a technology is assessed, but also what is then done with that assessment.

> ### Consensus conference clinical guidelines
> The National Institutes of Health (NIH) is the medical research agency of the United States. One function of the NIH is to organize conferences to bring together expert scientists and physicians to produce consensus statements on important and controversial topics in medicine. Consensus recommendations are based on publicly available scientific data. These consensus guidelines influence the practice of many physicians throughout the world.
>
> For example, the 2000 *NIH Consensus Statement* on Adjuvant Therapy for Breast Cancer recommends that the majority of women who have localized breast cancer be treated both with surgery and also with chemotherapy because of the small but statistically significant improvement in survival [9].

Often emerging technologies surface through reports in the literature of a series of cases. Once an existing technology has been accepted as standard clinical practice, its use can be tracked with data obtained from health registers and institutional and organizational databases, and by use of national administrative and financial data and post-marketing surveillance data. Often policy decisions about the use of existing technologies are then made by group judgment methods (e.g. **consensus conference**) [9]. Social and ethical issues should be considered throughout the development of a technology.

> ## Clinical preventive services guidelines
> **Recommendation on screening for HIV infection**
> Clinicians should assess risk factors for HIV infection by obtaining a careful sexual history and inquiring about injection drug use in all patients. Periodic screening for infection with HIV is recommended for all adolescents and adults at increased risk of infection. Early therapeutic intervention reduces the risk of clinical progression and mortality. Screening is recommended for all pregnant women. Treatment can significantly reduce rates of mother-to-child transmission. The US Preventive Services Task Force makes no recommendation for or against routinely screening non-pregnant adolescents and adults who are not at increased risk for HIV infection. The benefits of screening those without risk factors are too small relative to potential harms to justify a recommendation [4].

The US and Canadian Preventive Services Task Forces, which make recommendations about clinical preventive services, have devised a hierarchy of evidence upon which to base recommendations. The hierarchy has two determinations: quality of evidence and strength of recommendation [4]. The Task Forces categorize the overall *quality* on a three point scale divided into good, fair and poor. Good evidence is derived from well designed, well conducted studies in representative populations that directly assess effects on health outcomes. While fair evidence is sufficient to determine effects on health outcomes, the strength of the evidence may be limited by number, quality, or consistency of the studies, Poor evidence is considered insufficient to assess the effects on health outcomes due to a limited number of studies or flaws in the study design. The strength of recommendation is divided into five categories: there is good, fair, or insufficient evidence to support the recommendation that the test *not* be used in periodic health examinations, and there is fair or good evidence to support the recommendation that the test be used in periodic health examinations. This hierarchy has been applied by many technology assessors. Not all procedures apply to the periodic health examination, but the concept of using evidence to justify the strength of a recommendation is logical.

> ## Learn more about the Institute of Medicine Health Care Quality Initiative
> The report *Crossing the Quality Chasm: A New Health System For The 21st Century* was issued by the IOM in 2001. This report calls for comprehensive reform of the healthcare system to ensure that all patients receive quality, evidence based care [12].
>
> Committee on Quality of Healthcare in America. (2001). Institute of Medicine. *Crossing the Quality Chasm: A New Health System for the 21st Century*, Washington, D.C.: National Academy Press. http://www.nap.edu/catalog/10027.html.

The most recent **Clinical Preventive Services Guidelines** include recommendations for several tests we will examine later in this book [4]. For example, the current recommendation regarding breast cancer screening suggests that all women aged 40 and over undergo a mammogram every one–two years, with or without clinical breast examination. In contrast, screening for HIV is not universally recommended. According to the guide, only pregnant women and non-pregnant adults and adolescents who are at risk for HIV need to be regularly tested. All recommendations are made by reviewing the available evidence; a test, service, or immunization is recommended only when the data suggests that it will be effective.

Given that health technology assessment has an important effect on policy, it is important to track its use and application, and to ensure that health services delivered to patients are consistent with current professional knowledge. The Institute of Medicine (IOM) has undertaken a comprehensive effort to assess and improve the quality of healthcare throughout the United States. The first phase of this initiative began in 1996, and documented serious problems of the quality of healthcare delivered in the United States, concluding that the burden of harm conveyed by healthcare quality problems is staggering [10]. The following are some examples [11].

- Only 55% of patients in the USA receive care consistent with consensus guidelines.
- The delay between the discovery of more effective forms of treatment and the incorporation of these treatments into routine patient care averages 17 years.
- More than 18,000 Americans die every year as a result of heart attacks because they did not receive preventive medications, even though they were eligible to receive them.
- More Americans are killed every year as a result of medical errors than by breast cancer, AIDS or motor vehicle accidents.

In the second phase of the review, the IOM established a vision to transform the US healthcare system in order to close the gap between quality care and what exists today in practice. Recommendations include establishing healthcare systems where decision making is evidence based rather than based on a physician's training and experience, and shifting the view that patient safety is ensured by an individual's responsibility to "do no harm," to one where safety is an inherent property of the healthcare system as a whole [12]. The third phase of the review, currently ongoing, is focused at implementing these reforms on three levels, the environmental, at the level of healthcare organizations, and at the interface between clinicians and patients.

As we implement reforms to improve the quality of health systems, it is important to measure whether these reforms actually improve the health of our population. In the next chapter, we will look at various types of health data which are used to assess health status. We will use these measures to compare health status of populations throughout the world.

Homework

1. Two scientists want to know if a certain drug is effective against high blood pressure. The first scientist wants to give the drug to 1000 people with high blood pressure and see how many experience lower blood pressure levels. The second wants to give the drug to 500 people with high blood pressure and not give the drug to another 500 people with high blood pressure and see how many people in both groups experience lower blood pressure. Source: [13].
   a. What is the better way to test this drug?
   b. Why is it better to test the drug this way?

2. Find a news report describing a new health technology published in the past year.
   a. Based on this article, summarize which steps in the technology assessment process have been carried out for this technology.
   b. Given this, do you believe that the news report provides a balanced discussion of the potential promise and the potential limitations of this technology?

3. Dr. Maurice Hilleman died recently. A quote from his obituary stated, "I think it can be said without hyperbole that he was a scientist who saved more lives than any other modern scientist."
   a. What was Dr. Hilleman's contribution to medical science?
   b. How many lives per year are saved as a result of his work?
   c. Discuss Dr. Hilleman's work as an example of translational research.
      http://www.upenn.edu/almanac/volumes/v51/n29/obit.html

References

[1] Littenberg B. Technology assessment in medicine. *Academic Medicine: Journal of The Association of American Medical Colleges*. 1992 Jul; **67**(7): 424–8.

[2] Deber RB. Translating technology assessment into policy. Conceptual issues and tough choices. *International Journal of Technology Assessment in Health Care*. 1992 Winter; **8**(1): 131–7.

[3] Szczepura A, Kankaanpaa J. An Introduction to Health Technology Assessment. In: Szczepura A, Kankaanpaa J, eds. *Assessment of Health Care Technologies*. New York: John Wiley & Sons; 1996.

[4] *Guide to Clinical Preventive Services, 2006*. June 2006 [cited AHRQ Publication No. 06–0588]; Available from: http://www.ahrq.gov/clinic/pocketgd.htm

[5] Goodman C. A basic methodology toolkit. In: Szczepura A, Kankaanpaa J, eds. *Assessment of Health Care Technologies*. New York: John Wiley & Sons; 1996.

[6] Tan-Torres T. Technology assessment in developing countries. *World Health Forum*. 1995; **16**(1): 74–6.

[7] Cantor SB, Ganiats TG. Incremental cost-effectiveness analysis: the optimal strategy depends on the strategy set. *Journal of Clinical Epidemiology*. 1999 Jun; **52**(6): 517–22.

[8] U.S. Preventive Services Task Force. *Guide to Clinical Preventive Services*. 2nd edn. Alexandria, VA: International Medical Publishing; 1996.

[9] Adjuvant Therapy for Breast Cancer. *NIH Consensus Statement*. 2000 November 1–3; **17**(4): 1–35.

[10] Chassin MR, Galvin RW. The urgent need to improve health care quality. Institute of Medicine National Roundtable on Health Care Quality. *JAMA: The Journal of The American Medical Association*. 1998 Sep 16; **280**(11): 1000–5.

[11] Institute of Medicine. *The Chasm in Quality: Select Indicators from Recent Reports*. 2006 May 30 [cited 2007 May 28]; Available from: http://www.iom.edu/?id=14991.

[12] Institute of Medicine (U.S.). Committee on Quality of Health Care in America. *Crossing the Quality Chasm: A New Health System for the 21st Century*. Washington, D.C.: National Academy Press; 2001. 337 pages.

[13] National Science Board. 2008. *Science and Engineering Indicators 2008*. Two volumes. Arlington, VA: National Science Foundation (volume 1, NSB 08-01; volume 2, NSB 08-01A).

Health and economic data: a global comparison

As we consider the development of new technologies to improve health, it is important to step back and consider how we define health and how we assess the health of a population. We all have our own perceptions of disease; things like pain, fever, and symptoms of illness which can interfere with our normal activities, or reduce our ability to respond to stress and physical injury (Figure 3.1). The World Health Organization defines health to be "a state of complete physical, mental and social well being and not merely the absence of disease or infirmity [1]." In Chapters 3 and 4, we will see that the health status of a population is frequently correlated with economic measures such as income and health expenditures.

The focus of this book is the engineering of new health technologies to meet world health needs. The goal of bioengineering design is to apply advances in science to solve health problems in a way that meets resource constraints. Thus, to engineer new technologies we must understand both global health needs and global resource constraints. Both health needs and resource constraints vary dramatically throughout the world – a technology that solves a health problem in one part of the world, may not be a solution in a different part of the world.

In this chapter, we will develop metrics to assess the health and economic status of populations. We will use

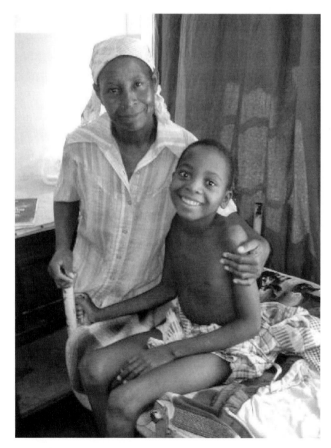

Figure 3.1. *A mother and her son smile after doctors at Chicuque Hospital in Mozambique saved his life. Absence of disease is only one component of overall health. Shannon Trilli, 2003. Used by permission of UMCOR (United Methodist Committee on Relief).*

these metrics to compare the health of different regions, as well as the ability of new technologies to improve health throughout the world. This will help give us a clear picture of where technologies have been effective, and where new technologies are needed.

Health data

How does the health of a population differ from an individual's health? In characterizing the health of a population, we must somehow pool together data about the health of the individuals that comprise that population. Scientists called **epidemiologists** specialize in the study of the health of populations. They calculate pooled figures such as infant mortality rates, numbers of deaths and causes, and immunization rates to develop a picture of the health of a population. In this chapter, we will examine why we need health data, what data we need, where we obtain these data, and how we use them to improve health.

The importance of health data became clear at the beginning of the twentieth century. From 1870 to 1900, biomedical science advanced more than it had in the previous three millennia. In this period, Darwin's concept of evolution was established; chemistry and microscopy were used to carry out field based research around the world. During this time, the means, transmission route, and causative agent of almost every important infectious disease were established. With this improved understanding of disease and how to control it, governmental health agencies were first established.

The **World Health Organization** (WHO) was established by charter of the United Nations after World War II. The WHO is headquartered in Geneva, and its mission is the "attainment by all peoples of the highest possible level of health [1]." The WHO serves a critical role, providing important health information to governments, including epidemiologic intelligence and data on world health problems, international standardization of vaccines, and reports of expert committees on health problems (Figure 3.2). Countries that are members of the WHO must provide certain information in regular reports, including information about disease outbreaks, the health of their population, and steps the country is taking to improve health. The WHO's website

Portrait of epidemiologist John Snow

In the 1840s, cholera outbreaks in London claimed many lives. At the time, most scientists believed that cholera was transmitted by breathing contaminated vapors. John Snow believed that cholera was spread through contaminated food and water but was unable to prove this theory.

In 1854, a cholera outbreak struck London. Snow began plotting the number of deaths by location throughout the city. At this time, London received its water supply from two companies, one which drew water from the Thames upstream of the city and the other which drew water downstream of the city. Snow noted that the concentration of cholera victims was higher in areas of the city supplied by water drawn downstream of the city, a location that was more likely to be contaminated by city sewage. Snow noticed a particularly high number of cases at the intersection of Cambridge and Broad streets – more than 500 people died of cholera over a ten day period. Snow convinced city officials to remove the pump handle that supplied water to this neighborhood and the epidemic was contained.

Snow's study was one of the first epidemiologic analyses, and it firmly established the value of health data in tracking and eliminating the spread of disease [2].

Source: Snow, J. *On the Mode of Communication of Cholera. 1855.*

Malaria epidemics in Africa 1997–2002

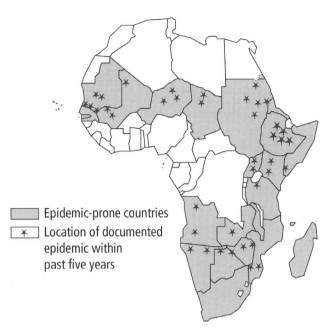

Epidemic-prone countries

Location of documented epidemic within past five years

Source: RBM data

Figure 3.2. *This map displays locations of documented malaria outbreaks. The WHO compiles data on a range of health related issues, such as malaria, ensuring that crucial health concerns are addressed and resources allocated effectively [6].*

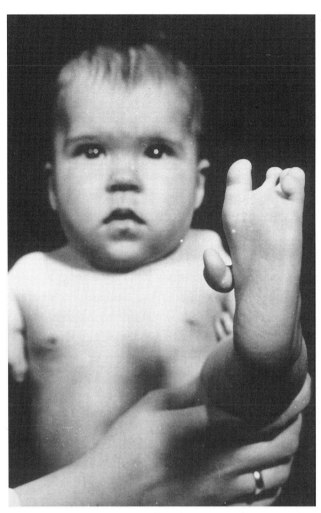

Figure 3.3. *Analysis of health statistics led to the removal of thalidomide from the market after it proved to be very harmful for pregnant women, as shown in this photo of a baby born with an extra appendage on the foot. Source: NCI/G. Terry Sharrer, Ph.D. National Museum of American History.*

(http://www.who.int/en/) includes much useful health data that we will refer to throughout the course.

How do societies use this type of health data? Health data can be used to provide an early warning system to identify emerging health problems. For instance, epidemiological data helped established the connection between rubella early in pregnancy and severe birth defects, even fetal death. As a result, rubella vaccines are now routinely administered during childhood, and cases of rubella associated birth defects have decreased. Another example involves thalidomide, a drug once used to treat morning sickness. Beginning in late 1959, cases of rare, severe birth defects involving the limbs and digits started to accumulate in high numbers (Figure 3.3). Given the sudden spike in the number of cases, investigators suspected a new drug might be the culprit; soon after, the defects were traced to thalidomide and in 1961 the drug was taken off the market [7]. In 1981, two diseases almost exclusively seen in older, immunocompromised people, Kaposi's sarcoma and *Pneumocystis carinii* pneumonia (PCP), began to appear in young, previously healthy, homosexual men [8]. As a result of these unusual appearances, the CDC launched an investigation into what appeared to be a new disease, later identified as AIDS. Health data can be used to help estimate the impact of health problems, such as the number of people affected by a disease and their ages and locations, to determine public policy to respond to a disease, to educate legislators and set societal priorities for healthcare funding, and to monitor progress toward goals so that interventions can be assessed objectively.

Given these uses, what types of health data should be collected? Health data can include data on the population of a country, such as the number of people, their age, sex, ethnic origin, and urbanization. They include vital statistics, such as the number of live births, and the number of deaths (including infant deaths) by sex, age, and cause. Such data are essential in order to accurately assess the health of populations and make decisions regarding health resources; unfortunately these data are lacking from many areas. Vital statistic data from many countries are not complete. In some countries deaths are recorded only in certain areas, while in others all areas are covered but not all deaths are recorded. Average rates of coverage vary widely, from only 10% in Africa, to over 90% in Europe [9].

World Health Report

Each year, the WHO publishes the *World Health Report*. This report provides data quantifying health throughout the world, tracks progress in improving world health and gives an overview of a specific health problem. The topics of past *World Health Reports* include the following.

2006: *Working Together for Health* calls for improvements in the health workforce throughout the world [3].

2005: *Make Every Mother and Child Count* calls for ways to improve maternal and child health throughout the world [4].

2004: *Changing History* calls for a comprehensive HIV/AIDS strategy [5].

Health data also include health statistics, such as the frequency of disease by type, severity and outcome. Certain epidemic-prone diseases are considered to be "reportable" diseases, meaning any incidence must be reported to health agencies. For example, the WHO maintains a list of reportable diseases. This allows for monitoring of potential outbreaks and prevention of epidemic spread. In the United States the Centers for Disease Control and Prevention (CDC) in Atlanta manages mandatory reporting to the WHO. The CDC also maintains a list of **nationally notifiable diseases** [10]. State health departments collaborate with the CDC to determine which diseases should be on this list, and the list varies slightly from state to state. State reporting is voluntary, although all states report diseases in compliance with WHO regulations. In addition to tracking reportable diseases, the WHO also maintains tumor registries, which compile epidemiological data regarding cancer cases. A final category of health data tracks health services, including the number and type of healthcare facilities, the number and qualifications of health personnel, health services and utilization rates, and costs and payment mechanisms.

Nationally notifiable diseases

The 2007 list of US nationally notifiable diseases includes 63 infectious diseases, including AIDS, anthrax, mumps, pertussis, plague, rubella, smallpox, tuberculosis, and typhoid fever [10].

The complete list can be found at: http://www.cdc.gov/epo/dphsi/phs/infdis2007.htm

Quantitative health measures

If we are to develop an accurate picture of health throughout the world, it is important to measure such health data in a quantitative manner. A number of health statistics have been found to be useful in assessing the health of a population. Here we examine some of the most important. **Incidence** refers to the number of *new* cases of a disease in a population over a period of time. We can calculate the annual incidence rate of a disease as follows [7].

$$\text{Annual incidence rate} = \frac{\text{Number of new cases of a defined condition in a defined population in one year}}{\text{Number in that population at mid-year of that same year}} \quad (3.1)$$

In contrast, the **prevalence** of a disease indicates the number of *existing* cases of the disease in a given population at a specific time. The prevalence of a disease is calculated as follows [7].

and prevalence of disease in populations? These data allow one to estimate the magnitude of health problems and to detect epidemics; one example is the "reportable disease" law discussed earlier. In 1951 the WHO

$$Point\ Prevalence = \frac{Number\ of\ cases\ of\ a\ defined\ condition\ in\ a\ defined\ population\ at\ a\ point\ in\ time}{Number\ in\ that\ population\ at\ same\ point\ in\ time} \tag{3.2}$$

While the definitions of incidence and prevalence appear similar, they are actually quite different and it is important to appreciate the distinction. When would incidence and prevalence of a disease differ? Consider a disease with a relatively short duration such as the flu – the annual incidence rate is much higher than the point prevalence, because while many people contract the flu each year, at any given time throughout the year they are not all sick. In contrast, for a disease with a relatively long duration, such as HIV/AIDS, the point prevalence can be much higher than the annual incidence rate. Why is it important to examine incidence

adopted the International Health Regulations (IHR), global legislation that requires all countries to notify the WHO of incidences of specific "quarantinable diseases." The first new case must be reported within twenty-four hours, all subsequent cases and deaths must be reported as well. This mandatory notification is currently required for cases of cholera, plague, and yellow fever [11]. However, owing to the renewed spread of old diseases and the rise of new ones, the WHO is currently working to broaden the powers of the IHR to cover any "public health emergency of international concern [12]."

Global outbreak alert and action

In 2000, the WHO formed a new infrastructure to organize world response to outbreaks of infectious disease. The Global Outbreak Alert and Response Network (GOARN) unites 130 existing agencies and networks throughout the world to work together to systematically detect and verify disease outbreaks, to provide real time alerts and to respond rapidly to contain outbreaks [13].

The effectiveness of GOARN relies on international cooperation. More than 800 people died in the SARS outbreak in 2002–3. World response to the outbreak was delayed because Chinese officials delayed full disclosure of initial cases of SARS.

Courtesy of WHO/Global Outbreak Alert and Response network http://www.who.int/csr/outbreaknetwork/en/

EXAMPLE 1: Incidence Rate

Using data obtained by the World Health Organization below, calculate the annual incidence rate of Pertussis (whooping cough) in New Zealand in 2002 and compare it to the annual incidence rate in 2001.

New Zealand – Population Data in Thousands		
	2002	2001
Live Births	54	54
Female 15-49 years	968	965
Pop. Less than 15 years	869	869
Pop. Less than 5 years	273	275
Surviving Infants	54	54
Total Population	3846	3815

New Zealand – Number of Reported Cases		
	2002	2001
Diptheria	1	0
Measles	21	65
Pertussis	1068	4143
Polio	0	0

Solution

For 2002:

Incidence $= 1,068 / 3,846,000$
$= 2.8 \times 10^{-4}$ or
$= 0.00028$
$=$ or about 3 in 10,000 people

For 2001:

Incidence $= 4,143 / 3,815,000$
$= 1.1 \times 10^{-3}$
$=$ or 0.0011
$=$ or 11 in 10,000 people (over 3 fold higher)

EXAMPLE 2: Mortality Rate

Based on the data available in this EXAMPLE, and EXAMPLE 1, which country has a lower infant mortality rate, New Zealand or India in 2002?

India – Population Data in Thousands		
	2002	2001
Live Births	25,221	25,477
Female 15-49 years	259,828	254,534
Pop. Less than 15 years	349,470	348,562
Pop. Less than 5 years	119,524	120,343
Surviving infants	23,793	24,032
Total Population	1,049,549	1,033,395

Solution

25,477,000 children were born in India in 2001 and 23,793,000 were surviving in 2002.
1,684,000 must not have survived.

Infant mortality rate $= 1,684,000 / 25,477,000$
$= 0.066$
$=$ or 6.6% infant mortality rate

54,000 children were born in New Zealand in 2001 and 54,000 were surviving in 2002. Although some may have died it is not evident in this data. As many as 499 could have died and it would not be apparent as this data has been rounded to the thousands place. If 499 children had died, it would represent 0.009 or a 1% infant mortality rate.

Once a potential outbreak is identified, preventative measures can be put in place to avert an epidemic. These include sending supplies and expert personnel to the disease site, as well as offering technical advice and sometimes initiating an epidemiological investigation. To facilitate information dissemination the WHO publishes the *Weekly Epidemiological Record* (*WER*); an essential tool for providing health personnel with information pertaining to outbreaks of communicable diseases [14]. Likewise the CDC publishes a morbidity and mortality weekly report (*MMWR*), available at www.cdc.gov/mmwr/.

The **mortality** rate in a population quantifies how many people have died. Many types of mortality rates are monitored; however, we will examine two in this text. The first is the crude death rate, often referred to as the mortality rate. The mortality rate is defined as follows.

$$Mortality\ Rate = \frac{Number\ of\ deaths\ in\ a\ defined\ population\ in\ a\ year}{Number\ in\ that\ population\ at\ mid\text{-}year\ of\ the\ same\ year} \tag{3.3}$$

Infant mortality refers to the number of deaths of persons under one year of age and is defined as follows.

$$Infant\ Mortality\ Rate = \frac{Number\ of\ deaths\ under\ 1\ yr\ of\ age\ in\ a\ defined\ population\ in\ a\ year}{Number\ of\ live\ births\ in\ that\ population\ in\ same\ year} \tag{3.4}$$

The infant mortality rate is often used as an indicator of how well a country's health system functions. Finally, health data can be used to estimate not only the occurrence of disease and death, but also the burden of disease. **Morbidity** refers to the degree or severity of a disease. We will use one measure of the burden of disease which combines the effects of both morbidity and mortality, the **disability adjusted life year** (DALY). DALYs measure the years of disability free life lost when a person contracts a disease. A DALY combines several elements, including the levels of mortality by age, the levels of morbidity by age and the value of a year of life at specific ages. For example, it is assumed that losing a year of life at age 20 is more serious than losing a year of life at age 90. While we will not learn to calculate DALYs (a complex calculation), we will use the DALY to compare the impact that different diseases have on different populations. Table 3.1 provides the DALYs lost per person associated with some common diseases and conditions in two regions of the world. In each case, you can think of a DALY as the average number of years of disability free life that an individual who contracts that disease or condition would lose.

Organizing and interpreting health data

The WHO collects vast quantities of data yearly regarding a multitude of health statistics. The challenge lies in organizing these data to create an overall picture of world health, identifying which diseases are most prevalent, where outbreaks are occurring, and how specific health concerns can be addressed. First, health data can suggest which diseases should be cause for great concern. For example, leading causes of mortality throughout the world help prioritize which diseases constitute the greatest threat to world health, allowing resources to be targeted effectively. HIV/AIDS, heart disease, and cancer are all leading causes of death (Table 3.2), and correspondingly many resources are devoted to improving their prevention, diagnosis, and treatment. [16]

Health data are also commonly used to learn more about the global impact of a particular disease. In 2004, the WHO devoted its entire annual report to the HIV/AIDS pandemic [5]. Data showed that although AIDS affects people throughout the world it does not do so equally; roughly 2/3 of those affected live in Africa (Figure 3.4). In order to effectively address HIV/AIDS, the WHO has called for an acceleration of

Table 3.1. *Average DALYs lost per person in North America and Africa for some common diseases and conditions [15].*

Disease/Condition	DALYs Lost /Person	
	North America	**Africa**
Stroke	9	10
Car accidents	28	37
Self-inflicted injuries	22	26
Violence	37	39
Lower respiratory infections	5	30
HIV	33	30

Table 3.2. *Leading causes of death by age [16].*

Leading causes of mortality among adults worldwide 2002						
Age: 15–59			**Age: 60+**			
Rank	**Cause**	**Deaths (000)**	**Rank**	**Cause**	**Deaths (000)**	
1	HIV/AIDS	2279	1	Ischemic heart disease	5825	
2	Ischemic heart disease	1332	2	Cerebrovascular disease	4689	
3	Tuberculosis	1036	3	Chronic obstructive pulmonary disease	2399	
4	Road traffic injuries	814	4	Lower respiratory infections	1396	
5	Cerebrovascular disease	783	5	Trachea, bronchus, lung cancers	928	
6	Self-inflicted injuries	672	6	Diabetes mellitus	754	
7	Violence	473	7	Hypertensive heart disease	735	
8	Cirrhosis of the liver	382	8	Stomach cancer	605	
9	Lower respiratory infections	352	9	Tuberculosis	495	
10	Chronic obstructive pulmonary disease	343	10	Colon and rectum cancers	477	

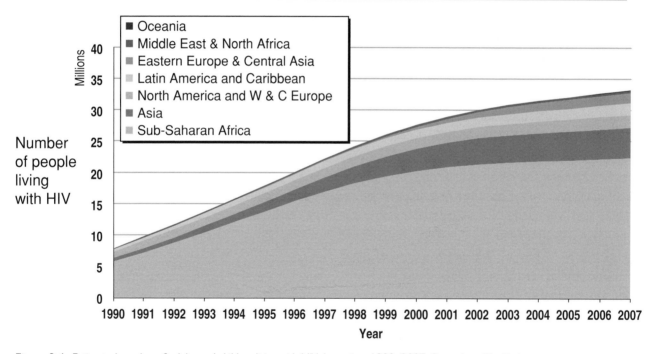

Figure 3.4. *Estimated number of adults and children living with HIV, by region, 1990–2007. Reproduced by kind permission of UNAIDS.*

both prevention and treatment programs in those areas hit hardest by the disease.

Health data can identify how to address health problems. Vital statistics can be used to track birth rates and identify areas that need improved prenatal care or other measures to prevent infant mortality. For example, Table 3.3 lists the leading causes of infant mor-

tality in the developing world. Three of seven are diseases that could be prevented with a simple vaccination: measles, pertussis (whooping cough), and tetanus. These data suggest that increased use of childhood vaccinations could have dramatic effects on infant mortality, and, motivated by this need, researchers are currently working to develop new technologies

Table 3.3. *Leading causes of infant mortality in the developing world [16].*

Causes	Numbers (000)
Lower respiratory infections	1856
Diarrhoeal diseases	1566
Malaria	1098
Measles	551
HIV/AIDS	370
Pertussis	301
Tetanus	185

to make it easier to transport and administer these vaccines.

Finally, health data can help identify the factors that contribute to morbidity and mortality. Epidemiologic studies help identify those factors which are associated with disease. **Association** is simply the statistical dependence between two or more events (e.g. the incidence of cancer rises with increasing age); association does not imply that one event causes the other [7]. However, understanding how certain factors are associated with health status can help us generate hypotheses about cause and effect, and can help us identify populations at most risk. A metric frequently used to characterize the strength of an association is **relative risk**. The relative risk is simply the risk of disease in a population that has been exposed to a certain factor divided by the risk

New Adherence Project . . . :
July 12, 2007
Tessa Swaziland

Now that the new volunteers have taken over most of the World Food Programme (WFP) duties, I've been free to work on the new adherence project with Tina. Some background info about outreach . . . Baylor sends doctors to Good Shepherd roll-out clinics. These are clinics that are located in rural villages and treat mild ailments. But once a month, Good Shepherd (the hospital in the Lobamba region of Swaziland) sends nurses to distribute to the HIV patients there. The Baylor doctors accompany them to see the patients, assess their health, and assess their drug regimen.

Good Shepherd uses a certain form to record the adherence of each patient. It basically has a place for counting the pills and then a place for a general assessment, but the adherence percentage is never calculated. Also, no recommendations concerning the adherence level are made. Tina gave me a new form to edit and test out in the field. This form makes it easy to calculate the adherence percent *and* make an appropriate recommendation (ie, continue ART, see an adherence counselor, refer to doctor, put in "high-risk adherence failure" group, etc.).

Before I went to the outreach clinics, I edited the sheet to incorporate the system they use now. I figured the fewer changes they had to make, the more likely they would be to change. The first day we tried it out was Tuesday. Tina explained to the

Good Shepherd adherence counselors what I was doing. One of them was happy to help me out, but the other one was initially very resistant. So I spent the first day filling the sheets out myself alongside them. This gave me a chance to get to know the new counselor better. At the end of the day, they looked at the sheets with me and gave me input about what they thought worked well and what needed to be changed.

Today, only one of the counselor was there. She was much more receptive and friendly, and she actually had to use the forms for the last ten or so patients since she ran out of her own forms. That gave her a chance to actually see what she thought about the new adherence sheet. Once she got the percentage calculations down, she seemed very pleased and asked for extra copies to take back with her to the hospital.

Table 3.4. *Relative risk of death associated with exposure to several factors [17–20].*

Groups Compared	Relative Risk of Death
Death from Lung Cancer: People who smoke *versus* People who do not smoke	RR = 15
Death from SIDS: Infants of mothers who smoke *versus* Infants of mothers who do not smoke	RR = 5
Death from Diarrheal Disease: Infants <1 year who received only powdered or cows milk *versus* who were exclusively breastfed	RR = 14
Death from Acute Respiratory Infection: Infants <1 year who received only powdered or cows milk *versus* who were exclusively breastfed	RR = 4
Death in a Car Crash: Sober driver traveling 80 mph *versus* Sober driver traveling 60 mph	RR = 32
Death in a Car Crash: Driving with a 0.21 g/dl blood alcohol concentration (b.a.c.) *versus* Driving with a 0.00 g/dl b.a.c.	RR = 30

in a population that has not been exposed to the same factor, and can be calculated as follows [7].

$$Relative\ Risk = \frac{Number\ of\ cases\ in\ exposed\ population/total\ number\ in\ exposed\ population}{Number\ of\ cases\ in\ unexposed\ population/total\ number\ in\ unexposed\ population} \quad (3.5)$$

A relative risk of 1 indicates that there is no risk associated with the factor, while a relative risk greater than 1 indicates an increased risk, and one less than 1 indicates a protective effect. Table 3.4 lists the relative risk of death associated with various factors.

Economic data

As we have seen, health data collected by the WHO show that health concerns differ dramatically throughout the world. For example, polio has been eradicated in the United States, but is a continuing concern in India. At the same time, diseases like AIDS threaten both rich and poor countries alike, requiring global cooperation to fight them. In understanding and addressing these differences, it is important to consider how the available infrastructure and economic resources devoted to healthcare vary throughout the world.

In characterizing populations, we will consider three economic measures which influence health status: the average annual per capita income, the average annual per capita health expenditure, and the human development index. The average per capita income is simply the average annual income per person in a population. Similarly, the average per capita health expenditure is the average amount spent on healthcare each year per person in a population. Typically, these values are reported in US dollars, accounting for currency exchange rates. This conversion doesn't take into account differences in price levels for goods and services between countries. Another approach is to convert currencies based on the cost of an equivalent basket of goods and services – for example, we could compare the cost of a loaf of bread in several different countries – this gives us a different kind of exchange rate which measures purchasing power parity (PPP). We find that prices are typically lower in a developing country, so that the same dollar will buy more in a developing country than in a developed country. Table 3.5 lists the ratio of the official exchange rate to the PPP conversion factor as determined by the World Bank for a number of different countries [21]. Note that this ratio is typically close to 1 for developed countries, but is significantly larger than

Queen II: June 20, 2007
Christina Lesotho

Today we visited the only public hospital in the capital city of Maseru. Before our visit, we were told by some that it may be one of the worst things we have ever seen. We were going to observe a group adherence counseling session aiming to promote adherence to HIV/AIDS anti-retroviral therapy (ARVs). This is supposed to be a technique that has been anecdotally successful in many cases, and one of our projects is to try to implement this at Baylor.

We spent some time in the Children's Ward where some of the physicians were doing their rotations. I entered warily. The smell was very unpleasant. Since I expected the worst, the fact that each child had his/her own hospital bed or crib was a surprise. The cleanliness of those beds was a bit questionable, but I was glad to see that each child seemed relatively comfortably situated. Most were very ill though, with complicated cases of TB; some looked very malnourished and wasted.

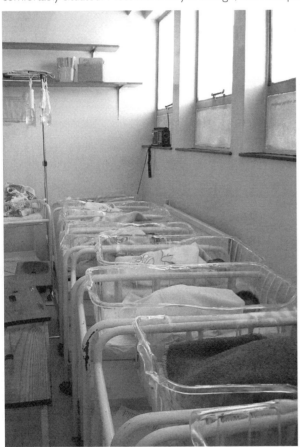

The site was very unpleasant, but I tried to smile at as many as I could. The face of one of the little girls will not leave me anytime soon. She has been paralyzed by a severe case of TB meningitis and spent at least five minutes straight just staring at me from her large crib. The stare was very blank, and she was not responsive to my waves or smiles. Her mother and all of the other mothers around seemed tired, yet very hopeful. Their love and care for the children seems very deep, and I am glad that the children are being treated.

Outside the HIV/AIDS clinic where the counseling was to take place, I read a small sign near the entrance stating that the building had been built by the Clinton Foundation in 2005. The building itself looked much older than 2 years and was very packed with patients. After the session, which was all in Sesotho, the nurse explained to us what went on and mentioned that ARVs have just recently become available in Lesotho. Before 2004, there were no ARVs available to HIV+ patients. This was completely unbelievable to me . . . almost 20 years after the beginning of the disease and no treatment in the hardest hit areas. She seemed very grateful for Clinton's efforts in negotiating drug prices and bringing them into Lesotho and she seemed genuinely interested in and appreciative of our work. She then said something that I did not at all expect to hear while abroad, "Thanks to your country and your people." This was perhaps the most surprising part of the day.

1 in developing countries. Even when income values and health expenditures in developing counties are corrected for PPP, we still find a large discrepancy in per capita wealth and health expenditures.

The Human Development Index (HDI) measures the average achievements in a country in three basic areas including health, education and income. Basically, HDI is the average of three indices [22]. The first is the life expectancy index, which depends on life expectancy at birth. The second is the education index, which depends on the adult literacy rate and the primary and secondary school enrollment rates. The third is the standard of living, which depends on the gross domestic product per capita, converted to equivalent purchasing power in the USA. Each year, the United Nations ranks countries according to these measures. Figure 3.5 shows a world

Table 3.5. *Ratio of exchange rate to purchasing power parity (PPP) conversion factor for several countries [21].*

Country	Exchange Rate/PPP Conversion Factor
United States	1
Canada	1.3
Switzerland	0.8
Haiti	4.1
Niger	4.2
Rwanda	6.0

Learn more about colonialism

Things Fall Apart by Chinua Achebe provides a fictional account of tribal life in Nigeria before and after European colonialism. It is one of the first novels to describe the effects of colonialism from the perspective of the colonized native people [23].

map indicating the HDI of each nation. Many countries with low HDI are located in Africa, Asia, Latin America and Eastern Europe. Often, these countries have experienced a history of **colonialism**.

In examining world health, it is useful to divide the world's countries into three groups: developed countries, developing countries, and least developed countries. Developed countries generally have diversified economies which rely on technology and, as a result, enjoy relatively high standards of living. In contrast, developing countries are characterized by a low per

capita income, underdeveloped infrastructure, and a low HDI. There are no universally accepted criteria which define a developing country; in fact, the United Nations allows each nation to determine whether it should be designated as a developing country [24]. A commonly used definition is to designate a country as a low income country if the per capita annual income is less than $400, and to designate those countries where the per capita annual income ranges from $400 to $4000 as middle income countries [25]. Developing countries have been previously referred to as **Third World countries**, although this terminology is not preferred today.

In 1971, the United Nations created a least developed country member category. Countries must apply for this

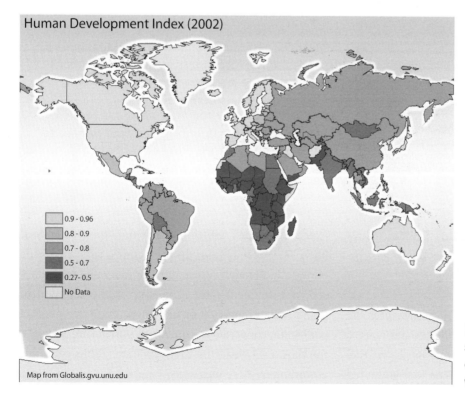

Human Development Index (2002)

0.9 - 0.96
0.8 - 0.9
0.7 - 0.8
0.5 - 0.7
0.27 - 0.5
No Data

Map from Globalis.gvu.unu.edu

Figure 3.5. *World map of HDI. Lighter shaded countries have a higher HDI than darker shaded countries. Source: Globalis, www. http://globalis.gvu.unu.edu/*

During the Cold War, the **Third World** was used to distinguish countries that were aligned with neither the West (the First World) or the East (the Second World). Over time, the Third World came to refer to poor nations which are not as industrialized as developed countries. The term Fourth World is sometimes used to describe the least developed countries.

status, and to qualify must have a low national income (<$900 per capita GDP), low levels of human capital development, and be considered economically vulnerable [26]. Originally, 25 countries were considered to be the least developed countries (LDCs). As of 2005, just over approximately 678 million people (11.3% of the world's population) live in the world's 50 least developed countries [27]. Figure 3.6 shows a map of the LDCs; note that many are located in Africa. Many LDCs have experienced periods of sustained economic decline.

It is eye-opening to contrast health and economic data in the LDCs and the rest of the world. The average per capita GDP in the LDCs is only $235; this figure is $24,522 on average for all developed countries. The average annual healthcare expenditure in LDCs is only $16 per person, whereas the average annual healthcare expenditure in high income countries is $1800 per person. Half the population in LDCs is illiterate. The average life expectancy in the LDCs is only 51 years, compared to 78 years in industrialized nations [28]. In African countries most affected by the global AIDS pandemic, life expectancy continues to decline – by 2010 the life expectancy in Botswana is expected to be only 27 years [29]. One child in 10 dies before his or her first birthday in LDCs; 40% of children under age five are underweight or suffer from stunted growth [30]. The mortality rate for children under five in LDCs is 151/1000 live births; in high income countries this mortality rate is only 6/1000 live births. A child born today in an LDC is more than 1000 times more likely to die of measles than one born in an industrialized country. Compounding these challenges, population growth in the LDCs is expected to triple by 2050 [28].

There are frequently strong correlations between economic data, the presence of health infrastructure, and health data. Figure 3.7 shows world maps which indicate (a) the per capita GNI at PPP, (b) the fraction of the population with access to safe drinking water, and (c) the average life expectancy at birth. Countries with low income tend to have much lower access to safe drinking water as well as a much shorter life expectancy. Moreover, the resultant burden of disease in these

Figure 3.6. *More than 10% of the world's population lives in the 50 least developed countries as shown on the map. These countries account for only 0.5% of the world's gross domestic product [27]. Reprinted with permission from the United Nations Conference on Trade and Development.*

We are just finishing up the third day in clinic, but it's probably only 8:00 a.m. where you're reading. Dr. Lowenthal has had Rachel and me just hanging out around the clinic and helping where we can – I think she just wants us to get a feel for the clinic before we actually start work.

I went for a couple hours to the government hospital to help out in triage, just taking vitals. There were a few differences between doing vitals here and doing vitals in Houston. It is difficult to communicate here because although the kids know a little English, they don't always want to speak it, and because many people only know Setswana. In Houston, I could get by with my scattered knowledge of medical Spanish and a little pantomiming; here, there's a whole lot of pantomiming because my Setswana needs some work. Lucky for me, most of the patients had been through the drill several times and knew what to do. Also, in Houston, I usually had to use the adult size blood pressure cuff, even on the kids, because everybody was so big. Here, there is rarely an adult who uses above a child-size/small-adult cuff. Most of the kids need infant-size. I got to see a lot many more patients today and to get a better idea of the types of patients that the Baylor International Pediatric AIDS Clinic in Botswana helps.

I also helped out in the pharmacy and got to see what typical drug regimens look like and how to fill prescriptions. I had worked in the BANA 2 trial pharmacy the day before, so I had an idea of what I was supposed to do. The BANA 2 trial is a study to see whether interrupting HAART (highly active antiretroviral therapy) at different intervals is as effective as or more effective than HAART without interruptions. This would be great for patients because drug regimens are a hassle as well as expensive for the government. There are hundreds of patients involved, and each has a large blue binder with his files, so there are a lot of stacks around. The trial patients get the same drugs (paid entirely by the government) as other patients and are on similar regimens.

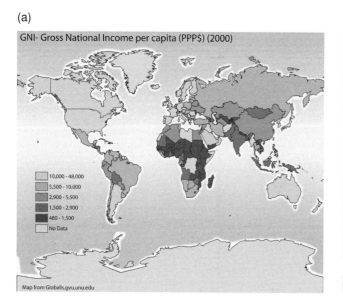
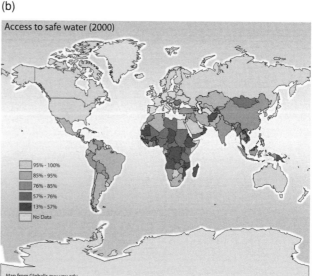

Figure 3.7. *The correlation between economic and health data is apparent in these maps of (a) income, (b) access to safe water, and (c) life expectancy. Source: Globalis, www. http://globalis.gvu.unu.edu/*

(c)

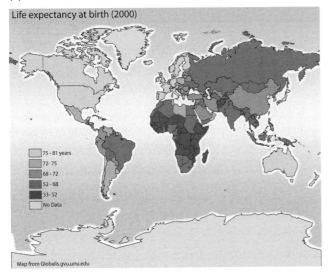

Life expectancy at birth (2000)

75 - 81 years
72- 75
68 - 72
52 - 68
33- 52
No Data

Map from Globalis.gvu.unu.edu

Figure 3.7. *(cont.)*

countries in turn further destabilizes the economy and political system. In fact, according to the WHO, between 1960 and 1994 a high infant mortality rate was a main predictor of state failure due to coups, civil war, and other unconstitutional changes in regime [28]. With this global background on health and economic data we now turn to examine the major health challenges throughout the world. In Chapter 4, we will examine the leading causes of death in developed and developing countries.

Homework

1. One function of the World Health Organization is to provide data on world health problems to member countries.

   a. Describe three ways in which health data are used to improve world health.
   b. Discuss the challenges of obtaining health data in the developing world.
   c. What is the goal of the "3 by 5" initiative of the WHO? Why is this important in developing countries? Describe how collection of health data is important in assessing whether the WHO has met their goal.

2. Incidence and prevalence.

   a. Explain the difference between *incidence* and *prevalence* of a disease.
   b. Which would you expect to be higher: the *incidence* of the flu or the *prevalence* of the flu? Why?
   c. Which would you expect to be higher: the *incidence* of HIV/AIDS or the *prevalence* of HIV/AIDS? Why?

3. The software Gapminder can be found at: http://tools.google.com/gapminder/ and provides a useful graphical tool to explore trends in health and demographic data throughout the world.

   a. Watch the video: http://www.gapminder.org/video/talks/ted-2007–the-seemingly-impossible-is-possible.html
   b. Using Gapminder, build a graph that shows the relationship between life expectancy at birth and per capita income from 1975 to 2004.
   c. Compare trends in life expectancy and income over time in the USA, China, Botswana, Malawi, Lesotho and Swaziland. What do you think is responsible for the differences that you observe?

4. Use the data in the chart below to answer the following questions.

Country	Gross domestic product (GDP) per capita	Life expectancy at birth for males	Total health expenditures per capita	Total health expenditures as % of GDP
United States	$34 637	74.3	$4499	13%
Canada	$27 956	76.6 years	$2534	9.1%
India	$1461	60 years	$71	4.9%
Angola	$1457	34.1 years	$52	3.6%

Source: World Health Organization.

a. Make a graph that shows the life expectancy at birth for males vs. the total health expenditures as a percentage of the GDP for these four countries. Include a title and axis labels.

b. List three reasons that life expectancy is lower in Angola than in Canada and the USA.

c. HAART is a highly effective treatment for HIV infection. Do you think that a poor citizen living in each of these four countries would have access to HAART? Why or why not?

5. Use the information in the chart below to answer the questions that follow.

More than 41,000 people are believed to have died. The earthquake destroyed approximately 20,000 homes and almost completely destroyed health facilities in the area. Immediately following the earthquake, the WHO cited the danger of outbreak of endemic diseases such as cholera, typhoid fever, malaria and leishmaniasis as a critical priority. Before the earthquake, the total population of Bam District was 240,000. The point prevalence of cholera was three per 100,000 population. The point prevalence of malaria was 109.1 per 100,000 population.

Country	Population	Gross domestic product (GDP) per capita	Total health expenditure per capita	Estimated annual incidence rate of tuberculosis per 100,000 population
Brazil	172,558,000	$7,548	$631	64
India	1025,095,000	$1,461	$71	178
Uganda	24,022,000	$932	$36	324

Source: World Health Organization.

a. For each of the three countries shown, calculate the approximate number of new cases of tuberculosis each year.

b. Make a graph that shows estimated annual incidence rate of tuberculosis (per 100,000 population) vs. total health expenditure per capita for these three countries. Include a title and labels.

c. For each of the three countries shown, calculate the percentage of the gross domestic product (GDP) that is spent on health expenditures.

d. Make a graph that shows percentage of GDP spent on health expenditures vs. GDP per capita for these three countries. Include a title and labels. Based on your knowledge of the US healthcare system, describe where you think the United States would lie on this graph.

6. In December 2003 an earthquake with magnitude of 6.7 on the Richter scale hit the city of Bam, Iran.

a. Calculate the approximate number of cases of cholera and malaria in Bam District before the earthquake.

b. Suppose a survey after the earthquake finds 43 cases of cholera in Bam District. How many times greater is the point prevalence of cholera compared to pre-earthquake levels? Include the change in population in your calculation.

c. In a post-earthquake survey, how many cases of malaria in Bam District would it take to represent a ten-fold increase in the point prevalence of malaria compared to pre-earthquake levels? Include the change in population in your calculation.

7. In 1994, the WHO published estimates of the expected number of new cases of TB as well as deaths due to TB expected throughout the world in 1990, 1995 and 2000. In addition, the WHO published estimates of the fraction of these cases and deaths that were attributed to HIV co-infection. The tables below summarize some of these data.

TB cases and deaths in Africa

Year	New cases of TB	Incidence rate per 100,000 population	New HIV attributed TB cases	Deaths due to TB	HIV attributed TB deaths
1990	992,000	191	194,000	393,000	77,000
1995	1,467,000	242	380,000	581,000	150,000
2000	2,079,000	293	604,000	823,000	239,000

TB cases and deaths in industrialized countries

Year	New cases of TB	Incidence rate per 100,000 population	New HIV attributed TB cases	Deaths due to TB	HIV attributed TB deaths
1990	196,000	23	6000	14,000	500
2000	211,000	24	26,000	15,000	2000

Source: World Health Organization (1994). Bulletin of the WHO.72(2):213–220.

a. Compare the trends predicted in the incidence rate of TB in Africa to that seen in industrialized countries between 1990 and 2000.

b. Calculate the predicted mortality rate of TB per 100 000 population in Africa in 1990 and 2000. Compare that to the mortality rate of TB predicted in industrialized countries in the same period.

c. Do you think that increases predicted in the number of new cases of TB in Africa represent demographic changes or epidemiologic changes? Justify your answer.

8. Recent violence in the Darfur region of Sudan has displaced large numbers of people, resulting in a major humanitarian crisis. The table below, from the United Nations Office for the Coordination of Humanitarian Affairs, lists mortality cases among internally displaced persons (IDPs) in one particular refugee camp in Sudan during April 2004, when the camp had a population of approximately 17,750 internally displaced persons.

a. Calculate the mortality rate in Kalma Camp during April 2004. Express your answer as "number of deaths per 1000 population per month."

b. At the current rate, how many times higher is the mortality rate in Kalma Camp than the baseline annual mortality rate in Sudan (9.59 deaths per 1000 population per year)?

Mortality cases amongst internally displaced persons (IDPs) in Kalma Camp during April 2004

April	Total deaths	Age group				Cause of death				
		<1 yr.	<5 yr.	5–15 yr.	>15 yr.	ADD	Measles	Malaria	MN	ARI
Total	114	11	74	21	8	78	7	10	8	11

ADD: Acute Diarrheal Diseases. ARI: Acute Respiratory Infections. MN: Malnutrition. http://www.reliefweb.int/rw/rwb.nsf/AllDocsByUNID/32e427b169a1004085256e8a004f2a1f

c. At the current rate, how many times higher is the mortality rate in Kalma Camp than the baseline annual mortality rate in the United States (8.44 deaths per 1000 population per year)?

d. Current estimates indicate that there are one million internally displaced persons in the region. Based on the statistics for Kalma Camp, estimate how many of these people can be expected to die of acute diarrheal diseases in the next three months.

e. Calculate the point prevalence of malaria in Kalma Camp in April 2004.

9. The table below shows the number of new cases of polio reported for several countries from 2000 to 2005.

per year. Which country had the highest incidence rate of polio in 2002?

b. Make a graph comparing the incidence of polio per year in Nigeria and Afghanistan from 2000 to 2004. Be sure to label both axes and include a title. Assume that the population did not change over this time period. Discuss any trends you observe in the data. Give some possible explanations for the trends that you see.

10. Using the data in the following two pages, answer the questions below.

a. Calculate the point prevalence of HIV/AIDS in the following countries: United States, Canada, India, Angola, Botswana.

Country or territory	Wild virus confirmed cases							Under investigation (2004)	Date of most recent confirmed case
	Total (01 Jan–08 Feb)								
	2000	2001	2002	2003	2004	2004	2005		
Pakistan	199	119	90	103	53	0	1		06-Jan-05
Sudan	4	1	0	0	123	0	1		01-Jan-05
India	265	268	1600	225	133	0	0		22-Dec-04
Saudi Arabia	0	0	0	0	2	0	0		17-Dec-04
Nigeria	28	56	202	355	782	0	0	3	10-Dec-04
Cameroon	0	0	0	2	13	0	0		25-Nov-04
CAR	3	0	0	1	30	0	0		21-Nov-04
Afghanistan	27	11	10	8	4	0	0		14-Nov-04
Guinea	0	0	0	0	5	0	0		08-Nov-04

http://www.polioeradication.org/content/general/casecount.pdf

a. Calculate the incidence of polio in India, Afghanistan and Nigeria in 2002. At that time the population of India was 1,045,845,226, that of Afghanistan was 28,513,677, and that of Nigeria was 129,934,911. Report the incidence as the number of cases per 100,000 population

b. Using data from all twelve countries, make a graph of life expectancy vs. health expenditure per capita. Include a title and labels.

c. Using data from all twelve countries, make a graph of infant mortality rate vs. health expenditure per capita. Include a title and labels.

d. Discuss any relationships or trends you observe in your graphs for parts b and c. Do you observe any other trends in the data listed for the twelve countries?

Brazil

CIA FACTBOOK (2003)

Population: 182,032,604 (July 2003 est.)
GDP per capita: $7,600 (2002 est.)
Life expectancy at birth: 71.13 years
Infant mortality rate: 31.74 deaths / 1000 live births
Fertility rate: 2.01 children born / woman (2003 est.)
People living with HIV/AIDS: 610,000 (2001 est.)

WHO STATISTICS (2000)

Total health expenditure per capita: $631
Health expenditure as percentage of GDP: 8.3 %
Out-of-pocket percentage of health expenditure: 38.5 %

 MEDECINS SANS FRONTIERES
DOCTORS WITHOUT BORDERS

2004 Activity Report : http://www.doctorswithoutborders.org/publications/ar/i2004/brazil.cfm

Angola

CIA FACTBOOK (2003)

Population: 10,766,471 (July 2003 est.)
GDP per capita: $1,600 (2002 est.)
Life expectancy at birth: 36.96 years
Infant mortality rate: 193.82 deaths / 1000 live births
Fertility rate: 6.38 children born / woman (2003 est.)
People living with HIV/AIDS: 350,000 (2001 est.)

WHO STATISTICS (2000)

Total health expenditure per capita: $52
Health expenditure as percentage of GDP: 3.6 %
Out-of-pocket percentage of health expenditure: 44.1 %

MEDECINS SANS FRONTIERES
DOCTORS WITHOUT BORDERS

2004 Activity Report: http://www.doctorswithoutborders.org/publications/ar/i2004/angola.cfm

Cameroon

CIA FACTBOOK (2003)

Population: 15,746,179 (July 2003 est.)
GDP per capita: $1,700 (2002 est.)
Life expectancy at birth: 48.05 years
Infant mortality rate: 70.12 deaths / 1000 live births
Fertility rate: 4.63 children born / woman (2003 est.)
People living with HIV/AIDS: 920,000 (2001 est.)

WHO STATISTICS (2000)

Total health expenditure per capita: $55
Health expenditure as percentage of GDP: 4.3 %
Out-of-pocket percentage of health expenditure: 66.3 %

MEDECINS SANS FRONTIERES
DOCTORS WITHOUT BORDERS

2004 Activity Report:
http://www.doctorswithoutborders.org/publications/ar/i2004/cameroon.cfm

Australia

CIA FACTBOOK (2003)

Population: 19,731,934 (July 2003 est.)
GDP per capita: $27,000 (2002 est.)
Life expectancy at birth: 80.13 years
Infant mortality rate: 4.83 deaths / 1000 live births
Fertility rate: 1.76 children born / woman (2003 est.)
People living with HIV/AIDS: 12,000 (2001 est.)

WHO STATISTICS (2000)

Total health expenditure per capita: $2,213
Health expenditure as percentage of GDP: 8.3 %
Out-of-pocket percentage of health expenditure: 16.8 %

Report by the US Office of Technology Assessment (1995)

"The health care system in Australia is pluralistic, complex, and only loosely organized. It involves all levels of government as well as public and private providers ... The government contribution is funded from general taxation revenues and a Medicare levy on taxable incomes ... For each health care technology included on the Medical Benefits Schedule, Medicare reimburses a proportion of the cost. If a technology is not included on the schedule, costs are typically paid by the patient; private insurance coverage is relatively limited."

For the complete report: http://www.wws.princeton.edu/cgi-bin/byteserv.prl/~ota/disk1/1995/9562/956204.PDF

Canada

CIA FACTBOOK (2003)

Population: 32,207,113 (July 2003 est.)
GDP per capita: $29,400 (2002 est.)
Life expectancy at birth: 79.83 years
Infant mortality rate: 4.88 deaths / 1000 live births
Fertility rate: 1.61 children born / woman (2003 est.)
People living with HIV/AIDS: 55,000 (2001 est.)

WHO STATISTICS (2000)

Total health expenditure per capita: $2,534
Health expenditure as percentage of GDP: 9.1 %
Out-of-pocket percentage of health expenditure: 15.5 %

Report by the US Office of Technology Assessment (1995)

"Under the Canadian constitution, health care is a provincial responsibility; the federal role is limited to health care financing, health protection, and environmental health ... Universal health insurance, administered by provincial governments on a shared-cost basis with the federal government, covers inpatient and outpatient care in hospitals, ambulatory care and, in some provinces, prescribed medication and appliances."

For the complete report: http://www.wws.princeton.edu/cgi-bin/byteserv.prl/~ota/disk1/1995/9562/956205.PDF

Botswana

CIA FACTBOOK (2003)

Population: 1,573,267 (July 2003 est.)
GDP per capita: $9,500 (2002 est.)
Life expectancy at birth: 32.26 years
Infant mortality rate: 67.34 deaths / 1000 live births
Fertility rate: 3.27 children born / woman (2003 est.)
People living with HIV/AIDS: 330,000 (2001 est.)

WHO STATISTICS (2000)

Total health expenditure per capita: $358
Health expenditure as percentage of GDP: 6.0 %
Out-of-pocket percentage of health expenditure: 11.0 %

Japan

Population: 127,214,499 (July 2003 est.)

GDP per capita: $28,000 (2002 est.)

Life expectancy at birth: 80.93 years

Infant mortality rate: 3.3 deaths / 1000 live births

Fertility rate: 1.38 children born / woman (2003 est.)

People living with HIV/AIDS: 12,000 (2001 est.)

WHO STATISTICS (2000)

Total health expenditure per capita: $2,009

Health expenditure as percentage of GDP: 7.8 %

Out-of-pocket percentage of health expenditure: 19.3 %

China

CIA FACTBOOK (2003)

Population: 1,286,975,468 (July 2003 est.)

GDP per capita: $4,400 (2002 est.)

Life expectancy at birth: 72.22 years

Infant mortality rate: 25.26 deaths / 1000 live births

Fertility rate: 1.7 children born / woman (2003 est.)

People living with HIV/AIDS: 850,000 (2001 est.)

WHO STATISTICS (2000)

Total health expenditure per capita: $205

Health expenditure as percentage of GDP: 5.3 %

Out-of-pocket percentage of health expenditure: 60.4 %

2004 Activity Report http://www.doctorswithoutborders.org/publications/ar/i2004/china.cfm

Sweden

CIA FACTBOOK (2003)

Population: 8,878,085 (July 2003 est.)

GDP per capita: $25,400 (2002 est.)

Life expectancy at birth: 79.97 years

Infant mortality rate: 3.42 deaths / 1000 live births

Fertility rate: 1.54 children born / woman (2003 est.)

People living with HIV/AIDS: 3,300 (2001 est.)

WHO STATISTICS (2000)

Total health expenditure per capita: $2,097

Health expenditure as percentage of GDP: 8.4 %

Out-of-pocket percentage of health expenditure: 22.7 %

Report by the US Office of Technology Assessment (1995)

"The high tax rate pays for extensive health and welfare benefits. All Swedes have compulsory health insurance that covers all health care, including outpatient and hospital services (except for some copayments for physician visits), home care, long-term and nursing care, and all equipment and aids for the disabled and handicapped … The Swedish health care system is decentralized … the Federation of County Councils plays a key role in health policy and structural and manpower issues."

For the complete report: http://www.wws.princeton.edu/cgi-bin/byteserv.prl/~ota/disk1/1995/9562/956209.PDF

Germany

CIA FACTBOOK (2003)

Population: 82,398,326 (July 2003 est.)

GDP per capita: $26,600 (2002 est.)

Life expectancy at birth: 78.42 years

Infant mortality rate: 4.23 deaths / 1000 live births

Fertility rate: 1.37 children born / woman (2003 est.)

People living with HIV/AIDS: 41,000 (2001 est.)

WHO STATISTICS (2000)

Total health expenditure per capita: $2,754

Health expenditure as percentage of GDP: 10.6 %

Out-of-pocket percentage of health expenditure: 10.6 %

Report by the US Office of Technology Assessment (1995)

"The most important institutions in the German health care system are the approximately 1,100 mandatory sickness funds … About 90 percent of the population are obligatory or voluntary members (or coinsured family members) of mandatory sickness funds, which operate as nonprofit statutory corporations. In addition, 45 private insurance companies offer health insurance … The services to be reimbursed by mandatory sickness funds are defined by law."

For the complete report : http://www.wws.princeton.edu/cgi-bin/byteserv.prl/~ota/disk1/1995/9562/956207.PDF

United States

CIA FACTBOOK (2003)

Population: 290,342,554 (July 2003 est.)

GDP per capita: $37,600 (2002 est.)

Life expectancy at birth: 77.14 years

Infant mortality rate: 6.75 deaths / 1000 live births

Fertility rate: 2.07 children born / woman (2003 est.)

People living with HIV/AIDS: 900,000 (2001 est.)

WHO STATISTICS (2000)

Total health expenditure per capita: $4,499

Health expenditure as percentage of GDP: 13.0 %

Out-of-pocket percentage of health expenditure: 15.3 %

Report by the US Office of Technology Assessment (1995)

"The organization and delivery of health care in the United States is a good reflection of the free market system … The delivery system is loosely structured … The government is the major purchaser of health care for older people and, along with the states, for some poor people. By and large, however, payments for health insurance and health care are private sector transactions. Access to health care is not universal, and even among those with health insurance, coverage is uneven … One of the most significant recent changes in the U.S. health care system is the growth in the number and variety of managed care plans … In the United States, substantial investment in health care R&D in the public and private sector has ensured a steady flow of technological innovations. These advances, many of which provide at least some benefit to some population of patients, are introduced into an environment in which explicit fiscal limits are unusual."

For the complete report: http://www.wws.princeton.edu/cgi-bin/byteserv.prl/~ota/disk1/1995/9562/956211.PDF

India

CIA FACTBOOK (2003)

Population: 1,049,700,118 (July 2003 est.)

GDP per capita: $2,540 (2002 est.)

Life expectancy at birth: 63.62 years

Infant mortality rate: 59.59 deaths / 1000 live births

Fertility rate: 2.91 children born / woman (2003 est.)

People living with HIV/AIDS: 3,970,000 (2001 est.)

WHO STATISTICS (2000)

Total health expenditure per capita: $71

Health expenditure as percentage of GDP: 4.9 %

Out-of-pocket percentage of health expenditure: 82.2 %

2004 Activity Report : http://www.doctorswithoutborders.org/publications/ar/i2004/india.cfm

11. The data in the table below were adapted from the *WHO World Health Report, 2005*. Based on the provided information for each area/region calculate the following.

    a. The total under five mortality rate for each region
    b. The percentage of under five deaths due to each of the six causes listed for each region (don't worry if they don't sum to 100%).

Area/region	Under 5 population (000)	Total number of under 5 deaths (000)	Number of under 5 deaths (000) due to:					
			Measles	**Malaria**	**Diarrheal diseases**	**Neonatal Causes**	**Acute Respiratory Diseases**	**Injuries**
Africa	110,944	4396	227	802	701	1148	924	76
Canada and USA	22,978	50	0	0	0	29	1	5
South East Asia	178,987	3070	103	12	552	1362	590	71
Europe (Low mortality states)	22,050	25	0	0	0	14	0	2

12. Using data obtained by the World Health Organization below, calculate the annual incidence rate of pertussis (whooping cough) in New Zealand in 2002 and compare it to the annual incidence rate in 2001.

New Zealand – population data in thousands							
	2002	**2001**	**2000**	**1999**	**1998**	**1990**	**1980**
Live births	54	54	54	54	54	58	50
Female 15–49 years	968	965	963	961	960	884	763
Pop. less than 15 years	869	869	867	863	858	786	832
Pop. less than 5 years	273	275	277	281	285	274	248
Surviving infants	54	54	54	54	55	57	51
Total population	3846	3815	3784	3752	3719	3360	3113
New Zealand – number of reported cases							
Diphtheria	1	0	0	0	1	0	1
Measles	21	65	65	106	164	–	–
Pertussis	1068	4143	4143	1046	153	91	0
Polio	0	0	0	0	0	0	0

The annual incidence rate of pertussis in the United States in 2001 was one in every 100,000 people and in 2002 was three in every 100,000 people. How does this compare to the rate and trend in New Zealand? Suggest contributing factors that may help to explain the difference.

More information about pertussis from the CDC can be found at http://www.cdc.gov/nip/publications/pink/pert.pdf

Sources of data for this exercise can be found at http://www.who.int/country/en/

13. Polio has been eradicated in the United States and many other countries because of immunization but is a continuing problem in India. Using the data below, calculate the incidence rate of polio in India in 2002.

mortality rate in Iraq according to the CIA is 5.84 deaths/1000, how many people could be expected to die at baseline rates out of the number who are displaced this year if the displacement has no effect on mortality? If the

India – population data in thousands							
	2002	**2001**	**2000**	**1999**	**1998**	**1990**	**1980**
Live births	25,221	25,477	25,779	26,074	26,307	26,117	23,517
Female 15–49 years	259,828	254,534	249,253	243,992	238,764	201,498	159,836
Pop. less than 15 years	349,470	348,562	347,158	345,236	342,821	309,227	265,551
Pop. less than 5 years	119,524	120,343	120,878	121,071	120,957	115,404	96,705
Surviving infants	23,793	24,032	24,261	24,434	24,521	23,759	20,294
Total population	1,049,549	1,033,395	1,016,938	1,000,161	983,110	846,418	688,856
India – number of reported cases							
Diphtheria	5,472	5,101	3,094	1,786	1,378	8,425	39,231
Measles	51,780	37,969	22,236	21,013	33,990	89,612	114,036
Pertussis	34,703	30,653	27,851	11,264	31,199	112,416	320,109
Polio	1,600	268	265	2,817	4,322	10,408	18,975
Tetanus (neonatal)	1,178	3,241	1,679	610	2,049	9,313	–
Tetanus (total)	–	8,880	6,694	2,125	6,705	23,356	45,948
Yellow Fever	–	0	–	–	–	–	–

14. Based on the data in problems 11 and 12, which country has a lower infant mortality rate, New Zealand or India, in 2002? What information did you use to determine this? Is New Zealand's infant mortality rate really zero?

15. War creates tremendous challenges in meeting the health needs of people near the conflict. Read the article entitled "Deadly comrades: war and infectious diseases" published in *The Lancet*, volume 360, pages s23–s24, 2002.

    a. According to Connolly, crude mortality rates over 60 times higher than baseline have been recorded when populations are suddenly displaced in temporary settlements due to war. Current estimates indicate as many as 900,000 displaced persons in Iraq. If the baseline adult

displacement does have an effect on mortality, use the highest rate Connolly has recorded to calculate the number of people that could be expected to die. What percent of the displaced population is the new number of possible deaths?

   Source of data for displaced persons http://www.who.int/disasters/country.cfm?countryID=28&DocType=2.

   Source of data for baseline mortality rate http://www.cia.gov/cia/publications/factbook/geos/cg.html

    b. If the population of Zaire in 1960 (now Democratic Republic of Congo) was 16.2 million, calculate the point prevalence of trypanosomiasis for the entire country during this war-torn period if the number of cases of this disease there were estimated at 40,000.

Source of data for population in 1960 http://lcweb2.loc.gov/frd/cs/zaire/zr_appen.html#table2

References

[1] *Constitution of the World Health Organization*. International Health Conference 1946 July 22; New York: World Health Organization; 1946.

[2] Crosier S. *John Snow: The London Cholera Epidemic of 1854*. CSISS Classics [cited; Available from: http://www.csiss.org/classics/content/8

[3] Chen L, Evans D, Evans T, Sadana R, Stilwell B, Travis P, *et al. The World Health Report 2006*: *Working Together for Health*. Geneva: World Health Organization; 2006.

[4] VanLerberghe W, Manuel A, Matthews Z, Wolfheim C. *The World Health Report 2005: Make Every Mother and Child Count*. Geneva: World Health Organization; 2005.

[5] Beaglehole R, Irwin A, Prentice T. *The World Health Report 2004: Changing History*. Geneva: World Health Organization; 2004.

[6] *The Africa Malaria Report 2003*: World Health Organization and UNICEF; 2003.

[7] Hennekens CH, Buring JE. *Epidemiology in Medicine*. Philadelphia: Lippincott Williams and Wilkins; 1987.

[8] Kasper D, Braunwald E, Fauci A, Longo D, Hauser S, Jameson JL. *Harrison's Principles of Internal Medicine*. 16th edn. New York: McGraw-Hill; 2005.

[9] Mathers CD, Fat DM, Inoue M, Rao C, Lopez AD. Counting the dead and what they died from: an assessment of the global status of cause of death data. *Bulletin of the World Health Organization*. 2005 Mar; **83**(3): 171-7.

[10] *Nationally Notifiable Infectious Diseases: United States 2007*. Centers for Disease Control and Prevention; 2007.

[11] *International Health Regulations (1969)*. Geneva: World Health Organization; 1983.

[12] *Revision of the International Health Regulations*. In: Assembly tWH, ed.: World Health Organiation; 2005.

[13] Global Outbreak Alert and Response Network. [Online] 2007 [cited]; Available from: http://www.who.int/csr/outbreaknetwork/en/

[14] WHO. *The Weekly Epidemiological Report (WER)*. 2007 May 25 [cited 2007 May 30]; Available from: http://www.who.int/wer/en/

[15] WHO. *Mortality: Revised Global Burden of Disease (2002) Estimates*. Geneva: World Health Organization; 2002.

[16] Beaglehole R, Irwin A, Prentice T. *The World Health Report 2003: Shaping the Future*. Geneva: World Health Organization; 2003.

[17] Ross RM. Estimating the standard mortality ration for lung cancer due to the confounding effects of smoking. *Chest*. 2004 October 24; **126**(4): 850S.

[18] Mitchell EA. SIDS: facts and controversies. *Medical Journal of Australia*. 2000; **173**(4): 173-4.

[19] Victora CG, Smith PG, Vaughan JP, Nobre LC, Lombardi C, Teixeira AM, *et al.* Evidence for protection by breast-feeding against infant deaths from infectious diseases in Brazil. *Lancet*. 1987 Aug 8; **2**(8554): 319-22.

[20] Peden M, Scurfield R, Sleet D, Mohan D, Hyder AA, Jarawan E, *et al. World Report on Road Traffic Injury Prevention*: World Health Organization; 2004.

[21] Swanson E, *et al. World Development Indicators*. Washington D.C.: The World Bank; 2006 April.

[22] Fukuda-Parr S. *Human Development Report 2004: Cultural Liberty in Today's Diverse World*. New York: United Nations Development Program; 2004.

[23] Achebe C. *Things Fall Apart*. New York: Knopf : Distributed by Random House; 1992.

[24] Definition of: developed, developing countries [cited 491; Available from: http://unstats.un.org/unsd/mispa/mi_dict_xrxx.aspx?def_code=491#top

[25] *Entering the 21st Century: World Development Report 1999–2000*. New York: World Bank; 2000.

[26] *The Criteria for the Identification of LDCs*: UN Office of the High Representative for the Least Developed Countries, Landlocked Developing Countries and Small Island Developing States; 2003.

[27] UNCTAD Secretariat. *Statistical Profiles of the Least Developed Countries, 2005*. Geneva: United Nations Conference on Trade and Development; 2005.

[28] Sachs J. *Report of the Commision on Macroeconomics and Health, Macroeconomics and Health: Investing in Health for Economic Development*. Geneva: World Health Organization; 2001.

[29] Steinbrook R. Beyond Barcelona – the global response to HIV. *The New England Journal of Medicine*. 2002 Aug 22; **347**(8): 553-4.

[30] UNCTAD. *3rd UN Conference on the Least Developed Countries, 2001*. Geneva: United Nations Conference on Trade and Development; 2001.

World health and global health challenges

4

In Chapter 3 we examined measures which are used to assess the health of populations and to guide efforts to improve health throughout the world. In this chapter, we examine the leading causes of death in the developed and developing worlds. Before reading this chapter, you should visit "A Tale of Two Girls" on the World Health Organization website (http://www.who.int/features/2003/11/en/) [1]. This brief presentation contrasts the lives of two baby girls, one born in Japan and the other in Sierra Leone, where the average life expectancy for women differs by 50 years.

Overview

We have seen that health data differ dramatically between developed countries and developing countries. In developing countries, infectious diseases, such as

Healthcare standards differ greatly between the developed and developing countries. The gap in resources is apparent in comparing Texas Children's Hospital in Houston, Texas (a), to a medical clinic in Afghanistan (b) [2]. Image (a) courtersy of Texas Children's Hospital. Image (b) from World Vision International/Geno Teofilo.

(a)

(b)

(a) (b)

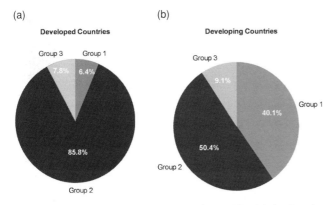

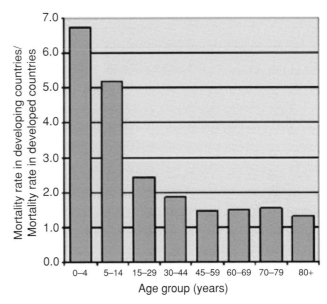

Figure 4.1. *Causes of death throughout the world in (a) developed countries and (b) developing countries.* **Group 1**: *Communicable diseases, maternal/perinatal conditions, nutritional deficiencies;* **Group 2**: *Noncommunicable diseases (cardiovascular, cancer, mental disorders);* **Group 3**: *injuries. Note that Group 1 causes account for a much larger portion of deaths in developing countries [2].*

Figure 4.2. *Ratio of the mortality rate in developing countries to that in developed countries for various age groups [2].*

HIV/AIDS, tuberculosis, malaria and measles, are an important cause of mortality. In contrast, mortality in developed countries is more commonly associated with chronic diseases, such as cancer and heart disease. Figure 4.1 shows the percentage of deaths caused by infectious diseases, noninfectious diseases and injuries in developed and developing countries [2]. In this figure, Group 1 causes of mortality include infectious diseases, maternal and perinatal conditions, and nutritional deficiencies. Group 2 causes of mortality include noncommunicable diseases such as cardiovascular disease, cancer, and mental disorders. Group 3 causes include both intentional injuries such as suicide, interpersonal violence and war as well as unintentional injuries such as motor vehicle accidents and accidental drowning. Throughout the world, Group 2 diseases account for most deaths; in developed countries they are responsible for more than 85% of deaths, while in developing countries Group 2 diseases account for just over half of all deaths. In contrast, Group 1 conditions account for only 6% of deaths in developed countries, but more than 40% of deaths in developing countries. Injuries account for less than 10% of deaths throughout the world.

Within a population, the mortality rate depends strongly on age. In general, mortality rate decreases from infanthood to childhood, and then increases with increasing age throughout adulthood. Mortality rates in developing countries are higher than those in developed countries for all age groups (Figure 4.2); these differences are greatest for infants, children and young adults – the mortality rate for children under five years of age is more than six times higher in developing countries than in developed countries [2].

The leading causes of mortality within a population also depend strongly on age. The incidence and mortality of cancer increase with increasing age, accounting for a large fraction of adult death. The infectious diseases pertussis, poliomyelitis, diphtheria, measles and tetanus are known as childhood cluster diseases and cause many deaths in children in developing countries [2]. In order to devise the most effective health interventions for different populations, it is important to understand how causes of morbidity and mortality vary with age and geographic region. In 1993, the Harvard School of Public Health began a collaboration with the World Bank and the WHO called the Global Burden of Disease Project to estimate the mortality and morbidity of diseases throughout the world. This study generated the most comprehensive and consistent set of estimates of mortality and morbidity by age, sex and region to quantify the burden of disease throughout the world [3]. The World Health Organization now estimates the Global Burden of Disease for the years 2000 and beyond.

Tessa and I were planning on working on some projects relating to nutrition.

Well, we've found a project. And it's not just any project.

We have been dubbed, by Carrie (one of my favorite doctors here), directors of the initiation and direction of the World Food Programme food distribution at the clinic. It's no small task; the World Food Programme (WFP) is very picky about who to give out food to, how much food to give out, and how often to give food out. They also want to make sure that every last kilogram of food is accounted for. So, Tessa and I are now in charge of creating a system to determine who is eligible to receive WFP food as well as organizing the quite complicated registration of each participant and then the doling out of food to each participant.

This is all quite tricky, as it seems the doctors and the administrative staff all have their own opinions about how the system should work and who should get food. It's also complicated because we have to make it so that our WFP registration and distribution does not interfere with the normal going-ons of the clinic.

But everything seems to be going quite well – we have developed a quite meticulous and well-thought-out plan for the program – a triple-check system with three different ledger books to be filled out – which we hope to initiate next Thursday. And, I must say, our plan would never have been such a meticulous and well-thought-out one, if it were not for Tessa's amazing planning skills and her relentless approach of not stopping until we had carefully looked at the program from every possible angle. Some day, I hope to be half as organized as she is. (Hey, I think it's already starting to rub off a little. :))

The food that we are giving out is called Corn Soy Blend (CSB) and is pretty much what you would assume it to be – a mix of corn flour and soy flour (in order to get a healthy mix of both carbohydrates and protein), but with the added elements of some fat, vitamins and minerals to make it more of a complete meal.

We will also be working on creating a hand-out for the WFP participants to describe different ways to prepare the CSB to make beverages, porridge, breads, cakes, soups and even cookies. The purpose of this, besides just teaching them how to use this perhaps somewhat foreign food source, is to allow us to show them how they can add certain items (such as a tablespoon of oil or an avocado or some milk) in order to make the meal even more nutritionally complete. This is important to ensure that our young patients have the necessary dietary requirements for normal growth and development.

For now, however, we are mostly concentrating on rolling out the WFP distribution itself. And even though I feel like our plan is pretty much bullet-proof (okay, maybe I'm giving us too much credit), there will definitely be quite a bit of difficulty in getting everyone to agree that our solution is best, and there could be some problems that other members of the staff will bring up that we haven't thought of yet. We will have to be ready to deal with those problems.

If all goes well, we hope to start training several members of the community to take over for us. It will be good for them because it gives them a job and will be great for us because it ensures that our program will continue after we're gone and also frees us up to work on new and equally exciting projects.

I sincerely hope that everything goes well. There are so many patients that I've seen already that need this food so badly. Malnutrition is extremely common in Swaziland. It is even more common in our clinic as people with HIV/AIDS have increased caloric needs for a number of reasons. For example, the body's metabolic rate increases in order to fuel the large immune system response to fight off the infection. This food is also crucial to our patients because many have living situations that are altered by the infection. Many live with their grandmothers (gogos) due to the death of both of their parents as a result of AIDS. These grandmothers often take care of many other children who have been orphaned by AIDS (In clinic the other day, a gogo came in while I was there that was taking care of 11 children.) They have great difficulty in providing food for so many hungry mouths. Also, if the parents are living with HIV/AIDS they are likely to be weakened so much by their disease that they are unable to hold a job and therefore cannot afford to buy enough food for their children.

In terms of keeping patients healthy, making sure they have enough food to meet their needs is just as important, if not more so, than giving them ARVs. In fact, if they are not getting enough food, their immune systems will have a lessened ability to fight the infection, which will allow the virus to spread more quickly and thus hasten the onset of AIDS.

What it boils down to is that most of the patients seen at the Baylor Clinic are in desperate need of food. We need our program to work so that WFP will continue to give us more food to feed these patients.

I'm keeping my fingers crossed.

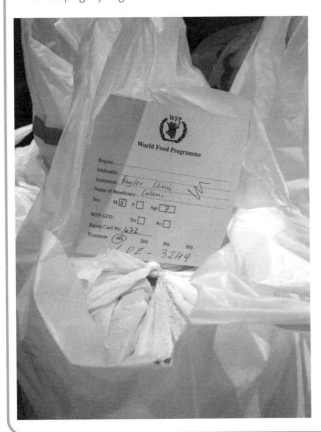

Data from the Global Burden of Disease study can be used to identify the top ten causes of mortality (Table 4.1) by age for developed and developing countries [2]. In the remainder of this chapter, we will examine the four leading causes of mortality by age group in more detail. We begin with the leading causes of child mortality.

Leading causes of mortality, ages birth to four years

More than ten million children under the age of five die every year throughout the world [4]. Ninety-eight percent of childhood deaths occur in developing countries – the number of children who die every year in developing countries is more than two times the number of children born each year in the USA and Canada combined [5]. Ninety percent of childhood deaths occur in just 42 countries, with nearly half occurring in sub-Saharan Africa and nearly one quarter in India [4]. Childhood mortality rates have dropped by nearly 50% throughout the world over the past 30 years (Figure 4.3) [5]. However, progress to further reduce childhood mortality has slowed, and rates of decline have begun to level off, particularly in African countries.

Past efforts to reduce childhood mortality have been most successful for children over the age of two months. As we will see in Chapter 8, worldwide childhood immunization campaigns have reduced childhood mortality substantially; these vaccinations prevent diseases which typically affect older infants and children. Today, more than 40% of deaths to children under five occur during the first 28 days of life [5]. This time is called the neonatal period, and more than four million babies die each year within their first month of life. The perinatal period refers to the period which extends from 22 weeks of pregnancy to the first seven days of life. One quarter

Table 4.1. *Top ten causes of death in developed and developing countries for three age groups [2].*

Ages 0-4	Developing Countries			Developed Countries	
Cause of Death	# Deaths	% of Total Deaths	Cause of Death	# Deaths	% of Total Deaths
Perinatal Conditions*	2,378,099	23.2%	Perinatal Conditions*	83,877	36.3%
Lower respiratory infections	1,701,383	16.6%	Congenital anomalies§	38,169	16.5%
Diarrhoeal diseases	1,597,647	15.6%	Lower respiratory infections	32,872	14.2%
Malaria	1,149,195	11.2%	Unintentional injuries#	15,486	6.7%
Measles	535,504	5.2%	Diarrhoeal diseases	11,940	5.2%
Congenital anomalies§	387,262	3.8%	Meningitis	9,603	4.2%
HIV/AIDS	356,500	3.5%	Neuropsychiatric conditions	4,791	2.1%
Pertussis	293,543	2.9%	Measles	4,712	2.0%
Unintentional injuries#	273,040	2.7%	Cardiovascular diseases	4,347	1.9%
Tetanus	198,236	1.9%	Non-communicable respiratory disease	3,514	1.5%
Protein-energy malnutrition	147,607	1.4%	Malignant neoplasms**	3,218	1.4%

* low birth wt, birth asphyxia/trauma, other
§ heart, spina bifida, down syndrome, anencephaly, other
# drownings, traffic accidents, fires, falls, poisoning, other

* low birth wt, birth asphyxia/trauma, other
§ heart, spina bifida, down syndrome, anencephaly, other
# drownings, traffic accidents, fires, falls, poisoning, other
** leukemia, lymphomas, liver cancer, other

Total Deaths	10,247,719	Total Deaths	230,861
Total Population	536,962,742	Total Population	81,206,312
Mortality Rate	1.9%	Mortality Rate	0.3%

Ages 15-44	Developing Countries			Developed Countries	
Cause of Death	# Deaths	% of Total Deaths	Cause of Death	# Deaths	% of Total Deaths
HIV/AIDS	1,826,460	24.2%	Unintentional injuries#	261,693	26.3%
Unintentional injuries#	1,212,096	16.0%	Cardiovascular diseases‡	172,194	17.3%
Cardiovascular diseases‡	596,038	7.9%	Malignant neoplasms**	129,897	13.0%
Tuberculosis	591,316	7.8%	Self-inflicted injuries	106,759	10.7%
Maternal conditions¥	488,346	6.5%	Digestive diseasesΦ	55,043	5.5%
Malignant neoplasms**	449,127	5.9%	Violence	52,996	5.3%
Self-inflicted injuries	375,684	5.0%	Neuropsychiatric conditions	49,202	4.9%
Violence	330,313	4.4%	HIV/AIDS	30,897	3.1%
Digestive diseasesΦ	247,284	3.3%	Tuberculosis	29,686	3.0%
Lower respiratory infections	246,949	3.3%	Lower respiratory infections	16,739	1.7%

road traffic accidents, fires, drownings, other
‡ ischaemic heart disease (IHD), cerebrovascular disease, other
¥ maternal haemorrhage, sepsis, other
** liver cancer, leukaemia, stomach cancer, other
Φ cirrhosis of the liver, other

road traffic accidents, poisoning, drownings, other
‡ ischaemic heart disease (IHD), cerebrovascular disease, other
** breast cancer, lung cancer, leukemia, other
Φ cirrhosis of the liver, other

Total Deaths	7,555,885	Total Deaths	996,707
Total Population	2,312,272,679	Total Population	597,682,683
Mortality Rate	0.3%	Mortality Rate	0.2%

Ages 45-59	Developing Countries			Developed Countries	
Cause of Death	# Deaths	% of Total Deaths	Cause of Death	# Deaths	% of Total Deaths
Cardiovascular diseases‡	1593447	27.3%	Cardiovascular diseases‡	569767	33.7%
Malignant neoplasms**	1051947	18.0%	Malignant neoplasms**	495519	29.3%
Unintentional injuries#	455323	7.8%	Unintentional injuries#	163055	9.6%
HIV/AIDS	386869	6.6%	Digestive diseasesΦ	116643	6.9%
Tuberculosis	362682	6.2%	Intentional injuriesΨ	89868	5.3%
Digestive diseasesΦ	340739	5.8%	Neuropsychiatric conditions	49873	2.9%
Chronic obstructive pulmonary disease	283883	4.9%	Lower respiratory infections	33114	2.0%
Intentional injuriesΨ	215023	3.7%	Diabetes mellitus	26317	1.6%
Diabetes mellitus	150126	2.6%	Chronic obstructive pulmonary disease	26179	1.5%
Lower respiratory infections	145994	2.5%	Tuberculosis	21530	1.3%

‡ ischaemic heart disease (IHD), cerebrovascular disease, other
** liver cancer, stomach cancer, lung cancer, other
# road traffic accidents, poisonings, falls, other
Φ cirrhosis of the liver, other
Ψ self-inflicted, violence

‡ ischaemic heart disease (IHD), cerebrovascular disease, other
# poisoning, road traffic accidents, falls, other
** lung cancer, breast cancer, colon cancer, stomach cancer, other
Φ cirrhosis of the liver, other
Ψ self-inflicted, violence

Total Deaths	5,844,812	Total Deaths	1,692,592
Total Population	600,316,766	Total Population	254,600,864
Mortality Rate	1.0%	Mortality Rate	0.7%

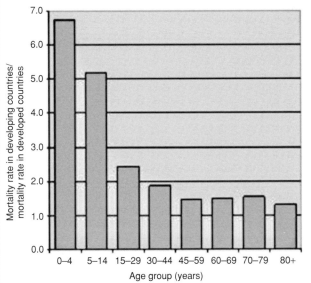

Sad Story: June 13, 2007
Kim Malawi

At St. John's yesterday on our rounds trying to find children to test, the nurses in the Pediatric Ward brought to our attention a four-month-old orphan.

The infant lives in a residential orphanage, which is very rare in Malawi. Orphans are usually taken in by a member of their extended family. To be in a residential orphanage means this little girl has nobody left. She was breathing extremely rapidly and with great difficulty. They had her on oxygen.

By the time we saw her, they had already done a rapid test, which came back positive. This indicates that the little girl's mother was certainly positive (and probably died of HIV) and that the baby was definitely exposed, if not infected. We managed to collect a dry blood spot sample to send to the lab, after a little difficulty. (She was so small and she was hard to put in the right position because of the oxygen connection.)

We're pretty sure she has PCP pnemonia, which is really, really bad (and really specific to HIV infected individuals) because she hadn't responded to 2 days of antibiotics. They put her on high-dose cotrimoxazole (the treatment for PCP) and hoped for the best. In this country, that's the best they can do. That's the limit of their options treatment-wise. It's really distressing that they have so few options.

She died overnight.

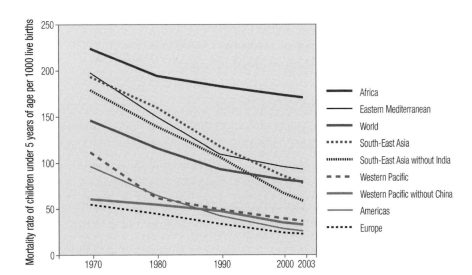

Figure 4.3. *Childhood mortality rates over time for several regions of the world. Used with permission from [5].*

of deaths to children under five occur during childbirth and the first week of life [5]. It has been estimated that 3.3 million babies are stillborn each year; these losses are not included in the estimates of the more than ten million infants and children who die each year throughout the world [5].

Tragically, most childhood deaths, both in infancy and childhood, still occur as a result of preventable and treatable causes such as inadequate care during pregnancy, unsanitary childbirth conditions, pneumonia (lower respiratory infection), diarrhea and malaria. The underlying cause of most of these deaths is poverty

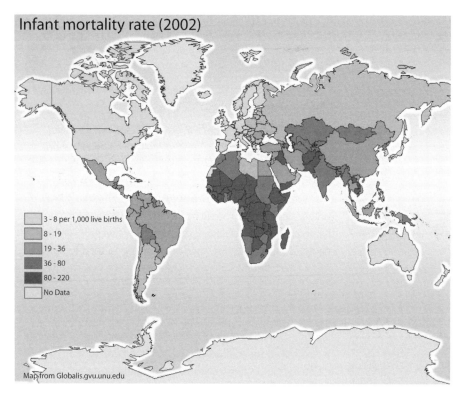

Infant mortality rate (2002)

3 - 8 per 1,000 live births
8 - 19
19 - 36
36 - 80
80 - 220
No Data

Map from Globalis.gvu.unu.edu

Figure 4.4. *Infant mortality rates throughout the world in 2002. Source: Globalis www. http://globalis.gvu.unu.edu/*

and the associated malnutrition, crowded and unsanitary living conditions, and lack of access to healthcare (Figure 4.4). Undernutrition and malnutrition contribute to more than half of childhood deaths. Unsafe drinking water and poor sanitation are responsible for nearly 90 % of deaths due to diarrheal diseases [4]. In the 42 countries where most childhood deaths occur, many children do not have access to healthcare. For example, 60 % of children with pneumonia failed to get the antibiotics they needed, and 70 % of children with malaria did not receive treatment [5].

In this chapter, we will consider the pathophysiology of the four leading causes of mortality for children under the age of five in greater detail, along with potential interventions to address these conditions. For children in developing countries, the four leading causes of death are as follows [2].

(1) **Perinatal conditions** (conditions during childbirth and the first seven days of life),
(2) **lower respiratory infections** (pneumonia),
(3) **diarrheal diseases**, and
(4) **malaria**.

For children in developed countries, the four leading causes of death are as follows [2].

(1) **Perinatal conditions**,
(2) **congenital anomalies** (birth defects),
(3) **lower respiratory infections**, and
(4) **unintentional injuries**.

Perinatal conditions

Pregnancy, childbirth and the seven days after is a particularly dangerous time, both for babies and their mothers. More than 2.5 million children die as a result of perinatal conditions and more than 500,000 women die as a result of complications of pregnancy and childbirth each year [6]. These perinatal mortality statistics probably underestimate the scope of the problem, since the vast majority of these deaths occur in developing countries where rates of vital registration are lowest. Only about one fourth of the world's births are registered and usually countries with the highest mortality rates have the lowest rates of vital registration [6].

Most deaths in the first week of life result from inadequate access to healthcare during pregnancy, during childbirth and immediately afterward. As a result,

(a)

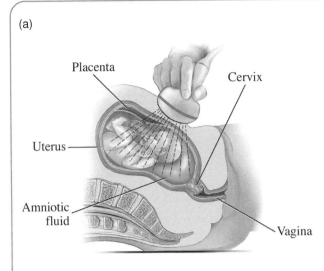

Placenta

Cervix

Uterus

Amniotic fluid

Vagina

(b)

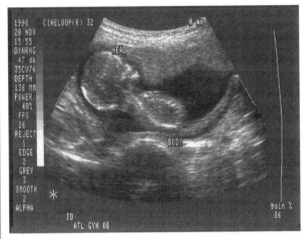

(c)

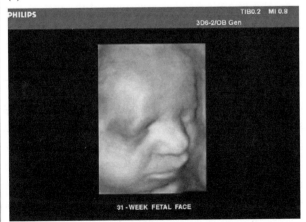

Figure (a) Medical illustration copyright © Nucleus Medical Art, all rights reserved. www.nucleusinc.com. (b) CDC/Jim Gathany. (c) Image provided courtesy of Philips Medical Systems.

Fetal ultrasound provides a window to examine gestational development. In developed countries ultrasound is routinely used to identify ectopic pregnancies (where the fertilized egg has implanted in the fallopian tube rather than the uterus), to determine the gestational age of the baby, to confirm the number of babies, to evaluate the baby's growth and provide early evidence of delays in growth which might require intervention, to study the placenta which provides nourishment to the baby and identify any problems, and to help identify congenital abnormalities.

In an ultrasound imaging procedure, a transducer which emits and receives ultrasonic waves is placed in contact with the mother's stomach (a). The transducer contains a quartz crystal called a piezoelectric crystal; when an oscillating electric current is applied to the piezoelectric crystal it changes shape rapidly and emits sound waves. Structures beneath the transducer reflect sound waves back toward the transducer; when these waves hit the crystal, it vibrates, producing an electric current. The time delay between incidence and detection depends on the speed of sound and the distance between the transducer and the reflecting surface. To provide a two dimensional image of a plane located in front of the transducer, we record the intensity of reflected sound as a function of time at each point along the transducer. Knowing the speed of sound in tissue, the time-dependent reflectance profile can be converted to yield the depth of reflective structures beneath each point of the transducer. These data can then be processed to yield an image of the structures underneath the transducer (b). Image resolution is determined by the frequency of the sound waves. Higher frequency waves have shorter wavelength, and thus better spatial resolution; however, higher frequency sound waves don't penetrate through as much tissue. Generally, fetal ultrasound uses sound waves with a frequency of 3–7.5 MHz.

A new technology, 3D ultrasound, creates volumetric images of the structures beneath the ultrasound transducer; these can be processed to display detailed surface renderings of the fetus (c). To obtain 3D images, sound waves are sent in at different angles, and the position of objects in the pyramid shaped volume beneath the transducer are calculated based on the reflectance profile.

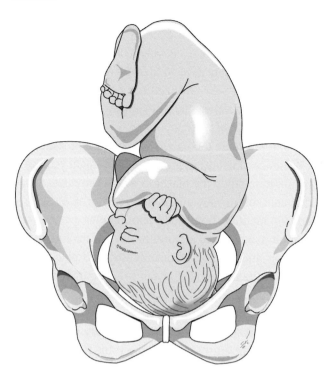

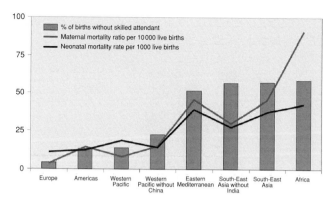

Figure 4.6. *Correlation between neonatal and maternal mortality rate and fraction of births without a skilled attendant. More than 529,000 women die in childbirth or shortly after due to severe bleeding, infection, high blood pressure and obstructed labor, most in developing countries. In developing countries, the consequences of pregnancy and childbirth represent the leading cause of death and disability for women of childbearing age. Used with permission from [5].*

Figure 4.5. *Cephalopelvic disproportion refers to the condition where the baby's head is too large to pass through the mother's bony birth canal. The resulting mechanical trauma can impede the progress of labor and can result in significant injury to both mother and child. Life ART image © 2009 Lippincott Williams & Wilkins. All rights reserved.*

many babies are born with low birth weight, many suffer asphyxia during birth or other birth trauma, and many acquire infections during childbirth; although these complications are frequently fatal, they are easy to prevent [6].

Proper nutrition during pregnancy and prenatal care can substantially reduce the risk of premature delivery and low birth weight. A delivery attended by a skilled healthcare worker (midwife, nurse or physician) dramatically reduces the risks of birth asphyxia and birth trauma [6]. Birth asphyxia occurs when the baby does not initiate and sustain normal breathing at birth. Asphyxia can occur in cases when the umbilical cord is wrapped around the baby's neck, or during a breech delivery. Skilled healthcare workers are trained to perform neonatal resuscitation to treat birth asphyxia. Birth trauma occurs during obstructed labor, when mechanical forces prevent descent of the baby through the birth canal. If the infant's head is too large to pass through the bony birth canal (Figure 4.5), or the baby presents in

a breech position (feet or bottom first), prolonged labor can result in severe trauma, both to mother and child.

Injuries to the baby can include intracranial hemorrhage, blunt trauma to internal organs such as the liver or spleen, and injury to the spinal cord or peripheral nerves – all of which can be fatal or result in lifelong disability [6]. Birth trauma is more common in developing countries. In these settings, a mother's growth may be stunted by malnutrition or young mothers may bear children before pelvic growth is complete; in these cases, the infant's head is frequently too large to pass through the birth canal.

The high perinatal mortality rate in developing countries is strongly related to conditions during childbirth. Most births in developing countries are not attended by a skilled healthcare worker; Figure 4.6 shows the correlation between neonatal and maternal mortality rates and the fraction of births which are unattended [5].

Unfortunately, there are no good screening tests to indicate those women who will need emergency medical care during childbirth and those who will not; thus, it is important that all births are attended by a trained healthcare worker. The effects of unattended birth extend beyond mortality; infants who survive traumatic birth frequently suffer lifelong disability such as cerebral palsy [5].

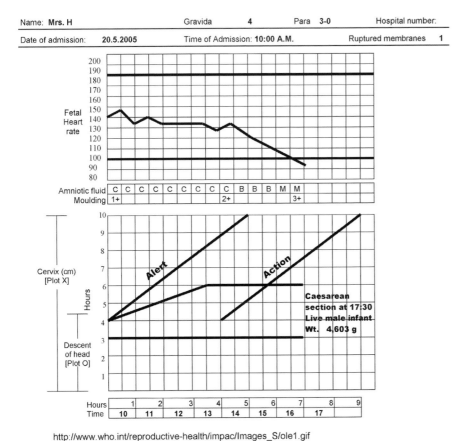

Name: **Mrs. H**   Gravida   **4**   Para   **3-0**   Hospital number:

Date of admission: **20.5.2005**   Time of Admission: **10:00 A.M.**   Ruptured membranes   **1**

Figure 4.7. *Partograph showing obstructed labor. Used with permission from [7].*

http://www.who.int/reproductive-health/impac/Images_S/ole1.gif

The WHO has developed an important tool for health-care workers to monitor and track the progress during labor – the partograph. Figure 4.7 shows a portion of a partograph, which graphs the progress of labor and helps anticipate the need for interventions, such as transfer to a hospital, before serious complications occur [6].

Infection is another important source of mortality during the perinatal period. Infections during the first week of life are usually acquired from exposure to organisms in the maternal genital tract during birth. Infants born prematurely are at higher risk for developing sepsis (bloodstream infection) as a result of this exposure. Sepsis is an extremely serious condition; left untreated, the fatality rate of sepsis is over 40% [6]. Because of high perinatal and neonatal mortality rates, in many cultures a child's birth is not celebrated until he or she has survived the first weeks of life and mother and infant are isolated until this period has passed. This practice can be helpful to the child because isolation reduces exposure to infectious agents in the environment. However, it can also result in delays in seeking medical care needed for infections acquired during childbirth.

Infections of the umbilical cord are a common problem in developing countries where many births occur in the home. The use of non-sterile instruments to cut the umbilical cord can result in infections of the cord; these infections can lead to sepsis. The non-profit organization PATH develops and disseminates new health technologies for low resource settings. To reduce the incidence of infection following childbirth, PATH has supported the development of a delivery kit (Figure 4.8) in Nepal to create a clean childbirth environment for home births. PATH evaluated the impact of a similar clean delivery kit in Tanzania, which resulted in a 13-fold reduction in umbilical cord infection [8].

Lower respiratory infections

Almost one million children die every year as a result of infections of the lower respiratory tract, the most

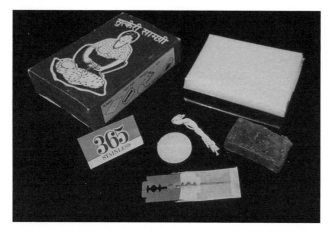

Figure 4.8. *To reduce the incidence of infection following childbirth, PATH has supported the development of a delivery kit in Nepal to create a clean childbirth environment for home births. PATH evaluated the impact of a similar clean delivery kit in Tanzania which resulted in a 13-fold reduction in umbilical cord infection. Courtesy of PATH.*

SARS

Severe acute respiratory syndrome (SARS) is a viral respiratory illness that frequently leads to pneumonia. SARS is caused by a previously unrecognized virus called the SARS associated coronavirus. Coronaviruses are a common cause of mild upper respiratory illnesses; the newly discovered virus causes a much more serious lower respiratory illness. Tests are currently underway to determine whether the SARS associated coronavirus responds to antiviral drugs.

SARS first emerged in Southern China in November of 2002 and quickly spread to more than 24 countries throughout the world. Between November 2002 and July 2003 more than 8000 people became ill with SARS and 774 died. By late July, 2003, the WHO declared the global outbreak to be over [11, 12].

---

serious of which is pneumonia [6]. Pneumonia is an infection of the lung, and represents a group of infections caused by multiple organisms, including viruses, bacteria and fungi. Pneumonia is a particularly serious infection because it can interfere with the ability to oxygenate blood within the lungs. Ordinarily, when we breathe, oxygenated air is drawn into millions of alveoli, tiny sacs within the lung at the tips of the airways. As shown in Figure 4.9, alveoli are surrounded by capillaries which bring in deoxygenated blood from the right ventricle via the pulmonary artery. Oxygen diffuses across the alveolar membrane where it binds to hemoglobin within red blood cells, and the resulting oxygenated blood is collected in the pulmonary vein, where it travels to the left side of the heart to be pumped throughout the body. In attempting to fight off a lower respiratory infection, the body's immune system produces fluid and pus that can fill the alveoli, interfering with crucial gas exchange. The symptoms of pneumonia include fever, cough, chest pain, and breathlessness. Treatment with antibiotics has since substantially reduced the mortality due to pneumonia; between 1936 and 1945 mortality due to pneumonia dropped 40% due to the widespread availability of antibiotics [9]. Current risk factors for child mortality due to pneumonia include living in poverty, low parental education level, environmental pollutants such as cigarette smoke or smoke from firewood burning in the house, low birth weight, malnutrition, and HIV/AIDS [10].

About half of pneumonias are produced by bacterial infection and about half by viral infection. The most common causes of bacterial pneumonia include *Streptococcus pneumoniae*, *Haemophilus influenzae*, *Staphylococcus aureus* and pertussis. The most common viral causes include respiratory syncytial virus (RSV), influenza virus, parainfluenza virus and measles [10]. **SARS** is an emerging cause of pneumonia.

Bacterial pneumonias are treated with antibiotics; the choice of antibiotic depends on the causative factor. Viral pneumonias usually resolve on their own without treatment. In severe cases, oxygen can be administered, and antiviral drugs can be used. Therefore, in diagnosing pneumonia it is particularly important to ascertain whether the cause is viral or bacterial.

In developed countries, the presence and location of pneumonia can be confirmed by a chest X-ray (Figure 4.10). A complete blood count is performed to determine whether the infection is likely to be bacterial or viral;

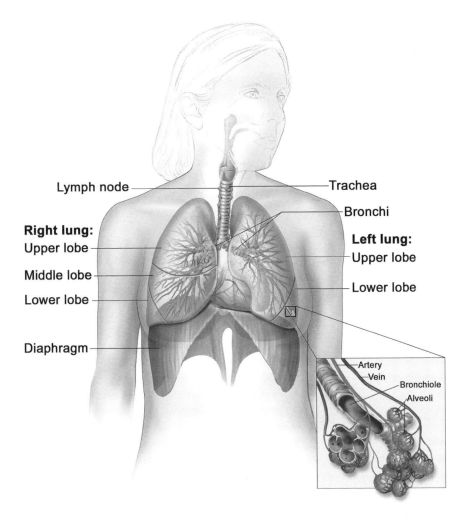

Lymph node

Trachea

Bronchi

Right lung:

Upper lobe

Middle lobe

Lower lobe

Left lung:

Upper lobe

Lower lobe

Diaphragm

Artery

Vein

Bronchiole

Alveoli

National Cancer Institute

Figure 4.9. *Alveoli are the site of gas exchange within the lung. In a lower respiratory infection, alveoli can fill with fluid and pus interfering with gas exchange. Source: © 2006 Terese Winslow, US Government has certain rights.*

as we will see in Chapter 8, the types of immune cells present in blood differ with a bacterial or viral infection. To determine the cause of bacterial pneumonia, sputum produced when the patient coughs is examined under the light microscope to examine bacterial shape and staining characteristics (Figure 4.11).

In some cases, fluid is obtained directly from the lungs using a long, skinny needle which is advanced into the area of lung consolidation under the guidance of computed tomography. This fluid is cultured to determine which bacterial organisms are present and to determine which antibiotics they are responsive to. In addition, blood is obtained and cultured as well. Recently, rapid tests have been developed to identify the cause of viral pneumonias from nasal swabs. These tests are

based on technology known as **direct immunofluorescence assays** (DFA) or enzyme-linked immunosorbent assays (ELISA). Accurate tests have been developed to detect influenza, RSV, parainfluenza, and other viruses [13].

In developing countries, diagnosis of pneumonia is made primarily on the basis of clinical signs and symptoms and sometimes a chest X-ray. Signs of particular importance include rapid breathing higher than 40 breaths/min in children over one year of age, chest indrawing, cyanosis (blue nailbeds), and poor feeding [10]. In mild pneumonia children will exhibit rapid breathing; chest indrawing is also present in moderate pneumonia. These symptoms are present in addition to cyanosis and poor feeding in severe pneumonia. In these

Direct immunofluorescence assays

To rapidly determine whether the cause of a patient's pneumonia is viral, a DFA test is used. DFA tests whether cells lining the upper respiratory tract are infected with viruses which cause many lower respiratory infections. If these cells are infected, it is assumed that the pneumonia is viral in origin.

In the test, nasal secretions are collected. A centrifuge is used to concentrate the cells into a small pellet. A drop containing these cells is placed on a slide and the cells are allowed to dry. The cells are immersed in alcohol to permeabilize the cell membranes. A solution containing monoclonal antibodies which bind tightly and specifically to viruses such as influenza, adenovirus, parainfluneza, and RSV is applied to the slide. These antibodies are coupled to a dye which shows bright green fluorescence when illuminated with blue light. The unbound antibody is washed from the slide. The slide is examined under a fluorescence microscope. The DFA test will stain all virus infected cells bright apple green [13].

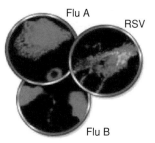

Result of DFA tests showing cells which stain positive for viral infection. The healthcare provider examines the cells under a fluorescence microscope. To yield accurate results, at least 20 cells must be examined. The right to use this figure has been granted by Millipore Corporation to Dr. Rebecca Richards-Kortum of Rice University. Any further reproduction or use of this figure must be authorized by Millipore Corporation.

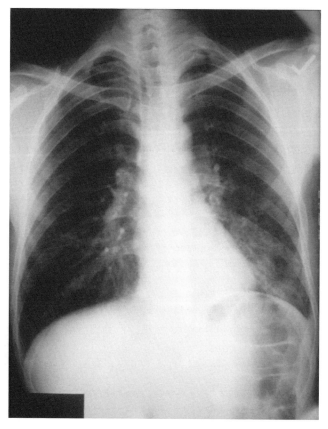

Figure 4.10. *Chest X-ray indicating pneumonia in the lower lobe of the left lung. Source: CDC/Dr. Thomas Hooten.*

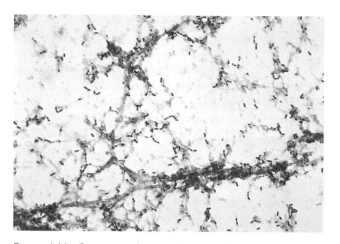

Figure 4.11. *Gram stain of sputum indicating* Streptococcus pneumoniae. *Source: CDC/Dr. Mike Miller.*

settings, maternal education regarding when to seek medical care for pneumonia is particularly important.

Because of the difficulty in discriminating between viral and bacterial pneumonias on the basis of symp-

toms, the WHO recommends treating all pneumonias in children in developing countries with antibiotics [10]. This conservative approach ensures that all children who need therapy receive it and has proven to

Antibiotic resistance

Widespread use of antibiotics began during World War II and led to dramatic reductions in the mortality of infectious diseases. Over the past 60 years, some bacteria have evolved resistance to commonly used antibiotics, and today, many physicians and scientists are concerned by the increasing difficulty of treating infections caused by antibiotic resistant strains of bacteria.

If you are admitted to the hospital in the USA today, you have a 5–10% chance of acquiring an infection in the hospital. Some 70% of bacteria that cause hospital acquired infections are resistant to at least one drug. Resistant infections are more difficult and expensive to treat and more than 90,000 people die each year in the US from hospital acquired infections, up from only 13,300 deaths in 1992. Antibiotic resistance is not just a problem in hospitals; recently, multi-drug resistant strains of *Staphylococcus aureus* have been found in locker rooms.

Antibiotics kill bacteria which are sensitive to the drug; this efficiently selects out those bacteria resistant to the drug. Widespread use of antibiotics has spurred evolutionary changes in bacteria that allow them to resist the action of drugs. For example, penicillin kills bacteria by attaching to and destroying a key part of the wall. In response, microbes can alter their cell walls to prevent binding, enabling them to evade the action of the drug. Since bacteria are single celled organisms with a small number of genes, even a single random gene mutation can greatly affect the ability to cause disease. Bacteria divide every few hours, so they can evolve rapidly and a mutation that helps a bacterium to survive drug exposure can quickly become dominant in that population. Complicating the problem, bacteria can trade genes by exchanging plasmids – in effect, one strain of bacteria can develop resistance to one or more drugs and then efficiently transfer this resistance to other strains of bacteria.

The problem of drug resistance is exacerbated when antibiotics are overused or when patients don't finish taking their prescription for antibiotics – drug resistant microbes can proliferate and be spread. Increases in antibiotic resistance have occurred in parallel with increased antibiotic use. Later in this chapter, we will see the worldwide problems associated with multi-drug resistant tuberculosis. Of particular concern today is the emergence of strains of bacteria resistant even to Vancomycin – the antibiotic of "last resort" for treating resistant infections [14, 15].

reduce mortality associated with pneumonia in developing countries. For example, in one trial in India, child mortality rate due to pneumonia was 13.5% when untreated, and only 0.8% when treated with antibiotics [10]. However, one danger is that more **resistant strains of bacteria** will emerge from overuse of antibiotics.

A number of vaccines can prevent infections which cause pneumonia. For example, the Hib vaccine protects against *Haemophilus influenzae*. The WHO recommends worldwide Hib vaccination; unfortunately, only about 10% of the world's children have been vaccinated [10]. Pneumococcal vaccine protects against *Streptococcus pneumoniae*. Yearly flu vaccination can protect against pneumonias caused by the influenza virus.

Diarrheal diseases

Diarrhea is a gastrointestinal disorder characterized by frequent watery stools. Diarrhea is caused by bacterial or viral infection of the gastrointestinal tract; the most common causes are infection with the *Escherichia coli* bacterium, Rotavirus and the *Vibrio cholerae* bacterium [16]. Diarrheal disease is uncommon in neonates who are often isolated and exclusively breastfed; instead it usually occurs in older infants and

Yay! I feel incredibly good about what I just did! I just finished creating a film! I'm so excited with having made it that I just need to write about it (get it all out) right away.

The film is only a short one, I'll admit, but it fulfills a very important purpose here at the clinic. You see, now that the World Food Program distribution has taken off, Tessa and I spend every morning (from 7:15 to noon) registering patients, filling out record books to keep track of the amount of food we give out, and, of course, handing out food. We also have to somehow get across to the patients, who have never done this before, what exactly is going on. This is rather difficult as we have only a few choice SiSwati vocabulary words and our patients' English varies greatly and is often non-existent. We have to explain who is eligible to receive food – people on ARVs are, people about to start ARVs aren't, etc. We have to try to explain to each patient that they must bring the little green card we give them back to the clinic every time they come and that they must also bring a plastic bag in which to carry their food in. And on top of that, as Corn Soy Blend requires a different preparation than the food they are used to, we must explain to them the steps necessary to cook it.

Originally, we planned on just having one of the translators explain this to the patients several times each day. This approach worked quite nicely at first, but on busier days in the clinic, it is sometimes impossible to find a translator, as they are all at work in the exam rooms translating for the doctors. And, on top of that, it seems that more times I ask them, the less the translators are willing to give the same speech over again. (If I had a lot of work to do and was asked to yell the same instructions three times a morning, I wouldn't be too happy either.)

It was this problem that sparked my idea of making a video. I thought that if I could make a video of the translators' speeches and show it several times a day in the waiting-room, I would be able to kill two birds with one stone: it would make our jobs easier by making us less dependent on the translators to get our message across, and it would make the translators happy because they would be freed from their monotonous task. And, I happened to have exactly the bare minimum in terms of equipment that I would need to make it: my picture camera (which has a video option) and a computer.

So, I set to work.

I decided that the first thing I would need to do was get the approval of the translators. I needed their help – as they would be my actors – and I wanted to get their advice on what the film should say so that it would be effective and would come across clearly to the Swazi people. I sat down with them in the lobby and we talked for quite some time. As a group, they were quite willing to help out, but when it came to acting, none of them was very ready to volunteer (I think they were shy – about appearing on film). It took some work, but I finally convinced three of them – Lulu, Nomsa and Bongiwe to do it.

Then, I sat down right in the lobby, and, with my little camera, filmed Lulu (who seems to be Queen Bee of the lot) explaining the rules and regulations of the program. And Lulu did not fail me; a born star (you can tell she thinks so), with stage voice and lines memorized, she did it perfectly in one take. The second bit didn't work out quite so well. I tried Nomsa at first. Nomsa, however, was a little bit too bashful for the big screen and kept forgetting what she was saying. Before I could attempt coaching her to success, she was called away to clinic duties. I was left with Bongiwe to finish the job. She, too, was shy, but Lulu urged her (somewhat forcefully) to conquer her fears and "just do it." And so she did. The next day, however, when putting the movie together, I realized that Bongiwe's voice was far, far too soft, and so I had to have Bongiwe try again. This time, we filmed in the board room in order to have fewer distractions. And boy was I surprised – I don't know if it was the room or the timing or what, but Bongiwe laid it out like a professional. Go Bongiwe!

So, now that I had my film clips, I just had to put it all together. I uploaded the clips on to my computer and then into a program called Windows Movie Maker, which is a very primitive film editing program. Then, I spent quite a while discovering the ins-and-outs of the program and experimenting with different video effects. I put the clips in sequence, trimmed off the beginning "and GO!" or "ACTION!" parts that I had accidentally filmed, and created title pages. Then, to involve Nomsa (after her film debut didn't work out as planned), I had her help me translate the titles into SiSwati.

Later, Sipho (pronounced See-po) came into the room and saw my project and was very excited about it. He told me that he had made a couple of movies in the past, so I, wanting input, had him watch my film. He loved it, but he had an idea to make it even better: what it really needed was a soundtrack!

I thought this was a wonderful idea, so we started looking for some music on my computer. But then, Sipho had another great idea: We needed to choose a song that our audience (Swazi patients from ages toddler to granny) would enjoy – and the best song to do that would certainly be an African one. And he knew just who to get it from! "Mlungisi," he told me, "has all the best African music."

So, we found Mlungisi in the filing room (he works at the clinic as a data clerk) and he was quite happy to have been thought of first when it came to good music taste. He helped us find a song that, he said "would be perfect." The song he found was called "Mama ka Sibongile" by the Soul Brothers. Sipho and Mlungisi were both very satisfied with the choice because, apparently, the Soul Brothers are one of the best loved groups in Southern Africa. I hurried back to my computer to try it out.

The song worked wonderfully – it was quite up-tempo and peppy and was mostly instrumental (which is nice because lots of lyrics would have muddled Lulu and Sibongiwe's words). I had to splice up the song according to the video segments and play with volume levels so that it would be louder during the titles and softer during the speaking parts. It took a while, but I got it eventually.

And, as no film is complete without the ending credits, I added those in too. Lulu, Sibongiwe, Mlungisi (whom I listed as musical director), and Sipho (whom I listed as technical support) were all quite excited to see their names scrolling boldly across the screen.

Finally, I had to transfer the movie from my computer to another and then burn it onto a DVD. Now this is where I would normally break down and want to give up (computers and I don't often get along), but wonderful, amazing Sipho stayed with me (an hour past the time he usually goes home) and helped me every step of the way. And though we encountered quite a few difficulties, with Sipho's help we were able to keep trudging forward and finally we had done it. We had our DVD.

With fingers crossed (so, so tightly), we walked downstairs and popped it into the DVD player. I think I might have even closed my eyes – I was really on pins and needles worrying that after all our work, something would go wrong.

But, when Sipho pressed play . . . there it was!!! My very short film in all its glory. I jumped in the air with excitement, and Sipho and I exchanged a very heartfelt high five.

I raced upstairs to find Carrie and Julia (two of the doctors who are my advisors-of-sorts) to have them come watch the film. On the way downstairs, Carrie stopped by Busi's office and asked if she would like to come see the film. (Busi is the Executive Director, so everything in the clinic must be approved by her.) From the first title screen, they all had huge smiles and were already talking about how impressed they were. After the film was over, they all congratulated me on my work. Busi said that she wished that I was here earlier so that I could have made a similar video of some of the patients speaking about living with HIV at an event they had at the clinic for the RED Campaign. She also asked me to make another video of the

support group's meeting on Saturday so that they can show it to other visitors. I felt quite honored by her request. The film wasn't exactly Oscar material, but it was a good first attempt. (Hey, I'm sure even Spielberg wasn't that good on the first go!) While watching the film on the TV, I saw quite a few things I would like to change. I thought about what I would do differently next time to make the film better. (Like using an actual video camera.;))

What I love most about the entire experience is how inspired I now feel to do more. I feel like I've just opened the door to so many possibilities. I've realized that, using film, I can express all of the messages that I wanted to with the card project (discussed in an earlier blog) – proper hand-washing, hygiene, using clean water, proper infant feeding, etc. – and I can do it in a way the patients (who often cannot read) will be able to understand much more easily. And, if I do make films about these subjects, it will be incredibly helpful to the doctors.

As it is, they must try to express all of these points to every patient they see. And, with so many patients, they are pressed for time. Perhaps with a detailed video, they would be able to really take these messages in. And, some of these basic messages are so important to get across. By using clean water, a mother can protect her baby from diarrhea (one of the major causes of infant mortality here). By adding just a teaspoon of oil to the baby's porridge, a mother can prevent her child from improper brain development and stunting. These and many other problems have relatively simple solutions. Simple behavioral changes can prevent the loss of a child's life.

I'm glad that I can do something that might actually help. I really can't wait to do more. And I love that I can't wait to do more.

children, particularly in situations where safe sources of drinking water are not available [4]. Most children can recover from these gastrointestinal infections; however, diarrhea can lead to loss of substantial body fluids, sodium, chloride and other electrolytes. The symptoms of dehydration include low blood pressure, fainting, sunken eyes and fontanelle (soft spot in baby's head) [17]. This dehydration can quickly lead to death, particularly in babies and malnourished children. Diarrhea leads to 2.2 million deaths per year throughout the world, most in children in developing countries [18]. This is the equivalent of 20 jumbo jets crashing with no survivors every day.

How does diarrhea produce such rapid fluid loss? Ordinarily, during digestion, food is mixed with large quantities of water in the stomach. As the mixture passes through the colon, 98% of this water is reabsorbed along with electrolytes and nutrients, leaving solid waste to be eliminated [19]. Bacterial or viral infection interferes with the processes that control fluid reabsorption, leading to frequent, watery stools. The loss of just 10% of body fluids can prevent maintenance of adequate blood pressure and is enough to cause death [17].

In the late 1960s, a simple new treatment was developed to prevent the fatal dehydration that frequently accompanies diarrheal disease. This inexpensive treatment, oral rehydration therapy, reduced mortality due to diarrhea from 4.6 million deaths per year to 1.5 million

deaths per year in 2000 [20]. In 1978, the development of oral rehydration therapy was called the "most significant medical advance of the century" by the prestigious medical journal *The Lancet* [21]. To understand how a simple mixture of one liter of boiled water, one teaspoon of salt and eight teaspoons of sugar can save so many lives, we must first understand some physiology of digestion.

The epithelial cells which line the colon are responsible for fluid reabsorption. Water moves from the lumen of the colon, across this epithelium and back into the blood vessels in response to osmotic gradients. In the upper gastrointestinal tract, the epithelium absorbs osmotically active products of digestion such as amino acids [16]. This reduces the tonicity of the solution in the lumen, and water passively leaves the lumen to equalize the osmotic pressure gradient. In the lower digestive tract, sodium is actively pumped out of the lumen, and again water follows.

There are several mechanisms by which colonic epithelial cells pump sodium out of the lumen. Toxins produced by bacteria which cause diarrhea (such as *E. coli* and *V. cholerae*) bind tightly to the luminal surface of the colonic epithelial cells and cause these cells to secrete chloride ions into the lumen; at the same time, these toxins interfere with one of the most important mechanisms of sodium reabsorption [16]. The reduced ability to pump sodium out of the lumen results in watery diarrhea (Figure 4.12).

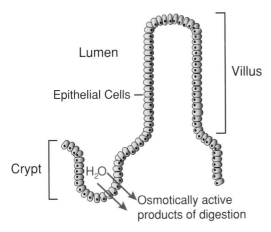

Figure 4.12. *Epithelium absorbs osmotically active products of digestion, reducing the tonicity of solution in lumen. Water leaves lumen due to osmotic pressure gradient. Toxins produced by bacteria interfere with the ability to reabsorb sodium and cause epithelium cells to secvete chlorine into the lumen, resulting in watery diarrhea.*

Cholera is associated with a particularly severe diarrhea because the *Vibrio cholerae* bacterium produces a toxin which binds so tightly to epithelial cells in the gut that it remains active until the epithelium regenerates itself, which occurs every five to seven days [16].

Simply providing water or even salt water to cholera victims does not prevent dehydration, it simply adds to the volume of diarrhea. However, in the 1950s and 1960s researchers discovered a new mechanism of sodium reabsorption in the colon which was coupled to the transport of glucose [16]. This coupled glucose-sodium transporter was found to be unaffected by the cholera toxin, and in the presence of glucose, sufficient sodium reabsorption can occur to prevent dehydration. This is the basis for oral rehydration

Cholera

Cholera is spread by drinking water or food contaminated with the cholera bacterium, *Vibrio cholerae*, or by coming in contact with feces of an infected person. Approximately one in 20 people infected with the cholera bacterium develop severe diarrhea. Without oral rehydration therapy, death can occur within a few hours. In countries with advanced water treatment and sanitation systems, cholera is rare. However, cholera can spread rapidly in crowded areas with inadequate sewage and water treatment. Today, cholera is prevalent in India, Bangladesh, Pakistan, Indochina, Indonesia, Afghanistan, Africa, South America and Mexico.

Outbreaks of violence or natural disasters often lead to conditions that cause cholera outbreaks. Beginning in April, 1994 the extremist Hutu militia slaughtered nearly one million Tutsis in Rwanda over a period of just 100 days. In July, the Tutsi led Rwandese Patriotic Front overthrew the genocidal Hutu government, resulting in a massive exodus of Hutus to the neighboring countries of Burundi and Zaire. More than one million Hutus entered Goma, Zaire over a three to four day period in July, 1994. Cholera spread rapidly in the refugee camp and at the height of the epidemic, more than 2000 people died each day. Over the course of the epidemic, more than 46,000 lives were lost. Ironically, it was the cholera outbreak and loss of life amongst the genocidaires which captured the attention of the western world and aid agencies [16, 22–24].

The book *We wish To Inform You That Tomorrow We Will Be Killed With Our Families* by Philip Gourevitch provides a compelling account of the factors that led to genocide in Rwanda, the events of 1994, and the world's response.

The WHO publishes a short brochure to guide aid workers who suspect a cholera outbreak:

http://www.who.int/topics/cholera/publications/en/first_steps.pdf

WHAT TO DO IF YOU SUSPECT AN OUTBREAK

- **Inform and ask for help**
- **Protect the community**
- **Treat the patients**

■ Inform and ask for help

The outbreak can evolve quickly and the rapid increase of cases may prevent you from doing your daily activities

- Inform your supervisor about the situation
- Ask for more supplies if needed (see **Box**)
- Ask for help to control the outbreak among and outside the community

Check the supplies you have and record available quantities

- ➡ IV fluids (Ringer Lactate is the best)
- ➡ Drips
- ➡ Nasogastric tubes
- ➡ Oral Rehydration Salt (ORS)
- ➡ Antibiotics (see **Table 2**)
- ➡ Soap
- ➡ Chlorine or bleaching powder
- ➡ Rectal swabs and transport medium (Cary Blair or TCBS) for stool samples
- ➡ Safe water is needed to rehydrate patients and to wash clothes and instruments

therapy. Researchers theorized that adding an oral rehydration solution containing both salt and sugar could enable sufficient sodium absorption to compensate for the effect of the toxins. This was confirmed in studies in the late 1960s in India and Bangladesh which showed that oral rehydration solutions containing glucose and sodium resulted in net reabsorption of fluid into the bloodstream of patients with cholera [16]. In 1975, the WHO and UNICEF agreed to promote a single oral rehydration solution containing 90 mm sodium, 20 mm potassium, 80 mm chloride, 30 mm bicarbonate, and 111 mm glucose [25]. Today, a packet of oral rehydration salts costs about ten cents (Figure 4.13).

While oral rehydration salts prevent dehydration, they don't reduce the volume of diarrhea. This is because the glucose present in these solutions is absorbed as fast as it is delivered [16]. Newer solutions under development may act both to prevent dehydration and to reduce diarrhea volume. These new solutions contains glucose polymers. These polymers are not instantly reabsorbed – instead they are slowly broken down into glucose in the intestine. This glucose then permits sodium reabsorption in the same manner as today's oral rehydration solutions. Solutions containing glucose polymers provide a continuous source of glucose to facilitate reabsorption, and can be thought of as a "glucose battery" [16].

While diarrhea leads to far fewer deaths in developed countries, it is still an important cause of childhood morbidity. Diarrhea is the number two cause of visits to the pediatric emergency room in the USA [16]. Although oral rehydration therapy is inexpensive, highly effective, and widely used throughout developing countries, many physicians in developed countries fail to use it to treat mild or moderate dehydration due to diarrhea. Instead, these physicians rely on rehydration using intravenous fluids, a method which is more painful and more expensive. In one survey, training directors in pediatric emergency medicine were asked to recommend treatment in ten hypothetical scenarios. In each case, treatment with oral rehydration salts was the appropriate therapy; yet, only 17.2% of directors believed this was the best therapy and only 6.7% of directors said they would use oral

(a)

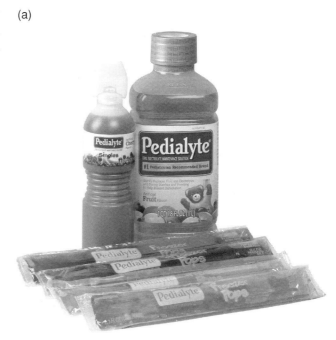

(b)

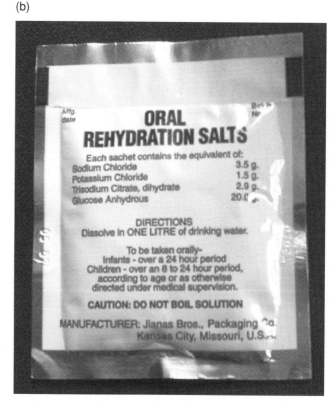

Figure 4.13. *Oral rehydration solutions contain both salt and sugar. (a) Pedialyte is one of several commercially available oral rehydration solutions in the USA (b). Photograph (a) is used with permission: Ross Products Division, Abbott Laboratories Inc., Columbus, Ohio: Pediatric Nutritional Product Guide © 2007. Photograph (b) is used with permission from Jianas Brothers Packaging Company.*

rehydration salts in every case [19]. The reluctance to use oral rehydration therapy in the USA is likely owing to early experience with the first commercially available solutions in the USA. A number of patients developed elevated sodium levels (hypernatremia); this occurred because the carbohydrate concentration was initially too high [19]. In addition, the initial product was dispensed in powder form and incorrect mixing exacerbated the problem. Most US physicians are still uncomfortable with oral rehydration therapy even though many clinical trials have shown that treatment with current solutions does not result in hypernatremia.

Vaccines to prevent infections such as rotavirus that cause diarrheal diseases are under active development and testing. Infection with rotavirus causes 29–45% of all deaths due to diarrheal diseases [26]. Rotavirus is found in every country and the incidence of rotavirus infection does not seem to go down as improvements are made in sanitation and hygiene [26]. Rotavirus is highly contagious; almost every child will have one infection with rotavirus before they are two years old (Figure 4.14) [27]. In the USA, 55,000 children are hospitalized each year due to rotavirus [27]. Rotavirus infection causes explosive, watery diarrhea and sometimes is accompanied by vomiting. Although oral rehydration therapy is effective at preventing dehydration associated with rotavirus infection, parents and caregivers often stop giving it owing to vomiting [28].

A vaccine called RotaShield designed to prevent rotavirus infection was approved by the US Food and Drug Administration in 1998. Prior to approval, clinical trials were carried in the USA, Venezuela, and Finland to test the safety and effectiveness of the rotavirus vaccine. These tests showed that the vaccine was 80–100% effective and found no evidence of statistically significant serious adverse side effects [29]. After the vaccine came into widespread use in the United States a small number of infants who had been vaccinated developed a severe complication, bowel obstruction, which can be fatal. This complication developed in about one out of every 10,000 infants vaccinated and as a result the manufacturer withdrew the vaccine from

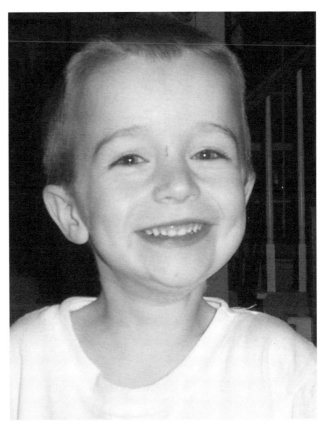

Figure 4.14. *One of the author's sons who developed a severe rotavirus infection as a baby in 1998, one month before the vaccine was commercially available.*

the market in 1999 [30]. The decision to withdraw the vaccine in the USA generated significant controversy. Researchers argued about whether it was ethical to continue trials of the rotavirus vaccine in developing countries and whether the high mortality associated with rotavirus infection could possibly justify the small risk of serious complications. It had been estimated that widespread use of the rotavirus vaccine in developing countries would result in 2000–3000 deaths per year due to vaccine related complications, but could potentially save the lives of 500,000 children per year [30]. We will revisit this debate in Chapter 9 when we consider the ethics of research involving human subjects.

In 2006, two new vaccines to prevent rotavirus completed large clinical trials in more than 130,000 infants in the USA, Latin America and Finland [31]. These vaccines are based on a different technology, less likely to

Malaria risk areas, 2005

Areas where malaria transmission occurs
Areas with limited risk
No malaria

This map is a visual aid only. It is not a definitive source of information about malaria endemicity

Source: ©WHO, 2005

Figure 4.15. *Areas of malaria risk as of 2005. Used with permission from [33].*

produce side effects, which we will examine in Chapter 8. Trial results show between 85–98% reduction in serious illness and 42–63% reduction in hospitalizations with no serious complications [31]. While these results are very encouraging, current estimates place the cost of the vaccine at approximately $100 per dose – too expensive for developing countries where rotavirus causes the most mortality [31].

Malaria

Malaria is a life threatening disease spread by mosquitoes which carry a parasite that can cause disease in humans. Forty percent of the world's population lives in malaria endemic countries (Figure 4.15), and more than 300 million cases of malaria occur each year throughout the world [32].

Each year on average, African children suffer between 1.6–5.4 episodes of malarial fever [34]. More than 1 million children under the age of 5 die as a result of malaria; the vast majority of these children live in sub-Saharan Africa [32].

Malaria carrying mosquitoes harbor a parasite that can be transmitted to humans when bitten. There are four species of malaria parasite that infect humans. The most deadly of these, *Plasmodium falciparum*, (Figure 4.16) is the species most commonly found in mosquitoes in sub-Saharan Africa [35].

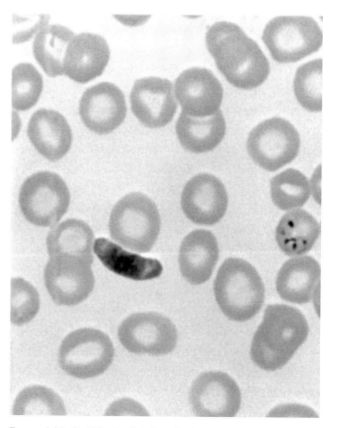

Figure 4.16. *Red blood cells infected with the malaria parasite* Plasmodium falciparum. *Diagnosis of malaria can be confirmed by examining red blood cells with the light microscope. Source: CDC/ Steven Glenn, Laboratory & Consultation Division.*

Malaria parasites are capable of evading the human immune system; they initially travel to the liver where they infect liver cells and multiply. The infected liver cell bursts, and the daughter parasites can then attach to the surface of red blood cells, where they consume the hemoglobin and divide. The red blood cell then ruptures, releasing more daughter parasites which can go on to destroy other red blood cells. Symptoms of malaria first appear 9–14 days after infection when daughter parasites are released from liver cells and include fever, headache, vomiting, and flu-like symptoms [32]. Malaria kills by destroying oxygen carrying red blood cells; the destruction of these cells can lead to severe anemia. The debris produced when red blood cells are destroyed can clog the capillaries that carry blood to the brain (cerebral malaria) depriving the brain of crucial oxygen supply [32]. Many children who survive cerebral malaria are left with permanent neurological problems, including blindness and epilepsy [34].

Pregnant women have increased susceptibility to malaria. The resulting maternal anemia can result in babies with low birth weight and malaria can be transmitted from mother to child across the placenta [36]. Insecticide treated bed nets offer a cost effective solution to prevent malaria in pregnant women and infants. Since malaria carrying mosquitoes bite predominantly at night, encouraging mothers and infants to sleep under mosquito nets treated with insecticide can reduce malaria incidence, the incidence of low birth weight babies and malaria deaths in children [36]. In Kenya in the 1990s, residents were bitten 60–300 times a year by malaria carrying mosquitoes before control measures were put in place [37]. Providing insecticide treated nets (Figure 4.17) for pregnant women and their babies to sleep under resulted in a 25% reduction in low birth weight babies and a 20% reduction in deaths in young children [34, 36]. Insecticide treated bed nets cost about $1.70; however, they must be retreated with insecticide each year, which costs about three to six cents per treatment [34]. Long lasting insecticide treated nets which are effective for up to four years are under development [34].

Several drugs are available to treat malarial infection. The least expensive of these are chloroquine

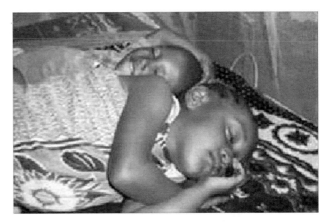

Figure 4.17. *Young children sleeping under an insecticide treated net. Courtesy of WHO.*

(13 cents/course), sulfadoxine-pyrimethamine (14 cents/course), and quinine ($2.68/seven day course) [37]. Unfortunately, past inappropriate use of antimalarial drugs has resulted in malaria parasites with high levels of resistance to drugs. The use of antimalarials on a large scale, given as monotherapies, introduced in sequence, and continued use even in the face of high levels of resistance has made the treatment of malaria more difficult [38]. Resistance to chloroquine is common in Africa and resistance to sulfadoxine-pyrimethamine is increasing [35]. New treatments for resistant malaria include artemisinin containing drugs such as artesunate. These drugs are derived from a Chinese plant; currently, it takes six to eight months to grow the plant, and another two to five months to process the plant and yield the drug [38]. Combination therapies including artemisinin cost about $1 per course. No resistance has been observed to these new therapies yet.

Congenital anomalies

About 2–3% of all children are born with a birth defect and more than 400,000 children die every year as a result [2, 6]. As the health of a population improves, the proportion of mortality attributed to congenital anomalies increases. In developing countries, birth defects account for less than 4% of childhood deaths, whereas in developed countries they account for 16.9% of childhood deaths [2].

Table 4.2 shows that the causes of congenital anomalies can be grouped into three categories: those caused

Table 4.2. *Cause and classification of selected birth defects [6].*

Cause	Classification	Example
Genetic	Chromosomal Single gene	Down syndrome Cystic fibrosis
Environmental	Infectious disease Maternal nutritional deficiency – folic acid	Congenital rubella syndrome Neural tube defects
Complex	Congenital malformations involving single organ system	Congenital heart disease

by chromosomal abnormalities and single gene defects, those caused solely by environmental exposure or nutritional deficiencies, and those with complex or unknown causes which may include interaction between several genes and possibly environmental factors [6]. Sporadic losses or rearrangements of genetic material have been estimated to affect 10% of conceptions and more than 90% of these events are incompatible with development and result in spontaneous abortion [6]. Infants born with such defects can suffer congenital malformations and mental retardation. Advanced maternal age is a risk factor, and increases particularly after age 35. As a result, such defects are relatively more common in developing countries, where the fraction of births to women over 35 years of age is 11–15% compared to only 5–9% in developed countries [6].

Environmental causes of birth defects include exposure to infectious agents, exposure to teratogens in the environment, as well as maternal nutritional deficiencies. Exposure to infectious agents during gestation such as malaria or rubella can result in serious birth defects. Congenital rubella syndrome results from maternal infection with rubella in early pregnancy. This syndrome includes blindness, deafness, cardiovascular defects and severe mental retardation [5]. Rubella epidemics occur every four to seven years in populations that have not been immunized and are of particular concern in developing countries [6]. Maternal consumption of alcohol, particularly binge drinking, can lead to **fetal alcohol syndrome**, which is characterized by altered facial features, fetal growth reduction, and cognitive defects [6].

Learn more about fetal alcohol syndrome

The Broken Cord by Michael Dorris chronicles Dorris' life as a single father after adopting a three year old child who suffered from fetal alcohol syndrome. Dorris was the first single male in America to adopt a child. His book was pivotal in encouraging Congress to pass laws regarding warnings about fetal alcohol syndrome on alcoholic beverage labels. The book won a National Book Critics Circle Award in 1989 [39].

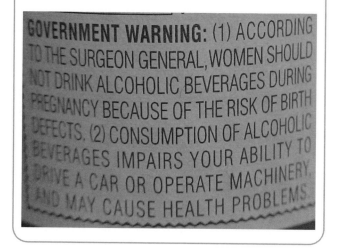

Several medicines are known to cause congenital anomalies. Thalidomide is still used to treat leprosy and HIV in developing countries and causes severe fetal limb and organ defects [6]. Some anticonvulsants can cause congenital heart defects [6]. Environmental pollutants such as heavy metals, pesticides and solvents, can also cause birth defects. For example, exposure to the fungicide methylmercury can cause central nervous system defects, mental retardation, cerebral palsy, deafness, and blindness [6]. In the 1970s, seed grain which had been treated with methylmercury was inadvertently used to make bread in Iraq, resulting in a number of fetal defects [40]. A range of nutritional deficiencies can also result in birth defects. Folate deficiency can cause neural tube defects (spina bifida, anencephaly) and iodine deficiency can cause mental retardation, hypothyroidism, goiter, and cretinism [6].

Unintentional injuries

In developed countries, unintentional injuries are the fourth leading cause of death for children under the age of five [2]. Unintentional injuries result in the death of approximately 15,000 children under the age of five each year in developed countries and 273,000 children in developing countries. In both settings, the single largest cause of fatal unintentional injuries to children include drownings, which kill 82,000 children each year and road traffic accidents, which kill 58,000 children each year worldwide [2]. We will examine technologies to prevent injuries in road traffic accidents later in this chapter.

> ### Learn more about preventing childhood deaths
>
> Implementing preventive interventions such as breastfeeding, insecticide treated materials, access to clean water, childhood vaccination, adequate nutrition, clean delivery, and therapeutic interventions such as oral rehydration therapy, antibiotics and anti-malarial drugs can reduce childhood death by more than 2/3. [42]
>
> Source: How many child deaths can we prevent this year? *The Lancet*, **362**: 65–71, 2003.

We saw in Chapter 1 that one of the UN Millenium Development Goals is to reduce childhood mortality by 2/3 by 2015 [41]. Is this possible with today's technology? A recent study estimated that 2/3 of **childhood deaths** could be prevented today if an essential set of currently available interventions which are feasible to implement in low income countries were made universally available to the populations in need [42]. The study authors conclude that, in most cases, we have technology available to reduce child mortality, but children continue to die because the technology does not reach them.

Mortality rates for children under five years of age have dropped by 1/3 between the late 1970s and the late 1990s [43]. Over this same time period, neonatal mortality rates have dropped much more slowly. Future progress in reducing childhood mortality worldwide will require more progress in reducing infant and neonatal mortality. The island country of Sri Lanka, located off the southern tip of India, provides an example of a successful approach. Sri Lanka has seen a dramatic drop in infant mortality rates over the past 50 years (Figure 4.18).

The infant mortality rate of 15 deaths per 1000 live births in Sri Lanka is much lower than would be expected on the basis of its per capita GDP at PPP of $4600 in 2006 [44, 45]. Reductions in infant mortality are a result of substantial investments in health infrastructure and education. Healthcare is provided free of charge and the female literacy rate is nearly 90%. The first maternal hospital was established in 1879, and training of midwives began in 1926. The government

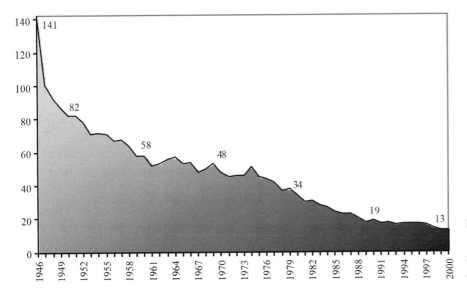

Figure 4.18. *Infant mortality rate (per 1000 live births) vs. time in Sri Lanka for the period 1946–2000. Courtesy and © World Bank.*

Thursday – the long awaited WFP launching day – finally arrived! It was a whirlwind, of which I remember little. It was soooo busy, and Dave and I were running about and registering people all day. It actually went surprisingly well, but there are quite a few kinks for us to work out for Monday. The biggest problem was that there were about 30 patients that walked in (i.e., they didn't have an appointment). We had to look each one up to determine whether they were eligible before we could register them and give them food. That kind of threw our system off a bit (and compounded our struggle to communicate with some of the Saswati-speaking women), but it really wasn't too difficult.

The greatest part of the day was actually giving the women and children their food. Most of them looked so happy to receive it. Some of them looked confused. And a few of them really didn't seem like they needed it all. They were wearing nice clothes and seemed significantly less excited about the food than the others (although they still wanted it). It's impossible to know for sure though because most of them dress up when they come into the clinic. I mean, we can't say, "Um, no. I don't think you need the food. Your outfit is too nice." Of course not! There is really no way for us to know who is really poor and who isn't. The majority of them are in dire need, though, so this food distribution program is really crucial for most of them.

I was able to interact a lot more with the patients than I had before, but because we were trying to register everyone and give out food and look people up all at once, I didn't get to talk to very many of them for long. At the very end of the day, things slowed down, and Dave and I were able to talk to the women and play with the children a bit. One boy was so cute and had so much energy. He just kept walking around, climbing the stairs, following people out to the street . . . I couldn't tell who he belonged to for awhile. He seemed to be friends with everyone.

Friday was equally busy. Dave and I presented on WFP and did boring stuff like fill out forms and fill bags of food. That afternoon, Eileen (another doctor and Dave's mentor) took us to pick up a cabinet for the kitchen she is creating at the Mbabane hospital. They offered to deliver it right then, so we accompanied her to the children's ward, where the kitchen was located. Surprisingly, I ran into someone I knew from the clinic. He was there with his nine-year-old daughter who was suffering from toxoplasmosis. I'm reading a book called *And the Band Played On* right now about the beginning of the HIV/AIDS epidemic. Many of the first men to suffer and die from AIDS came down with this exact disease. It is a disease that normally only infects cats, but when a person's immune system is down, they can contract it as well. If a person doesn't receive the medication he or she needs, he or she will die from it. The poor girl was clearly not receiving the treatment she needed. As Eileen hunted down a doctor from the hospital, Dave and I hung out in the children's ward.

A group of women beckoned to me and started asking me all sorts of questions. After the usual exchange (where we're from, what we do, what our family is like, why they want to go to the USA, etc.), we started talking about why their children were sick. Mostly, they all had diarrhea and vomiting. I asked them if they boiled their water. They explained to me that they do occasionally, but most of the time electricity is too expensive (or fire wood is too hard to gather) to boil water. I knew that saying, "Well, if you boiled your water every time, then your child probably wouldn't be sick" would be way too critical and sweeping of a statement. So I kept my mouth shut. They clearly knew that they should boil their water but felt that they were unable to. I wondered what could be done to make it easier for them to do so. My mind wandered to Rice's Engineers Without Borders group and thought that would be a good problem for them. Behavior change is so difficult. If something can be made automatic so that it is incorporated into their daily routine without adding any extra work, only then is it likely to make a difference.

Another interesting behavior-change problem we discussed dealt with the rapid spread of HIV/AIDS throughout Swaziland. I don't remember how the topic came up, but they were telling me angrily how there are so many people who have AIDS yet still have sex. They say everyone knows that sex spreads the disease, but people continue to sleep with multiple partners, often without protection. They also told me that girls usually start having sex when they are 13!

I asked them if they all knew each other from home or if they had met there in the hospital. One woman responded that they had met there and had become very close. They were there to support each other and comfort each other when one of their children dies. She said it so matter-of-factly. Most of them had already lost at least one child.

As we walked out, Eileen explained to us how frustrating it was to come to this hospital. It always ran out of medicines, and as a result, people died. People that could easily be saved died because the hospital had no medicine to give them. I thought of the still, silent figure of the nine-year-old with toxoplasmosis. She was so skinny and looked way younger than nine. I wondered if my friend's daughter would be dead next week when he came to work, and I felt sick to my stomach. I am so lucky to live where there are clean hospitals with more medicine than they need. I am so lucky that not every family I know has lost a child.

began to promote institutional deliveries in 1948, and today more than 90% of deliveries occur in healthcare institutions. Since 1989, the partogram has been used to monitor labor, improved methods of resuscitating asphyxiated babies have been introduced, and neonatal tetanus has been eliminated by vaccinating mothers and by using aseptic procedures during delivery. Future challenges include providing equitable care throughout all provinces in the nation, particularly those affected by conflict [46].

Developed countries also face challenges associated with infant mortality. In the USA, infant mortality increased in 2002 to seven per 1000 live births for the first time in more than 40 years. Most of the rise occurred in the neonatal period and was related to the excess contribution of prematurity. While the causes of premature birth are not well understood, risk factors include maternal age, multi-fetal pregnancies, stress, smoking, and obesity [47].

Leading causes of mortality, ages 15–44

Worldwide, more than 8.5 million people between the ages of 15 and 44 die each year, 88% in developing countries [2]. The death of a young adult can leave a family without a breadwinner and children without a parent, resulting in far ranging social consequences. The leading causes of mortality for young adults differ substantially for those in developed and developing countries. For persons aged 15–44, the four lead-

ing causes of mortality in the developing world are as follows [2].

(1) **HIV/AIDS**,
(2) **unintentional injuries**,
(3) **cardiovascular diseases,** and
(4) **tuberculosis**.

In the developed world, for persons aged 15–44, the four leading causes of mortality are as follows [2].

(1) **Unintentional injuries**,
(2) **cardiovascular diseases**,
(3) **cancer** (malignant neoplasms), and
(4) **self-inflicted injuries**.

In developing countries, more than 1/3 of mortality in this age group is a result of two infectious diseases: HIV/AIDS and tuberculosis. In contrast, more than 1/3 of mortality in this age group in developed countries occurs as a result of injuries, both unintentional and self-inflicted. Here, we briefly review the epidemiology and pathophysiology of each of these causes of death.

HIV and AIDS

Globally, the burden of HIV/AIDS is staggering. HIV/AIDS is the leading cause of death among people aged 15–44 [2]. Approximately 33 million people are living with HIV/AIDS worldwide and more than 20 million people have been killed by the disease. In 2003 alone,

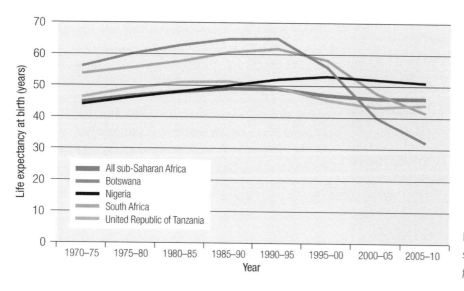

Figure 4.19. *Life expectancy over time in selected African countries. Used with permission from [48].*

three million people died of AIDS and five million others became infected with HIV. The disease has been particularly devastating to the African continent. Two-thirds of those living with AIDS are in Africa, where one in 12 adults has HIV/AIDS. AIDS has reduced gains in life expectancy in sub-Saharan Africa which peaked in the late 1980s (Figure 4.19) [48].

There is worldwide concern over growing HIV/AIDS epidemics in eastern Europe and central Asia. In the United States today, it has been estimated that there are currently between 0.8–1.2 million people living with HIV/AIDS and between 30,000 and 40,000 new HIV infections per year occur in the USA, most in African Americans and Hispanics [48]. The annual cost to treat HIV/AIDS is approximately $15 billion in the USA [49].

The clinical course of HIV/AIDS is characterized by initial infection with the HIV virus (Figure 4.20). HIV is spread by sexual contact with an infected person, by sharing needles for drug injection with someone who is infected, or by transmission from mother to child during pregnancy, childbirth or breastfeeding. The resulting acute infection can initially produce mononucleosis-like symptoms, including fever, sore throat, headache, muscle ache and nausea. This is followed by viral dissemination, eliciting an HIV-specific immune response. The virus continues to replicate, and begins to destroy an important component of the host's immune system, the CD4+ lymphocyte [50].

The rate of progression of HIV during the asymptomatic, latent period is correlated with the number of viruses present per unit volume of blood (viral load).

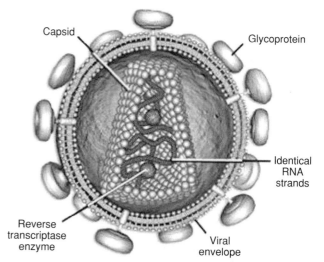

Figure 4.20. *Anatomy of the HIV virus. The virus is spherical in shape and approximately one ten-thousandth of a millimeter in diameter. The core of the virus contains RNA and enzymes called reverse transcriptase, enveloped within a protein coat. An outer lipid membrane contains several proteins important in fusing to host immune cells. There are two HIV virus types, HIV-1 is the predominant type found in the US, while HIV-2 predominates in Africa. There are several sub-types of HIV-1. Reprinted with permission from Michael W. Davidson at Florida State University.*

During the latent phase the viral load remains low (Figure 4.21), but as disease progresses the viral load increases and CD4+ T lymphocytes decrease. Eventually the disease reaches the point where the resulting immunologic dysregulation produces AIDS. AIDS is characterized by a series of opportunistic infections and cancers, including *Pneumocystis carinii* pneumonia (PCP) and Kaposi's sarcoma [50]. As more CD4+ lymphocytes are destroyed, the patient becomes

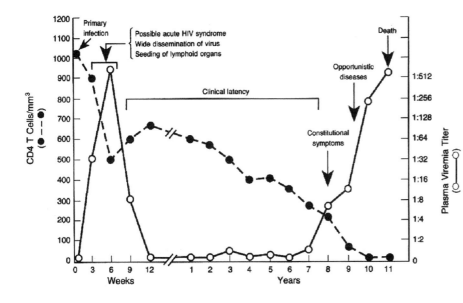

Figure 4.21. *The HIV viral load is high in the two to four weeks following infection. During the latent period, the viral load drops and remains low, but levels of CD4+ lymphocytes decrease, ultimately progressing to AIDS. With AIDS, viral loads increase and lymphocyte levels continue to drop. Pantaleo, G., Graziosi, C., Fauci, A. (1993) The Immunopathogenesis of Human Immunodeficiency Virus Infection. Mechanisms of Disease. 328 (S): 327–335.* © 1993. Massachusetts Medical Society. All rights reserved.

HIV screening of blood donations

In 1985, it became mandatory to test donated blood for the presence of antibodies to HIV. Before screening, more than 10,000 individuals in the USA acquired HIV/AIDS as a result of a transfusion; many of these individuals were hemophiliacs. Today, because of good screening procedures, the risk of acquiring HIV from a blood transfusion in the USA is estimated to be between 1/725,000 and 1/835,000.

Donated blood is screened for the presence of antibodies to HIV, which usually develop within two to eight weeks following infection. An enzyme linked immunosorbent assay (**ELISA**) test is used to screen blood for the presence of HIV antibodies. The ELISA principle was discovered in 1960 and is used to test many bodily fluids, including blood, urine and saliva, for the presence of antibodies, hormones, and proteins.

In an ELISA test for HIV antibodies, a small plastic well is coated with partially purified, inactivated HIV antigens. The sample to be tested is then incubated in the well; any antibodies against HIV will bind to the antigen and become immobilized. Unbound antibodies are then washed away. A second antibody which binds to all human antibodies is then added. The second antibody is conjugated to an enzyme designed to react with dye added in the third step. Again, unbound antibody is washed away. In the third step, dye is added; in a positive test, the enzyme present on the captured antibody acts on the dye, producing a color change. The enzyme acts as an amplifier, so that even if only a few antibodies have been captured, the test generates a large enough color change to be detected. The optical absorption of the fluid is directly proportional to the concentration of the captured antigen [51, 52].

increasingly susceptible to infections. Without treatment, within ten years of infection, 50% of patients will develop clinical AIDS, 40% will develop illness associated with HIV, and 5–10% will remain asymptomatic [49]. Left untreated, the average AIDS patient dies within one to three years [50].

The history of HIV/AIDS likely unfolded throughout your lifetime. In 1981, the Centers for Disease Control and Prevention (CDC) reported cases of pneumo-

nia caused by an unusual organism – *Pneumocystis carinii* – in five previously healthy, homosexual men and cases of a rare cancer – *Kaposi's sarcoma* – in 26 previously healthy, homosexual men. In 1981–2, it was noted that these findings were also associated with IV drug use, recipients of blood transfusions, and hemophiliacs. In 1983, the HIV virus was isolated, and in 1984 this virus was shown to be the causative agent of AIDS. In 1985, a simple blood test was developed

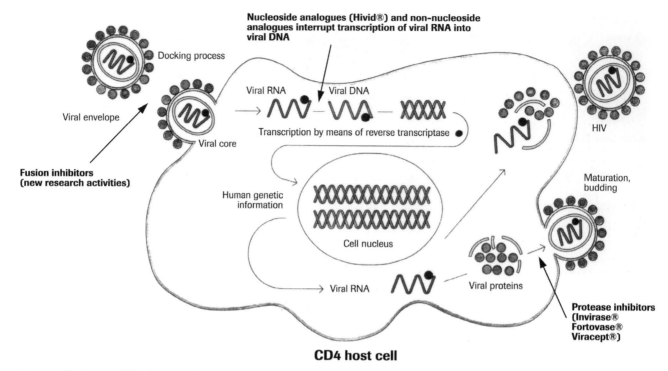

Figure 4.22. *Steps in HIV infection and reproduction, indicating opportunities for drug intervention. Used with permission from Roche.*

to detect the virus – the **ELISA** test [51]. Initially AIDS was a rapidly fatal disease; however, in 1996 a breakthrough new treatment was developed – highly active anti-retroviral therapy (HAART). Although **HAART** cannot cure AIDS, it has dramatically increased the lifespan of HIV infected persons, changing the management of HIV/AIDS from that of a fatal disease to that of a chronic, lifelong disease like diabetes [53].

To understand how HAART works, we need to know more about the details of how the virus replicates within host immune cells. Figure 4.22 illustrates the process which occurs when the HIV virus infects CD4+ lymphocytes. The viral gp120 protein binds tightly to the CD4 receptor on T lymphocytes [50]. The gp41 viral protein helps the membranes of the two cells fuse, enabling the HIV RNA to enter the host cell [50]. The virus then uses the reproductive machinery of the host cell to reproduce. Viral RNA is transcribed to double stranded DNA by means of reverse transcriptase enzymes contained within the virus [50]. With the help of a viral enzyme, the viral DNA is then randomly incorporated into the host DNA [50]. This allows the host cell to produce new copies of viral RNA and viral coat proteins. Viral

protease enzymes process these viral proteins so that the RNA and proteins can assemble into mature virus particles which then bud off the host cell membrane [50].

HAART drugs act to block several steps in the viral reproduction process. One type of HAART drug inhibits the reverse transcriptase enzymes that enable replication of the viral RNA. Without reverse transcriptase, the virus cannot replicate, preventing further infection of other cells. This enzyme is specific to HIV so that inhibition does not affect other normal cellular processes [50]. Two classes of reverse transcriptase inhibitor therapies are currently used in patients – nucleoside reverse transcriptase inhibitors which were first approved to treat patients in 1987 and non-nucleoside reverse transcriptase inhibitors which were first approved in 1996 [54]. Another type of HAART drug inhibits HIV proteases, so that the viral product cannot leave the infected cell to spread to other cells. HIV proteases are distinct from mammalian proteases, so their inhibition does not affect other normal cellular processes [50]. Protease inhibitors were first approved for use in humans in 1995 [54]. Another strategy under active investigation is the development of drugs which inhibit the initial fusion of HIV to the cell membrane. This is an area of active new research. Fusion inhibitors were approved for use in humans in the USA in 2003, but only for patients who have failed other therapies [55]. Other drugs under development include those that may block integration of viral DNA into host genome [50].

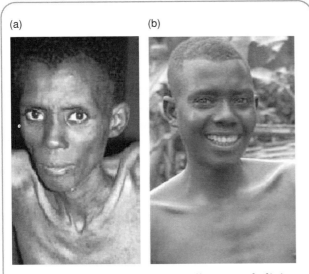

Joseph Jeune is one of the 33 million people living with HIV/AIDS today. The picture on the left (a) was taken at his home in coastal Haiti in March of 2003, when he was suffering from advanced AIDS. The picture on the right (b) shows Joseph six months later, after having started **HAART** and treatment for tuberculosis. (Source: WHO Report, 2004.)

HIV can rapidly mutate, so that it can quickly develop resistance to either type of reverse transcriptase inhibitor or protease inhibitor if given alone. However, resistance develops much more slowly when patients take all three types of drugs simultaneously. This combination therapy is what is known as highly active anti-retroviral therapy (HAART), and represents the most significant advance yet in the treatment of HIV/AIDS. The use of HAART greatly reduces the risk of disease progression and deaths due to AIDS. The clinical impact of HAART is dramatic, the use of anti-retroviral therapeutics decreased death rates due to HIV/AIDS by up to 80% in Europe and the Americas (Figure 4.23)

[48]. Figure 4.24 shows the probability of survival for patients who acquired an HIV infection in the years 1986–96 (pre-HAART period) and the years 1997–8 (HAART period) [56]. Long term survival has significantly increased as a result of this important advance.

While HAART dramatically prolongs survival, it does not cure HIV/AIDS. The goal of HAART is to reduce **viral loads** to an undetectable level, which is usually attained although the virus is not eliminated. Patients must have their viral load and CD4 counts tested every few months to monitor for resistance, which can develop to individual drugs within the HAART regimen. When resistance develops, then the drug regimen must be changed or the typical patient will develop AIDS within three years. New combinations of drugs should include at least two drugs that the patient has not taken before. There are more than 20 approved anti-retroviral drugs [57]. HAART regimens are complex – patients must take 2–13 pills a day and they must continue HAART for the rest of their lives [53]. Many of the HAART medicines have serious side effects.

It has been estimated that there are 6.5 million people living with HIV/AIDS in developing countries in need of HAART; 90% of these people can be found in just 34 countries [48]. Figure 4.25 documents the discrepancies between the need for and the availability of anti-retroviral therapy (HAART) throughout the world.

In the Americas, HAART is available to meet over 3/4 of the need, while in Africa drugs are available to meet only 10% of the need. As a result in late 2003, the WHO launched the 3 × 5 Initiative, a program whose goal was to provide anti-retroviral (ARV) treatment for three million people by 2005 [59]. However, by June of 2005, roughly only a third of this goal was reached (Table 4.3) [60]. The 3 × 5 program is part of a greater global effort to provide universal HIV prevention and treatment as a basic human right, with an expansion goal of universal access to ARV treatment by 2010 [61].

Developing effective tools to prevent future HIV infections has proven to be very difficult. There is active research to develop vaccines to prevent and/or control HIV infection. Pre-clinical work in animals is

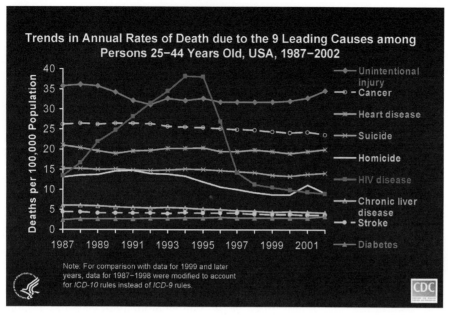

Figure 4.23. *Annual rates of death due to leading causes of death in the USA among persons aged 25–44 years from 1987 to 2002. Courtesy of CDC.*

Viral load tests are widely used to monitor patients on HAART to: stratify patients for therapy, to predict their expected clinical course and the onset of AIDS, and to establish the kinetics of HIV production and destruction.

The viral load test measures the amount of HIV RNA in a sample (note: each virus carries two copies of HIV RNA). There are two types of viral load test: (1) branched DNA (bDNA) test, which sets off a chemical reaction so that HIV RNA gives off light, and (2) RT-PCR, which uses an enzyme to multiply viral RNA by about a million fold. Test results are given in copies per ml; where low results are <500/ml and high results are >40,000/ml. RT-PCR can identify as few as 40/ml, while bDNA can measure as few as 50/ml.

In a bDNA test, concentrated viral RNA is "captured" using complementary oligonucleotides which are bound to the wells of a microtiter plate. Any captured viral RNA is next hybridized to branched oligonucleotides, which then are hybridized to alkaline phosphatase (AP) labeled probes. These AP labeled probes bind with a chemiluminescent substrate that allows for direct measurement of viral RNA concentration by light emission intensity [53, 58].

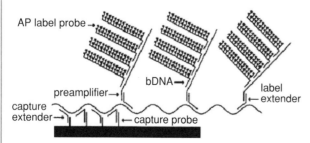

*Figure from ML Collins et al., A branched DNA signal amplification assay for quantification of nucleic acid targets below 100 molecules/mL, Nucleic Acids Research, 1997, Vol. **25**, No. 15, 2979–2984, by permission of Oxford University Press.*

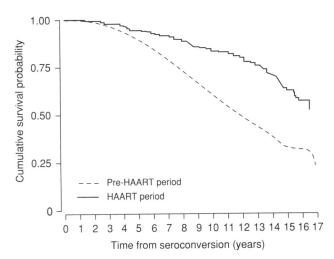

Figure 4.24. *The long term probability of survival following an HIV infection acquired after the development of HAART is much greater than that for patients who acquired an HIV infection pre-HAART. Reprinted from* The Lancet, *Vol.* **355**, *The CASCADE Collaboration, Survival after introduction of HAART in people with known duration of HIV-1 infection, page 1, 2000, with permission from Elsevier.*

transmission (MTCT) is also an area of targeted educational efforts; with 700,000 children infected by MTCT each year [48]. In the absence of ARV intervention, 15–30% of infected mothers transmit HIV during pregnancy and delivery, and 10–20% through breastfeeding [48]. Anti-retroviral drugs drastically reduce transmission. In Africa, less than 5% of women and neonates who need interventions to prevent MTCT receive them [48].

Efforts to prevent the spread of HIV are complicated by the fact that individuals may not realize they are infected; more than 90% of HIV positive individuals do not know they are infected [62]. The reality is that stigma and discrimination – at home, at work, in healthcare settings, etc. – continue to deter people from having an HIV test, particularly if there is fear of violence, social isolation, and no guarantee of treatment availability for those who test positive.

promising and we will examine these vaccines in detail in Chapter 8. Another approach is to educate and counsel patients about the risks of HIV infection and the steps they can take to reduce their risk, an approach that proved very successful in the USA for homosexual men. Globally, the most common means of transmission of HIV infection is unprotected sexual intercourse between men and women [48]. Transmission rates are as high as 45% without treatment. Prevention of mother-to-child

Unintentional injuries

Unintentional injuries are the second leading cause of death for people in developing countries between the ages of 15–44, killing more than one million in this age group each year [2]. In developed countries, unintentional injuries are the leading cause of death, and more than 260,000 people aged 15–44 die each year as a result [2]. The leading cause of fatal unintentional injury worldwide is road traffic accidents (Figure 4.26),

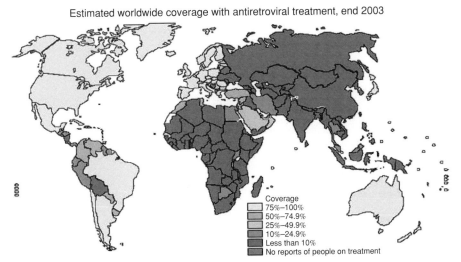

HIV/AIDS: THE TREATMENT GAP

Estimated worldwide coverage with antiretroviral treatment, end 2003

Coverage
- 75%–100%
- 50%–74.9%
- 25%–49.9%
- 10%–24.9%
- Less than 10%
- No reports of people on treatment

Figure 4.25. *In many countries where HIV prevalence is highest, fewer than 10% of those infected have access to HAART. Used with permission from [48].*

I spent most of yesterday trying to collect voice recordings of the adherence counseling sessions at the clinic. One of our projects is to develop a group ARV adherence counseling module. We wanted to know what is being said in the individual counseling sessions that the clinic currently has. The doctors said that this would be a good idea since they were finding that many of the patients did not know enough about their condition and ARVs when they came to them to start treatment even after finishing the adherence counseling sessions. One of the problems is that the current module places too little emphasis on educating the patient, but rather focuses mostly on counseling.

In one of the screening sessions, a mother came in with her eight-year-old girl to have the child tested. The girl was obviously ill – she appeared weak and was not expressive. Her eyes were sunken in and her cheeks and ankles were swollen. One of the doctors told me that these were signs of severe malnourishment – most likely due to protein deficiency. Many times when I've been around Basotho children, they just stare at you without any expression on their faces. Then, when you smile or wave at them, all of a sudden their faces uncover joyous smiles that nearly overwhelm you and give you an amazing sense of happiness and pleasure. When I smiled at this young girl who was being tested, she also revealed to me that same sweet smile, yet it seemed that it took effort. Another interesting thing about Basotho children is that they can be surprisingly stoic, which is a characteristic not often seen in children. They hardly flinch or show any emotion on their faces when they get their fingers pricked during HIV testing. This child was no exception.

The mother was also with the young girl, but she explained to the social worker that she did not want to be tested because she was afraid of knowing what the result might be. The social worker explained that it was worse not to know her status since the virus would continue to replicate in her body until she ended up with AIDS and then would die. At this explanation, she became even more afraid and was persuaded into being tested.

There is about a five minute wait for getting the results of an HIV rapid test. While we were in the room, I saw several cold sores on the mother's mouth and was reminded of a powerpoint I had seen on some common symptoms of HIV infection. While we were waiting, I was anticipating that both the mother and child would be HIV positive. As we looked at the test results, the mother's test result had two lines, while the child's only had one. My face flushed with heat as I saw the result – only the mother came out positive. The mother did not show emotion when she was told her status – she remained calm and was quiet as she was referred to Senkatana clinic, which is the HIV clinic that treats adults. Maybe a positive test result is not as shocking for the Basotho seeing as they live in a country where 1/4th of the population is infected with HIV.

Table 4.3. *ARV therapy coverage in low and middle income countries, June 2005. Courtesy of WHO, 3 × 5 Initiative Progress Report, June 2005.*

Geographical Region	Number of people receiving ARV therapy	(low estimate – high estimate)	Estimated need	Coverage
Sub-Saharan Africa	500 000	(425 000 –575 000)	4 700 000	11%
Latin America and the Caribbean	290 000	(270 000 –310 000)	465 000	62%
East, South and South-East Asia	155 000	(125 000 –185 000)	1 100 000	14%
Europe and Central Asia	20 000	(18 000 – 22 000)	160 000	13%
North Africa and the Middle East	4 000	(2 000 –6 000)	75 000	5%
Total	970 000	(840 000 –1 100 000)	6.5 million	15%

Figure 4.26. *Road traffic accidents are a leading cause of death throughout the world. Used with permission from NHTSA. CDC/Gwinnett Country Police Department.*

which kill more than 500,000 people aged 15–44 every year; 90% of road traffic deaths occur in low income and middle income countries [43]. In developing countries most deaths involve pedestrians, bicyclists, motorcyclists and occupants of buses, while in developed countries most fatalities are occupants of cars [43]. Road traffic injuries are also a major cause of severe disability; 50 million people are injured each year in traffic accidents [63]. While traffic fatality rates have decreased in developed countries, traffic fatalities are expected to rise 65% worldwide between 2000 and 2020 [63].

In the United States, traffic fatality rates have decreased steadily over the past thirty years (Figure 4.27), yet automobile accidents still represent the leading cause of potential years of life lost. In 2005, 43,443 Americans were killed in road traffic accidents and 2,699,000 Americans suffered injuries in road accidents. Persons between the ages of 16 and 20 suffered the highest mortality rate. The fatal accident rate of male drivers is almost three times higher than for females. Mortality rates for motorcycles are even higher, with a death rate per mile nearly 40 times higher than that of automobiles. In the USA, 39% of fatal crashes are related to alcohol use and the deadliest time on the road is midnight to 3 a.m. Saturday and Sunday [64].

Successful prevention of road accidents and resulting death and injury has focused legislation, education and engineering to reduce conditions which lead to crashes and keep occupants safer during crashes. Throughout the United States, laws regulate maximum driving speed on roadways. All 50 states in the USA have laws requiring seat belt usage and the use of child safety restraints up to a certain age [65]. Twenty states require motorcycle drivers to wear helmets, and another 27 states require helmet usage with some provisions for exceptions; three have no requirements [66]. In 45 states it is illegal to drive with a blood alcohol concentration in excess of 0.08 g/dl [67]. New cars sold in the USA must

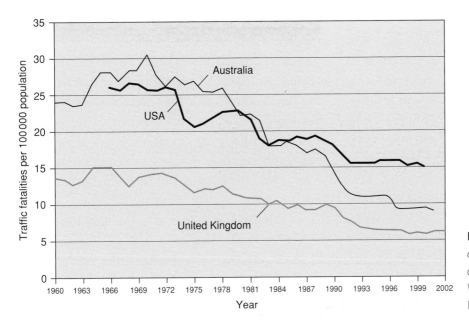

Figure 4.27. *Traffic fatality rates have decreased in many developed countries over the past three decades. Courtesy of WHO, World Report of Road Traffic Injury Prevention, 2004.*

be manufactured to meet certain safety standards which protect occupants in the event of a crash.

To understand how these laws reduce the number of road accidents and the number and severity of accident-related injuries we need only consider Newton's laws. Newton's second law tells us that that the force, F, acting on an object is a product of its mass, m, and its acceleration, a:

$$F = ma. \tag{4.1}$$

Acceleration describes how rapidly velocity, v, changes with time t, and can be described as the rate of change of velocity with time:

$$a = \frac{dv}{dt}. \tag{4.2}$$

We can approximate the acceleration in a crash as the change in velocity divided by the time over which it occurs:

$$a = \frac{initial\,velocity}{time\,to\,come\,to\,rest}. \tag{4.3}$$

In a crash, velocity decreases to zero in a very short time leading to large accelerations and large forces. If these large forces are transmitted to passengers, fatal injuries can result. There are primarily two ways to reduce the forces that impact passengers – we can (1) reduce the initial velocity of impact and (2) extend the time that it takes the passenger to come to rest.

Reducing initial crash velocity. High velocity at initial impact leads to higher forces in crashes. This can be caused by traveling at excessive speeds. Figure 4.28 shows the risk of pedestrian death as a function of impact speeds above 40 km/h. Excessive speed contributes to 30% of deaths due to road accidents in developed countries, and 50% in developing countries [63]. Drivers traveling at higher speeds have less time to stop and avoid the crash or slow and reduce the speed of impact.

When drivers can anticipate a crash, they have time to react and begin to brake in order to reduce the velocity of initial impact. Factors that slow driver reaction times can increase the likelihood of injury and death. A number of factors slow driver reaction time, including alcohol use, cell phone use, and poor visibility [63].

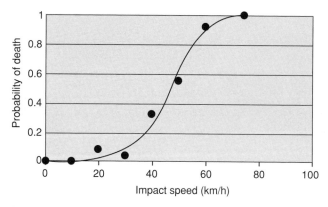

Figure 4.28. *Probability of death for pedestrians as a function of the velocity of impact. Courtesy of WHO,* World Report on Road Traffic Injury Prevention, *2004.*

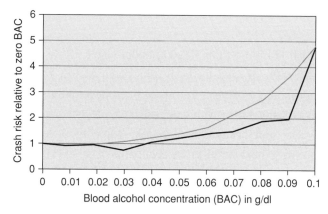

Figure 4.29. *The risk of crash as a function of blood alcohol concentration relative to a driver with zero blood alcohol concentration. Redrawn from* World Report on Road Traffic Injury Prevention, *2004.*

In addition, driver inexperience can contribute to poor judgment and slowed reaction times.

Alcohol impaired drivers have 17 times increased risk of being involved in a fatal crash relative to drivers who have zero blood alcohol. The effects of alcohol increase risks more in younger, less experienced drivers. Figure 4.29 shows the risk of crash as a function of blood alcohol concentration. A study carried out in the USA in the 1960s established the basis for the 0.08 g/dl limit on blood alcohol content; this study showed that at this level, drivers had a two fold increase in the risk of crash. New data indicate that the original study probably underestimated the risk of crash, and that the limit should be reduced to 0.05 g/dl for all drivers and 0.02 g/dl for young drivers. Data indicate that raising the minimum legal drinking age reduces the rates of alcohol related crashes. Enforcement of these laws is also important; a low expectation of getting caught increases the risk of alcohol related crashes. Alcohol also increases the risk of pedestrian death in traffic accidents; studies show that 50–60% of pedestrians killed in traffic accidents had been drinking [63].

Videos of crash tests

These crash test videos contrast what happens to occupants who are restrained with seat belts to those who are not wearing belts. In all three crashes, the crumple zone slows the deceleration of the passenger compartment [70–72].

Unbelted occupants rapidly accelerate into the dashboard:
http://www.regentsprep.org/Regents/physics/phys01/accident/nobelt.htm

Belted occupants are held with the passenger compartment:
http://www.regentsprep.org/Regents/physics/phys01/accident/withbelt.htm

Airbags can help protect unbelted occupants in the front seat, but unbelted rear occupants are not protected:
http://www.nhtsa.dot.gov/staticfiles/DOT/NHTSA/Communication%20&%20Consumer%20Information/Multimedia/Associated%20Files/crashdum2.ram

Airbags can help provide additional protection beyond seatbelts:
http://www.accidentreconstruction.com/movies/5thper.mov

This video illustrates the importance of child restraints. During the crash, a child seated in his unbelted mother's lap effectively serves as an air bag for his mother:
http://www.nhtsa.dot.gov/staticfiles/DOT/NHTSA/Communication%20&%20Consumer%20Information/Multimedia/Associated%20Files/crashdum3.ram

---

The use of mobile phones is of growing concern; at any given daylight moment in the USA it is estimated that 10% of drivers are using a cell phone [68]. Driver reaction times increase by 0.5–1.5 seconds when talking on the phone. The risk of crash increases four times when using a mobile phone; the equivalent effect of driving with a blood alcohol concentration of 0.09 g/dl. Some 35 countries have now banned use of hand held phones while driving. Hands free phones present less risk than hand held phones, but still distract drivers [63]. Only three states, New York, New Jersey and Connecticut, and the District of Columbia, ban the use of hand held mobile phones while driving [68].

Increased visibility can enable time for driver reaction and reduce crash risk. The use of daytime running lights reduces car crashes by 10–15% [63]. High mounted brake lights reduces rear end crashes by 15–50% [63].

Finally, inexperienced drivers have increased risk of fatal crashes. In the USA and Canada, graduated drivers license systems have reduced crashes of new drivers by 9–43% [63].

Extending time for passengers to come to rest. If we lengthen the time that it takes for passengers to come to rest during a crash we can also reduce the risk of injury and death. The front end of new cars are designed to crumple in a controlled manner, allowing the passengers additional time to decelerate and directing the energy absorbed in the crash away from the passenger compartment. In order for crumple zones to protect occupants, they must be wearing seat belts. Seat belts restrain occupants so that they remain in the passenger compartment; in addition, seat belts stretch during an impact, further slowing deceleration of passengers [69]. In a frontal collision, occupants not wearing seat belts will fly forward and decelerate rapidly when they come into contact with the dashboard, steering wheel or windshield. Air bags provide an additional cushion which slows the deceleration of occupants. Airbags inflate rapidly upon frontal impact; as occupants move forward, they strike the inflated airbag which slows their deceleration. In the USA, a national car assessment program, begun in 1978, tests the safety of a car's design – determining crash-worthiness in certain types of crashes – which consumers can consider when buying cars [63]. **Videos of crash tests** conducted by the National Highway Traffic Safety Administration provide a dramatic illustration of the effects of these safety measures.

The use of seat belts reduces the risk of death in a crash by 40–65%. Seat belts are most effective in roll overs and rear and frontal collisions. Numerous studies have shown that seat belt usage rates increase in areas where they are required by legislation and where these laws are enforced. Seat belt usage rates are much lower in developing countries, where half of all vehicles may lack functioning seat belts. When combined with seatbelts, air bags have been shown to reduce the risk of fatal injury by 68%. The use of child restraints, such as car safety seats, reduces infant death by 71% and toddler death by 54%. Mandatory child restraint laws in the USA have led to a 13% increase in usage rates and a 35% reduction in fatal injuries [63].

Head injuries are a major cause of death for users of two wheel vehicles. This is of growing concern in Asia, where two wheelers are commonly used as family vehicles. Non-helmeted users are three times more likely to sustain head injuries in a crash of a two wheeler [63]. When the skull receives an impact, crucial nerves and blood vessels can be torn, leading to bleeding within the brain and neurologic damage. Helmets contain a layer of foam which is designed to crush upon impact. In a crash the foam collapses and extends the time over which the head comes to rest, thus reducing the force of impact. Helmet laws reduce fatal injuries by 20% and serious head injuries by 45%. Mandatory helmet laws in the USA have reduced injuries to motorcycle drivers by 20–30% [63].

Cardiovascular disease

Cardiovascular diseases are the second leading cause of death for people aged 15–44 in developed countries and the third leading cause of death in this age group in developing countries [2]. Worldwide, more than 768,000 people aged 15–44 die every year as a result of cardiovascular diseases [2]. The most common cause of fatal cardiovascular disease in this age group is **ischemic**

Profile of bioengineers Albert King and Lawrence Patrick

The College of Engineering at Wayne State University has been a pioneer in the field of impact biomechanics since 1939. Two researchers in particular, Lawrence Patrick and Albert King, have made a number of significant contributions to the field of automotive safety. Dr. Patrick, who assumed the position of director of the Biomechanics Research Center in 1965, was known as a courageous researcher, volunteering himself for a number of different impact tests in the name of scientific advancement [73]. In the image below, reproduced here courtesy of the College of Engineering, Wayne State University, Dr. Patrick is shown preparing to ride the crash sled; a task usually reserved for cadavers and dummies. Dr. Patrick's research led to the development of the Wayne State Tolerace Curve (an index of head injury tolerance) as well as many automotive design improvements including the airbag. In 1976, Dr. Albert King took over the position of director, carrying on Patrick's legacy. Dr. King's research attempts to numerically model human bone structure from head to foot in order to predict injuries that might result from a crash. These models are verified in the laboratory experimentally using both crash test dummies and cadavers and then used to help design safer cars. Dr. King's work has contributed to the safety regulations and design specifications of vehicles, yielding in safer steering columns, high penetration resistance windshields, tempered glass side windows, three-point belt restraint systems, and airbags.

Source: http://www.eng.wayne.edu/page.php?id=266

Courtesy of the College of Engineering, Wayne State University.

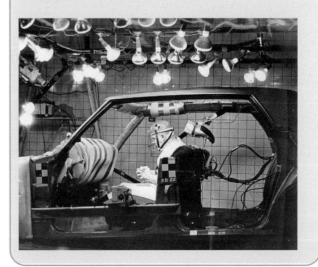

Ischemia refers to the condition where an organ does not receive sufficient oxygen because of inadequate blood flow. Ischemic heart disease occurs when the heart does not receive adequate oxygen due to decreased blood flow in the coronary arteries.

In the United States more than 12 million people have ischemic heart disease [51]. This disease causes more deaths, disability and economic cost than any other illness in the USA. The risk factors for developing ischemic heart disease include a positive family history, diabetes, hyperlipidemia (high cholesterol), hypertension (high blood pressure), and smoking [51].

heart disease, which kills more than 286,000 people aged 15–44 every year [2]. Cerebrovascular disease is the second leading cause of cardiovascular mortality in this age group, killing 159,000 people aged 15–44 yearly [2]. Here, we will consider the pathophysiology of both ischemic heart disease and cerebrovascular disease.

Ischemic heart disease is also called coronary artery disease (CAD) because it develops in the coronary arteries which supply blood to the heart. Throughout your lifetime, cardiac muscle cells in the heart contract rhythmically in order to supply the rest of your body with oxygenated blood. In fact, your body's total volume of blood is pumped through the heart each minute. In order to maintain its function, the heart itself requires a continuous supply of oxygenated blood. The coronary arteries are responsible for supplying blood to heart muscle. Figure 4.30 shows the two main coronary arteries, the right main coronary artery and the left main coronary artery. These arteries branch directly from the aorta, and are about 3 mm in outer diameter at their maximum point. They branch and encircle the heart to supply oxygenated blood throughout.

In coronary artery disease, the lumen of the coronary arteries becomes obstructed in a process known as atherosclerosis (Figure 4.31). As the atherosclerotic narrowing obstructs blood flow in the coronary arteries, the oxygenated blood flow to the heart muscle is decreased. The most common symptom of CAD is angina, or chest pain. A typical angina patient is a man

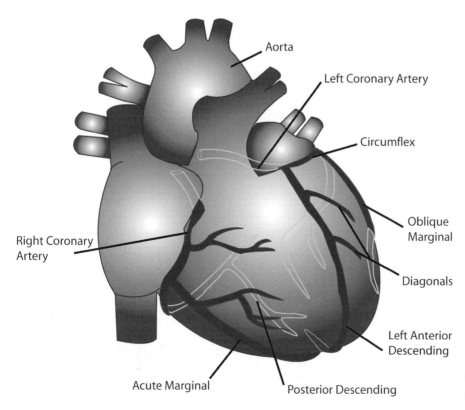

Figure 4.30. *The coronary arteries and their branches supply all areas of the heart muscle with blood.*

Figure 4.31. *The series demonstrates the process and growth of an atherosclerotic plaque, and the development of a blood clot at the site of a ruptured plaque.*

over 50 years of age or a woman over 60 years of age who suddenly experiences a sense of heaviness, pressure, squeezing, smothering or choking which is localized to the chest and may radiate to the left shoulder or both arms [51]. Angina attacks typically last from two to five minutes. When atherosclerosis initially develops, the residual coronary artery circulation provides sufficient oxygenated blood to perfuse the heart. However, as the blockage grows to occupy more than 75% of the lumen diameter, during periods of exertion there is an insufficient supply of oxygenated blood to the heart and angina results [51]. As the degree of coronary blockage increases, patients may develop unstable angina, with severe pain of increasing frequency and, as the obstruction occupies more than 80% of the lumen, angina at rest.

The lumen of a blood vessel is lined by a specialized layer of endothelial cells. These endothelial cells serve an important role; they control the transport of many substances between blood and the surrounding tissue. When this layer of cells is damaged, as can occur with high blood pressure or due to high serum nicotine levels, cholesterol from the blood can deposit at the site of injury. The blood vessel reacts to this injury by producing large amounts of fibrous tissue; this fibrous cap covers the lesion. The center of the lesion generally contains dead cells and cholesterol debris and is known as the necrotic core. Some atherosclerotic plaques remain stable over time and simply reduce the size of the lumen available for blood flow. However, in some plaques the fibrous cap ruptures and the material in the necrotic core is exposed to the blood within the lumen. These plaques are known as unstable plaques.

The material within the necrotic core is very thrombogenic and can rapidly cause a blood clot to form. The blood clot can lead to complete occlusion of the coronary artery, so that none of the tissue downstream from the blockage receives oxygenated blood. Without rapid treatment, the heart muscle supplied by the blocked artery will die. This event is known as a myocardial infarction, or heart attack. In the USA, approximately 30% of patients do not survive a first heart attack; more than half die before reaching a hospital. If the patient survives the myocardial infarction, the dead heart muscle is replaced with a fibrous scar. The scar tissue does not contract in the same way as healthy cardiac muscle, and the function of the heart can be severely compromised [51].

The diagnosis of ischemic heart disease is usually made by listening to the patient's history. Physical examination may reveal other disorders, including lipid disorders, hypertension, and diabetes. Diagnostic tests performed include an electrocardiogram to measure the electrical activity of the heart at rest and during activity (stress test), and **coronary arteriography**. In coronary arteriography, X-rays are used to take a picture of the lumen of the coronary arteries, also known as an angiogram. A radio-opaque dye is injected into the coronary circulation, and X-ray movies of the flow of the dye are obtained. For 50% of patients, their first symptom of CAD is a heart attack [51].

Treatment of ischemic heart disease includes giving drugs such as nitrates or calcium channel agonists; these drugs increase the supply of blood to the heart by increasing the diameter (dilating) of the coronary arteries [51]. Other drugs, such as beta blockers, can be given to reduce the oxygen demand of the heart. While these drugs can relieve symptoms of CAD, they do not reduce the coronary artery blockage.

Treatments that increase coronary artery blood flow include drugs which dissolve the blood clot precipitating a myocardial infarction, such as tissue plasminogen activator [74]. In addition, there are two invasive procedures which directly treat blockages in coronary arteries – coronary artery bypass grafting (CABG) and percutaneous transluminal coronary angioplasty (PTCA) [51]. Figure 4.32 provides an overview of these

Coronary arteriography

Coronary arteriography is a non-surgical diagnostic procedure used to evaluate patients who exhibit chest pain. In this procedure, which is also called cardiac catheterization, thin catheters are threaded through blood vessels and into the heart (a). Contrast dye which absorbs X-rays is injected into the catheters to allow physicians to see inside the heart and the vessels using X-rays. This procedure is often used to determine if there are any blockages in the arteries surrounding the heart (b).

(b)

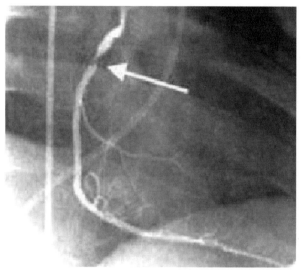

(a)

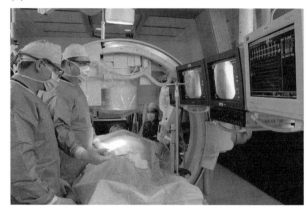

*(a) Photo courtesy of Ozarks Medical Centre. (b) Texas Heart®
Institute.*

two therapies, which we will consider in much more detail in Chapter 15.

Some 159,000 persons between the ages of 15–44 die each year as a result of cerebrovascular disease, more commonly known as a stroke [2]. Cerebrovascular disease is the most prevalent neurologic disorder in the USA. The vast majority of strokes (approximately 87%) are caused by ischemia and resulting infarction [75]. Ischemic strokes are a consequence of atherosclerosis in arteries which supply oxygenated blood to the brain (Figure 4.33). An ischemic stroke occurs when a blood vessel supplying the brain is blocked, depriving tissue downstream of oxygenated blood. This blockage can occur in a manner similar to that which occurs in a heart attack. If an atherosclerotic lesion within a brain vessel ruptures, blood in the vessel can come in con-

tact with thrombogenic material within the plaque. A clot quickly forms in the vessel, closing off all blood flow. A stroke can also occur when a large clot formed elsewhere in the bloodstream breaks loose and lodges in a blood vessel within the brain. Additionally, vessels within the brain can undergo transient vasoconstriction or spasm, resulting in stroke.

Because the occlusion forms suddenly, a stroke is characterized by the abrupt onset of a focal neurologic deficit. The maximal deficit typically occurs within hours of vessel occlusion. Often, patients awaken with neurologic deficit following a completed stroke. A small fraction of patients (approximately 15%) experience warning signs before a stroke called transient ischemic attacks (TIA) [75]. While a TIA has traditionally been defined as a neurological deficit that resolves within

(a)

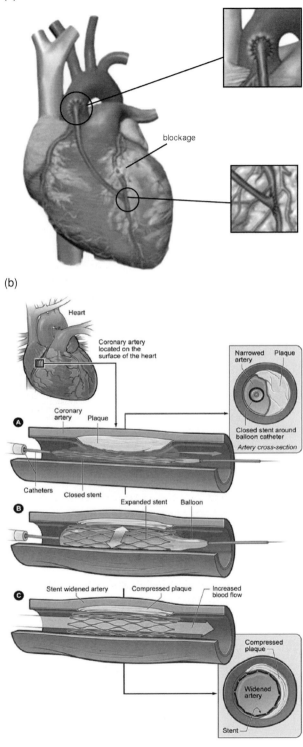

(b)

Figure 4.32. *Atherosclerotic blockages which cause coronary artery disease can be treated by (a) grafting a vessel to bypass the blockage – CABG. Alternatively, (b) a balloon-tipped catheter can be advanced to the site of the obstruction. Inflating the balloon increases the size of the lumen, enabling increased blood flow to heart – PTCA. Sources: (a) The MetroHealth System, Cleveland Ohio; (b) National Heart, Lung, and Blood Institute as part of the National Institutes of Health and the US Department of Health and Human Services.*

24 hours, imaging studies have revealed that nearly 2/3 of patients with a diagnosis of TIA have actually suffered an ischemic stroke. As a result, a new definition has been proposed that is based on tissue evidence of infarction rather than a set time interval [76].

Diagnosis of cerebrovascular disease is frequently based on the patient's history. Physical examination will include imaging studies such as computed axial tomography (CT) scan, magnetic resonance imaging (MRI) and MR angiography (MRA) to examine blood flow in the brain (Figure 4.34). Treatment of cerebrovascular disease includes the use of drugs to dissolve blood clots, or thrombolysis. Rehabilitation of stroke patients includes physical therapy, occupational therapy and speech therapy to provide encouragement and instruction for the patient and their family to overcome the resulting neurologic deficit [51].

Tuberculosis

More than 600,000 people between the ages of 15 and 44 die each year as a result of tuberculosis (TB) [2]. Worldwide in 2004, nine million new cases of TB occurred, 80% in just 22 countries; in this same year, TB caused two million deaths, 98% in the developing world [77, 78]. It is estimated that two billion people – *1/3 of the world's population* – are presently infected with TB bacilli, the causative agent of the disease. The global burden of TB continues to grow by 1% a year, primarily due to rapid increases in Africa (Figure 4.35). It has been estimated that TB will kill another 35 million people in the next 20 years if the current situation does not change [78].

TB is a bacterial infection of the lungs caused by *Mycobacterium tuberculosis* [51]. Not all people who are infected with TB have symptoms of the disease; most have what is known as latent TB. In latent TB, the patient's immune system has walled off the TB bacilli within a thick waxy coat to form granulomas. Latent TB can thus lie dormant for years before becoming active and causing symptoms (Figure 4.36) [51]. Only 5–10% of people with normal immune systems who are infected with TB will go on to develop active TB [79].

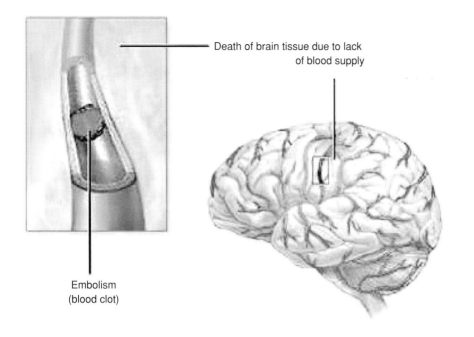

Death of brain tissue due to lack of blood supply

Embolism
(blood clot)

Figure 4.33. *In an ischemic stroke, blockage of a cerebral blood vessel results in reduced blood supply to the brain. Emboli (moving blood clots) that develop in the heart can cause the blockage, with the embolus traveling through the bloodstream and becoming stuck in a small artery in the brain.* © 2008 A.D. A.M. Inc.

A.D.A.M.

Just a quick update on what has been going on with the WFP project this past week . . .

We want to be sure that there will be someone to continue the WFP food distribution after Dave and I leave. So we are trying to find volunteers among the clinic patients. Before we could bring volunteers in to train, we had to find a way to cover their transportation costs (and perhaps a little extra as an incentive to volunteer). I managed to get a local grocery store to donate food vouchers. This is enough to cover the next month, and while it is very exciting, I need to see if we can get enough (perhaps from other grocery stores) to cover a few months more. It is difficult to get funds here at the clinic (even if it is a measly 400 rand – about 55 dollars – a month). At least now that we have the vouchers, we can bring in some women and start training them. The process of deciding on the trainees was interesting as well. I basically got a

list from the clinic social worker of women she thought would be appropriate for the job (and who were in desperate need of food). During the process of choosing, I asked her if these women were literate. The ability to read and write is clearly crucial to being able to register the patients. Unfortunately she didn't know, so now we basically have to bring them in one at a time and see how capable they are. One woman today actually approached me and asked about volunteering. She had completed high school and lived close by. She also seemed like she would be easy to train and like she would be capable of training other women after I leave. Despite all these factors, the clinic staff (the social worker and several other women met to discuss whom I should train) decided that I should try the other women first since their need was much greater. At first I wasn't sure about the decision, but the more I think about it, the more I like how they made the decision. It shows how much the people here care about their patients. And if these women don't work out, I think there will be enough time to train someone more capable.

(a)

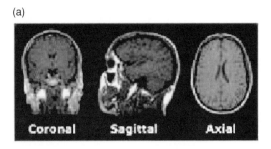

(b)

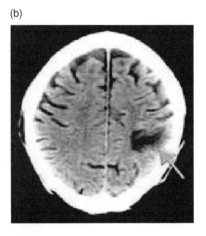

(c)

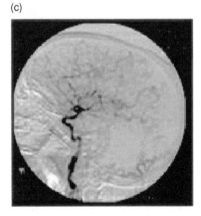

Figure 4.34. *MRI (a), CT scan (b), and MR angiography (c). These tests can be used to diagnose stroke. MRI uses magnetic fields and CT scans use x-rays to produce 3-D images. MRA is a special type of MRI used to see the blood vessels in the neck or brain. Used with permission from the Internet Stroke Center (http://www.strokecenter.org/).*

However, a much higher fraction of people with weak immune systems who are infected with TB will develop active TB. For this reason, TB is particularly problematic in people with weak immune systems, including

Skin and serum tests for TB [51, 84]

The Mantoux tuberculin skin test, also known as the purified protein derivative (PPD) test, is the most accurate and thus preferred type of skin test. For the test, tuberculin – protein derived from tubercle bacilli that have been killed by heating – is injected between the layers of the skin, usually on the forearm. Two to three days later, swelling at the injection site is assessed. Whether a reaction is classified as positive depends on the diameter of the swelling, as well as the person's risk factors for TB. Most people with TB infection have an immune reaction to the tuberculin, as it resembles the tubercle bacilli causing their infection.

The QuantiFERON-TB Gold (QFT-G) blood serum test – approved by the FDA in 2005 – measures a blood sample's response to the presence of TB proteins. QFT-G can be used in all circumstances in which tuberculin skin tests are used. Additionally it may be used with persons who test false positive with a tuberculin skin test, e.g. those vaccinated against TB with Bacille Calmette Guerin (BCG) and those who undergo serial evaluation for TB.

NPR Story: a new plan to combat HIV/AIDS and tuberculosis simultaneously
http://www.npr.org/rundowns/segment.php?wfId=1520699

those with AIDS, babies, and young children. In fact, people with AIDS are 100 times more likely to develop active TB once infected; HIV positive individuals have a 50% lifetime risk for developing active disease compared to a 5–10% lifetime risk for HIV negative individuals [80, 81]. The result is an escalation of TB incidence in countries with a high prevalence of HIV (Figure 4.37) [77]. Furthermore, TB is the leading cause of death among HIV positive individuals, accounting for 13% of AIDS deaths worldwide [82].

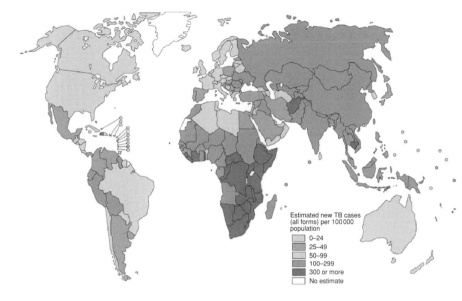

Figure 4.35. *Based on surveillance and survey data, there were an estimated 8.9 million new cases of TB in 2004 (140 per 100,000), including 3.9 million (62 per 100,000) new smear positive cases. Courtesy of WHO Report 2006:* Global Tuberculosis Control – Surveillance, Planning, Financing.

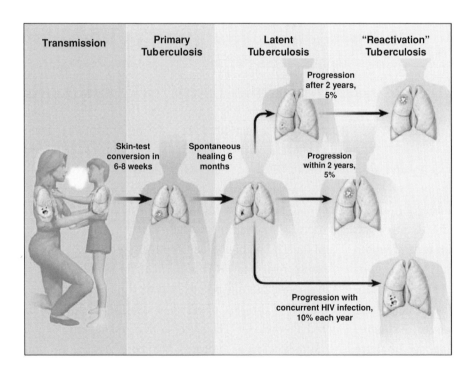

Figure 4.36. *Transmission, progression, and reactivation of tuberculosis. The risk of reactivation of latent tuberculosis is much higher for HIV-infected individuals. Used with permission from [79]. Copyright © 2003 Massachusetts Medical Society. All rights reserved.*

In active TB, the bacteria usually affect the lungs. Symptoms of active pulmonary TB include a persistent bad cough, chest pain, coughing up blood or sputum, weight loss, weakness, chills, fever, and night sweats [79]. Hence TB is spread through the air from a person with active infection to another. If untreated, each person with active TB infects on average 10–15 people every year [83]. Left untreated, active pulmonary TB results in lung tissue necrosis. This severe lung damage leads to hypoxia and death within five years in 65% of all cases. Additionally, highly fatal TB can result from bacilli dissemination to and destruction of a host of organs, such as the spleen, liver, kidneys, bone marrow and central nervous system [51].

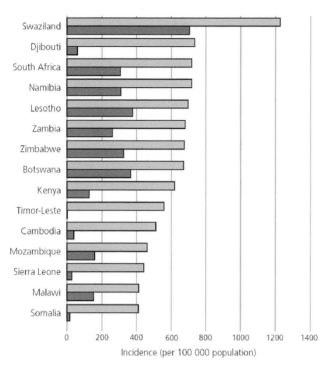

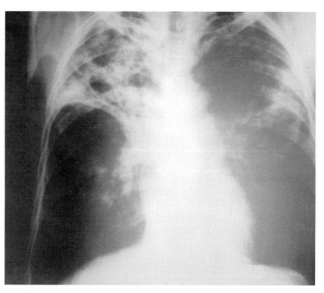

Figure 4.38. *This chest X-ray shows the presence of a bilateral pulmonary infiltrate, and a cavity like lesion. The diagnosis is advanced tuberculosis. Source: CDC.*

Figure 4.37. *Fifteen countries with the highest estimated TB incidence rates per capita (all ages, all forms – light gray bars) and corresponding incidence rates of HIV infected TB in adults aged 15–49 (dark gray bars), 2004. Courtesy of WHO* Report 2006: Global Tuberculosis Control – surveillance, planning, financing.

Latent infection of TB is indicated by a positive **skin or serum test** with a negative chest X-ray [84]. However, in HIV positive persons with TB infection or disease, skin tests often elicit no reaction due to their compromised immune response and sputum tests are often read as negative due to the unusual, difficult to diagnose strains of TB that HIV infected persons tend to contract [51]. Nevertheless, annual TB screening is recommended for these high risk individuals. In all patients suspected of TB infection or disease, diagnosis of active TB is made by a chest X-ray (Figure 4.38) which shows nodules in the lungs, and is confirmed by positive sputum culture for *M. tuberculosis* [51].

It is important to treat patients for both latent and active TB. Drugs which cure TB have been available since the 1940s. Patients with a latent TB infection are treated with the antibiotic isoniazid; this treatment will prevent them from developing active TB. Active TB can almost always be cured by taking several antibiotics in combination, including isoniazid, rifampin,

ethambutol and pyrazinamide. Active TB patients remain contagious during the first few weeks of antibiotic therapy and it is important for them to stay isolated for several weeks until they are no longer infectious. After several weeks of treatment, symptoms subside and patients begin to feel better. However, to completely cure the TB infection, patients must take drugs for six months. If patients discontinue medication before this period, it is much more likely that the TB will relapse and the TB bacilli will develop resistance [51].

Drug resistant TB is a particularly serious form of the disease and must be treated with special medicines. Because of the risk of resistant disease, poorly supervised therapy for TB is worse than no therapy at all. A simple clinical protocol has been shown to dramatically increase TB cure rates. Directly observed treatment short course (DOTS) is the process of administering TB medicine daily under the watch of a healthcare worker. The use of DOTS can achieve cure rates of up to 95% even in the poorest countries. A six month supply of DOTS costs only $10 [82]. More than 20 million TB patients worldwide have been treated with DOTS since 1995. However, approximately 20% of the world's population still does not have access to DOTS [77].

Prevalence of MDR-TB among new TB cases, 1994-2002

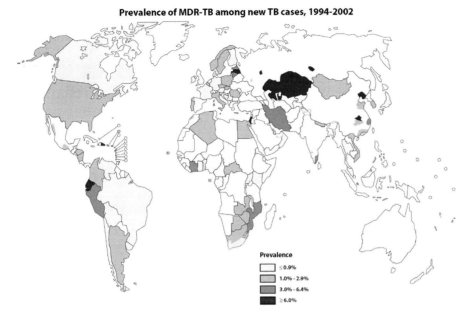

Prevalence
- ≤ 0.9%
- 1.0% - 2.9%
- 3.0% - 6.4%
- ≥ 6.0%

Figure 4.39. *Estimated prevalence of MDR-TB, i.e. resistance to two or more TB drugs, amongst new TB cases ranged from 0% (Andorra, Cambodia, Iceland, Luxembourg, Malta, New Zealand, Oman, Scotland, Slovenia, and Switzerland) to 14.2% (Kazakhstan and Israel). The median was 1.1%. Ten locations had an MDR prevalence higher than 6.5%. Courtesy of WHO / Anti-Tuberculosis Drug Resistance In the World; Report No. 4, 2004.*

Multi-drug resistant TB (MDR-TB) is TB disease that arises from bacilli resistant to at least isoniazid and rifampicin, the two most powerful anti-TB medicines. MDR-TB is present in all countries surveyed by WHO and is a growing concern [82]. The prevalence of resistance in new TB cases is as high as 21.7% (7% median) for a single drug and as high as 17.8% (2.2% median) for two drugs [85]. Today, more than 450 000 new cases of MDR-TB occur each year. The problem is of particular concern in the Russian Federation, India, and China, where up to two-thirds of new MDR-TB cases occur (Figure 4.39) [83].

Cancer

Cancer is the third leading cause of death in developed countries for persons aged 15–44, killing more than 129,000 people each year. Worldwide, more than 580,000 people aged 15–44 die of cancer each year. The leading causes of cancer death for people in this age group include liver cancer (68,000 deaths/year), leukemias (65,000 deaths/year), stomach cancer (58,000 deaths/year) and breast cancer (57,000 deaths/year) [2]. Nearly half of all men in the USA and one-third of women in the USA will develop cancer at sometime in their lives [86].

Cancer is a complex group of diseases, all of which are characterized by uncontrolled cell growth. Cancer cells usually form an abnormal mass called a tumor (some blood cancers like leukemia are exceptions) [86]. The tumor grows more rapidly than the surrounding normal tissue and can damage adjacent normal structures. Tumors are characterized as benign or malignant based on the ability of tumor cells to break away from the main lesion and spread (metastasize) to other parts of the body. Benign tumors cannot metastasize, and are usually not life threatening. Malignant tumors can spread to distant sites; it is this metastasis that is responsible for more than 90% of deaths caused by cancer [87]. Figure 4.40 shows the types of cancer which cause the greatest number of deaths in men and women in the United States.

A cancer cell develops as a result of non-lethal damage to DNA; this damage can accrue as a result of environmental exposure to carcinogens, hereditary defects or a combination of both [86]. In any case, the cancer cell multiplies and unless eradicated by the host immune system, a tumor arises from the expansion of a single progenitor cell that has incurred genetic damage.

As tumors grow and become more aggressive, tumor cells can detach from the primary tumor, and secrete enzymes to degrade the matrix of connective tissue which surrounds them. This process is known as local invasion and enables the tumor to infiltrate the organ where it originated. In some cases, tumor cells migrate

In Clinic with Amy and Clement: June 6, 2007
Dave Swaziland

I spent today making presentations, reading about nutrition, and checking up on some patient files for Dr. Eric. And, while that was interesting, all that sitting at a desk and staring into a computer all day can get quite tiresome. So at the end of the day, I decided to change things up a bit – get a little reality check – and step into a Dr.'s room and observe a check-up.

The room I chose was Dr. Amy's.

I absolutely love Amy, by the way. She, very kindly, offered to take Tessa and I to Mlilwane this weekend, which we of course accepted, and I got to know her pretty well along the way. I also had the pleasure of meeting her daughter, Molly – a darling two-year-old with curly blonde hair who is quite afraid of strangers (she covers her eyes with her hands whenever she meets one) but is quite talkative and friendly once she warms up to you. Molly accompanied us for the ride in her carseat. Before hitting the road for good, we had to make an emergency stop at Amy and Molly's in order to make about 50 tunafish sandwiches for a meeting she had to go to afterwards (for which she had to be a good little Swazi wife and bring a meal – despite the fact that, unlike the rest of the wives that would be attending just as wives, she, as a doctor, was actually going to present). Oh, and I plan on trying her special tuna fish sandwich recipe which involves apple slices as soon as possible.

So, I went in to Dr. Amy's room and asked her if it would be alright for me to observe, and she said she was quite happy to have me come in. She was just starting her last check-up of the day with an adorable four-year-old boy named Clement.

Clement is HIV positive. However, because the Baylor Clinic has been able to provide him with free ARVs for a year now he is now quite healthy. His CD4 count has risen since he began ARV treatment from around 500, which is quite low, to a now very normal level of about 1100.

Clement came in today with his gogo (grandmother) because of a cough he has had for about four weeks now. Amy treated him about two weeks ago with antibiotics, but it failed to help. The reasons for this, as Amy told me, could be that the prescribed antibiotic wasn't the right one, the cough could be caused by a virus rather than bacteria, or that it could be tuberculosis. Apparently, Swaziland holds the record right now for the country with the highest percentage of its population infected with tuberculosis. Because of the prevalence of the disease here, it was very important for Amy to find out if this was Clement's problem and treat it as quickly as possible. So, in order to check for TB, she had to administer a skin test.

A skin test, unfortunately, involves a needle, and when Amy reentered the room with needle in hand . . . you should have heard Clement's screams! I had no idea that a child that small could produce a noise that big. Or, quite frankly, that unnerving.

I was a bit shell-shocked, but, of course, to Amy, the seasoned pediatrician, Clement's screams were nothing out of the ordinary. She proceeded without hesitation. It took the combined effort of Gogo, Amy, and even Lulu (the translator) to hold Clement's thrashing body still long enough for Amy to inject him for the test.

After the shot (which was given with a needle so small that I doubt he was able to even feel it much) Clement continued to cry a bit, but it was a more resigned "Oh, I guess that didn't hurt too badly but I'm still mad at you for making me do something I didn't want to do" kind of cry.

Besides the screaming part, I had a fantastic time observing. Amy did a wonderful job of teaching me about all kinds of things like what she looks for during a check-up, what different symptoms mean, and how she uses the medical records. In Clement's case, it was the medical records that Amy and I found most interesting. While showing me the medical records, she noticed a very strange doctor's note: Apparently, Clement's house had recently burnt down.

And, apparently, somehow adorable little Clement did it. Poor guy!

2007 Estimated US Cancer Deaths

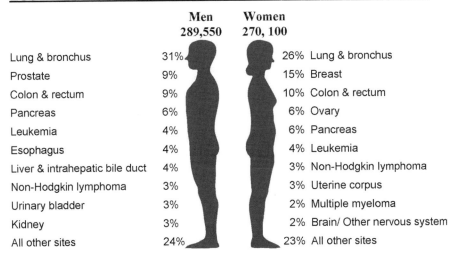

	Men 289,550	Women 270,100	
Lung & bronchus	31%	26%	Lung & bronchus
Prostate	9%	15%	Breast
Colon & rectum	9%	10%	Colon & rectum
Pancreas	6%	6%	Ovary
Leukemia	4%	6%	Pancreas
Esophagus	4%	4%	Leukemia
Liver & intrahepatic bile duct	4%	3%	Non-Hodgkin lymphoma
Non-Hodgkin lymphoma	3%	3%	Uterine corpus
Urinary bladder	3%	2%	Multiple myeloma
Kidney	3%	2%	Brain/ Other nervous system
All other sites	24%	23%	All other sites

Figure 4.40. *Leading causes of cancer death, by organ site, in the United States, estimated for 2007. Adapted with permission of the American Cancer Society, Inc. All rights reserved.*

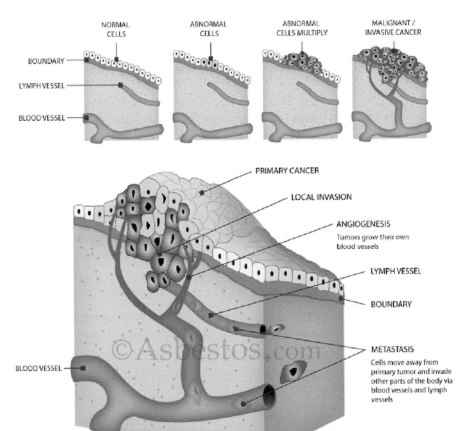

Figure 4.41. *Illustration of the major steps of cancer formation and metastasis. Courtesy of Asbestos.com.*

to blood vessels or lymphatic vessels. From there, tumor cells can circulate throughout the body, lodge in distant organs and form a metastasis [88]. Figure 4.41 illustrates the steps that occur during tumor formation and metastasis.

The most effective therapy to treat cancer is surgical removal. Generally, the patient's prognosis is excellent if all of a tumor can be resected. However, in many cases, cancers are not diagnosed until there has been extensive tumor growth and metastasis, preventing complete

Five-Year Relative Survival Rates by Stage at Diagnosis

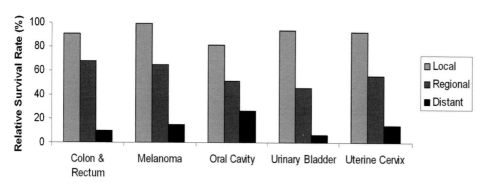

Figure 4.42. *The five year relative survival rate for different cancers in the United States by stage (local, regional, distant) at the time of diagnosis, estimated for 1992–1999. Local: cancer confined to the organ of origin. Regional: cancer that (1) has extended beyond the organ of origin into surrounding organs or tissue, (2) involves regional lymph nodes, or (3) both. Distant: cancer that has spread to parts of the body remote from the primary tumor. Adapted with permission of the American Cancer Society, Inc. All rights reserved.*

surgical removal. In these cases, chemotherapeutic agents or radiation therapy or a combination of the two, generally after surgery removing the majority of the tumor mass, can be used to kill elusive tumor cells. Because early lesions are so much easier to treat, improvements in early detection can play a large role in improving survival rates for cancer patients. Methods for the early detection of cancer are as varied as the sites in which they develop. Screening for early cancers generally occurs as a part of population based programs and frequently leads to a biopsy of the suspected tumor site for microscopic examination [89, 90].

Figure 4.42 shows the five year survival rate for colorectal cancer, skin cancer, oral cancer, bladder cancer, and cervical cancer as a function of the stage of the tumor when it is initially detected. When a tumor is detected at a stage when only local invasion has occurred, the five year survival rates are on average over 90%. If regional metastasis has occurred, five year survival rates drop to 50–70%. However, when metastasis to distant organ sites has occurred, the five year survival rates are only 5–30%. See also Figure 4.43. Later in this chapter, we will consider the pathophysiology of lung cancer in more detail. One reason that lung cancer is so deadly is that it is frequently not detected until distant metastasis has occurred.

Self-inflicted injuries

The fourth leading cause of death for people aged 15–44 in developed countries is suicide – worldwide, more than 480,000 people between the ages of 15 and 44 years take their own lives, nearly as many as die from cancer in this age group [2].

In 2003, the USA reported a suicide rate of 10.8/100,000 people. The rate was 17.6/100,000 for men and 4.3/100,000 for women [91]. Firearms are used in nearly 60% of suicides; hanging and drug overdose are the second leading means of suicide for men and women respectively. Alcohol intoxication is associated with 25–50% of suicides.

The causes of suicide are complex – including interactions between personal, family, community and societal problems. The major risk factors include ready availability of weapons, inadequate social problem solving skills, abuse of alcohol and drugs, psychiatric illness, and affective, personality and other mental disorders. Other contributing risk factors include social adjustment problems, serious medical illness, living alone, recent bereavement, personal history of suicide attempt, divorce or separation, and unemployment.

Several of the risk factors for suicide could be screened for in the physician's office. In fact 50–66% of all suicide victims visit a physician within one month

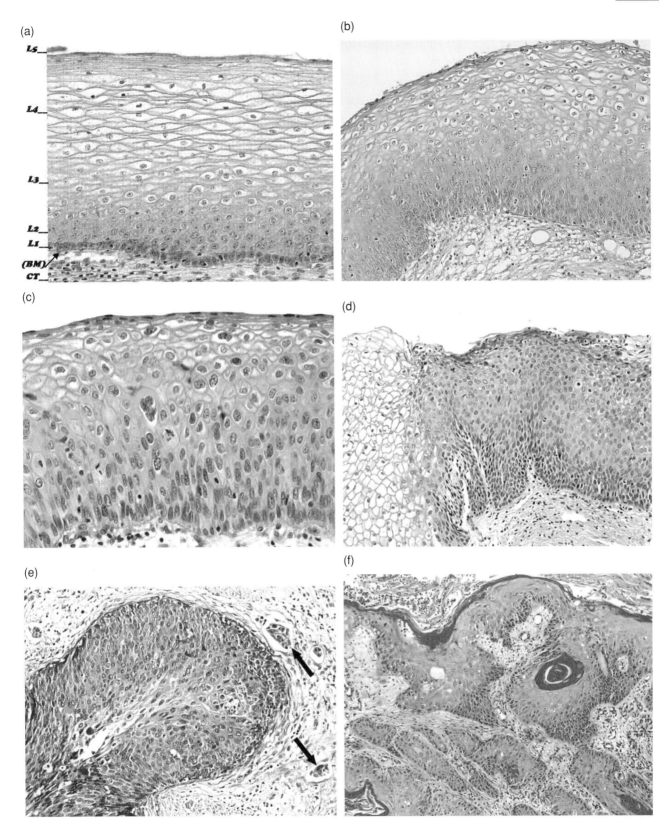

Figure 4.43. *Biopsies taken to detect the presence of cervical cancer and pre-cancer, viewed as histological sections. The above are examples of (a) normal tissue; early stage precancerous changes, (b) CIN I, (c) CIN II, (d) CIN III; (e) micro-invasive cancer; and (f) cancer. Used with permission from the International Agency for Research on Cancer.*

before the event and 10–40% of victims visit their physician in the preceding week, suggesting the potential for screening and prevention. However, it is difficult to identify who is at risk. Direct questioning, including general questions about sleep disturbance, depressed mood, guilt and hopelessness, has been found to have low yield. Similarly, different survey instruments have not been shown to be good at predicting what will happen [49].

Leading causes of mortality, ages 45–60

For persons aged 45–60, the four leading causes of death in the developing world are as follows [2].

(1) **Cardiovascular diseases**,
(2) **cancer** (malignant neoplasms),
(3) **unintentional injuries**, and
(4) **HIV/AIDS**.

In the developed world, the four leading causes of death in this age group are as follows [2].

(1) **Cardiovascular diseases**,
(2) **cancer** (malignant neoplasms),
(3) **unintentional injuries**, and
(4) **digestive diseases**.

Cardiovascular diseases

In both developed and developing countries, cardiovascular diseases are the leading cause of death for people aged 45–59, killing more than two million people in this age group every year [2]. Ischemic heart disease is the single leading cause of cardiovascular death in this age group, and is responsible for one million of these deaths [2]. Cerebrovascular disease is the second leading cause of cardiovascular death, killing nearly 625,000 people between the ages of 45 and 59 years worldwide [2].

Cancer

Cancer is the second leading cause of death in both developed and developing countries for people aged 45–59 [2]. More than 1.5 million people of this age die every year as a result of cancer. Lung cancer is by far the leading cause of cancer death in this age group, killing

263,000 people aged 45–59 each year throughout the world [2]. After lung cancer, stomach cancer (185,000 deaths/year), liver cancer (179,000 deaths/year) and breast cancer (148,000 deaths/year) account for most cancer mortality among this age group [2].

Lung cancer is the leading cause of cancer death in men in the United States. It was estimated that 89,510 males and 70,880 females in the USA would die of lung cancer in 2007 [92]. Only 16% of patients with lung cancer survive five years or more after the original diagnosis [92]. The survival rate of lung cancer is so low because it is usually not detected until it is at a very advanced and untreatable stage. Only 16% of patients are diagnosed with localized disease, when treatment is most effective [92]. The most important risk factor for the development of lung cancer is smoking. Patients who actively smoke increase their risk of developing lung cancer by 13-fold, while patients exposed to passive smoke increase their risk by 1.5 times [51].

The signs and symptoms of lung cancer include coughing, chest pain, difficulty breathing, and recurrent pneumonia [92]. A chest X-ray can document advanced lung cancer. The diagnosis of lung cancer is usually confirmed by a biopsy, obtained under the guidance of CT or obtained through a bronchoscope [51].

Small lung tumors can be removed surgically, usually resulting in good prognosis. Larger tumors which are confined to the lung are usually treated with chemotherapy or radiation therapy and surgery. Metastatic lung tumors are treated with a combination of chemo- and radiation therapy [51].

It is usually difficult to detect lung cancer at an early stage when it is most easily treatable. By the time patients experience symptoms, the disease has usually spread to other organs [51]. A number of technologies have been tested to determine if they are useful to screen the general population for early lung cancer. However, clinical trials of chest X-ray and sputum cytology, (two of the most promising technologies), have not proven adequate to screen for early disease [92].

Unintentional injuries

Unintentional injuries are the third leading cause of death in both developed and developing countries for people between the ages of 45 and 59 years, and are responsible for 618,000 deaths per year in this age group [2]. Road traffic accidents are by far the leading cause of death by unintentional injury in this age group, accounting for more than 222,000 deaths per year [2].

HIV/AIDS

The fourth leading cause of death in developing countries is HIV/AIDS, which kills 386,000 persons aged 45–59 in developing countries each year [2].

Digestive diseases

The fourth leading cause of death in developed countries for persons aged 45–59 is digestive diseases. Worldwide, 456,000 people in this age group die each year as a result of digestive diseases [2]. Cirrhosis of the liver is by far the most common fatal digestive disease, killing 250,000 people each year between the ages of 45 and 59 [2].

The liver is the largest organ in the body and performs a number of crucial physiologic functions (Figure 4.44) [93]. All blood that leaves the stomach and intestines passes to the liver through the hepatic portal vein; in addition, blood from the peripheral tissues enters the liver via the hepatic artery [93]. This blood contains nutrients as well as drugs and foreign substances that have been ingested. The liver metabolizes fat and glucose for energy storage and helps to remove toxic substances from the blood, transferring wastes to the kidney to be excreted [93]. The liver is the body's largest chemical factory, producing bile to help absorb fats, proteins that regulate blood clotting, and immune agents [93]. Loss of liver function can produce severe disease and death.

In cirrhosis, normal liver is replaced with scar tissue as a result of chronic injury, interfering with liver function [94]. There are several causes of cirrhosis, the two most common are chronic alcoholism and viral hepatitis infection [94].

Alcoholic cirrhosis usually develops after more than a decade of heavy drinking. Heavy drinkers include those who drink between 8–16 ounces of hard liquor per day; about 1/3 of those who drink this much will develop cirrhosis within 15 years [95]. However, some social drinkers will develop alcoholic cirrho-

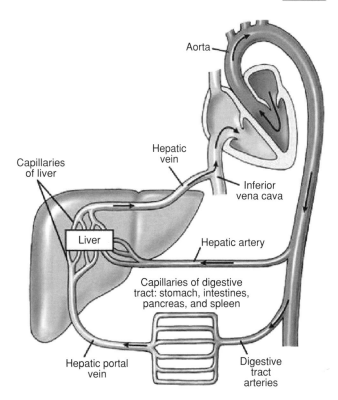

Figure 4.44. *The liver plays a central role in many important metabolic processes. Copyright © 2004 by Pearson Education, Inc. Reprinted with permission.*

sis; this is much more likely in women for reasons that are not well understood. Alcohol injures the liver by blocking protein, fat and carbohydrate metabolism [94].

Hepatitis is a viral infection which can also produce cirrhosis. In the United States, hepatitis C is the most common cause of infectious hepatitis; however, worldwide, hepatitis B is the most common cause [94]. Both infections cause liver inflammation and injury that can lead to cirrhosis over several decades. Acute hepatitis B infection leads to chronic hepatitis in approximately 1–5% of patients, a fraction of whom will develop cirrhosis [51]. Acute hepatitis C becomes chronic in 85–90% of patients. Approximately 20–50% of these patients will develop cirrhosis [51].

The symptoms of cirrhosis include exhaustion, loss of appetite, nausea, vomiting blood, weakness, weight loss, and abdominal pain [94]. Patients bruise and bleed easily and become highly sensitive to medicines with increasing loss of liver function.

Table 4.4. *Top ten causes of morbidity by age in developed and developing countries [2].*

Ages 0-4	Developing Countries				Developed Countries	
Cause of Disability	# DALYs	% of Total DALYs	Cause of Disability		# DALYs	% of Total DALYs
Lower respiratory infections	60,236,694	14.3%	Low birth weight		1,337,103	10.3%
Diarrhoeal diseases	55,543,335	13.2%	Congenital heart anomalies		1,327,944	10.2%
Low birth weight	44,997,007	10.7%	Birth asphyxia and birth trauma		1,310,241	10.1%
Malaria	42,244,474	10.0%	Lower respiratory infections		1,138,923	8.8%
Birth asphyxia and birth trauma	33,133,613	7.9%	Diarrhoeal diseases		537,438	4.1%
Measles	18,618,263	4.4%	Iodine deficiency		495,378	3.8%
Protein-energy malnutrition	14,718,970	3.5%	Mental retardation, lead-caused		467,625	3.6%
Congenital heart anomalies	12,851,427	3.1%	Meningitis		367,281	2.8%
Pertussis	12,264,915	2.9%	Down syndrome		367,109	2.8%
HIV/AIDS	12,181,146	2.9%	Asthma		259,845	2.0%
Total DALYs	420,827,539		Total DALYs		13,003,994	
Total Population	536,962,742		Total Population		81,206,312	

Ages 15-44	Developing Countries				Developed Countries	
Cause of Disability	# DALYs	% of Total DALYs	Cause of Disability		# DALYs	% of Total DALYs
HIV/AIDS	59,382,428	13.3%	Unipolar depressive disorders		10,484,105	13.7%
Unipolar depressive disorders	35,978,376	8.1%	Alcohol use disorders		6,308,519	8.3%
Tuberculosis	19,848,813	4.5%	Road traffic accidents		3,804,331	5.0%
Road traffic accidents	19,151,250	4.3%	Self-inflicted injuries		3,144,909	4.1%
Violence	14,923,499	3.3%	Violence		2,285,286	3.0%
Self-inflicted injuries	12,189,495	2.7%	Bipolar disorder		2,209,104	2.9%
Schizophrenia	12,074,350	2.7%	Drug use disorders		2,037,084	2.7%
Bipolar disorder	10,977,659	2.5%	Schizophrenia		1,944,628	2.5%
Alcohol use disorders	10,664,330	2.4%	Ischaemic heart disease		1,916,252	2.5%
Hearing loss, adult onset	9,186,758	2.1%	Hearing loss, adult onset		1,523,616	2.0%
Total DALYs	445,613,527		Total DALYs		76,416,610	
Total Population	2,312,272,679		Total Population		597,682,683	

Ages 45-59	Developing Countries				Developed Countries	
Cause of Disability	# DALYs	% of Total DALYs	Cause of Disability		# DALYs	% of Total DALYs
Ischaemic heart disease	12,050,270	7.6%	Ischaemic heart disease		5,286,352	11.3%
Cerebrovascular disease	10,212,640	6.4%	Cerebrovascular disease		3,123,891	6.7%
Cataracts	9,735,678	6.1%	Unipolar depressive disorders		3,006,141	6.4%
Unipolar depressive disorders	8,374,876	5.3%	Hearing loss, adult onset		1,884,097	4.0%
Chronic obstructive pulmonary disease	7,287,981	4.6%	Trachea, bronchus, lung cancers		1,770,453	3.8%
Hearing loss, adult onset	6,891,572	4.3%	Chronic obstructive pulmonary disease		1,762,640	3.8%
HIV/AIDS	6,703,167	4.2%	Osteoarthritis		1,545,250	3.3%
Tuberculosis	6,504,260	4.1%	Cirrhosis of the liver		1,438,096	3.1%
Vision disorders, age-related	4,787,811	3.0%	Diabetes mellitus		1,388,386	3.0%
Diabetes mellitus	4,045,375	2.5%	Alcohol use disorders		1,086,926	2.3%
Total DALYs	159,380,182		Total DALYs		46,615,959	
Total Population	600,316,766		Total Population		254,600,864	

The diagnosis of cirrhosis is made by a liver biopsy done through a needle. It is currently not possible to reverse the liver damage associated with cirrhosis [51]. Discontinuing use of alcohol can prevent further damage as can treatment of hepatitis with medicines to boost immune response. In advanced cases, liver transplant is currently the only treatment which can restore liver function.

Leading causes of morbidity and mortality

In this chapter, we have considered the leading causes of mortality throughout the world. In examining global health, it is important to consider both causes of mortality and morbidity. As we saw in Chapter 3, the number of DALYs lost to disease measures the combined effects of morbidity and mortality. Table 4.4 shows the ten diseases which result in the greatest number of

Table 4.5. *Grand challenges in global health From [96]. Reprinted with permission from AAAS.*

GOAL: To improve childhood vaccines GC #1 Create effective single-dose vaccines that can be used soon after birth GC#2 Prepare vaccines that do not require refrigeration GC#3 Develop needle-free delivery systems for vaccines
GOAL: To create new vaccines GC#4 Devise reliable tests in model systems to evaluate live attenuated vaccines GC#5 Solve how to design antigens for effective, protective immunity GC#6 Learn which immunological responses provide protective immunity
GOAL: To control insects that transmit agents of disease GC#7 Develop a genetic strategy to deplete or incapacitate a disease-transmitting insect population GC#8 Develop a chemical strategy to deplete or incapacitate a disease-transmitting insect population
GOAL: To improve nutrition to promote health GC#9 Create a full range of optimal, bioavailable nutrients in a single staple plant species
GOAL: To improve drug treatment of infectious diseases GC#10 Discover drugs and delivery systems that minimize the likelihood of drug resistant micro-organisms
GOAL: To cure latent and chronic infections GC#11 Create therapies that can cure latent infections GC#12 Create immunological methods that can cure chronic infections
GOAL: To measure disease and health status accurately and economically in developing countries GC#13 Develop technologies that permit quantitative assessment of population health status GC#14 Develop technologies that allow assessment of individuals for multiple conditions or pathogens at point-of-care

disability free years of life lost by age for developed and developing countries. In comparing Table 4.1 to Table 4.4, it is evident that the leading causes of mortality are not necessarily the same as the leading causes of morbidity. For example, unipolar depressive disorders are the leading cause of morbidity in the developed world and the second leading cause of morbidity in the developing world for ages 15–44. Despite the prevalence of these disorders and their significant contribution to the worldwide burden of disease, they are not listed as a leading cause of mortality. In assessing the state of health throughout the world it is important to consider not only the causes of death, but also those conditions, such as mental illness, that significantly impact a person's quality of life.

Global health challenges

Can technology address the health challenges that are faced by citizens in the developing world? In January 2003, the Bill and Melinda Gates Foundation announced a $200 million medical research initiative to solve medical issues identified as "grand challenges" in global health [96]. This initiative is designed to encourage scientific and technological solutions to diseases that disproportionately affect the developing world. The initiative seeks scientific or technical innovations that remove a critical barrier to solving an important health problem in the developing world, with a high likelihood of global impact and feasibility [96]. The Foundation solicited ideas for Grand Challenges in May 2003 and received 1048 submissions from scientists and institutions in 75 countries [96]. The Foundation's Scientific Board heard proposals in August 2003 and identified 7 long range goals and 14 grand challenges (Table 4.5). These are heavily oriented toward infectious disease, largely because infectious diseases account for the most profound discrepancies between advanced and developing economies, and because the causes of infectious diseases are well known and scientists can more easily understand and address technical and scientific obstacles to progress [96].

> ### Bioengineering and Global Health Project
>
> **Project task 1: Define a public health problem facing a particular country or region of the world**
>
> You may select your topic from a wide range of heath issues – you must simply demonstrate that the chosen issue significantly and adversely affects the lives of people in the country or region you have selected. Write a one-page summary of your disease including: epidemiology, prevalence, and incidence in the region you have selected; pathophysiology of the disease; physical signs and symptoms of the disease.

Homework

1. List the four leading causes of death in developed and developing countries for the following age groups.
   a. Ages 0–4
   b. Ages 15–44
   c. Ages 45–59
   d. Compare the differences in causes of mortality in these settings. What is responsible for the differences?
   e. Compare the leading causes of morbidity to the leading causes of mortality in each case. What differences do you note?

2. Road traffic accidents are a leading cause of death for young people.
   a. How do motorcycle helmets work to save lives?
   b. Why do factors which slow driver reaction time lead to increased crash frequency and crash severity? Name three factors which slow driver reaction times.
   c. Compare trends in motor vehicle related mortality rates in developed and developing countries over the past ten years. What factors do you think contribute to these differences?

3. Cholera can produce severe diarrhea. The associated fluid loss can lead to dehydration and death if untreated.
   a. How does cholera produce such severe fluid loss?
   b. What is oral rehydration therapy and how does it prevent dehydration associated with severe diarrhea?

4. The AIDS pandemic is a worldwide problem. An estimated 33 million people are living with HIV/AIDS and over 20 million deaths have been associated with this disease.
   a. Which component of the immune system is selectively targeted and destroyed by HIV?
   b. Sketch a plot showing (1) the viral load and (2) blood count of the cell type specified in part (a), over the time course of the disease. On the plot identify the acute phase of the infection, the latent period, and full-blown AIDS.
   c. What is the approximate length of the latent period? Can HIV be transmitted during this time?
   d. Draw a diagram which indicates the process of retrovirus replication inside a human cell.
   e. Combination drug therapies have been successful in suppressing viral levels. What is the name of the current treatment strategy and why is it so effective?
   f. There are several potential strategies for preventing retrovirus replication. On your diagram for part d, draw arrows to indicate the three stages targeted in current combination therapies. Name the type of each type of inhibition.
   g. Discuss the WHO 3 × 5 initiative and comment on the current progress and challenges associated with this effort.

5. Download "Drug Resistance and Malaria" by Peter Bloland, published by the WHO: http://www.who.int/csr/resources/publications/drugresist/malaria.pdf. Using this reference as your guide, answer the following questions.
   a. According to the report, which species of *Plasmodium* parasites have developed drug resistance?
   b. Which region of the world is experiencing the most significant problem with drug resistance?

c. What would be the yearly cost of treating every malaria infection on Earth with the least expensive single agent anti-malarial? (Note: use the WHO's low end estimate of global incidence, and assume that one treatment per patient will suffice.)

d. What is the yearly cost for the least expensive combination therapy?

e. In one sentence each, describe three strategies to prevent anti-malarial drug resistance.

6. The use of combinations of anti-retroviral drugs has proven remarkably effective in controlling the progression of human immunodeficiency virus (HIV) disease and prolonging survival, but these benefits can be compromised by the development of drug resistance. Resistance is the consequence of mutations that emerge in the viral proteins targeted by anti-retroviral agents. In the United States, as many as 50% of patients receiving anti-retroviral therapy are infected with viruses that express resistance to at least one of the available anti-retroviral drugs [Source: *NEJM* **350**:1023–35, 2004]. One new technology developed to decrease the development of resistance is described in the following NPR report: http://www.npr.org/templates/story/story.php?storyId=5554167. How might this particular development reduce the risk of drug resistance in the developing and developed world?

7. Access the following CDC *Weekly Report*: Emergence of *Mycobacterium tuberculosis* with Extensive Resistance to Second-Line Drugs – Worldwide, 2000–2004: http://www.cdc.gov/mmwr/preview/mmwrhtml/mm5511a2.htm. Answer the questions that follow.

a. Define MDR and XDR tuberculosis.

b. In 2004, which two regions of the world had the highest percentage of TB isolates classified as multi-drug resistant?

c. In 2004, which countries had more MDR isolates classified as XDR than any other?

d. How does one attempt to treat MDR-TB? (Hint: refer to the *Weekly Report*'s Editorial Note.)

8. The Framingham Heart Study was a monumental project not only for cardiovascular disease, but for all of science, health, and medicine. Answers to the following questions may be found at the study's website: http://www.nhlbi.nih.gov/about/framingham/.

a. What was the initial purpose of the Framingham study?

b. List 5 definite risk factors for heart disease, and the year in which they were found to be associated with an increased risk.

9. Complete the interactive tutorial on Coronary Artery Bypass Graft (CABG) surgery: http://www.nlm.nih.gov/medlineplus/tutorials/coronaryarterybypassgraft/htm/index.htm.

a. What non-surgical therapies are alternatives to CABG?

b. What vessels are used to form the "bypass"? What are the specific side effects associated with harvesting these vessels?

10. Cancer is the second leading cause of death in the United States and annually costs the healthcare system more than 100 billion dollars.

a. What type of cancer is responsible for the greatest number of deaths worldwide?

b. Why is the mortality rate so high for this type of cancer?

c. Describe the stages of malignant tumor formation and metastasis.

References

[1] *A Tale of Two Girls*. World Health Organization; 2003.

[2] WHO. *Mortality: Revised Global Burden of Disease (2002) Estimates*. Geneva: World Health Organization; 2002.

[3] Lopez AD. *Disease Control Priorities Project. Global Burden of Disease and Risk Factors*. New York Washington, D.C.: Oxford University Press; World Bank; 2006.

[4] Black RE, Morris SS, Bryce J. Where and why are 10 million children dying every year? *Lancet*. 2003 Jun 28; **361**(9376): 2226–34.

[5] WHO. *The World Health Report 2005: Make Every Mother and Child Count*. Geneva: World Health Organization; 2005.

[6] Bale JR, Stoll BJ, Lucas AO. Institute of Medicine (US). Committee on Improving Birth Outcomes. *Improving Birth Outcomes : Meeting the Challenges in the Developing World.* Washington, D.C.: National Academies Press; 2003.

[7] Mathai M, Sanghvi H, Guidotti RJ. *Managing Complications in Pregnancy and Childbirth: a Guide for Midwives and Doctors.* Geneva: World Health Organization; 2000.

[8] PATH. *Delivery Kit.* Seattle: PATH; 2006.

[9] PBS. *RX for Survival: Pneumonia.* 2006 March [cited 2007 May 30]; Available from: http://www.pbs.org/wgbh/rxforsurvival/series/diseases/pneumonia.html

[10] Cashat-Cruz M, Morales-Aguirre JJ, Mendoza-Azpiri M. Respiratory tract infections in children in developing countries. *Seminars Pediatric Infectious Diseases.* 2005 Apr; **16**(2): 84–92.

[11] CDC. *Frequently Asked Questions About SARS.* 2005 [cited; Available from: http://www.cdc.gov/ncidod/sars/faq.htm

[12] CDC. *Fact Sheet: Basic Information About SARS.* 2005 [cited; Available from: http://www.cdc.gov/ncidod/sars/factsheet.htm

[13] CHEMICON. *Respiratory Viral Screen Direct Immunofluorescence Assay.* Temecula, CA; 2001.

[14] Liaison OoCaP. *The Problem of Antibiotic Resistance.* Bethesda: National Institute of Allergy and Infectious Diseases; 2006.

[15] Lewis R. The rise of antibiotic-resistant infections. *FDA Consumer.* 1995; **29**(7).

[16] Field M. Intestinal ion transport and the pathophysiology of diarrhea. *Journal of Clinical Investigation.* 2003 Apr; **111**(7): 931–43.

[17] Nutrition Committee CPS. Oral rehydration therapy and early refeeding in the management of childhood gastroenteritis. *The Canadian Journal of Pediatrics.* 1994; **5**(1): 160–4.

[18] WHO. *Global Water Supply and Sanitation Assessment 2000 Report.* Geneva: World Health Organization and United Nations Children's Fund; 2000.

[19] Ulrickson M. Oral rehydration therapy in children with acute gastroenteritis. *Jaapa.* 2005 Jan; **18**(1): 24–9; quiz 39–40.

[20] Victora CG, Bryce J, Fontaine O, Monasch R. Reducing deaths from diarrhoea through oral rehydration therapy. *Bulletin of the World Health Organization.* 2000; **78**(10): 1246–55.

[21] Water with sugar and salt.*The Lancet.* 1978 Aug 5; **2**(8084): 300–1.

[22] Coordinating Center for Infectious Diseases/Division of Bacterial and Mycotic Diseases. *Cholera.* Atlanta: Centers for Disease Control and Prevention; 2005.

[23] Gourevitch P. *We Wish To Inform You That Tomorrow We Will Be Killed With Our Families: Stories From Rwanda.* 1st edn. New York: Farrar, Straus, and Giroux; 1998.

[24] WHO Global Task Force on Cholera Control. *First Steps for Managing an Acute Outbreak of Diarrhoea.* Geneva: World Health Organization; 2004.

[25] WHO/UNICEF. *Oral Rehydration Salts: Production of the New ORS.* Geneva: World Health Organization; 2006.

[26] Parashar UD, Gibson CJ, Bresse JS, Glass RI. Rotavirus and severe childhood diarrhea. *Emerging Infectious Diseases.* 2006 Feb; **12**(2): 304–6.

[27] National Center for Infectious Diseases Respiratory and Enteric Viruses Branch *Rotavirus.* 2006 [cited; Available from: http://www.cdc.gov/ncidod/dvrd/revb/gastro/rotavirus.htm

[28] Ahmed FU, Karim E. Children at risk of developing dehydration from diarrhoea: a case-control study. *Journal of Tropical Pediatrics.* 2002 Oct; **48**(5): 259–63.

[29] Cunliffe NA, Bresee JS, Hart CA. Rotavirus vaccines: development, current issues and future prospects. *Journal of Infect.* 2002 Jul; **45**(1): 1–9.

[30] Weijer C. The future of research into rotavirus vaccine. *BMJ.* 2000 Sep 2; **321**(7260): 525–6.

[31] Maugh TH, Kaplan K. 2 Vaccines Sharply Decrease Grave Diarrhea Causes. *Los Angeles Times.* 2006 January 5.

[32] *Roll Back Malaria. What is Malaria?* Geneva: World Health Organization; 2004.

[33] Roll Back Malaria, World Health Organization, UNICEF. *World Malaria Report 2005.* Geneva: World Health Organization; 2005.

[34] *Roll Back Malaria. Children and Malaria.* Geneva: World Health Organization; 2004.

[35] *Roll Back Malaria. Malaria in Africa.* Geneva: World Health Organization; 2004.

[36] *Roll Back Malaria. Malaria in Pregnancy.* Geneva: World Health Organization; 2004.

[37] National Center for Infectious Diseases Division of Parasitic Diseases. *Malaria Facts.* Atlanta: Centers for Disease Control and Prevention; 2004.

[38] *Roll Back Malaria. Facts on ACTs: January 2006 Update.* Geneva: World Health Organization; 2006.

[39] Dorris M. *The Broken Cord.* 1st edn. New York: Harper & Row; 1989.

[40] Greenwood MR. Methylmercury poisoning in Iraq. An epidemiological study of the 1971–1972 outbreak.

Journal of Applied Toxicology. 1985 Jun; **5**(3): 148–59.

[41] *Health in the Millenium Development Goals*: Millenium Development Goals, targets and indicators related to health. World Health Organization; 2004.

[42] Jones G, Steketee RW, Black RE, Bhutta ZA, Morris SS. How many child deaths can we prevent this year? *The Lancet.* 2003 Jul 5; **362**(9377): 65–71.

[43] Beaglehole R, Irwin A, Prentice T. *The World Health Report 2003: Shaping the Future.* Geneva: World Health Organization; 2003.

[44] Aturupane D, Deolalikar AB. *Attaining the Millenium Development Goals in Sri Lanka: How Likely and What Will It Take To Reduce Poverty, Child Mortality and Malnutrition, and to Increase School Enrollment and Completion?* The World Bank; 2005 October 7.

[45] Central Intelligence Agency. *The World Factbook 2007* [cited 2007 May 30]; Available from: https://www.cia.gov/library/publications/the-world-factbook/index.html

[46] WHO. *Report of a Regional Consultation.* Improving Neonatal Health in South-East Asia Region; 2002 April 1–5; New Delhi, India. World Health Organization; 2002.

[47] Howse JL, Caldwell MC. The state of infant health: is there trouble ahead? *America's Health: State Rankings 2004.* Minnetonka, MN: United Health Foundation; 2004.

[48] Beaglehole R, Irwin A, Prentice T. *The World Health Report 2004: Changing History.* Geneva: The World Health Organization; 2004.

[49] U.S. Preventive Services Task Force. *Guide to Clinical Preventive Services.* 2nd edn. Alexandria, Virginia: International Medical Publishing; 1996.

[50] Beal J, Orrick J, Alfonso K, Rathore M, eds. *HIV/AIDS Primary Care Guide.* Florida/Carribean AIDS Education Training Center; 2006.

[51] Kasper D, Braunwald E, Fauci A, Longo D, Hauser S, Jameson JL, eds. *Harrison's Principles of Internal Medicine.* 16th edn. New York: McGraw-Hill; 2005.

[52] Glick BR, Pasternak JJ. *Molecular Biotechnology: Principles and Applications of Recombinant DNA.* 3rd edn. Wahington D.C.: ASM Press; 2003.

[53] Hoffman C, Rockstroh JK, Kamps BS, eds. *HIV Medicine 2006.* 14th edn. Paris: Flying Publisher; 2006.

[54] US FDA. *Drugs Used in the Treatment of HIV Infection.* 2007 May [cited 2007 May 30]; Available from: http://www.fda.gov/oashi/aids/virals.html

[55] FDA notifications. FDA approves Fuzeon, the first fusion inhibitor. *AIDS Alert.* 2003 Jun; **18**(6): 78–9.

[56] Survival after introduction of HAART in people with known duration of HIV-1 infection. The CASCADE Collaboration. Concerted Action on SeroConversion to AIDS and Death in Europe. *The Lancet.* 2000 Apr 1; **355**(9210): 1158–9.

[57] *Guidelines for the Use of Antiretroviral Agents in HIV-1-Infected Adults and Adolescents.* US Department of Health and Human Services. October 10, 2006.

[58] *VERSANT HIV-1 RNA 3.0 Assay (bDNA): Summary of Safety and Effectiveness.* US Food and Drug Administration; 2002.

[59] WHO, UNAIDS. *Treating 3 Million by 2005: Making it Happen, The WHO Strategy.* Geneva: The World Health Organization; 2003.

[60] WHO, UNAIDS. *Progress on Global Access to HIV Antiretroviral Therapy: An Update on "3 by 5".* Geneva: World Health Organization; 2005 June 29.

[61] WHO. *Access to HIV Treatment Continues to Accelerate in Developing Countries, but Bottlenecks Persist, says WHO/UNAIDS report.* Press release. Geneva: World Health Organization; 2005.

[62] PBS. *RX for Survival: HIV/AIDS.* 2006 March [cited 2007 May 30]; Available from: http://www.pbs.org/wgbh/rxforsurvival/series/diseases/hiv_aids.html

[63] Peden M, Scurfield R, Sleet D, Mohan D, Hyder AA, Jarawan E, *et al.*, eds. *World Report on Road Traffic Injury Prevention.* Geneva: World Health Organization; 2004.

[64] National Center for Statistics and Analysis. *Traffic Safety Facts 2005.* Washington D.C.: National Highway Traffic Safety Administration; 2005.

[65] *Traffic Safety Facts: Occupant Protection.* Washington, D.C.: National Highway Traffic Safety Administration; 2006.

[66] *Traffic Safety Facts: Motorcycles.* Washington, D.C.: National Highway Traffic Safety Administration; 2005.

[67] *Traffic Safety Facts: Laws,.08 BAC Illegal per se Level.* Washington D.C.: National Highway Traffic Safety Administration; 2004.

[68] Glassbrenner D. *Traffic Safety Facts Research Note: Driver Cell Phone Use in 2005 – Overall Result.* Washington D.C.: National Highway Traffic Safety Administration; 2005 February.

[69] Ross M, Patel D, Wenzel T. Vehicle Design and the Physics of Traffic Safety. *Physics Today.* 2006 January: 49–54.

[70] National Transportation Safety Board. *Crash Tests Without Seatbelts.* Regents Exam Prep Center.

[71] *Lap Held Infant in Sedan Front Seat: 24 mph Front-into-barrier Crash.* National Highway Traffic Safety Administration.

[72] National Transportation Safety Board. *Crash Tests with Seatbelts.* Regents Exam Prep Center.

[73] Larry Patrick, pioneer auto safety researcher: 1920–2006. Wayne State University 2006 [cited 2007 June 1]; Available from: http://www.eng.wayne.edu/news.php?id=425

[74] de Boer MJ, Zijlstra F. Treating myocardial infarction in the post-GUSTO era. A European perspective. *PharmacoEconomics.* 1997 Oct; **12**(4): 427–37.

[75] American Heart Association. *Heart Disease and Stroke Statistics – 2007 Update.* Dallas, TX: American Heart Association; 2007.

[76] Shah KH, Edlow JA. Transient ischemic attack: review for the emergency physician. *Annals of Emergency Medicine.* 2004 May; **43**(5): 592–604.

[77] *Global Tuberculosis Control: Surveillance, Planning, Financing.* Geneva: World Health Organization; 2006.

[78] *Global TB Fact Sheet 2005.* Geneva: World Health Organization; 2005.

[79] Small PM, Fujiwara PI. Management of tuberculosis in the United States. *New England Journal of Medicine.* 2001 Jul 19; **345**(3): 189–200.

[80] Centers for Disease Control and Prevention. *The Deadly Intersection Between TB and HIV.* 2007 March 8 [cited 2007 May 30]; Available from: http://www.cdc.gov/hiv/resources/factsheets/hivtb.htm

[81] WHO Regional Office for South-East Asia. *TB/HIV.* 2006 April 27 [cited 2007 May 30]; Available from: http://www.searo.who.int/en/Section10/Section2097/Section2129.htm

[82] *Tuberculosis.* April 2006 [cited 2007 February 15]; Fact Sheet No. 104. Available from: http://www.who.int/mediacentre/factsheets/fs104/en/

[83] *Tuberculosis Facts 2006.* Geneva: World Health Organization; 2006.

[84] *Questions and Answers About TB.* Department of Health and Human Services, Centers for Disease Control and Prevention; 2005.

[85] *Anti-tuberculosis Drug Resistance in the World, Report No. 3.* Geneva: WHO/IUATLD Global Project on Anti-Tuberculosis Drug Resistance Surveillance; 2004. Report No.: WHO/HTM/TB/2004.343.

[86] *Detailed Guide: What is Cancer?* February 6, 2006 [cited 2007 February 16]; Available from: http://www.cancer.org/docroot/CRI/content/CRI_2_4_1x_What_Is_Cancer.asp?sitearea=

[87] Sporn MB. The war on cancer. *The Lancet.* 1996 May 18; **347**(9012): 1377–81.

[88] Hanahan D, Weinberg RA. The hallmarks of cancer. *Cell.* 2000 Jan 7; **100**(1): 57–70.

[89] Cotran RS, Robbins SL, Kumar V. *Robbins Pathologic Basis of Disease.* 5th edn. Philadelphia: Saunders; 1994.

[90] Frappart L, Fontaniere B, Sankaranarayanan R. *Histopathology and Cytopathology of the Uterine Cervix – Digital Atlas.* International Agency for Research on Cancer; 2004.

[91] Hoyert DL, Heron MP, Murphy SL, Kung H-C. *Deaths: Final Data for 2003.* National Vital Statistics Report. 2006 April 19; **54**(13).

[92] Society AC. *Cancer Facts and Figures 2007.* Atlanta: American Cancer Society; 2007.

[93] Silverthorn DU. *Human Physiology: An Integrated Approach.* 2nd edn. Upper Saddle River, N.J.: Prentice Hall; 2001.

[94] *Cirrhosis of the Liver.* 2003 December [cited 2007 February 17]; Available from: http://digestive.niddk.nih.gov/ddiseases/pubs/cirrhosis/

[95] *Cirrhosis: Many Causes.* Brochure. New York: American Liver Foundation; 2002.

[96] Varmus H, Klausner R, Zerhouni E, Acharya T, Daar AS, Singer PA. Public health. Grand Challenges in Global Health. *Science.* 2003 Oct 17; **302**(5644): 398–9.

5

Healthcare systems: a global comparison

In Chapter 4, we examined the health challenges facing our planet. Solving these problems will require creative scientific and economic approaches, given the vast differences in healthcare spending, particularly between developed and developing nations. Within each country, the **health system** brings together human resources, physical infrastructure, healthcare technologies, and economic resources in an attempt to improve the health of the population in a way that is fair and responsive to the needs of the citizens (Figure 5.1) [1].

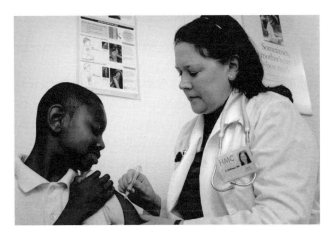

Figure 5.1. *All the components of a nation's health system must work together effectively to ensure that quality healthcare is available to all citizens. CDC/Judy Schmidt Update.*

> **Health system**
>
> A health system is the collection of human resources (doctors, nurses, healthcare workers), physical infrastructure (hospitals, health clinics), healthcare technologies (vaccines, MRI machines), and economic resources (healthcare expenditures) devoted to improving the health of the population.

New technologies have the potential to address many global health challenges only if they are engineered to meet the economic and cultural constraints faced by health systems. In this chapter we will examine the different types of health systems found throughout the world and will compare the performance of health systems in different countries.

Every nation, whether it has many healthcare resources or only a few, must make decisions about how to use those resources to best serve its population. To make these decisions, important societal questions must be answered. How much money are we willing to spend to gain a year of life? Which procedures should we pay for so that we get the most bang for the buck? One way to help answer these questions is to rank health interventions according to the ratio of economic cost to heath benefit. For example, we can divide the cost of providing a healthcare technology by the number of quality adjusted years of life (QALYs) gained by using a

Table 5.1. *A comparison of the costs per quality adjusted life year for some common health related technologies in the USA [2].*

Therapy	Cost per QALY
Motorcycle helmets, seat belts, immunizations	Cost saving
Anti-depressants for people with major depression	$1000
Hypertension treatment in older men and women	$1000–$3000
Pap smear screening every 4 years (vs none)	$16,000
Driver's side airbag (vs none)	$27,000
Chemo in 75 yo woman with breast CA (vs none)	$58,000
Dialysis in seriously ill patients hospitalized with renal failure (vs none)	$140,000
Screening and treatment for HIV in low risk populations	$1,500,000

technology. The cost per QALY gained provides a way to compare the relative benefit of healthcare spending associated with different interventions. Table 5.1 compares the cost per quality adjusted life year for several health interventions in the United States, ranking these from least to most expensive.

We see that some interventions actually generate cost savings (e.g. motorcycle helmets and childhood immunizations), while other interventions are significantly more costly (screening and treatment for HIV infection in low risk populations). Which of these interventions is our health system willing to pay for? In developed countries, we find that most interventions which cost $100,000 or less per quality adjusted year of life gained are generally made available through private and government sponsored health insurance plans.

A tale of two children

The actions of health systems directly impact the lives of individual citizens, and the result of a health system which performs poorly can be measured in terms of death, suffering, despair, and impoverishment [1]. As an example, we begin our global tour of health systems

with a look to the state of Oregon in July of 1987. At this time, the Oregon state constitution required a balanced state budget; any surplus revenue was required to be returned to taxpayers [3]. The state was faced with rapidly increasing Medicaid expenditures to provide healthcare for its poor, uninsured citizens. Health expenditures in Oregon were growing at a rate substantially higher than revenues. In an attempt to control expenditures while providing the greatest health benefit to its population, in 1987, the Oregon legislature made the decision to end state funded coverage of organ transplants for poor citizens through Medicaid in favor of funding prenatal care. They reasoned that the state government typically funded ten transplants per year, at a cost to the state of approximately $1.1 million annually. If these funds were instead used to support prenatal care, 25 infants would be saved each year [3].

In August of 1987 Coby Howard, a seven year old boy from Oregon, developed leukemia. He required a bone marrow transplant to treat his leukemia; under the new Oregon rules he was denied coverage. Coby's mother appealed to the Oregon legislature, but coverage was still denied. His mother began a media campaign to raise the $100,000 required for his treatment [4]. She raised $70,000, but Coby died in December, 1987. Ira Zarov, an attorney for a patient in similar circumstances, noted that Coby was "forced to spend the last days of his life acting cute" before the cameras [3].

At the same time, David Holliday, a two year old Oregon boy, also developed leukemia. His mother moved to Washington state, where the family lived in their car. In the state of Washington, Medicaid covered transplants with no minimum residency requirement, and David was able to receive treatment for his leukemia [3]. These disparities in access to care illustrate the difficult issues faced by a healthcare system that struggles to provide fair access to care in the face of limited resources. In Chapter 6, we will come back to the story of Coby Howard and learn how the State of Oregon reformed its healthcare system as a result. In this chapter, we will first gain a global understanding of how health systems function.

Health systems

Before comparing health systems throughout the world, we must first develop an understanding of the components of a health system, the different types of health systems, and learn how to assess the performance of health systems. With this introduction, we will then consider examples of several different health systems, comparing the situation in the United States, Canada, the United Kingdom, China, India, and Angola.

The primary goal of a health system is to provide and manage resources to improve the health of the population. The secondary goal of a health system is to ensure that good health is achieved in a way that is fair; health systems should protect citizens against the unpredictable and often high financial costs of illness. In many of the world's poorest countries, most people have to pay for medical care out of their own pockets, often at a time when they can least afford it. In these settings, illness is frequently a *cause* of poverty. Prepayment, through health insurance, leads to greater fairness in funding healthcare needs [1].

Within each country, the health system reflects historical trends in both economic development and political ideology [5]. Although there are many different types of health systems, they all provide four important functions [1]. (1) Health systems generate the human resources, physical infrastructure and knowledge base necessary to provide healthcare. (2) They provide healthcare services, through primary and preventive care clinics, hospitals, and specialized tertiary care centers, generally operated by a combination of governmental agencies and private providers. (3) Health systems raise and pool the economic resources necessary to pay for the healthcare of a population. Sources of resources may include tax revenues, mandatory social insurance programs, voluntary private insurance programs, charity, personal household income and, in many developing countries, foreign aid. (4) Finally, health systems provide stewardship for the healthcare system, setting and enforcing rules which patients, providers and payers must follow. The ultimate responsibility for stewardship of a country's health system lies with its government [1].

Types of health systems

The level of government intervention in a health system varies dramatically from country to country. Politically, health systems can be classified into four main categories of increasing levels of government intervention – entrepreneurial, welfare oriented, comprehensive, and socialist [5]. Health systems also vary dramatically based on income level. An entrepreneurial health system is strongly influenced by market forces, with some government intervention; the United States is the only industrialized nation to have an entrepreneurial health system. In contrast, in a welfare oriented system, the government mandates health insurance for all workers, often through intermediary private insurance agencies; Canada, South Korea, and Germany all are developed nations characterized by welfare oriented health systems. Comprehensive health systems provide complete coverage to 100% of the population almost completely through tax revenues. Many countries which initially developed welfare oriented systems evolved to comprehensive systems following World War II. Great Britain, Spain, and Slovenia provide examples of comprehensive health systems. In a socialist health system, all health services are operated by the government, and theoretically, are free to everyone. Cuba is an example of a middle income developing country which has a socialist health system [5]. Table 5.2 lists examples of each political classification of health system according to three per capita income levels: high income industrialized (per capita income >$11,116 USD), middle income developing (per capita income in the range $906–11,115 USD), low income developing (per capita income <$905 USD) [51].

Performance of health systems

The primary goal of a health system is to improve the health of a population. In Chapters 2 and 3, we compared the performance of health systems in different countries using health data such as average life expectancy at birth, infant mortality rate and total per capita health expenditures. We saw that, in general, as health expenditures rise, the health of the population

Table 5.2. *National health system political topologies versus gross national income (GNI) per capita, 2004 or latest year available.*

	Entrepreneurial	**Welfare oriented**	**Comprehensive**	**Socialist**
High Income Developed	United States	France, Germany, Netherlands, Canada, South Korea	Japan, United Kingdom, Spain, Ireland, Norway Denmark, Sweden, Iceland, Luxembourg, Slovenia, Czech Republic	
Middle Income Developing – Upper	South Africa, Uruguay	Argentina, Chile, Mexico, Turkey, Poland, Romania, Malaysia	Brazil, Costa Rica	
Middle Income Developing – Lower		Colombia, Paraguay, Peru, Nicaragua, Phillipines, Thailand, Sri Lanka, China	Ukraine	Cuba
Low Income Developing	Uganda, Vietnam, India, Nepal, Bangladesh, Burma/Myanmar	Ghana, Nigeria, Tanzania, Kyrgyzstan		North Korea

Source: Adapted from [5]; updated with data from the World Bank, WHO, PAHO, and IMF [51–54].

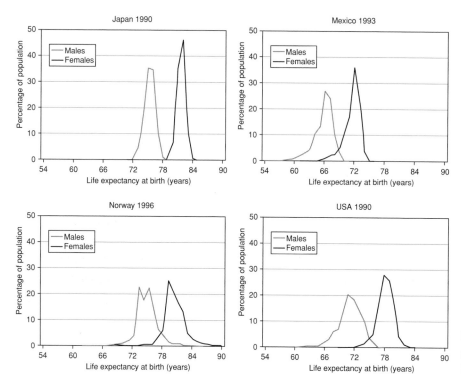

Figure 5.2. *Distribution of life expectancy at birth for men and women. Courtesy of WHO World Health Report 2000.*

increases. However, gains in health do not always accrue equally to all citizens.

Figure 5.2 shows the distribution of life expectancy at birth for men and women in four countries [1]. In all countries, the life expectancy is higher for women. Health is distributed most equally in Japan, which shows the most narrow distribution. In the USA and Mexico, the distributions are much wider, indicating inequities in the health of the population. We also have seen that gains in spending do not always result in

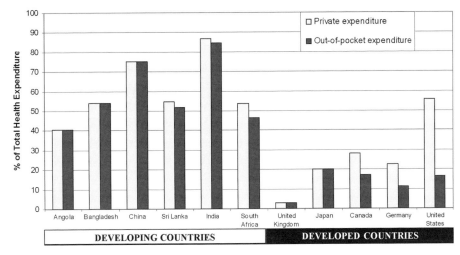

Figure 5.3. *The fraction of total health expenditures paid from private sources and fraction paid as out of pocket expenses for several countries. Countries are ranked in order of increasing per capita health expenditures from left to right. Data are from 1997 [1].*

gains in health. The United States has the largest per capita health expenditure in the world but this does not translate into the longest life expectancy or the lowest infant mortality rates [6]. In many countries, the actual performance of the health system falls short of its potential given the resources invested. As we will see, this failure is often disproportionately borne by poor citizens.

The secondary goal of a health system is to improve health in a way that is fair to all citizens. In some health systems, individuals who face serious illness must make a choice between forgoing care or being forced into poverty as a result of attempting to pay for care [1]. Within other health systems, economic resources are pooled to ensure that the risk of paying for healthcare is borne by all members of the pool. Health systems use a variety of mechanisms to pool resources, including general taxation, mandated contributions to social health insurance programs, and voluntary private health insurance contributions [5]. All of these mechanisms ensure that individuals prepay a portion of their healthcare expenses. Out of pocket healthcare spending is defined as that spending which occurs at the time healthcare is delivered. Some health systems, particularly those in developing countries, rely heavily on out of pocket spending; this restricts access to only those who can afford it. In nearly half of low income countries, citizens must pay more for than 50% of their healthcare expenditures out of their own pockets; the same is true

in only 18% of middle income countries and 5% of high income countries [1].

Figure 5.3 compares the fraction of total health expenditures which are paid from private funds for several countries; the fraction of health expenditures paid as out of pocket expenses is also shown. The countries are ranked from left to right in order of increasing per capita health expenditures. The fraction of health expenditures paid from private, non-governmental funds is higher in developing countries, and the vast majority of these are paid out of pocket. The fraction of expenditures paid from private sources is lowest in the comprehensive healthcare system of the United Kingdom, increases slightly in the welfare-oriented systems of Canada, Japan and Germany, and increases dramatically in the entrepreneurial US health system. However, the fraction of health expenditures paid out of pocket is below 20% in all these developed countries [1].

Stewardship to ensure fairness is an important function of a health system. In developing countries, infrastructure to implement large insurance programs to pool risk is often simply non-existent [1]. In India, for example, some large health insurance schemes currently have no legal status [1]. In other developing countries, health ministries often do not enforce regulations and public providers charge patients an extra fee for care. This "informal charging" can severely restrict access to care – a recent study in Bangladesh indicated that unofficial fees are 12 times higher than official payments [1].

Table 5.3. *Overview and comparison of the health systems in several countries using important measures such as life expectancy, infant mortality rate, and health expenditures per capita.*

Country	Healthcare system	Total healthcare expenditure as % of GDP*	Health expenditure capita per (US $)*	Life expectancy at birth (years)([0–9])[†]	Infant mortality rate (per 1000 live births)([0–9])[†]
Angola	A fragmented and unstable public healthcare system with limited resources that relies heavily on international aid. Privatized care is available to high-income sector.	1.9	$26	40	154
China	Few, but improving, options for public health insurance in rural and urban areas, with programs suffering from financial limitations and accommodating migrating workers.	4.7	$70	73	23
India	Limited public healthcare is funded by tax revenue, community financing and out-of-pocket sources. Private insurance covers only 10% of population.	5.0	$31	63	56
Japan	Workers can enroll in insurance programs through jobs; others join national health insurance plan.	7.8	$2823	83	3
United Kingdom	National Health Service provides free care. Optional private insurance is available for those wanting treatment outside the state system.	8.1	$2900	79	5
Canada	Limited, but universal, coverage through the provincial government, which acts as the sole insurer. Supplemental private insurance can cover dental services, drug plans, etc.	9.8	$3038	81	5
France	Mandatory health insurance provided by Social Security, funds universal care, private supplemental coverage can fill gaps.	10.5	$3464	80	4
Germany	Government approved health insurance plans are financed by employer and employee contributions. Some high income workers buy private insurance.	10.6	$3521	79	4
United States	Costs of care for senior citizens and poor are funded by state and federal governments. Most citizens are covered by employer or personal financed insurance. About 47 million people lack coverage.	15.4	$6096	78	7

Data: World Health Statistics 2007 [10], World Health Organization (*2001; †2005); Brown, Barry. "In Critical Condition: Health Care in America Canada's Way," San Francisco Chronicle. 14 Oct 2004, c-1.

Table 5.3 provides an overview of the health systems in some developed and developing countries and quantifies their performance using several metrics – health expenditures, life expectancy, and infant mortality rate. In the next section, we consider several of these health systems in greater detail, including the USA, the UK, Canada, India, China, and Angola. We will see that these countries have made very different choices about how much to spend on healthcare and how to provide these resources.

US healthcare: an entrepreneurial health system

In the entrepreneurial health system of the United States, healthcare funding comes from three primary sources: private insurers, the government, and individuals. Many Americans have private health insurance which pays for most healthcare costs [11]. Private insurers include conventional insurance plans, as well as health maintenance organizations (HMOs). What happens to those Americans who don't have health insurance? If they meet certain income guidelines, they are eligible for government sponsored insurance. Government funding comes from three primary sources: Medicare, Medicaid, and State Children's Health Insurance Program (SCHIP). While the eligibility rules and coverage vary from state to state, generally the state government pays a portion of the costs for medical coverage and the federal government matches the rest. As a result, state governments have been forced to make difficult choices about which procedures are covered by their limited Medicaid budgets. Earlier in this chapter we saw an example of such a choice in the State of Oregon. Within the USA, individuals who have private or government sponsored insurance are usually required to pay for a fraction of their healthcare costs, and uninsured individuals must pay for their entire costs.

In the year 2000, the USA spent approximately $1.3 trillion dollars on healthcare [11]. Figure 5.4 summarizes where this money came from. The single largest share of these costs is borne by the government, covering about 45% of all healthcare expenditures. About 40% of healthcare costs are paid for from private sources, most by private insurers. Individuals bear

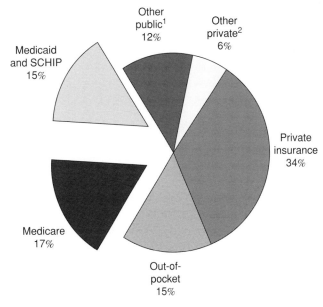

Figure 5.4. The Nation's Health Dollar, 2000. *Medicare, Medicaid, and SCHIP account for one-third of national health spending [11].*
[1] *Other public includes programs such as workers' compensation, public health activity, Department of Defense, Department of Veterans Affairs, Indian Health Service, and State and local hospital subsidies and school health.*
[2] *Other private includes industrial in-plant, privately funded construction, and non-patient revenues, including philanthropy. Note: Numbers shown may not sum to 100% owing to rounding.*

about 15% of these costs. These figures represent an aggregate across the country. For example, although Medicaid pays for approximately 41% of all births in the United States, there is substantial variation from state to state (Figure 5.5) [12].

Where does the money go? Figure 5.6 shows what these healthcare dollars buy in the USA. The largest fraction of money pays for hospital care (32%); other large pieces of the pie include physician services (22%), prescription drug costs (9%), and administrative costs (6%).

Interestingly, healthcare spending in the USA is concentrated mostly on a few very sick individuals. More than 1/4 of US health spending is devoted to 1% of the population, and approximately 1/2 of US health spending is devoted to 5% of the population [13]. This trend has been stable over many years.

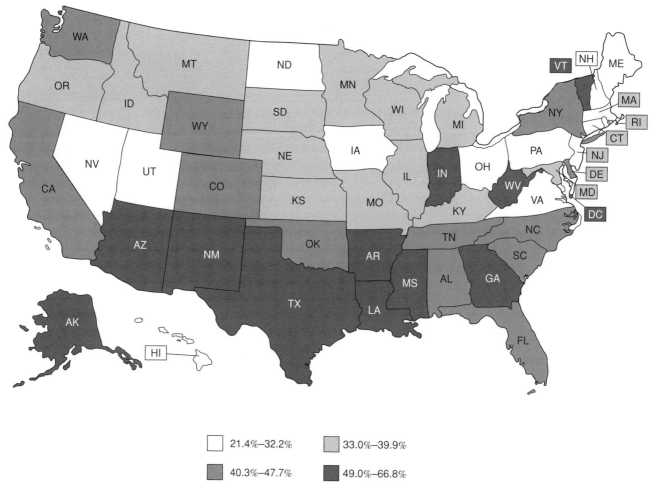

Figure 5.5. *Births in the USA financed by Medicaid as a percent of total births by state, 1998 [12]. This information was reprinted with permission from the Henry J. Kaiser Family Foundation. The Kaiser Family Foundation, based in Menlo Park, California, is a nonprofit, private operating foundation focusing on the major healthcare issues facing the nation and is not associated with Kaiser Permanente or Kaiser Industries.*

For the past four decades, the USA has consistently devoted a higher fraction of its GDP to healthcare costs, and this discrepancy has been steadily growing. In 2004, more than 15% of the GDP was devoted to health expenditures in the USA, compared to only 8% in Japan [15]. In 2004 the USA devoted more than $4400 per person to healthcare in the USA, compared to just over $1900 per person in Japan [16]. By almost any measure, the USA spends more for healthcare than any other country in the world.

Despite the substantial resources devoted to healthcare in the USA, access to care is far from equal, and disparities in access are increasing. Figure 5.7 shows the sources of health coverage for people under 65 years of age in the USA as a function of time. Over

the past 25 years, rates of private insurance coverage have dropped from 83% to 69%, while the fraction of uninsured has risen from 10% to 18%. In 2004, there were more than 45.8 million Americans with <u>no</u> health insurance [18]. There is wide regional variation in the number of uninsured, with the southern and western USA showing the highest fraction of uninsured individuals [18]. During this time, state programs to provide health insurance to low income children have resulted in a decline in the fraction of children without health insurance. From 1988 to 2005, the fraction of children without insurance declined from 13.1% to 11.6% [19]. Since the inception of SCHIP in 1998, an additional 5.1 million children have gained access to care [20].

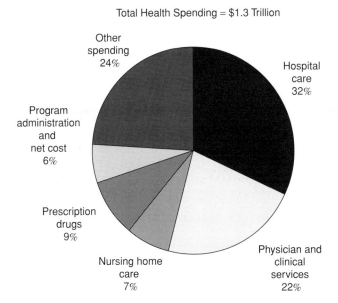

Total Health Spending = $1.3 Trillion

Figure 5.6. The Nation's Health Dollar, 2000. *Hospital and physician spending accounts for more than half of all health spending. Other spending includes dentist services, other professional services, home health, durable medical products, over-the-counter medicines and sundries, public health, research and construction [14].*

During this same time the character of private insurance plans in the USA changed dramatically, moving from primarily conventional insurance plans to primarily health maintenance organizations (HMOs), preferred provider organizations (PPOs) and point of service (POS) plans, as shown in Figure 5.8. In 1988, nearly 3/4 of private insurance plans were conventional, and only 25% were HMOs. In 2005, only 3% of private insurance plans were conventional, and more than 95% were HMOs, PPOs and POS plans. Initially, many of these HMOs were not-for-profit plans; however, today nearly 2/3 of HMOs are for-profit organizations [21]. The motivation for this change was largely to attempt to control rising healthcare costs in the USA. In Chapter 6, we will examine whether this was successful.

How does the system of healthcare financing in the USA affect availability of care and access to care? How does the US system compare to that available in the rest of the world? In the next sections, we compare the US financing system to that of the welfare oriented system of Canada, the comprehensive system of the UK, and examine the situation in three developing countries: India, China, and Angola.

Canadian healthcare: a single payer system

While healthcare costs in the entrepreneurial system of the United States are paid by multiple sources (entrepreneurial system), Canadians have developed a welfare oriented, single payer system, in which a single insurer (the government) is billed for medical services. Health expenditures within the public sector are financed by taxes paid to the provincial and federal governments and account for approximately 70%

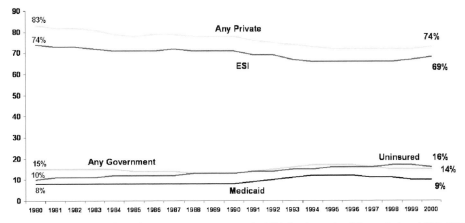

Figure 5.7. *Sources of health insurance coverage in the USA for the under 65 population, 1980–2000. Over the past two decades, private coverage has declined, public coverage has remained relatively constant, but the number of Americans with no health insurance has increased [17]. Notes: ESI – Employer Sponsored Insurance. Any Private includes ESI and individually purchased insurance. Any government includes Medicare for the disabled population. Source: Tabulations of the March Current Population Survey files by Actuarial Research Corporation, incorporating their historical adjustments, Centers for Medicare and Medicaid Services.*

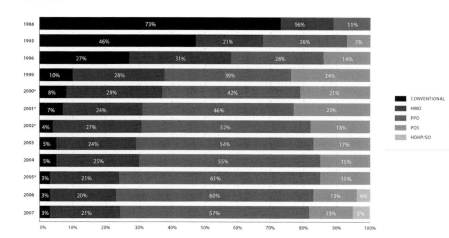

Figure 5.8. *Health plan enrollment in the USA by plan type, 1988–2001. Over the 1990s, managed care grew from about a quarter of employees to the vast majority [22]. This information was reprinted with permission from the Henry J. Kaiser Family Foundation. The Kaiser Family Foundation, based in Menlo Park, California, is a nonprofit, private operating foundation focusing on the major healthcare issues facing the nation and is not associated with Kaiser Permanente or Kaiser Industries.*

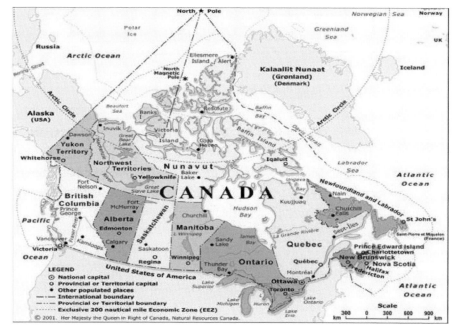

Figure 5.9. *Canada's provinces. Used with permission from the* Atlas of Canada.

of total health expenditures. The remaining 30 % of total health expenditures are paid for by the private sector, with private health insurance and out-of-pocket payments accounting for the largest portion. Public coverage within the Canadian system is characterized as narrow, but deep; it is described as narrow because coverage is limited to hospital and physician services, and deep due to the lack of any kind of fee to the user. Private insurance largely applies to dental and vision services, as well as some prescription drug plans [23].

The Canadian system is characterized by five principles: comprehensiveness, universality, portability, accessibility, and public administration. All 13 Canadian provinces and territories (Figure 5.9) have different health systems, providing some degree of local control, with the Provincial government as the sole insurer. In the Canadian system, patients can choose their own doctors, and physicians work on a fee-for-service basis where fees are capped at a maximum value by the government [23].

Before 1947, the Canadian healthcare system was much like the current US system. However, in 1947 the province of Saskatchewan implemented North America's first universal hospital insurance plan. British Columbia and Alberta followed suit in 1949 and 1950, respectively [23]. In 1957, the Canadian federal government adopted the Hospital Insurance and Diagnostic Services Act, which stated that the federal government

Table 5.4. *Some useful indicators for comparing health systems in various countries. OECD Health Data 2004; A Comparative Analysis of 30 Countries, 4th Edition. Paris, Organization for Economic Cooperation and Development, 2004.*

Disease mortality indicator relative rankings, 2000 (overall OECD rankings)				
	Malignant neoplasms	**Cerebrovascular diseases**	**Respiratory system diseases**	**Ischaemic heart diseases**
Australia	2 (8)	4 (5)	4 (12)	2 (11)
Canada	4 (15)	1 (2)	3 (10)	3 (12)
France	5 (18)	2 (3)	2 (8)	1 (3)
Sweden	1 (2)	5 (11)	1 (4)	4 (16)
United Kingdom	6 (20)	6 (18)	6 (25)	6 (22)
United States	3 (14)	3 (4)	5 (22)	5 (21)

would reimburse the provinces for approximately 50% of the cost of acute hospital care and diagnostic services [24, 25]. By 1961, agreements were in place with all of the provinces and 99% of Canadians had access to the freehealth care services covered by the legislation [25].

In 1962, Saskatchewan introduced full blown, universal medical coverage for physician services. Shortly afterward, in 1966, the federal government offered cost sharing to those provinces whose plans met criteria for comprehensiveness, portability, public administration and universality. By 1972, all Canadians were guaranteed access to essential medical services [23].

Throughout the 1970s and 1980s, medical costs rose substantially; however fees to doctors did not rise at the same rate. Some Canadian physicians began to bill patients themselves for the additional cost. In 1984 the Canada Health Act outlawed extra billing in an attempt to enforce the principal of accessibility. In the 20 years since the passage of the Canada Health Act, healthcare reform was marked by a period of extreme cost cutting, followed by a period of increasing public expenditures. While Canadians continue to enjoy universal access to healthcare, there is growing concern that rising health expenditures threaten the fiscal sustainability of the system [23].

How does the Canadian health system compare to that of the USA? Costs on average are lower in Canada;

Canada spends 9.9% of its GDP on healthcare, compared to 15.2% in the USA [8]. The Canadian system is far simpler than that of the USA; the legislation for Canadian Medicare is only eight pages long, compared to the 35,000 pages of US Medicare legislation. In terms of citizen's satisfaction with their own healthcare system, results from both countries are mediocre, with only 46% of Canadians and 40% of Americans reporting that they are fairly or very satisfied [26]. When comparing performance of national health systems, one must carefully select health status and health system indicators. In terms of disability adjusted life expectancy, Canada ranks 12th while the USA ranks 24th [1]. The infant mortality rate in Canada is five deaths per 1000 live births, while that of the USA is seven deaths per 1000 live births [9].

While these indicators provide a picture of the overall health of a population, they are subject to influence by environmental factors and do not measure a health systems' actual performance. The performance of a health system is better evaluated with indicators that measure processes or outcomes of care. Table 5.4 provides examples of such indicators, assessing how each country ranks in terms of treating each disease [23]. According to many indicators, Canada's health system fares well, relative to other industrialized nations.

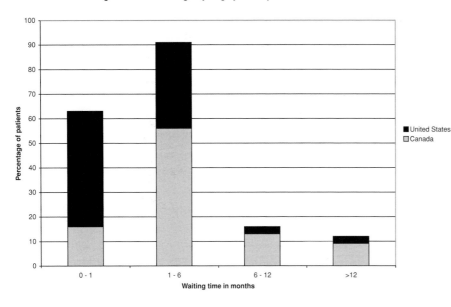

Figure 5.10. *Comparing waiting times in the USA and Canada for elective surgery. Data:* Siciliani, L., Hurst, J. *OECD Health Working Papers no.7.* Explaining Waiting Times Variations for Elective Surgery Across OECD Countries. Paris, Organization for Economisc Cooperation and Development, 2003.

Problems with the Canadian health system

Despite its many achievements, the Canadian health system currently faces a number of challenges. Waiting times for medical services are frequently longer in the Canadian health system than in the USA; 57% of Canadians wait more than four weeks to see a specialist physician, compared to 23% of Americans [27]. Canadians wait an average of 47 days for an MRI, while Americans wait an average of three days [28]. The increased waiting time for non-emergency surgeries is one of the most controversial features of the Canadian Medicare system. Figure 5.10 compares the waiting times for elective surgery in the USA and Canada, and shows that 22% of Canadians must wait more than six months. Why are waiting times so long? While a lack of equipment, health personnel, and finances are often cited as a major cause, challenges related to coordinating access and delivery of care, as well as maintaining up-to-date waitlists also impact waiting times [27].

More recently, Canada has seen an emergence of for-profit care; in exchange for an extra fee, facilities will provide more rapid access to Medicare insured services, such as diagnostic imaging. While some argue that this violates the principle of universality, others argue that long waiting lists are a violation of patients' rights. The Canadian Supreme Court's narrow ruling in the case of *Chaoulli v. Quebec* has fundamentally challenged the Canadian healthcare system, finding that lengthy waiting times violate the rights granted to individuals under the Quebec charter of human rights and freedoms [29]. The ruling effectively declares that it is unconstitutional for Quebec to prevent its citizens from obtaining private insurance and gave Quebec one year to make its Medicare policy consistent with the ruling. In December 2006, the National Assembly of Quebec passed Bill 33; a controversial piece of legislation amending its healthcare policy. The new legislation allows for the creation of private clinics that receive public funds in exchange for offering total hip and knee replacement, and cataract surgery, the elective procedures with some of the longest waiting times [30]. These private clinics will be allowed to directly bill patients for any fees not covered by Medicare. In addition, the new legislation allows patients to obtain private insurance to cover the costs of these specific medical procedures [31]. While proponents of the bill claim that the introduction of public and private partnerships will alleviate problems in the public sector, opponents argue that it moves dangerously close to a two-tier system of health care that has the potential to destroy the single payer model [32]. It remains to be seen how the legislation will impact Quebec, and whether or not other provinces will follow its lead.

Comprehensive system in the UK

Healthcare in the United Kingdom is provided through a large, publicly owned and financed health system and a smaller, but parallel private health system [33]. The public health system, headed by the National Health Service (NHS), was founded on the principle of collective responsibility and equality of access. While the structure of the NHS has evolved over time, it continues to provide universal health coverage for a broad range of acute and primary care services, and on a much more limited basis, coverage for dental and ophthalmic services and pharmaceuticals [34].

As of 2004, 86% of total health expenditures were funded by public sources. The remaining 14% of private financing represents a combination of out-of-pocket payments and private health insurance [35]. In contrast to the Canadian healthcare system, the UK does not prohibit its citizens from obtaining private health insurance for those services provided through the NHS. As of 2003, 11.2% of the population in the UK had some form of private insurance [36].

Since its inception in 1948, the NHS has been through many iterations of organizational restructuring. Initially, the NHS was designed around the idea of command and control, with orders being passed from the central government down to local officials, through a hierarchical system of administrators [34]. Physicians were salaried employees of the government, or in the case of general practitioners, independent contractors. This system did not account for changing demographics and served to heighten geographical disparities [37].

By the mid 1980s the NHS was faced with increasing demand for services and severely constrained resources. It was accused of being rigid, overcentralized, and lacking of incentives for innovation [38]. In 1991, the Conservative government headed by Prime Minister Margaret Thatcher introduced sweeping changes to establish an internal market. Under the new system, local health authorities and some general practitioners became "purchasers" responsible for assessing the needs of their patient population and purchasing needed services. Hospitals, ambulance services, and long term care facilities became the "providers" in the form of National Health Service trusts [39]. Each trust essen-tially functioned as a non-profit organization with its own system of management [37].

In separating purchasers and providers, the government aimed to create a competitive environment that would improve patient care [38]. While observers credit the internal market with providing a means for more rational planning of services, greater flexibility, and improved cost consciousness, concerns were raised about inequities of care [37, 39].

With the election of the Labour Party in 1997, the policy direction of the NHS again changed course. Major changes included the requirement that all general practitioners organize into geographically based Primary Care Groups (PCGs), with each serving approximately 100,000 patients [37]. As of 2002, PCGs have been given the option to reorganize as Primary Care Trusts, organizations that control over 75% of the NHS budget, in an effort to give greater control to the patients and front line staff [40].

Similar to the Canadian health system, the National Health Service of the United Kingdom has found itself battling lengthy patient waiting times for more than 20 years. While much progress was made in the early 1990s, waiting times slowly began to creep up in 1997 (Figure 5.11) [41]. In 2000, the Labour government introduced a new plan promising that by December 2005, maximum waiting times would not exceed three months for outpatient appointments and six months for inpatient appointments, a goal that was largely met through the creation of elective surgery treatment centres as well as targeted initiatives to reduce waiting times in orthopaedics and opthalmology [42].

Buoyed by its initial success, the government launched an ambitious plan in 2004 to further reduce all waiting times to 18 weeks by 2008, as measured from the time of first referral to hospital admission. In order to accomplish this goal, the government has embraced the idea of patient choice; the creation of a quasi-market in which patients will be given information on waiting times and in turn choose their own provider. In theory, patients will use their power of choice to select the hospital with the lowest waiting time, thereby equalizing waiting times across all hospitals. While there is some evidence to suggest that patients will utilize their power

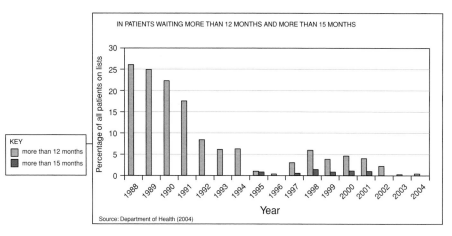

Figure 5.11. *Waiting times for inpatient appointments in the National Health Service [41].*

Tessa's Final Blog: July 19, 2007 Tessa Swaziland

Working here made me realize how important access to and quality of healthcare are, how difficult it can be to implement a seemingly simple project, and how there are always going to be unforeseen challenges no matter how well you plan, so it is important to be adaptable and constantly aware of what is going on.

I have no idea what exactly I will be doing five, ten, or 25 years from now, but after working here, I am certain that whatever it is, it will be somehow related to inequalities in healthcare. There are so many resources and so much wealth in the world, and so much of it (even when it is intended to go to the sick and the poor) never makes it to those who most need it.

For example, a few months ago, there were hundreds of packages of plumpy-nut sent to Swaziland. Plumpy-nut is a food supplement that is used to treat malnourished children on an outpatient basis, and it has been very successful in other countries that have started using it. It is supported and promoted by many organizations, including the World Health Organization. The doctors at the clinic were elated to hear the good news until the Ministry of Health declared that they needed their own specialists to look at plumpy-nut before it could be distributed. Knowing the snail-pace at which everything happens here, the elation was quickly replaced with extreme frustration. So, the plumpy-nut sat here in Swaziland for months, only a few hundred kilometers from the starving, dying children, and there was nothing the doctors could do about it. Luckily a conference of sorts was held here with many international organizations, and one of their first questions was, "Why doesn't Swaziland get plumpy-nut here?!" Once they were held accountable for their behavior, the Ministry took action, and now things are looking up. The frustrating thing is that there usually isn't anyone holding them accountable, and in the future, a similar situation could occur again.

So I guess I will leave it at that. This internship has been eye-opening, to say the least. As much sadness and pain and needless suffering as I have encountered here, I have also seen just as many successes in the healthcare community, and I leave here feeling both hopeful and determined to stay an active member of this global community that strives to transcend political, social, and economic boundaries in an attempt to relieve the needless suffering that exists in the world.

to choose, it is hard to predict whether or not more rapid access to care will take priority over proximity of the hospital and reputation [42].

Healthcare in the developing world: India, China and Angola

We now turn to the developing world, and examine some of the challenges faced by three countries: India, China and Angola.

Healthcare in China

Over the past half century, the People's Republic of China has undergone rapid economic and demographic transitions. Following the introduction of sweeping government reforms in 1978, China's gross domestic product grew at an average annual rate of 9.4% through 2003, and is projected to continue growing at 7% per year. This rapid economic growth brought urbanization, industrialization, and rural to urban migration; changes that have in turn led to a new and emerging set of health issues. Wealthy, aging, urban residents are more likely to suffer from noncommunicable diseases such as cardiovascular disease and diabetes, whereas their poor and rural counterparts are much more susceptible to the communicable diseases typical of developing countries [43].

According to a 2000 WHO survey of 190 countries, China ranked fourth worst in terms of fairness of financial contribution to health expenditures; a measure of financial health equality [1]. As of 2002, only 29% of the population have some form of health insurance and over 58% of healthcare spending came from out of pocket expenses [44]. The Chinese healthcare system is in a state of transition as it struggles to address the needs of an economically and geographically disperse population and a growing demand for healthcare. To better understand these demands, it is interesting to examine the evolution of the Chinese healthcare system over the past half century.

Healthcare in a centrally planned economy

With the founding of the People's Republic of China in 1949, health was viewed as a welfare undertaking, with the government providing medical insurance and free preventive services to the majority of the population [43]. All health facilities, from large urban hospitals to countryside clinics were government owned and physicians were direct employees of the state. In rural areas, healthcare was provided through the Cooperative Medical System and the local communes; institutions that owned the land, distributed the harvest, and provided social services [44]. Throughout the 1950s, the government launched massive campaigns to wipe out infectious diseases such as schistosomiasis; this is a disease contracted by wading in water contaminated with worms that live in the bodies of freshwater snails. Efforts to control the disease included drying out swamps and building drainage ditches to eliminate the snails' habitat and organizing teams of workers to lance the snails with sharpened chopsticks [45, 46].

With the start of the Cultural Revolution in 1965, Chairman Mao Zedong exercised his authoritarian rule. In an effort to seize power from the purportedly privileged medical establishment and return control to the people, physicians with formal training were sent to the countryside and medical schools and specialty departments in hospitals were shut down [46]. To meet the health demands of the rural population, thousands of young peasants were recruited to become primary healthcare workers known as "barefoot doctors." With three to six months of training in traditional Chinese and Western medicine, the barefoot doctors returned to the countryside where they worked in the fields and provided basic healthcare services to the people in their community. By 1975, there were an estimated one million barefoot doctors in China [45].

The overall success of the barefoot doctor program is difficult to assess and remains controversial. Throughout the 1960s and 1970s, 90% of China's rural population had access to basic healthcare [45, 47]. Further, between 1952 and 1982 infant mortality fell from 200 to 34 deaths per 1000 live births and life expectancy increased from 35 to 68 years [44]. While these improvements lend support to the program's goal of providing basic healthcare to the masses, Zedong's reforms are criticized for having adversely affected the quantity and quality of medical education and health services throughout the country [48].

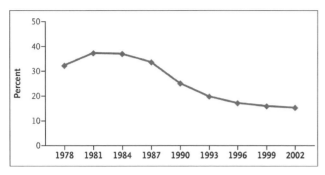

Figure 5.12. *The percentage of total healthcare expenditures borne by the Chienese government vs. time. Declining public healthcare spending in China has led to increased privatization in the system.* [44] © 2005 Massachusetts Medical Society. All Right Reserved.

Healthcare in a free market economy

In 1978, China's move toward a free market economy dealt a serious blow to the healthcare system. In an effort to privatize the Chinese agricultural economy, the commune system was dismantled, leaving 900 million rural citizens without insurance or access to basic healthcare [44]. A call for government decentralization placed the responsibility for healthcare funding at the provincial level, in effect favoring the wealthier, urban municipalities that could levy higher taxes. Additionally, public healthcare funding was significantly reduced (Figure 5.12), encouraging healthcare facilities and providers to privatize their services. While the government implemented price controls for basic healthcare services, high tech procedures and new pharmaceuticals went unregulated, providing an incentive for physicians to raise revenue through the use of costly and often times needless interventions [43, 44].

While private health insurance is available to those who can afford it, the Chinese government has recently taken steps to make social health insurance more widely available to urban and rural citizens. Started in 1998, the Urban Employee Basic Medical Insurance System (UEBMIS) allows workers to pay into a private medical savings account for outpatient expenses. When this money is exhausted, participants may pull from a risk-pooling fund for up to four times the average monthly wage, with expenses beyond this paid out-of-pocket. While this program provides 110 million citizens with some degree of coverage, participants are at risk of incurring large expenses in the event of catastrophic illness [43].

Today, the Chinese healthcare system is managed through 12 different ministries and administrations, divided into urban and rural branches, and is characterized by a high degree of inequality. In the urban system, facilities range from large regional medical centers to small community hospitals. By comparison, the rural health facilities range from county hospitals and Traditional Chinese Medicine facilities to small village clinics. As of 2003, 89% of rural villages had access to clinics that could provide basic treatment and preventive services; however, the heavy reliance on out-of-pocket spending still prevents many citizens from using these facilities [43].

In 2002, the New Rural Cooperative Medical Scheme (NRCMS) was started in an effort to provide some degree of social insurance to the rural population. The voluntary program allows citizens to buy in at affordable prices and pool funds to pay for inpatient care or catastrophic illness. Because participation is voluntary observers worry that only those people at high risk will choose to participate, threatening the financial sustainability of the program. Further, the program has caused concern due to its lack of portability. Over the next 15 years, 300 million rural dwellers are expected to migrate to urban areas and in turn lose any coverage they had through the NRCMS. As the government works to scale up the urban and rural health insurance schemes it will be necessary for them to consider the changing demographics [43].

India

The Indian healthcare system is at an important crossroads. Due to economic progress and improvements in healthcare, mortality and fertility rates are decreasing, and as communicable diseases come under better control, degenerative diseases of older age are emerging. However, reliance on private spending for healthcare in India is among the highest in the world; out-of-pocket payments at the point-of-care account for nearly all of this. More than 40% of Indians must borrow money or sell assets when hospitalized [7].

The transformation taking place in India has led to geographic disparities in both health spending and health outcomes. Southern and western states have

Figure 5.13. *The diverse demographics of India have created regional disparities in health outcomes and health spending.*

Table 5.5. *Fraction of population with access to selected services. Healthcare coverage in India varies dramatically from state to state [7].*

State	Prenatal care	Institutional deliveries	Immunization rates
India[†]	28% (2–95%)	34% (5–100%)	54% (3–100%)
Kerala	85%	97%	84%
Gujarat	36%	46%	58%
Bihar	10%	15%	22%

[†]Country-wide average (range across states.)

better health outcomes, higher spending on health, greater use of health services, and a more equitable distribution of services (Figure 5.13). Table 5.5 compares the availability of prenatal services, the fraction of babies born in healthcare institutions, and childhood immunization rates for India overall and for three states, illustrating the wide disparity in access to health services. The percentage of women with access to prenatal care ranges from a low of 2% in some states to a high of 95% in others, while the fraction of children receiving recommended immunizations ranges from a low of 3% to a high of 100% [7].

Obviously, a country in the midst of such transition faces unique healthcare delivery and financing challenges. These challenges include determining how the government can work most effectively with private health providers, and can fairly test new health financing systems. At the same time, the government must deal with emerging new infectious diseases, make HIV drugs affordable throughout the country, and develop strategies to increase the number of trained healthcare workers. The solutions to these problems will be neither formulaic nor uniformly the same for all of India.

Angola

At the other end of the spectrum is Angola – a country moving from crisis to recovery. The 27 year long civil war between UNITA rebels and government forces ended in April, 2002. In this conflict more than one million people died, out of a total population of only 13 million. An additional four million Angolans fled, many to neighboring countries. With the end of the war, 3.8 million Angolans have returned to their areas of origin. Many people have precarious access to food, and more than 70% of the people live on less than 70 cents per day [49].

The United Nations responded to these needs through the UN World Food Programme, which provides food to an average of 1.7 million people per month. Under this program another 740,000 people receive rations through food-for-work programs. While these programs provide some essential services, they do not address the significant infrastructure needs within Angola – more than 500 roads need reconstruction, many key bridges are unstable, and millions of landmines scatter the countryside. In addition, corruption is a substantial problem within the country. Angola produces 900,000 barrels of oil per day, but the resources from the sale of this oil are not used to address the needs of the population. Massive corruption has undermined donor confidence [49].

The overall public health situation in Angola is critical. As a result of the lack of healthcare resources, more than one in four Angolan children die before

A chronologically capricious smattering of topics (BMS, KITSO, outreach, chatting with locals), July 11, 2007 Rachel Botswana

About two weeks ago Lindsay and I followed Marape to KITSO training in Kanye. Kitso is Setswanan for "knowledge" and is an acronym for Knowledge Innovation and Training Shall Overcome. It is a government training program for health professionals involved with HIV/ AIDS care. The Baylor Botswana team of doctors is responsible for traveling to sites such as Kanye and giving PowerPoint lectures along with pretests and post tests to determine the efficacy of the teaching. Our KITSO took place in a small conference room of a lodge and we covered topics such as CDC categorization of AIDS levels, nutrition, and ethics and law surrounding AIDS cases to an audience of about 40. I found the last topic the most interesting as there are many issues about disclosure and the children's right to know, autonomy over taking drugs, employers' responsibilities to employ infected people and the government's responsibility to protect the infected people's rights. Marape was an excellent presenter, throwing in funny anecdotes such as his poking fun at the traditional Setswanan wild spinach dish's right to being called a vegetable as it has all its nutrients boiled, dried, mashed and simmered out before it hits the table. Lindsay and I filled in for the normal intern, namely starting up the presentation, stapling, and grading post tests. (Basically we were there just to learn about how this program operates.)

Last Friday I followed doctors Chelsea and Jeff to Outreach training in Kanye, not too far from where KITSO training was. Jeff presented to a small room of about ten doctors an interesting PowerPoint on MDR and XDR TB and how to treat and prevent it. The presentation's approach to treating MDR TB I found to be revealing of Botswana's strong commitment to healthcare. According to a Peace Corp coordinator for Botswana I met recently, the Botswanan budget's biggest expenditures are on Health and Education (lets compare that to the US, shall we?). This is especially so in contrast to what I read about in "Mountains Beyond Mountains" which dealt largely with how MDR TB was being inadequately treated in Haiti and Russian prisons. Speaking of TB, this brings up one of the many letdowns that the Meditech system has had since implementation, including a several day "down time" in which the system was offline. About 4 months ago the internet wire connecting the BOTUSA (Botswana USA) TB lab to Meditech was cut by a lawn mower.

After the presentation Chelsea and one of the Nigerian doctors there saw a couple of the more difficult pediatric AIDS patients. The hospital in Kanye doesn't have a pediatric clinic or a pediatrician, so it is only on these weekly Outreach days that a Baylor pediatrician can aid in the diagnosis of the child and prescription of which ARVs. We saw a five year old girl who was failing second line and needed to be moved onto new medication before her viral loads went up and her CD4s went down too low. Through chart analysis of clinical data (VL and CD4 and previous lines) and calling the last Baylor doctor who saw her, Chelsea helped in deciding which drugs to move her on next- a very important decision considering once put on these new

3rd line drugs (which are often more toxic/ more side effects), these are the only drugs they have to take the rest of their lives. If the patient fails on them – by not taking them correctly and building up a resistance – then they are out of options for treatment. She filled out a special request form to get the 3rd line drugs because they are strictly regulated by the government in order to stop wide prescribing and hence resistance building. Unfortunately, as Chelsea informed me, sometimes the approval process for these drugs can take up to three months which is sometimes too long to wait for some kids. Luckily for this little girl, through the connection to the Baylor Clinic COE, this special order can be expedited.

age five. Measles – a completely preventable disease – claims 10,000 Angolan children per year. UN Agencies have conducted vaccination campaigns, holding National Immunization Days, with more than seven million children vaccinated for measles and five million children vaccinated against polio. The UN is working to implement routine immunization programs in Angola, but the country must still cope with the challenge of building a complete health system in the face of enormous constraints [50].

Bioengineering and Global Health Project
Project task 2: Evaluate current technology meant to address the issue

Assess the effectiveness and efficacy of current devices or strategies employed to prevent, diagnose or treat the disease. What are the limitations? Are current solutions available only to a limited population due to cost or other factors? Could alternate materials or approaches be used to reduce the overall cost? Is there another strategy/design that has never been tried? Write a one-page summary of the following. (1) Current methods for prevention and their limitations; are these available in the region you have selected? (2) Current methods for diagnosis and their limitations; are these available in the region you have selected? (3) Current treatments and their limitations; are these available in the region you have selected?

Homework

1. Define each of the following types of health systems (one sentence each). For each health system, give an example of one country which has that type of health system.
   a. Entrepreneurial health system.
   b. Welfare oriented health system.
   c. Comprehensive health system.
   d. Socialist health system.
   e. Compare and contrast the fraction of health expenditures which are paid for from private health spending and out-of-pocket health spending in entrepreneurial and welfare oriented health systems in developed and developing countries.

2. What are the five principles of the Canadian healthcare system? What does it mean for a country to have a two-tier system of healthcare? Contrast the degree to which the USA and Canada have two-tier systems of healthcare. How might this differ in a developing country?

3. Discuss three advantages and three disadvantages of the single payer Canadian system of healthcare financing compared to that of the multi-payer US system of healthcare financing.

4. The US has an entrepreneurial health system including three federal/state run programs: Medicare, Medicaid, and CHIP.
   a. Describe the population covered by each of these three systems (be specific to your state when applicable):
      i. Medicare,
      ii. Medicaid,
      iii. CHIP.
   b. Those not covered by the three programs listed above usually receive insurance through employer sponsored insurance or by buying insurance on their own. However, it is estimated that 24.6% of the Texas population is uninsured, this large population includes 21% of children, many working adults, and a growing population of adults that earn > $50,000–75,000/year. Read the following report and list four reasons for this large uninsured population: http://www.window.state.tx.us/specialrpt/uninsured05/

5. It is sometimes said that "illness is a cause of poverty."
   a. Explain what is meant by this saying in the context of what we have already learned. You may find it useful to use the specific example of malaria in your answer.
   b. How do health systems attempt to protect citizens from the financial burdens associated with illness?

6. The "absolute poverty line" is the threshold below which people lack the resources to meet basic needs

for healthy living, and have insufficient income to provide the food, shelter and clothing needed to preserve health. According to the 2005 guidelines, what is the absolute poverty line for a family of four living in the United States?

References

[1] Musgrove P, Creese A, Preker A, Baeza C, Anell A, Prentice T. *The World Health Report 2000: Health Systems: Improving Importance*. Geneva: World Health Organization; 2000.

[2] Neumann PJ. *25 Years of Cost-Utility Analyses: What Have We Learned?* University of Minnesota Carlson School of Management; 2003.

[3] Egan T. Rebuffed by Oregon, patients take their life-or-death cases public. *The New York Times*. 1988 May 1.

[4] Kolata G. Ideas & Trends; Increasingly, life and death issues become money matters. *The New York Times*. 1988 March 20.

[5] Roemer MI. National health systems throughout the world. *Annual Review of Public Health*. 1993; **14**: 335–53.

[6] Brown B. *In Critical Condition: Health Care in America Canada's Way: What a Universal Health Care System Delivers, Good and Bad*. San Franccisco Chronicle. 2004 October 14.

[7] Peters DH, Yazbeck AS, Sharma RR, Ramana GNV, Pritchett LH, Wagstaff A. *Better Health Systems for India's Poor: Findings, Analysis, and Options*. Washington, D.C.: The World Bank; 2002.

[8] Chen L, Evans D, Evans T, Sadana R, Stilwell B, Travis P, *et al. The World Health Report 2006: Working Together for Health*. Geneva: World Health Organization; 2006.

[9] Beaglehole R, Irwin A, Prentice T. *The World Health Report 2004: Changing History*. Geneva: World Health Organization; 2004.

[10] *World Health Statistics 2007*. Geneva: World Health Organization; 2007.

[11] CMS Office of the Actuary National Health Statistics Group. The Nation's Health Dollar, CY 2000, Medicare, Medicaid, and SCHIP account for one-third of national health spending. *Program Information on Medicare, Medicaid, SCHIP, and Other Programs of the Centers for Medicare and Medicaid Services*. Centers for Medicare and Medicaid Services; 2002.

[12] Kaiser Family Foundation State Health Facts. Table 4.31 Births Financed by Medicaid as a Percent of Total Births by State, 2005. *An Overview of the US Health Care System: Chart Book*. Centers for Medicare and Medicaid Services; 2007.

[13] William W Yu & Trena M Ezzati-Rice – AHRQ. Table 1.8, Concentration of Health Spending, 1987–2002. *An Overview of the US Health Care System: Chart Book*. Centers for Medicare and Medicaid Services; 2007.

[14] CMS Office of the Actuary National Health Statistics Group. The Nation's Health Dollar, CY 2000: Hospital and physician spending accounts for more than half of all health services. *Program Information on Medicare, Medicaid, SCHIP, and Other Programs of the Centers for Medicare and Medicaid Services*. Centers for Medicare and Medicaid Services; 2002.

[15] OECD Health Data. Percent of GDP Spent on Health Care by OECD Country, 1960–2004. *An Overview of the US Health Care System Chart Book*. Centers for Medicare and Medicaid Services; 2007.

[16] OECD Health Data. Per Capita Spending on Medical Services by OECD Country, 1980–2004. *An Overview of the US Health Care System Chart Book*. Centers for Medicare and Medicaid; 2007.

[17] Actuarial Research Corporation. Health Insurance Coverage for the Under 65 Population, 1980–2004. *An Overview of the US Health Care System Chart Book*. Centers for Medicare and Medicaid; 2007.

[18] Census Bureau. Table 6.9 The Uninsured by State. *An Overview of the US Health Care System Chart Book*. Centers for Medicare and Medicaid Services; 2007.

[19] Actuarial Research Corporation. Table 4.34 Health Insurance Coverage of Children, 1988–2005. *An Overview of the US Health Care System Chart Book*. Centers for Medicare and Medicaid Services; 2007.

[20] Center for Medicaid and State Operations. Table 4.35 State Children's Health Insurance Program, Spending and Enrollment, 1998–2004. *An Overview of the US Health Care System Chart Book*. Centers for Medicare and Medicaid Services; 2007.

[21] Kaiser Family Foundation. Table 1.16 HMO Enrollment by Ownership Status, 1981–2004. *An Overview of the US Health Care System Chart Book*. Centers for Medicare and Medicaid Services; 2007.

[22] The Kaiser Family Foundation and Health Research Educational Trust. Table 5.9 Private Health Insurance Enrollment by Plan Type, 1988–2005. *An Overview of the US Health Care System Chart Book*. Centers for Medicare and Medicaid Services; 2007.

[23] Marchildon GP. *Health Systems in Transition: Canada*. Copenhagen: World Health Organization Regional

Office for Europe on behalf of the European Observatory for Health Systems and Policy; 2005.

[24] Health Canada. *Canada Health Act: Federal Transfers and Deductions.* 2007 [cited 2007 March 9]; Available from: http://www.hc-sc.gc.ca/hcs-sss/medi-assur/transfer/index_e.html

[25] Canadian Economy Online. *Key Economic Events: 1957 – Advent of Medicare in Canada: Establishing Public Medicare Access.* 2007 [cited 2007 March 8]; Available from: http://www.canadianeconomy.gc.ca/english/economy/1957medicare.html

[26] Blendon RJ, Kim M, Benson JM. The public versus the World Health Organization on health system performance. *Health Affairs (Project Hope).* 2001 May–Jun; **20**(3): 10–20.

[27] *Waiting for Health Care in Canada: What We Know and What We Don't Know.* Ottawa: Canadian Health Institute for Health Information; 2006.

[28] *Medical Imaging in Canada, 2004.* Ottawa: Canadian Health Institute for Health Information; 2005 January 13.

[29] Krauss C. In blow to Canada's heath system, Quebec law is voided. *The New York Times.* 2005 June 10.

[30] Blackwell T. Quebec's hybrid medicare: clinics to get public money and charge fees. *National Post.* 2007 January 29.

[31] Bordeau S. Canada: After "Chaoulli" Bill 33 opens doors to private clinics and private insurance in Quebec. *Mondaq Business Briefing.* 2006 July 28.

[32] Auger PL, *et al.* Two-tier medicine in Quebec is a slippery slope: Quebecers should be alarmed at shift away from single payer model. *The Gazette.* 2007 February 1.

[33] Tuohy CH, Flood CM, Stabile M. How does private finance affect public health care systems? Marshaling the evidence from OECD nations. *Journal of Health Politics, Policy and Law.* 2004 Jun; **29**(3): 359–96.

[34] *Health Care Systems in Transition: United Kingdom.* Copenhagen: European Observatory on Health Care Systems; 1999.

[35] *OECD Health Data 2006 – United Kingdom.* Paris: Organization for Economic Co-operation and Development; 2006.

[36] Foubister T, Thomson S, Mossialos E, McGuire A. *Private Medical Insurance in the United Kingdom.* Copenhagen: WHO on behalf of the European Observatory on Health Systems and Policies; 2006.

[37] Greengross P, Grant K, Collini E. *The History and Development of The UK National Health Service:*

1948–1999. London: Produced by the Health Systems Resource Centre on behalf of the UK Department for International Development; 1999 July.

[38] Enthoven AC. Internal market reform of the British National Health Service. *Health Affairs (Project Hope).* 1991 Fall; **10**(3): 60–70.

[39] *The NHS from 1988 to 1997.* [cited 2007 March 14]; Available from: http://www.nhs.uk/England/AboutTheNhs/History/1988To1997.cmsx

[40] *The NHS from 1998 to the Present.* [cited 2007 March 14]; Available from: http://www.nhs.uk/England/AboutTheNhs/History/1998ToPresent.cmsx

[41] *An Independent Audit of the NHS Under Labour (1997–2005).* London: King's Fund; 2005.

[42] Lewis R, Appleby J. Can the English NHS meet the 18-week waiting list target? *Journal of the Royal Society of Medicine.* 2006 Jan; **99**(1): 10–3.

[43] *A Health Situation Assessment of the People's Republic of China.* Beijing: United Nations Health Partners Group in China; 2005 July.

[44] Blumenthal D, Hsiao W. Privatization and its discontents – the evolving Chinese health care system. *The New England Journal of Medicine.* 2005 Sep 15; **353**(11): 1165–70.

[45] Valentine V. *Health for the Masses: China's Barefoot Doctors.* National Public Radio. 2005 November 4.

[46] Hesketh T, Wei XZ. Health in China. From Mao to market reform. *BMJ (Clinical research ed.).* 1997 May 24; **314**(7093): 1543–5.

[47] Riley NE. China's population: new trends and challenges. *Population Bulletin.* 2004 June; **59**(2).

[48] Song F, Rathwell T, Clayden D. Doctors in China from 1949 to 1988. *Health Policy and Planning.* 1991; **6**(1): 64–70.

[49] Eisenstein Z. WFP says Angola moving from crisis to recovery. *ReliefWeb.* 2003 October 23 [cited 2007 June 1]; Available from: http://www.reliefweb.int/rw/rwb.nsf/0/2677aa9c9581d10749256dc900024d77?OpenDocument&Click=

[50] USAID. Angola – Complex Emergency Situation Report #1 (FY 2004). *ReliefWeb.* 2004 January 7 [cited 2007 June 1]; Available from: http://www.reliefweb.int/rw/rwb.nsf/AllDocsByUNID/737e00912f1ee31285256e14006a913d

[51] *Data & Statistics: Country Classification.* The World Bank Group, 2007. [cited 2007 August 17] http://go.worldbank.org/K2CKM78CC0

[52] *Data & Statistics: Country Groups*. The World Bank Group, 2007. [cited 2007 August 17] http://go.worldbank.org/D7SN0B8YU0

[53] *World Health Statistics 2007: Core Health Indicators*. The World Health Organization. [cited 2007 August 17] http://www.who.int/whosis/database/core/

core_select.cfm?strISO3_select=afg&strIndicator_select= nha&intYear_select=latest&language=english

[54] *Health Systems Country Profiles*. The Pan-American Health Organization. [cited 2007 August 17] http://www.paho.org/English/DPM/SHD/HP/ Health_System_Profiles.htm

Healthcare costs vs. time: trends and drivers

In Chapter 5, we saw that countries have developed many different types of health systems to attempt to meet the health needs of their population in an equitable manner. Over the past 50 years, these systems have dramatically improved health in many countries. At the same time, the fraction of the world's economy devoted to healthcare has more than doubled; world health expenditures rose from 3% of world GDP in 1949 to 7.9% in 1998. In most countries, recent growth in healthcare expenditures has consistently outpaced economic growth. Figure 6.1 shows the growth in health expenditures over time for several developed countries. Rising healthcare expenditures have been a particular problem for the United States, which has by far the most expensive health system in the world [1].

In 2004, medical costs in the United States increased at a rate of 7.9%, nearly four times that of growth in wages and over three times the rate of inflation [2]. Is this growth in health spending expected to continue? Can such growth be sustained? Do advances in

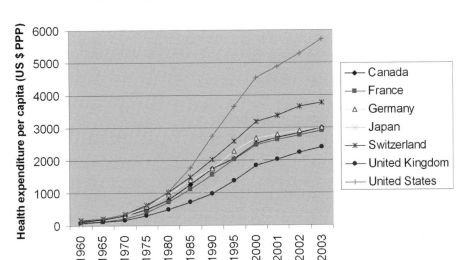

Figure 6.1. *Growth in per capita health expenditures in several countries. Based on OECD Health Data 2005, OECD 2005, www.oecd.org/els/health/data*

bioengineering fuel this growth or can new technologies be developed to help slow or reverse increases in health spending? As new medical technologies are developed and adopted, it is particularly important to consider the impact they may have on health spending.

In this chapter, we will examine trends in healthcare spending over time in the United States and other industrialized countries. We will explore which factors have contributed most to increases in health spending and we will examine three attempts to reform the entrepreneurial US health system to deal with rising costs. Finally, we will consider ways in which new medical technologies can be developed to improve the performance of health systems and reduce healthcare costs.

The challenge of rising healthcare costs (2005 data)

- 23% of Americans report trouble paying medical bills; 61% of these people have health insurance [4].
- 50% of all bankruptcy filings in the USA are partly a result of medical expenses [3].
- 29% of Americans have delayed or failed to seek needed healthcare because of concerns about cost [4].
- 70% of uninsured Americans cite cost as the main reason they do not have insurance [4].
- Insurance premiums rose by 9.2%, five times the rate of inflation. The average annual premium for an employer sponsored health plan for a family of four is nearly $11,000 [3].
- Workers are now expected to pay more of the costs for health insurance and pay more out of pocket for their own care. From 2000 to 2005, the average employee contribution to health insurance has increased by 143% and average out-of-pocket costs have increased by 115% [3].
- Annual healthcare spending in the USA is 4.3 times the amount spent on national defense [3].
- At the current rate of growth, Medicaid is projected to run out of funds in 2019 [5].

Costs in the United States

Total per capita health expenditures in the USA were $6102 in 2004 [6]. Health expenditures in the USA consumed 15.3% of the 2004 GDP; the next closest country was Switzerland, with health expenditures totaling 11.6% of their GDP or $4077 per capita [6]. Healthcare costs have increased and are projected to continue increasing. Adding to the problem is the fact that the US economy has not grown at the same rate as healthcare spending. By 2015 it is projected that healthcare costs will rise to $4 trillion, representing 20% of the GDP – one fifth of the total economy [7]. Richer nations spend more per capita on health expenditures than do countries with lower incomes, but even when this is taken into account, the United States spends substantially more on healthcare than other comparable countries (Figure 6.2) [8].

Why have healthcare costs in the US increased so much? Figure 6.3 shows the annual rate of growth for overall health spending in the USA, as well as growth rates for different types of health services. The annual growth rate peaked in 2001 and appears to have stabilized at around 8% per year. Growth in prescription drug spending and hospital outpatient spending are the two most rapidly growing sectors of healthcare spending [9].

Most health economists believe that six major factors have contributed to the increase in healthcare costs in the USA: (1) an aging population, (2) increasing prescription drug costs, (3) increasing costs of technology and increasing rates of technology utilization, (4) increased administrative costs, (5) a growing shift from non-profit to for-profit healthcare providers and (6) an increase in the number of under-insured and uninsured individuals.

Aging population

Over the past 50 years, average life expectancy has increased from 68 to 78 years in the United States. The aging of the population has contributed significantly to increases in healthcare spending. The prevalence of chronic diseases, such as cancer and coronary artery disease, are much higher in individuals over the age of 50. The elderly account for much of healthcare

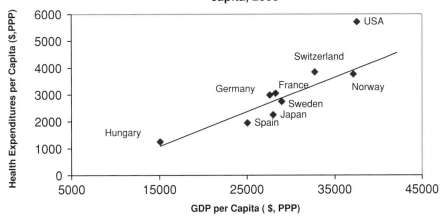

Health expenditures and gross domestic product (GDP) per capita, 2003

Figure 6.2. *Health expenditures per capita vs GDP per capita, 1998. The United States spends substantially more than other developed countries. From [8]. Used with permission from the* Annals of Internal Medicine.

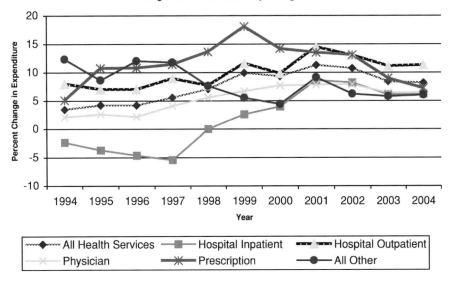

Annual growth rate for health spending in U.S.

Legend: — All Health Services — Hospital Inpatient — Hospital Outpatient — Physician — Prescription — All Other

Figure 6.3. *Annual rate of growth for overall health spending in the USA, as well as growth rates for different types of health services [9].*

spending in the United States – 40% of short term hospital stays, 25% of prescription drug use, and 58% of all health expenditures [10]. Health expenditures for individuals over the age of 75 average five times those of expenditures for individuals between the ages of 25–34 [8]. How will healthcare costs change as the population continues to age? The "baby boomers" are expected to place unprecedented strains on the healthcare system as they get older. The US government provides health insurance for senior citizens and Americans with disabilities through the Medicare program. Enrollment in Medicare nearly doubled from 1970 to 2000, and further increases are projected. By the year 2030, more than one in five US citizens are projected to be enrolled in Medicare, compared to the almost one in

eight now enrolled (Figure 6.4) [11]. This effect will be felt most in 2011–2030, and will result in the greatest single demand the USA has ever faced for long term care [10].

Increasing prescription drugs costs

The cost of prescription drugs represents the fastest growing category of health spending. From 1994 to 2005 prescription drug costs rose more than 12% per year on average, while other sectors of health spending grew by only 6–7% per year [12]. Over the past decade, the fraction of health spending devoted to prescription drugs has almost doubled from 5.6% of total health expenditures in 1993 to 10.1% of total expenditures in 2004 [13]. Today, more people take medicines – from

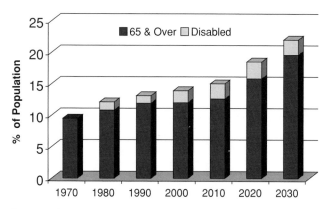

Figure 6.4. *Percentage of the US population eligible for Medicare from 1970 to 2030. The fraction eligible increases rapidly from 2000 to 2030 due to the aging US population [11].*

1994–2005 the number of prescriptions purchased increased by 71%, while the population grew by only 9% [12]. The costs to develop new drugs have risen dramatically over the last decade, and this has also contributed to increasing healthcare costs. Rising prescription drug prices are a particular challenge in the United States; drugs tend to be more expensive in the USA because the government does not set drug prices or regulate profits of drug companies as many other countries do. Compounding these challenges, direct marketing of drugs to the general population has resulted in increased usage in the USA; drugs advertised to the general public tend to be newer and more expensive [14].

The aging population is expected to further fuel increases in prescription drug costs. Today, 91% of senior citizens take prescription medicines on a regular basis [14]. Prior to 2006, Medicare did not cover the costs of prescription drugs for senior citizens, and more than 25% of seniors report they did not purchase medicines they were prescribed due to high prices. In January of 2006, Medicare instituted a new prescription drug benefit; coping with these additional costs will put further strain on the US healthcare system [14].

Increasing rates of technology utilization

Expensive new technologies are another factor that help drive healthcare costs higher. Examining data in the USA, we find that from 2001 to 2002 new technology was responsible for 22% of the increase in healthcare costs [15]. Much of this growth in spending is asso-

ciated with expensive radiologic imaging systems. For example, $175,000 X-ray machines are being replaced with more useful, but substantially more expensive, X-ray computed tomography (CT) machines, which cost in excess of $1 million each [15].

In addition, as new technologies which improve diagnostic or therapeutic outcomes are introduced they are used more frequently than the older technologies, so that even if they cost less there is the potential to increase total health spending [15]. For example, high-tech therapies for atherosclerosis, the leading cause of heart attacks, improved considerably throughout the 1990s. Balloon angioplasty is a much less invasive alternative than open heart surgery; as the technology associated with balloon angioplasty became safer and more effective and the costs of performing balloon angioplasty dropped, the utilization of this technology increased. From 1990 to 1998, four times more balloon angioplasties were performed in patients aged 65–74 [15]. This expansion of treatment allows more people to be treated, but the increase in utilization negates the potential cost savings associated with a less expensive procedure. Since newer technologies do not completely substitute for older technologies, their use is likely to increase cost rather than to save cost. The growth of an aging population, who can now be treated more intensively, with new technologies and new pharmaceuticals is a powerful combination which leads to increased cost.

Finally, there has been a recent move to **directly market** high-tech healthcare procedures to US consumers in an attempt to create a market for these expensive, and sometimes unproven, new technologies [15].

> **Direct marketing of medical technologies**
> The follow website provides an example of the direct marketing of healthcare technology to the consumer: http://www.ew1.org/ew1/EW1ataGlance/tabid/551/Default.aspx [16].

While medical technologies have contributed to increasing healthcare costs, in some cases, new technologies led to decreased spending. For example, some technologies have simplified medical procedures

Table 6.1. *The 2002 per capita health expenditures and average inflation adjusted growth in per capita health expenditures from 1992 to 2002 [18].*

Country	2002 per capita health expenditures ($PPP)	Average real annual growth per capita, 1992–2002
Canada	$2931	2.2%
Germany	$2817	2.0%
Japan	$2077	3.6%
Korea	$996	6.7%
Mexico	$553	2.2%
Switzerland	$3446	2.5%
Turkey	$446	9.2%
United Kingdom	$2160	3.8%
United States	$5267	3.3%
OECD Median	$2193	3.4%

so that they can be performed on an outpatient basis rather than necessitating a costly hospital stay. Many chemotherapy drugs are now given in outpatient clinics, eliminating expensive hospital visits. In the future, cost effective technologies which can address chronic diseases of the elderly have the most potential to reduce costs. A low cost cure for Alzheimer's disease could significantly reduce nursing home costs, among other expenses [17]. A new vaccine to prevent cervical cancer has recently been approved by the US Food and Drug Administration and has the potential to reduce costs. We will consider this technology in detail in Chapter 10.

Costs in other countries

Costs have risen in other industrialized countries at similar rates. Table 6.1 shows the average annual growth adjusted for inflation in per capita health expenditures from 1992 to 2002 for several different countries. Average real annual growth in the USA was 3.3% compared to the median growth of 3.4% in all Organisation for Economic Co-operation and Development (OECD) countries. When healthcare costs consistently outpace inflation, societies must make choices about whether to con-

tinue to devote increased resources to healthcare or to curtail access to healthcare. Health systems throughout the world struggle with this issue. Developing countries often face different challenges associated with rising health costs. For example in 2006, rapid inflation associated with the fragile economy of **Zimbabwe** led to a crisis in the availability of healthcare in that country [19].

Hyperinflation in Zimbabwe severely limits access to healthcare

In March of 2006, the rate of inflation in Zimbabwe approached 1000%. The costs of medicine have doubled or tripled every few months, pushing the cost of medical care beyond the reach of most citizens. In May of 2006, government hospitals raised consultation fees from one third of a cent to about $10, a 300,000% increase. Healthcare workers report that many AIDS patients have stopped taking anti-retroviral drugs provided through government health programs because of the increased cost, raising concerns that drug resistant strains of HIV would emerge [19]. Learn more about it:

Zimbabweans pay dearly for cost of healthcare by Craig Timberg, *Washington Post*, May 11, 2006. Available:

http://www.washingtonpost.com/wp-dyn/content/article/2006/05/10/AR2006051002308.html

Why are costs higher in the USA?

While healthcare costs have risen in most industrialized nations, the level of per capita health expenditures in the USA is still dramatically higher than in any other industrialized country. Why is the level of health spending in the USA disproportionately higher? Figure 6.5 compares several measures of health spending and health resources for the USA, Denmark, the United Kingdom, and Sweden [20]. Shown are the number of MRI and CT scanners, physicians, and hospital beds per person, the fraction of the population employed in the health sector, national health expenditures as a fraction

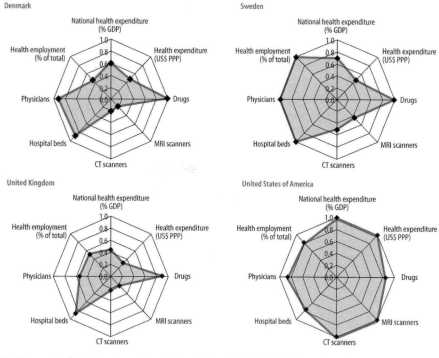

Figure **6.5**. *Comparison of several measures of health spending and health resources for the USA, Denmark, the United Kingdom, and Sweden. Each value has been normalized by the highest value reported for the four countries. Used with permission from [20].*

of the GDP, total health spending per capita, and total spending on prescription drugs. Each value has been normalized by the highest value reported for the four countries. The USA has the highest health expenditures, and also the highest number of MRI and CT scanners. On every other dimension, the USA ranks near the top as well. Denmark, the United Kingdom and Sweden report per capita health expenditures which are less than half those in the USA; these countries have substantially fewer MRI scanners and CT scanners than the USA. All countries report similar levels of prescription drug spending.

In addition to having a high number of physicians per capita, the USA has a higher ratio of specialists to primary care physicians than many developed countries [21]. Specialists are more likely to perform high cost procedures than primary care physicians, leading to increased costs. For example, in the USA the number of cardiac bypass procedures performed in 2003 was 161 per 100,000 population, compared to only 98 per 100,000 in Canada [22]. Similar patterns of usage are found for other high cost procedures (Table 6.2).

What has led to the rapid spread of medical innovation in the United States? In part, it may be due to the lack of regulatory constraints in the USA, where payments to physicians are directly related to the number of high tech procedures performed and hospitals must recoup huge investments in expensive new technologies. Clearly, new technologies have led to improvements in health, but overuse of these technologies may lead to increased cost. A recent analysis of regional variations in Medicare costs throughout the United States supports this view. Figure 6.6 shows the average Medicare cost per enrollee for the USA and for several different cities [21]. On average, Medicare spending in Miami was more than two fold higher than that in Des Moines. These intriguing data suggest several important questions. What factors are responsible for these large differences? Do Medicare enrollees in high cost areas receive better quality of care? Analysis indicates that the costs are primarily related to the predominance of specialists and hospital beds in the high cost regions [21]. Interestingly, increases in the quantity and cost of care do not seem to correlate with increases in quality of care.

Table 6.2. *Comparison of numbers of procedures performed for several industrialized countries [22–24].*

Country	Coronary angioplasty procedures per 100,000 population
United States	426.4 (2003)
Germany	301.6 (2004)
Canada	167.4 (2001)
United Kingdom	113.4 (2004)
Netherlands	92.6 (2003)
Country	**Number of heart transplants performed in 2005 per 100,000 population***
United States	0.72
Canada	0.54
Germany	0.44
Netherlands	0.23
Country	**Number of kidney transplants performed in 2005 per 100,000 population***
United States	5.6
Canada	3.6
Germany	2.6
Netherlands	2.5

* Calculated using mid year population estimates as provided in the 2005 *CIA World Factbook*.

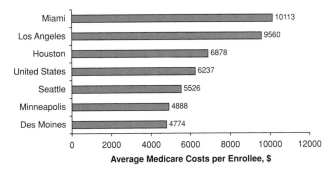

Figure 6.6. *Average Medicare costs per enrollee by hospital region, 2001. From [21]. Used with permission from the* Annals of Internal Medicine.

Disparities in practice patterns and costs have arisen in part because the United States lacks a nationally coordinated policy on technology assessment. Technology assessment programs which set standards of appropriate care and link them to payment systems which only reimburse for appropriate care have the potential to limit inappropriate technology diffusion and cost escalation [21].

Increasing administrative costs

The introduction of managed care insurance plans in the 1980s initially slowed growth of costs through a number of factors. HMOs resulted in slower adoption of technology, and reduced hospital cost growth through controls and limitations on care [25]. However in some cases restrictions on services were so extreme that they resulted in a consumer backlash. For example, several HMOs would cover only 24 hour hospital stays following vaginal childbirth. This procedure came to be known as "drive through deliveries." Many physicians felt that premature discharge of mothers and infants led to worse health outcomes, and the resulting consumer backlash led Congress to mandate that insurance plans provide coverage for minimum 48 hour hospitalization stay following uncomplicated vaginal deliveries [26]. In response to consumer backlash against HMO insurance plans that were perceived to overly restrict access to healthcare, plans with less tightly managed care were introduced, including PPOs and POS plans. Such plans give patients greater flexibility, but they also cost more.

We have seen that the fraction of health expenditures paid for from private sources is much higher in the USA than in other industrialized countries. The administrative burden associated with the complex, multi-payer US system leads to a higher fraction of the total healthcare budget which must be devoted to administrative costs as compared to welfare oriented and comprehensive systems which usually are characterized by a single payer system. In the USA, approximately 25–30% of the total healthcare budget is devoted to administrative overhead, and 27% of US healthcare workers do "mostly paperwork" [17, 27]. In contrast, Canada spends only 10–15% of the total healthcare budget for administrative costs [17, 27]. The problem continues to grow; from 2002 to 2004, administrative costs of private health

**Teen Club: July 15, 2007
Kim Malawi**

Saturday, I came to the clinic for the monthly Teen Club that they do. It's a really cool program.

In the morning, the kids get medical care – either a doctor or a nurse visit, depending on what they need. Then they get a free lunch. And, then, there is a variety of games for them to play.

I saw patients with Dr. Anjalee, which was interesting. I met a kid who got switched off first-line therapy, not because he failed adherence and built up a resistance, but because he was experiencing side effects. (Peripheral Neuropathy, to be specific.) But he was doing a really great job of following the more complicated second-line regimen. (He had a huge jump in his CD4 count, which is really nice to see!)

Lunch was a typical Malawian meal, except that it had rice instead of nsima. (Which is good, because I like rice more than nsima.) I ate with Amanda and Jenny, two nursing students from Seattle who have been around for a while.

After everyone is done eating, the kids form a huge circle and start chanting/singing in Chichewa. It took us quite a while to figure out what was going on, but we decided that they must have been introducing the newcomers. They dragged in a group

of five English people, all of whom were very nice (Patty, Simon, Katherine, Sophie and Chris.) Then, they dragged in Amanda, Jenny and me. It was fun and a little embarrassing.

Then, we had the kids divide up based on what activity they wanted to do. (They could choose between football, volleyball, song and dance, and arts and crafts.) I ended up playing football with them. It was really fun, but I'm so bad! I haven't played in years, and when I did play I was a goalie. So, my fieldwork is really quite terrible. The kids were nice, though, and didn't laugh at me (much.)

It was really nice to see these HIV-positive kids be so happy and healthy and excited. But it was sad to realize that the kids who were alive and playing were fewer than those who had died of the disease by this time. (75% of children born with HIV will die by 7 years of age if they don't receive treatment.)

insurance in the USA rose at an average of 16% per year, well above the average increase in health costs during that same period [8]. Publicly funded programs in the USA had much lower overall administrative costs. In 2002, the federal Medicare program spent about 3% of its total budget on administrative costs. The federal and state Medicaid program devoted 6.7% of its total budget to administrative costs, compared to 12.8% of revenues for private insurers [28].

Learn more about the problem of uninsured individuals in Texas

http://www.utsystem.edu/hea/codered [31]

Number of under-insured and uninsured individuals

Stephen Ayres noted in his book *Health Care in the United States*, that the USA is the "only country in the developed world, except for South Africa, that does not provide healthcare for all its citizens" [29]. In 2005, 46.6 million Americans, nearly 16% of the population, were without health insurance and the number is growing [30]. In some states, the number is even larger; 25% of Texas were without health insurance in 2005, the highest percentage of uninsured in the nation [31].

An increase in the number of under-insured and uninsured individuals can contribute to rising health costs. Frequently, individuals without health insurance will turn to the emergency room for routine care, which leads to increased expenditures [31]. In addition, uninsured individuals may delay seeking care until health problems have become serious and may be more expensive to treat. As healthcare costs continue to rise, the number of uninsured and under-insured are expected to rise [3]. As healthcare costs rise, employers stop offering insurance to some employees. Employees may decline employer-sponsored insurance because they cannot afford their part of the premium. In addition, rising health costs have forced many states to reduce the number of individuals eligible for Medicaid [32]. If increases in the price of health insurance continue to exceed income growth, health insurance will become unaffordable to more people.

Healthcare reform in the USA

Given these trends in healthcare costs, how can society responsibly allocate a limited number of healthcare dollars to address an increasing need for healthcare? Let's go back to Oregon in 1987, where Coby Howard's death resulted in widespread media coverage and forced Oregon to respond to the rise in healthcare costs. At this time, John Kitzhaber, a former ER physician, was president of the Oregon state senate. Kitzhaber later went on to become Governor of Oregon. In 1989, the state of Oregon set a goal of providing universal health insurance coverage for all of its citizens [33]. At the time only 42% of low income Americans in need of health insurance were covered by Medicaid [34].

Oregon took two steps to ensure universal insurance coverage. First, Oregon passed a bill which mandated private employers provide insurance for employees. Employers would be required to offer basic coverage to permanent employees working at least 17.5 hours per week or to make a contribution to a new state fund to help pay for coverage of the uninsured [35]. There was substantial business opposition to Oregon's "play or pay" employer mandate and it never received the federal waivers needed for implementation.

Second, Oregon expanded Medicaid to provide coverage for all people in the state living below the federal poverty line. Governor Kitzhaber reasoned that Oregon could not afford to pay for every medical service for every person, but that Oregon could expand insurance to

Table 6.3. *Health services ranked near the cut-off line for the Oregon Health Plan in 1999 [36].*

Oregon Health Plan initial rankings				
Rank	Treatment	Benefit (arbitrary units)	Duration	Cost
371	Tooth capping	0.08	4 years	$38
372	Surgery for ectopic pregnancy	0.71	48 years	$4000
376	Splints for TMJ	0.16	5 years	$98
377	Appendectomy	0.97	48 years	$5700

cover all its citizens if it was willing to ration care. While Oregon never received the federal permission necessary to mandate insurance coverage by private employers, it took a bold new approach to Medicaid reform [34].

The governor appointed a Health Services Commission which attempted to rank different health services according to their cost-effectiveness. In their first try they prioritized 1600 different health services. For each health service, they divided the cost of the treatment by the product of the total benefit provided by that health service and the expected duration of the benefit [36]:

Priority rating
$$= \frac{Cost\ of\ treatment}{Net\ expected\ benefit \times Duration\ of\ benefit}. \quad (6.1)$$

While this process yielded an objectively prioritized list of health services, many of the rankings were counter-intuitive. Table 6.3 shows some of the health services ranked closely together. You can see that some life saving procedures (surgery for ectopic pregnancy or appendectomy) were ranked similarly to inexpensive interventions with relatively minor benefits (tooth capping). Public reaction to the prioritized list was very negative [36].

The Commission went back to the drawing board and divided 709 condition/treatment pairs into 17 categories [34]. They first ranked categories according to net benefit. For example, the highest priority was given to services which provided treatment of acute life-threatening conditions where treatment prevents imminent death

Table 6.4. *Health services ranked near the cut-off line for the Oregon Health Plan in 1999. Those treatments above the line were covered, while those below were not [34].*

Oregon Health Plan, 1999		
Rank	**Diagnosis**	**Treatment**
570	Contact dermatitis and atopic dermatitis	Medical therapy
571	Symptomatic urticaria	Medical therapy
572	Internal derangement of knee	Repair/medical therapy
573	Dysfunction of nasolacrimal system	Medical/surgical treatment
574	Venereal warts, excluding cervical condylomata	Medical therapy
575	Chronic anal fissure	Medical therapy
576	Dental services (e.g. broken appliance)	Complex prosthetics
577	Impulse disorders	Medical/psychotherapy
578	Sexual dysfunction	Medical/surgical therapy
579	Sexual dysfunction	Psychotherapy

with a full recovery. Category 14, one of the lowest priority groups, included services which provided repeated treatment of nonfatal chronic conditions with minimal improvement in quality of well being with short term benefit [36]. Then services were separately ranked within each of the 17 categories.

These rankings were used to decide which healthcare services would be covered as part of Medicaid. Each legislative session, the legislature would decide how much money was available to allocate to the new Oregon Health Plan (OHP). Given the available resources, they calculated how many services on the ranked list they could afford. A line was drawn at this point, and the Oregon Health Plan would cover all services above the line, and no services below the line [34]. Table 6.4 shows services ranked near the cut-off line for the OHP in 1999.

The OHP was one of the first health plans implemented where cost effectiveness data were used to justify which services were excluded. Was the plan successful? The plan did not result in widespread rationing

of healthcare as many of its critics feared. The number of services actually excluded is very small and their medical value is marginal [34]. The current benefit package is now more generous than the state's old Medicaid system. In particular, coverage for transplants, the issue that gave rise to the initial controversy, is now more generous. Even the line separating which services are covered and which are not is rather fuzzy. The OHP pays for all diagnostic visits even if their treatment is not covered, and physicians frequently use this as a loophole to obtain coverage for services below the line. The OHP has not produced significant savings. During its first five years of operation, the OHP saved 2% compared to what would have been spent on the old Medicaid program [34].

Despite these limitations, there are many success of the OHP. Most notably, insurance coverage was significantly expanded. More than 600,000 previously uninsured residents are now covered under the OHP. From 1992 to 1997, the fraction of Oregon residents without health insurance dropped from 17% to 11%, and the number of uninsured children dropped from 21% to 8% [34, 37]. As a larger segment of the population had access to healthcare, the number of ER visits were reduced, and the number of low birth weight infants dropped [37]. How did the state pay for this? The additional coverage did not come from the savings from rationing, but instead the state raised additional revenues through a cigarette tax, and reduced costs by moving Medicaid recipients into managed care plans [34].

The attempts to reform healthcare in Oregon illustrate an important political paradox – the more public the discussions about priority setting, the harder it is to ration services to control costs [34]. Despite this difficulty, the state of Oregon significantly improved access to healthcare throughout the 1990s.

> **The Oregon Health Plan today**
> You can learn more about it at: http://www.npr.org/news/specials/medicaid/index.html

What is the status of **the Oregon Health Plan today**? Like many states, Oregon is facing a weak economy [38].

Given limited resources, the Oregon Senate established a special committee on the OHP to examine which citizens should be provided health insurance. Currently citizens qualified for the plan are ranked according to their need for insurance and their ability to pay. Those highest on list are the first to get services, while those at the bottom of the list will be the first cut if adequate resources are not available. Those given highest priority include poor pregnant women and children in families with incomes less than twice the federal poverty level. The next group includes adults at 50% of federal poverty line, followed by adults at 50–100% of federal poverty line, medically needy citizens (defined as those with a limited income and high medical expenses), and finally pregnant women and children in families with incomes up to 225% of the federal poverty line [39].

US healthcare reform

While reforms in Oregon are broadly viewed as successful, national attempts to reform healthcare in the USA have not been as successful. Employers have shifted an increasingly larger fraction of health insurance costs to their employees and rates of uninsured continue to climb.

Clinton plan

In response to rising healthcare costs, President Clinton assembled a task force to develop a plan for national health reform after he took office in 1992. The American Health Security Act of 1993 proposed a series of broad health reforms. Although it was ultimately not adopted by Congress, the plan attempted to address many of the challenges outlined in this chapter. The proposed plan would have guaranteed comprehensive healthcare coverage for all Americans regardless of health or employment status, including coverage for investigational treatments as part of approved research trials. The plan proposed to control costs through increased competition in the healthcare market, and through reduced administrative costs. Standards and guidelines were to be developed for practitioners in order to increase quality and further reduce costs. A National Health Board appointed by the President with Senate consent would

set national standards and oversee the administration of the new health system by the states. States would be responsible for ensuring that all eligible individuals enrolled in a health plan that delivered the guaranteed comprehensive benefit package. States were to establish one or more regional health alliances which would offer a variety of health plans that provide the comprehensive benefits package. The regional health alliances could be non-profit corporations or state agencies that would negotiate with private insurance companies to determine plans available in the region. These would include both traditional fee-for-service insurance plans as well as managed care plans. Employees of large corporations would obtain coverage through their employer unless the employer chose to participate in the regional health alliance. Individuals covered by most government sponsored health plans such as Medicare would continue to receive coverage through these plans. However, persons covered by state Medicaid programs would participate in plans offered through the regional health alliances. Government employees and employees of small businesses would obtain coverage through the regional health alliances. The plan was proposed to be financed by payroll taxes [40].

There was sustained and heated national debate about the merits and dangers of the Clinton plan. Interest groups on both sides of the issue spent large sums of money on television ads. One series of ads opposed to the plan noted that the Clinton plan would deprive families of their choice of doctor, while ads in support of the plan claimed that unless the plan was passed, millions of Americans would not have access to healthcare. A study by the University of Pennsylvania indicated that more than half of all the television ads sponsored by interest groups were misleading [41]. In the end, the plan was not adopted, nor were any of the other compromise plans offered during the debate about healthcare reform.

Situation today

Today, the USA continues to struggle with the complex issue of how to provide equitable access to healthcare for all citizens. While there is no "magic bullet," a number of possible strategies have been proposed. One suggestion is to devote more attention and resources to preventative medicine, including closer management of chronic diseases like heart disease or diabetes, which would theoretically decrease spending on expensive interventions like hospitalization. Another suggested reform is the adoption of high deductible insurance plans coupled with health savings accounts. In return for paying a higher insurance deductible, suggested to be at least $1000; policyholders would be able to invest money in a tax-free health savings account that could then be used to pay for medical costs. Employers could pass savings from lower premiums to employees, who would spend more wisely since they would pay out of pocket [32].

How can technology reduce costs?

Throughout this chapter, we have examined the challenges associated with rising healthcare costs. We found that, in many cases, new medical technologies have contributed to increases in healthcare costs. Increased spending associated with medical technologies arises in part due to the costs of the technology itself (e.g. expensive CT imaging systems) and in part because more providers and patients want access to new technologies which substantially improve the efficacy of a procedure or reduce its invasiveness. As a result, increasing use of a new technology may result in increased spending even if the new technology is cheaper than the current standard of care. Rigorous technology assessment should be used to establish consensus practice guidelines to ensure that new technologies are not inappropriately over utilized. In our next unit, we will examine the process of developing new medical technologies. We will consider how bioengineers can explicitly use health technology assessment to develop new medical technologies that make the best use of available healthcare resources.

Homework

1. Consider who pays for healthcare in the USA.
   a. What fraction of the US population currently does not have health insurance?
   b. List three differences between an HMO and conventional health insurance.
   c. What fraction of US healthcare dollars is spent on administrative costs? How does this differ in Canada?

2. Discuss the role that each of the following factors has played in contributing to increases in healthcare costs in the USA over the past decade.
   a. Physician income.
   b. Aging population.
   c. New technology development.
   d. Prescription drugs.

3. Cardiovascular disease is the leading cause of death in the USA. High blood pressure is a significant risk factor for both myocardial infarction and stroke. High blood pressure can be treated with drugs, but these can be expensive. A recent clinical trial involving more than 33,000 patients compared the efficacy of three different types of blood pressure reducing drugs.
   • Chlorthalidone, a diuretic (usual dose 25 mg/day).
   • Lisinopril, an ACE inhibitor, for which there is a generic available (usual dose 5 mg/day).
   • Amlodipine (Norvasc), a calcium channel blocker for which no generic is available (usual dose 5 mg/day).

   [Major outcomes of high risk hypertensive patients randomized to angiotensin-converting enzyme inhibitor or calcium channel blocker vs. diuretic, *JAMA* 12-18-02, **288**; 23. http://jama.ama-assn.org/cgi/reprint/288/23/2981.pdf]

   a. How much would a one year supply of each of the three drugs cost if you buy your prescriptions online at Walgreens (www.walgreens.com) and do not have health insurance? Present your answer as a bar graph.
   b. Listen to this story from NPR: http://www.npr.org/templates/story/story.php?storyId=1540336.
   c. What is the annual cost for these three drugs if you purchase them in Canada at http://www.canadianprescriptiondrugstore.com/? Add a second series of bars in your graph to compare them to US costs.

4. A commonly used anti-depressant, Zoloft (sertraline), recently became available as a generic drug. The most common regimen is 50 mg, taken daily. Find out how much a year's supply of Zoloft and the generic costs for the following.

   a. A person without insurance using the Walgreen's online pharmacy:
      i. Zoloft,
      ii. Sertraline (generic).
   b. A Canadian using the online Canadian Prescription Drugstore:
      i. Zoloft,
      ii. Sertraline (generic).
   c. Based on your calculations and the NPR report below, speculate on the reasons for the price differences you identified. http://www.npr.org/templates/story/story.php?storyId=1540336

5. Read the following report:
   http://www.npr.org/programs/atc/features/2002/may/uganda/
   Next, listen to this update on the story:
   http://www.npr.org/templates/story/story.php?storyId=6915566

   a. How is this system similar to the Oregon Health Plan?
   b. Evaluate this co-op system as a model for the rest of Africa by answering the following questions.
      i. Assess this health system in terms of access and likely outcomes for the town in Uganda that was described.
      ii. Only about 6% of the population of Uganda has AIDS; in Botswana and Zambia the prevalence is >2× greater. How might the growing AIDS epidemic impact participation in a healthcare co-op?
      iii. What sort of AIDS related care do you think community co-ops would decide to cover?

References

[1] Bureau of Labor Education. *The US Health Care System: Best in the World, or Just the Most Expensive?* Orono: University of Maine; 2001 Summer.

[2] Smith C, Cowan C, Heffler S, Catlin A. National health spending in 2004: recent slowdown led by prescription drug spending. *Health Affairs (Project Hope)*. 2006 Jan–Feb; **25**(1): 186–96.

[3] National Coalition on Health Care. *Facts on Health Care Costs*. 2004 [cited 2007 March 16]; Available from: http://www.nchc.org/facts/cost.shtml

[4] Brodie M, Hamel E, Deane C, Gutierrez C. *The Public on Health Care Costs*. Kaiser Family Foundation; 2005.

[5] Colliver V. In Critical Condition: Health Care in America. *San Francisco Chronicle*. 2004 October 11.

[6] OECD. OECD in Figures 2006–2007. *OECD Observer*. 2006(Supplement 1).

[7] Borger C, Smith S, Truffer C, Keehan S, Sisko A, Poisal J, et al. Health spending projections through 2015: changes on the horizon. *Health Affairs (Project Hope)*. 2006 Mar–Apr; **25**(2): w61–73.

[8] Bodenheimer T. High and rising healthcare costs. Part 1: seeking an explanation. *Annals of Internal Medicine*. 2005 May 17; **142**(10): 847–54.

[9] Strunk BC, Ginsburg PB, Cookson JP. Tracking healthcare costs: declining growth trend pauses in 2004. *Health Affairs (Project Hope)*. 2005 Jan–Jun; Suppl. Web Exclusives: W5-286–W5-95.

[10] Savage GT, Campbell KS, Patman T, Nunnelley LL. Beyond managed costs. *Health Care Management Review*. 2000 Winter; **25**(1): 93–108.

[11] Social Security Administration Office of the Actuary. Table 4.7 Medicare Beneficiaries as a Share of the U.S. Population, 1970–2030. *An Overview of the US Health Care System Chart Book*. Centers for Medicare and Medicaid Services; 2007.

[12] *Prescription Drug Trends Fact Sheet*. Kaiser Family Foundation; 2006 June.

[13] Table 2: National Health Expenditures Aggregate Amounts and Average Annual Percent Change, by Type of Expenditure: Selected Calendar Years 1960–2005. *NHE Web Tables*: Centers for Medicare and Medicaid Services; 2005.

[14] Woo A, Ranji U, Lundy J. *Prescription Drug Costs*. Kaiser Family Foundation; 2005 August 26.

[15] Goetghebeur MM, Forrest S, Hay JW. Understanding the underlying drivers of inpatient cost growth: a literature review. *The American Journal of Managed Care*. 2003 Jun; 9 Spec. No. 1: SP3–12.

[16] Heart Hospital of Austin. *Executive Wellness at a Glance*. 2006 [cited 2007 May 30]; Available from: http://www.ew1.org/ew1/EW1ataGlance/tabid/551/Default.aspx

[17] Mehrotra A, Dudley RA, Luft HS. What's behind the health expenditure trends? *Annual Review of Public Health*. 2003; **24**: 385–412.

[18] Anderson GF, Hussey PS, Frogner BK, Waters HR. Health spending in the United States and the rest of the industrialized world. *Health Affairs (Project Hope)*. 2005 Jul–Aug; **24**(4): 903–14.

[19] Timberg C. Zimbabweans pay dearly for cost of healthcare. *Washington Post*. 2006 May 11.

[20] Musgrove P, Creese A, Preker A, Baeza C, Anell A, Prentice T. *The World Health Report 2000: Health Systems: Improving Importance*. Geneva: World Health Organization; 2000.

[21] Bodenheimer T. High and rising healthcare costs. Part 3: the role of healthcare providers. *Annals of Internal Medicine*. 2005 Jun 21; **142**(12 Pt 1): 996–1002.

[22] OECD Health Data 2006. Table 2.14 Medical Technology and Use of High-Technology Medical Procedures by OECD Country, 2004. *An Overview of the US Health Care System Chart Book*. Centers for Medicare and Medicaid Services; 2007.

[23] The Organ Procurement and Transplantation Network. *Transplants by Donor Type*. [Database] 2006 [cited April 4, 2007]; Available from: http://www.optn.org/latestData/rptData.asp

[24] Canadian Organ Replacement Register. Table 1A. Transplants by Organ and Donor Type, Province of Treatment, Canada (Number). *e-Statistics Report on Transplant, Waiting List and Donor Statistics, 2005 Cumulative Report*. Canadian Institute for Health Information; 2006.

[25] Baker L. Managed care, medical technology, and the well-being of society. *Topics in Magnetic Resonance Imaging*. 2002 Apr; **13**(2): 107–13.

[26] Hyman DA. What lessons should we learn from drive-through deliveries? *Pediatrics*. 2001 Feb; **107**(2): 406–7.

[27] Woolhandler S, Campbell T, Himmelstein DU. Costs of healthcare administration in the United States and Canada. *New England Journal of Medicine*. 2003 Aug 21; **349**(8): 768–75.

[28] Bodenheimer T. High and rising healthcare costs. Part 2: technologic innovation. *Annals of Internal Medicine*. 2005 Jun 7; **142**(11): 932–7.

[29] Ayres SM, American Library Association. *Health Care in the United States: the Facts and the Choices*. Chicago: American Library Association; 1996.

[30] *The Number of Uninsured Americans is at an All-time High*. Washington D.C.: Center on Budget and Policy Priorities; 2006 August 29.

[31] *Code Red: The Critical Condition of Health in Texas*. Online: Task Force for Access to Health Care in Texas; 2006 April 17.

[32] Gutierrez C, Ranji U. *U.S. Health Care Costs*. Kaiser Family Foundation; 2005 September.

[33] Bodenheimer T. The Oregon Health Plan–lessons for the nation. First of two parts. *New England Journal of Medicine*. 1997 Aug 28; **337**(9):651–5.

[34] Oberlander J, Marmor T, Jacobs L. Rationing medical care: rhetoric and reality in the Oregon Health Plan. *Cmaj*. 2001 May 29;**164** (11): 1583–7.

[35] *The Oregon Health Plan: An Historical Overview*. Salem: Oregon Department of Human Services, Office of Medical Assistance Programs; 2006 July.

[36] Hadorn DC. The Oregon Priority Setting Process. In: Hadorn DC, ed. *Basic Benefits and Clinical Guidelines*. Boulder: Westview Press; 1992.

[37] Kitzhaber J. *A Reforming Governor's Perspective*. National Health Policy Conference; 2002 January 16; Washington, D.C.; 2002.

[38] Koffman W. Oregon Health Care Program in Jeopardy. *NPR's Morning Edition*: National Public Radio 2003.

[39] Colburn D. Health Plan 'Decision Day' Ends, But Senators Don't Act. *The Oregonian*. 2003 June 1.

[40] Baker N. Health care reform: summary of the Clinton administration's health reform plan: American Health Security Act of 1993. *Journal of Health and Hospital Law*. 1993 Oct; **26**(10): 289–95.

[41] Bok D. *The Great Health Care Debate of 1993–94*. Public Talk. 1998(2).

The evolution of technology: scientific method, engineering design, and translational research

In previous chapters, we learned about the epidemiology and pathophysiology of the leading causes of death throughout the world. We saw that both the causes of mortality and the availability of healthcare resources vary widely throughout the world. In this chapter, we will examine the process of scientific discovery and how it can lead to new medical technologies that benefit both individual patients and whole populations. Figure 7.1 outlines the process of developing a new medical technology. We begin with the science of understanding a disease. First, the etiology, or cause, of the disease must be identified. Next we examine how the disease produces symptoms, a process called pathophysiology, and how those symptoms can be detected and treated. We can use our understanding of the etiology and pathophysiology of a disease to design new healthcare technologies to diagnose, treat or prevent the disease. This process of design is referred to as bioengineering.

New healthcare technologies must be tested to determine whether they are safe and effective. Generally, researchers first carry out pre-clinical trials, using cell or tissue samples or animal models, to determine whether a technology is both effective and safe. Frequently the results of these trials suggest ways to improve a technology before it is ready to proceed to clinical trials with human subjects. New technologies which show promising pre-clinical results can proceed to clinical trials. These experiments must be carefully designed to ensure that the rights of patients are preserved. If a new technology shows promising results in clinical trials, then it can move from the research laboratory to general practice.

Figure 7.1 illustrates a vector which originates at the laboratory bench and terminates at the patient's bedside. Progress in medical research depends on moving new scientific ideas forward along this vector; research designed to advance new ideas from bench to bedside is known as translational research. Although the vector is shown as a straight line, we will see that progress is usually not made in a linear manner; more frequently the vector in Figure 7.1 can be thought of as a two-way street, with new scientific ideas as well as incremental improvements in technologies often originating from clinical observations. Translational researchers seek to make progress along this path by addressing the connections between the evolution of a new scientific idea, its translation into new technologies, the resultant improvements necessary to ensure their safety and efficacy, and the appropriate use of these technologies to improve health.

In the next few chapters we will examine the process of medical technology development in detail, using case

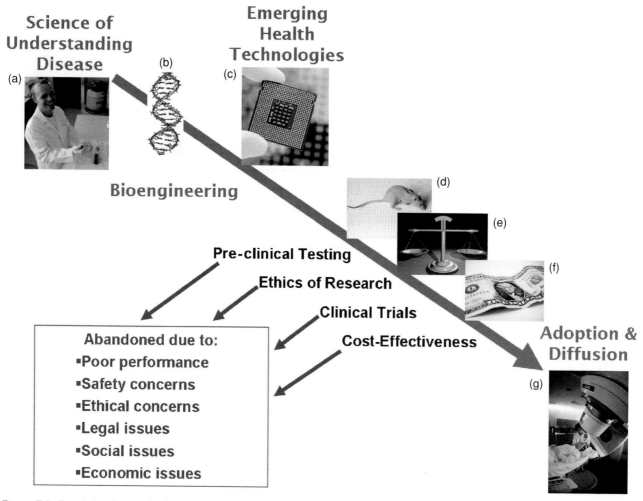

Figure 7.1. *Translational research: the process of developing a new medical technology. Sources as for Figure 2.1.*

studies centered around the prevention of infectious disease (Chapter 8), the early detection of cancer (Chapter 10), and the treatment of heart disease (Chapter 12). We begin by examining the process of technology development in detail. As we examine the technology development process, we will also profile several researchers who have made important contributions to translational research.

The process of discovery
The scientific method

The process of scientific discovery begins with a question. Why is the sky blue? Why do leaves turn color in the fall? Why does ice float (Figure 7.2)? As we seek to better understand natural phenomena, we develop new scientific knowledge. Science is the body of knowledge about natural phenomena which is well founded and testable. The purpose of scientific study is to discover, create, confirm, disprove, reorganize, and disseminate statements that accurately describe some portion of the physical, chemical, or biological world. Scientists have developed a methodic approach to guide them in the search for new scientific knowledge. This inquiry driven process is termed the scientific method, and it can be used to answer questions that are very basic in nature (what controls cell division?) as well as those that are more applied (why do tumor cells grow rapidly?). Often the distinction between basic and applied science is blurry, and advances in basic

Table 7.1.

Steps of the scientific method	Example
Pose a question	Why won't my car start?
Generate a hypothesis	The battery is dead
Design experiments to test the hypothesis	Try turning on the lights in the car
Carry out the experiments; analyze data	The lights turn on, therefore the battery is not dead
Revise the hypothesis if necessary	Perhaps I am out of gas

Figure 7.2. *An example of a simple scientific question is: why does ice float? Photograph courtesy of Dr. David A. Atkins.*

science can rapidly and unexpectedly lead to new technologies.

Table 7.1 gives an overview of the steps in the scientific method. After posing a question about a phenomenon that is not yet understood, scientists study work that has been done in this area. Based on this work they formulate a hypothesis to explain the phenomenon. The hypothesis represents an educated guess that answers the question originally posed and is consistent with observations made thus far. Formulation of the hypothesis is a critical step, because it serves as a guide for the rest of the inquiry process. You can think of a hypothesis as a mental model of how a process works. A good hypothesis will predict which variables affect a phenomenon, how these variables affect the phenomenon, as well as which variables are irrelevant.

Using this mental model as a guide, experiments are then designed to predict how the system will respond in ways that have not yet been tested. Where possible, in these experiments the effects of different variables are examined one at a time. In biological research, scientists often compare the response of two samples to different conditions. The control group and the experimental group are subject to identical conditions with one exception. Differences in the response of the two groups are attributed to the variable that was changed. The groups can be groups of cells, groups of animals, or groups of human subjects. Because of biological variability, the number of specimens in each group needs to be large enough to ensure that observed differences can be attributed to true differences in response and not just specimen-to-specimen variability. In Chapter 13, we will examine how to choose the number of subjects in a study to ensure statistically meaningful results. The results of the experiments are then analyzed to determine whether or not they are consistent with the hypothesis. If the results are not consistent, the hypothesis must be revised and new experiments are designed to test the revised hypothesis. Over time, if many experiments are found to be consistent with a given hypothesis then it becomes accepted as a theory or scientific law.

Engineering design

Advances in basic science can lead to inventions that improve our lives, but this is not the goal of the scientific method. Engineering is the profession that makes this connection. Thomas Tredgold defined engineering as the "art of directing the great sources of power in nature for the use and convenience of man." Vannevar Bush carried it further, saying, "engineering...in a broad sense...is applying science in an economic manner to the needs of mankind" [1]. Engineers design new products or processes in response to a particular need. Table 7.2 outlines the steps of the engineering design method. The first step is to identify a need. For example, cars that emit fewer pollutants would benefit the environment. Tools to detect cancers at their earliest stage would reduce cancer mortality.

Once a need has been identified, the next step in the engineering design method is to define the problem. The

Table 7.2.

The steps in the engineering design method	Example
Identify a need	Point of care diagnostic to measure HIV viral load in low resource settings
Define the problem (generate design specification)	Solution must be low cast and deliver results within minutes
Gather information	Reasearch current technologies, Review scientific findings in related areas
Develop solutions	Disposable microfluidic cartridge (PCR on a Chip)
Evaluate solutions	Carry out pre-clinical trials, clinical trials, Does design meet specification?
Communicate results	Journal article, patent, marketing brochure

problem definition consists of a carefully considered list of requirements that a solution must meet in order to be useful. This is sometimes referred to as a design specification. The specification considers both what the product should do as well as how much the product can cost. Once the specifications have been fully developed, engineers gather information and use this information to design alternative solutions. These solutions are evaluated to assess how well they meet the product specifications. Optimal solutions are selected and results are then communicated.

Differences between science and engineering

One of the most important fundamental differences between science and engineering lies in the goal of the work. In scientific research, the goal is simply to acquire new knowledge or understanding. In engineering, the goal is to use scientific knowledge to solve a particular problem. Clearly, there is much overlap between the two endeavors, and one can think of research and development as being part of a continuum, ranging from basic scientific inquiry to applied engineering design.

Translational research

Translational research operates across the continuum of basic scientific research, engineering design and clin-ical research. Progress along this continuum requires close interaction between scientists, engineers, and clinicians from a variety of different backgrounds; however, initiating and sustaining these interactions presents a number of challenges. The increasing depth of scientific knowledge within individual fields has made interdisciplinary collaboration more challenging. Many universities and medical schools are organized around individual basic science departments such as biology and chemistry, which sometimes operate as individual "silos." To address this need, the US National Science Foundation has emphasized the creation of new integrative graduate training programs which emphasize research and education at the interface between disciplines [2]. At the same time, the growing complexities of performing clinical research, have made it more difficult to connect advances in interdisciplinary science and engineering to clinical medicine. The US National Institutes of Health has recently established several new programs designed to fuse these fields together in new academic research and training programs designed explicitly to support clinical and translational research [3].

The field of biomedical engineering sits at the center of these opportunities. Advances in science drive new engineering designs to solve problems of humankind. As a discipline, the field of engineering initially built on advances in physics research, then later on research in chemistry, leading to the fields of mechanical engineering, electrical engineering and chemical engineering. More recently, engineers have capitalized on advances in biology, giving rise to the field of bioengineering. Biomedical engineers frequently collaborate with basic scientists and clinicians in an effort to develop new medical technologies. Biomedical engineers play an integrative role in the process of translational research illustrated in Figure 7.1.

Biomedical engineering and how it impacts human health

What is biomedical engineering?

The goal of biomedical engineering is to improve human health by integrating advances in engineering, biology,

As a bit of background, the ABC pump is how my senior design team refers to our project.

It's a device to accurately and precisely provide doses of liquid oral medication to children. It also assesses compliance/adherence with an attached counting mechanism.

I brought it with me to Malawi and am showing it to the doctors here (a little at a time) and asking for comments about it.

Dr. Jean thinks it's a really great idea. Her concerns are that it might not be durable enough and that patients might mess with the counting mechanism because they think they're supposed to have used a certain number of doses.

I think the durability problem will be taken care of when it's manufactured out of plastic. And we might be able to address the compliance issue by facing the numbers of the counter into the device so that only the doctor can take it out and check it.

I'll show it around more today.

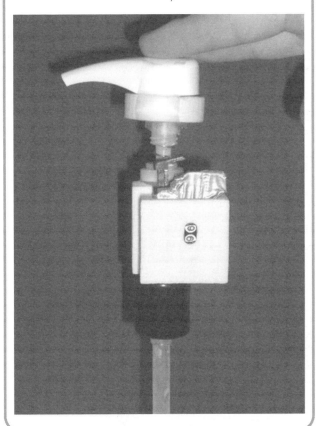

chemistry, physics and medicine. Biomedical engineers use engineering techniques to study living systems and better understand normal physiology and disease. Biomedical engineers also design new medical devices to diagnose and treat human disease. Bioengineering is closely related to the field of biotechnology – the use of living systems to make or improve new products. Advances in biotechnology are frequently targeted toward improving human health and bioengineering designs frequently rely on advances in biotechnology.

Major areas of bioengineering

The field of bioengineering draws on advances in many distinct areas of science and engineering. For example, designing improved imaging technologies such as portable magnetic resonance imaging systems draws heavily on advances in physics and electrical engineering. Designing and manufacturing new vaccines to prevent infectious diseases like influenza utilizes advances in chemical engineering, computational modeling, and biotechnology. Designing effective treatments to repair cartilage damaged as a result of injury or arthritis requires advances in both mechanical engineering and biology. Because the field of bioengineering is so diverse, it has been divided into several different sub-fields, which reflect both the different medical applications, the fundamental scientific underpinnings, and the analytic and experimental techniques which provide the basis of the sub-field. Some of the important areas of bioengineering include: (1) **biomedical imaging**, (2) **biosensors and bioinstrumentation**, (3) **biomechanics**, (4) **biomaterials and drug delivery**, (5) **tissue engineering and regenerative medicine**, (6) **biosystems engineering and physiology**, and (7) **molecular and cellular engineering**. In the following sections, we provide a brief overview of each sub-field, highlighting current challenges and the state-of-the-art in each area.

Areas of bioengineering
Biomedical imaging

Visualizing internal changes in anatomy and physiology can provide valuable diagnostic information. Advances in biomedical imaging over the past 100 years have

Profiles of translational innovation – Huda Zoghbi, M.D.

Courtesy of Photographer
Agapito Sanchez, Jr., Baylor
College of Medicine.

On a shelf in Huda Zoghbi's office at the Baylor College of Medicine sits a photo of Zoghbi and a group of people in her lab. The students, scientists, and technicians in the picture stand around a big cake celebrating the successful completion of Seung-Yun Yoo's Ph.D. thesis. The silver frame that surrounds the photo, a gift from Yoo, bears the inscription, "I love you, mom."

The affectionate engraving reflects the commitment that Zoghbi feels toward her students, her colleagues and collaborators, her friends, her family, and her patients. "It's all about relationships," Zoghbi says of her career. The people around Zoghbi provide her with the support, guidance, and inspiration to carry out her research – exploring diseases that lay waste to the human nervous system.

Zoghbi wasn't always dedicated to a life of science. "In high school I liked biology," she says. "But I loved literature. I fell in love with Shakespeare, with poetry." She hoped to major in literature and write poetry – but her mother put a stop to that. "She said that literature could be my hobby," says Zoghbi. "But she wanted me to go to medical school, graduate, and open a clinic – to keep my life simple," she laughs.

The budding poet listened to her mother, and she did well in medical school. "I enjoy anything as long as it's intellectually stimulating," Zoghbi says. After graduating, she went to Baylor, where she intended to go into pediatric cardiology. Until she did a rotation in neurology. "I just loved it," she says. "Neurology grabbed me because of how logical it is. You observe the patient, analyze her symptoms, and work backward to figure out exactly which part of the brain is responsible for the problem. You solve anatomical riddles by paying attention to details, listening to the patient, and then putting all the information together. It's like a puzzle." At the same time, Zoghbi found herself fascinated by disorders that affect the brain. "The brain is so vulnerable – so many things can go wrong," she says.

In the second year of her residency, Zoghbi encountered a very puzzling patient. The girl had been a perfectly healthy child, playing and singing and otherwise acting like a typical toddler. At the age of two, she stopped making eye contact, shied away from social interactions, ceased to communicate, and started obsessively wringing her hands. "She made a huge impression on me," says Zoghbi. What caused this sudden neurological deterioration? she wondered. And why was the child normal for so long?

The girl, it turns out, was a victim of Rett syndrome, a disorder that occurs primarily in females. Girls with this rare neurodevelopmental disorder develop normally for about 6 to 18 months and then start to regress, losing the ability to speak, walk, and use their hands to hold, lift, or even point at things.

With the help of the volunteers at the clinic, Zoghbi identified a half dozen girls who had Rett syndrome. All had been given the wrong diagnosis. She wanted to help these children, but in the end Zoghbi decided that she could do more for her patients at the lab bench than in the clinic. "Seeing these girls was so frustrating," she says. "I couldn't handle having to walk in, give the parents the bad news, and walk out."

So Zoghbi set aside her clinical work to devote herself to research. The switch was a bit frightening at first because, Zoghbi says, she "knew less molecular biology than the technicians" in the lab. But after three years of training, Zoghbi acquired the skills she needed to track down the genes that underlie neurological disorders.

Zoghbi wanted to start by studying Rett syndrome, but her scientific colleagues advised against it. "A lot of people told me I was ridiculous, that I would be wasting my time," she says. Many physicians doubted that Rett was a unique syndrome. "One very famous neurologist from a top institution said we were just putting a new name on an old diagnosis: cerebral palsy," says Zoghbi. And even if Rett were a "new" disease, it was a sporadic disease so rare – striking 1 child in 10,000 to 15,000 – that finding the gene responsible would be almost impossible. "After a while," says Zoghbi, "I stopped telling people I was working on Rett."

And for a time she did stop working on Rett – or at least put it on a back burner while she studied another disease, spinocerebellar ataxia type 1 (SCA1). This neurological disorder strikes later in life, when people reach their 30s or 40s. Patients with SCA1 experience a deterioration of balance and coordination that renders them unable to walk or talk clearly, or eventually to even swallow *or breathe. To help her track down the gene that causes SCA1, Zoghbi first identified a large family that seemed particularly prone to the disorder. She went from house to house in rural Montgomery, Texas – 50 miles north of Houston – to meet and examine members of the family. Of the 200 people she visited, 60 had the disease. "I felt so sad for the family," says Zoghbi. "They were very poor. They had no plumbing, no bathrooms. And they had no idea what was wrong with them. They thought they had rickets because they couldn't walk so well.*

"When I cloned the gene," she says, "I called them first."

After her success with SCA1, Zoghbi turned her attention back to Rett because, she says, "nobody could tell me not to." And 16 years after she saw her first patient with Rett syndrome, Zoghbi and her collaborators identified the gene responsible for the disorder.

"It wasn't all success," Zoghbi says of the journeys that led her to these genes. "I had more negative data than you can imagine." But she stuck with it because of the patients. "I worked with these families, I got to know them, and I wanted to help in some way," says Zoghbi. "When you see the plight of these patients and their families, how can you quit? They hadn't given up on me. How could I give up on them?"

Along the way, Zoghbi relied on the support of her collaborators, her husband, and even her children. "I've always involved my kids in what I'm doing," she says – and that includes her science. "I once submitted a manuscript to Science or Nature and was anxiously awaiting the verdict," says Zoghbi. "And my daughter said, 'If I were you I would have sent it to Nature Genetics. They're always so good and efficient with their reviews.' She was 13 at the time."

And Zoghbi still keeps in touch with the families who let her into their lives. "When they call, I'm happy that I have something to tell them," she says. "Maybe we don't yet have a treatment. Maybe we don't yet have a cure. But every six months, I can say we're another step closer."

"In 10 years," says Zoghbi, "I hope we'll have drugs to slow the progression of neurodegenerative diseases." In a way, she'd just be returning a favor. "For me," says Zoghbi, "I wouldn't be where I am without my patients."

Reprinted with permission of [4].

Advances in BME

Many medical devices that are part of routine clinical care today – angioplasty catheters to treat heart disease, magnetic resonance imaging systems, fiber optic endoscopes for minimally invasive surgery, artificial skin to treat burns, kidney dialysis machines, replacement hip joints, pacemakers, and the heart–lung machine – were developed and refined through collaborations between clinicians and engineers [5].

Today, the heart–lung machine is used to take over the function of blood circulation and oxygenation during more than 600,000 open heart surgeries each year in the USA. Mortality rates for many open heart procedures now approach 1%. Yet, the development of the heart–lung machine was fraught with setbacks that continued throughout the 1950s and 60s, and required close collaboration between pioneering clinicians and engineers.

The development of the heart–lung machine integrated advances in basic science to understand circulatory physiology and prevention of blood clotting together with engineering advances to pump blood without damaging red blood cells while supplying adequate oxygenation and removal of blood waste products.

The surgeon John Gibbon was inspired by clinical observations of patients who did not survive heart surgeries to build the first heart–lung machine in 1937. Animal trials at that time showed the device could sustain heart and lung function, but procedural mortality was high. Following World War II, he resumed research; complications with his prototype led him to a collaboration with Thomas Watson and other engineers at IBM in 1946. Around the same time, Forest Dodrill, a surgeon at Wayne State University, and engineers at General Motors developed a pump to take over circulation during heart surgery. The first successful left-sided heart bypass was performed using this device to support circulation while the patient's lungs were used to oxygenate the blood. In 1953, Gibbon performed the first successful human open-heart surgery with artificial circulation.

Despite these successes, early use of heart–lung bypass machines led to frequent complications, and many patients did not survive the procedures. An alternative procedure was developed in the early 1950s by C. Walton Lillehei called controlled cross-circulation. In this technique, the circulation of a patient's parent or close relative with the same blood type was used to temporarily support that of the patient.

Dr. John Kirklin led efforts to develop and test a modified heart–lung machine at the Mayo clinic in the 1950s. He wrote: "The Gibbon pump oxygenator had been developed and made by the IBM Corporation and it looked quite a bit like a computer. Dr. Dodrill's heart-lung machine had been developed and built for him by General Motors and it looked a great deal like a car engine . . . Most people were very discouraged with the laboratory progress. The American Heart Association and the NIH had stopped funding any projects for the study of the heart-lung machine because it was felt that the problem was physiologically insurmountable . . . The electrifying day came in the spring of 1954 when the newspapers carried an account of Walt Lillehei's successful open heart operation on a small child. In the winter of 1954 and 1955 we had 9 surviving dogs out of 10 cardiopulmonary bypass runs . . . We had earlier selected 8 patients for intracardiac repair . . . We had determined to do all 8 patients even if the first 7 died. All of this was planned with the knowledge and approval of the governance of the Mayo clinic. Our plan was then to return to the laboratory and spend the next 6–12 months solving the problems that had arisen in the first planned clinical trial of a pump oxygenator . . . Four of our first 8 patients survived, but the press of the clinical work prevented our even being able to return to the laboratory with the force that we had planned." These pioneering successes spurred many clinical and research labs to initiate open heart programs and led to the technologies that we rely on today[6].

(a)

(b)

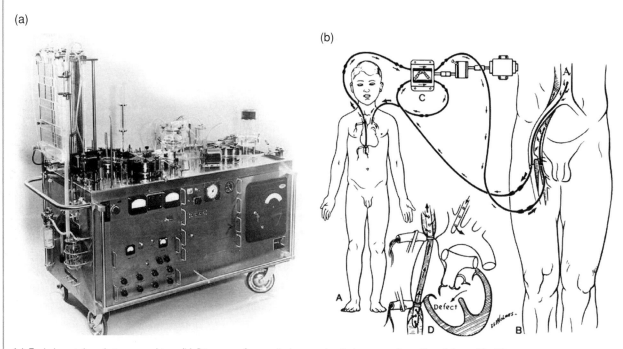

(a) Early heart–lung bypass machine. (b) Diagram of controlled cross-circulation procedure. Part (a) modified from http://mayoclinic.org/tradition-heritage-artifacts/images/68-1-1g.jpg. By permission from Mayo Foundation for Medical Education and Research. All rights reserved. Part (b) Courtesy of the Journal of the American Surgeons.

Learn more about it

A history of the bioengineering development of a number of medical devices can be found at:

http://bluestream.wustl.edu/WhitakerArchives/glance/heartlung.html

An excellent account of the history of cardiac surgery can be found at:

http://cardiacsurgery.ctsnetbooks.org/cgi/content/full/2/2003/3

revolutionized the ability to peer within the body; today physicians routinely use electromagnetic or acoustic radiation to produce high resolution, three-dimensional (3D) images of the structural and molecular changes associated with disease. Biomedical engineers play an important role in developing imaging hardware, new contrast agents, and software to construct, display and analyze images obtained with a variety of imaging modalities, including: X-ray/CT, MRI, nuclear/PET, ultrasound and optical imaging (Figure 7.3).

X-ray/CT

With the discovery of X-rays in the 1890s, the field of radiology was born. In the early 1900s, two dimensional images of skeletal structure were created by passing X-rays through the body and using photographic film to record a map of transmitted X-ray intensity, based on the strong absorption of X-rays by calcium in bone. In some of these early systems, patients themselves held the film cassette as the image was recorded (Figure 7.4) [9].

By using radio-opaque contrast agents, which can be administered orally or intravenously, a wide range of anatomic structures can now be imaged using X-rays. The upper digestive tract (the esophagus, stomach, and small intestine) can be visualized on X-ray after swallowing a barium-containing contrast agent (Figure 7.5); this is useful for diagnosing cancers, ulcers, and some swallowing problems. Blood vessels are routinely imaged following injection of iodine-containing contrast agents. As we saw in Chapter 4, the perfusion

of the heart via the coronary arteries can be visualized in real time in an X-ray imaging procedure called angiography; angiography is used to diagnose atherosclerosis, a process which can give rise to a heart attack.

In the 1970s, advances in computing technology made it possible to compute three dimensional images from a series of two-dimensional X-ray images acquired as the X-ray source and detector are rotated around the patient (Figure 7.6a). This computed tomography, or CT, exam images a slice approximately 1 cm in thickness; within this slice, a lateral spatial resolution approaching 0.25 mm can be obtained.

MRI

Magnetic resonance imaging (MRI) can provide high resolution, 3D images of soft tissues deep within the body by using our knowledge of the basic physical properties of water. Atomic nuclei which have an odd number of protons (e.g. hydrogen), possess a net magnetic moment (Figure 7.6a); MRI images are formed by measuring the interaction of these nuclei with an external magnetic field. To obtain an MRI image, a patient is placed inside the bore of a strong magnet; typical magnetic field strengths used in clinical MRI systems are 0.5–2.0 Tesla (10,000–40,000 times stronger than the Earth's magnetic field). MRI images essentially provide a spatial map of tissue water content by imaging the interaction between the imposed magnetic field and hydrogen atoms in water-containing tissues. The nuclei of these hydrogen atoms act as tiny magnets; when an external magnetic field is applied, the magnetic moments of the hydrogen atoms tend to align parallel or anti-parallel to the imposed magnetic field, with a slightly larger fraction aligning parallel to the applied field. When magnetized, these atoms can absorb and emit radiofrequency energy at a resonant frequency, which depends on the strength of the applied magnetic field.

In order to create images of tissue, the MRI machine (Figure 7.6b) delivers low energy pulses of radiofrequency (RF) energy to perturb the aligned hydrogen atoms. As the excited hydrogen atoms relax back to their original orientation, they emit RF energy at

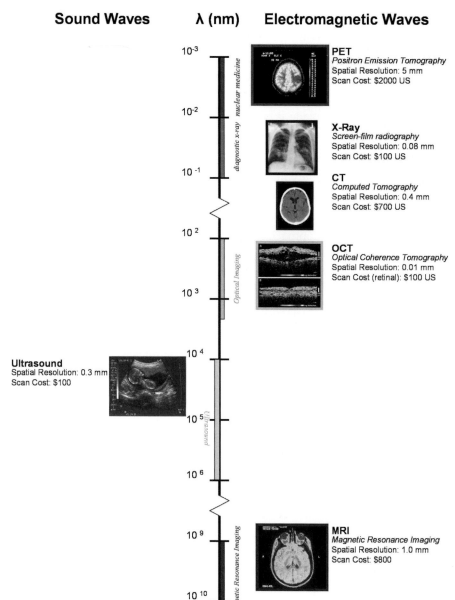

Sound Waves   **λ (nm)**   **Electromagnetic Waves**

PET
Positron Emission Tomography
Spatial Resolution: 5 mm
Scan Cost: $2000 US

X-Ray
Screen-film radiography
Spatial Resolution: 0.08 mm
Scan Cost: $100 US

CT
Computed Tomography
Spatial Resolution: 0.4 mm
Scan Cost: $700 US

OCT
Optical Coherence Tomography
Spatial Resolution: 0.01 mm
Scan Cost (retinal): $100 US

Ultrasound
Spatial Resolution: 0.3 mm
Scan Cost: $100

MRI
Magnetic Resonance Imaging
Spatial Resolution: 1.0 mm
Scan Cost: $800

Figure 7.3. *Biomedical imaging relies on the use of acoustic or electromagnetic radiation to visualize tissues within the body. The resolution of an imaging modality depends on the wavelength of the radiation used in the procedure. Sources: PET, Dr. Giovanni Di Chiro. Neuroimaging Section. National Institute of Neurologic; X-ray, National Cancer Institute; CT, photo courtesy of Philips Medical Systems; ultrasound, CDC/Jim Gathany; OCT, Courtesy of Dr. John Nolan; MRI, Courtesy of Dr. Leon Kaufman, University of California, NCI.*

the resonant frequency which can be detected to build up a 3D image proportional to the water content of tissue. Location of tissue within the image is encoded by applying a small gradient in the magnetic field which varies with spatial position. The frequency of RF energy emitted then depends on spatial location. The spatial resolution which can be obtained depends on the strength and uniformity of the applied magnetic field. A typical resolution of 0.5 mm can be obtained using commercial MRI systems; more expensive research systems use a field strength of up to 12 Tesla to achieve a

50 micron spatial resolution. These systems can weigh more than 5 tons and typically cost over $1 million per Tesla.

PET

In the 1950s, radiologists developed a new type of medical imaging based on the use of weakly radioactive (as opposed to radio-opaque) contrast agents. In a positron emission tomography (PET) scan, a contrast agent containing a radioactive isotope which emits positrons is given; a commonly used contrast agent is

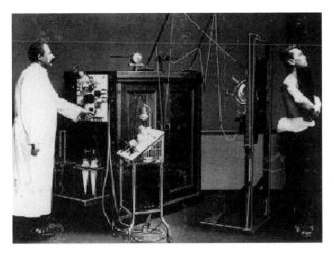

Figure 7.4. *An early X-ray system. Used with permission from [9].*

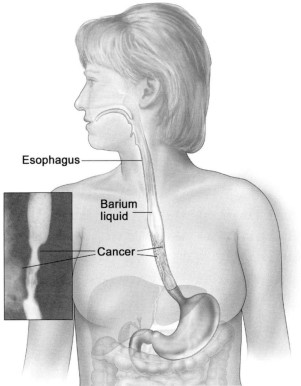

Esophagus

Barium
liquid

Cancer

National Cancer Institute

Figure 7.5. *Visualization of the upper digestive tract on X-ray after swallowing a radio-opaque contrast agent containing barium. © 2005 Tevese Winslow, US government has certain rights.*

(a)

Dipole magnetic field of
spinning proton

Magnetic moment of
spinning proton

Spinning protons in the *absence*
of an external magnetic field

Spinning protons in the *presence*
of an external magnetic field

(b)

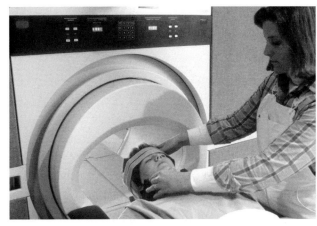

2-^{18}F-fluoro-2-deoxy-D-glucose (^{18}FDG), a glucose-based radiopharmaceutical. These radioisotopes are produced in a particle accelerator called a cyclotron and must then be included in the contrast agent very quickly before they decay. Thus, a major disadvantage of PET

Figure 7.6. *(a) Aligned magnetic moments after the application of an external magnetic field. (b) An MRI machine. Used with permission from Stanford Office of Communications and Public Affairs.*

technology is that all the facilities needed to produce the contrast agents must be located in close proximity to the imaging site. There, the contrast agents are ingested or injected and a gamma camera is used to detect the weak radiation emitted. In the patient, the contrast agent is taken up by the target tissue and enables 3D imaging of the metabolic activity of that region or tissue. When positrons emitted by the contrast agent collide with electrons in the tissue, a pair of photons is produced, which travel collinearly, but in opposite directions. An array of scintillation detectors detects the nearly simultaneous arrival of these photons. Because the photons travel along the same line, they give information about their original location. Computational algorithms can be used to reconstruct a three dimensional map of the tissue radioactivity using the detector locations at which many coincident photons are detected.

The spatial resolution of PET scans is limited by the size of the scintillation detectors; PET scanners record 2D images from a slice thickness of about 1 cm; within the slice, the lateral spatial resolution is approximately 6–8 mm (0.25 mm in CT). Newer time-of-flight PET scanners record the time difference (typically a few picoseconds) between the photon pairs, and can use this information to increase the resolution to approximately 4–5 mm. Because of the limited resolution of PET, and the fact that the image contains signal only from the regions of tissue which take up contrast agent, combined PET/CT imaging devices are frequently used to obtain images. The CT image provides anatomic reference information which is helpful in interpreting the functional imaging present in the PET scan.

Ultrasound

In ultrasound imaging, high frequency sound waves are used to create two-dimensional (2D) images of tissue beneath the surface of the skin. Ultrasound waves can easily penetrate through tissues with high water content; a portion of the incident ultrasound wave is reflected at borders between tissue with different densities. To produce an ultrasound image (Figure 7.7), an acoustic transducer is placed in contact with the tis-

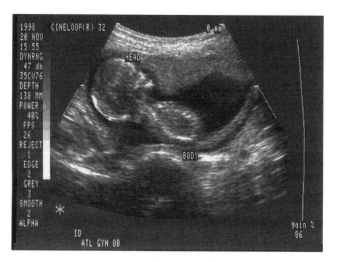

Figure 7.7. *A fetal ultrasound. NCI/Jim Gathany.*

sue to be imaged; a special gel is used to reduce the amount of ultrasound energy reflected from the tissue surface (impedance matching). Following an incident pulse of ultrasound energy, a detector in the ultrasound probe measures the strength of reflected ultrasound as a function of time. Using the speed of sound in tissue, these data can be used to calculate a map of differences in tissue density as a function of depth beneath the transducer. A typical slice thickness imaged with ultrasound is 1 mm, with in-plane lateral resolution of 0.5 mm. The resolution of ultrasound is directly related to the frequency of the soundwaves used to acquire the image. As the frequency increases, the wavelength of the sound waves decreases, yielding higher spatial resolution. For example, transabdominal obstetric ultrasound imaging systems use sound waves of 3.5–5.0 MHz frequency to achieve images with a typical spatial resolution of 0.5–1.0 mm. To image coronary arteries for intravascular ultrasound, higher frequencies (10–40 MHz) are used to yield 0.05–0.1 mm resolution; the trade-off is that tissue penetration decreases and scattering from blood increases at higher frequencies so that tissue can be imaged to a depth of only about 5–10 mm.

Recently, several companies have developed very small, portable ultrasound imaging systems which may dramatically expand access to ultrasound imaging in low resource settings (Figure 7.8). For example, SonoSite has developed a complete 2D ultrasound

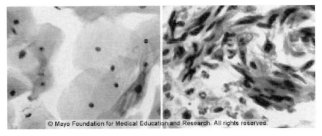

Figure 7.9. *A Pap smear examines cells from the cervix to screen for cervical cancer. In this Pap smear, normal cells appear on the left, while abnormal cells appear on the right. Copyrighted and used with permission from Mayo Foundation for Medical Education and Research, all rights reserved.*

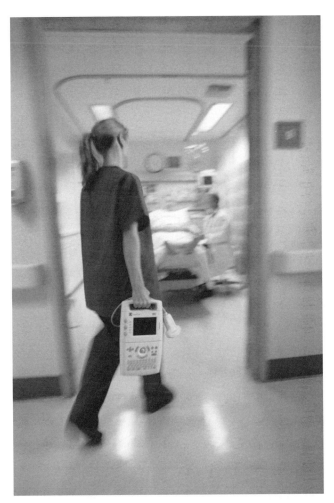

Figure 7.8. *A portable ultrasound system. Product photograph reprinted with permission from SonoSite; SonoSite® 180PLUS are trademarks owned by SonoSite, Inc.*

system which weighs less than six pounds; a variety of transducers can be used with the device to image at frequencies in the range 2–10 MHz; current prices begin at around $10,000.

Optical

The invention of the light microscope over 400 years ago provided the first view of the inner structure of single cells. The major underpinnings of cell biology and histopathology resulted from observations of tissues and cells through the light microscope. Today, advances in low cost optical technologies including lasers, LEDs, fiber optics and sensitive photodetectors promise another fundamental change in perspective in

cell biology and diagnostic medicine. As we will see in Chapter 10, the use of optical microscopy, both in the pathology lab and in the physician's office, has led to dramatic reductions in the mortality and incidence of cervical cancer in every country in which screening has been implemented. In Pap smear screening for cervical cancer (Figure 7.9), optical microscopy is used to examine cell scrapings from the uterine cervix; women whose smears contain abnormal cells are referred for a follow up procedure called colposcopy. A colposcope (Figure 7.10a) is essentially a microscope used to view the uterine cervix through a speculum at approximately 10× magnification. Based on changes in the color, surface pattern and vascularity observed through the colposcope, a practitioner will determine whether a biopsy is needed to determine whether a cervical cancer or precancer is present.

Used in the laboratory, the spatial resolution of optical microscopy is governed by the diffraction limit of light; typically structures as small as 500 nm can be resolved. Antibody targeted stains can be used to visualize molecular and genomic changes in cells in the lab. Low magnification optical microscopes have been used for decades to observe the surface of accessible tissues such as skin, and other mucous membranes to examine the surface of tissues with sub mm spatial resolution. The development of lasers and optical fibers led to great expansion of optical imaging in medicine in the early 1980s. We will see in Chapter 15 that small fiber optic microscopes are used to visualize structures in hollow organs such as the colon, the bladder, the lungs and the peritoneal cavity in order to provide diagnostically

(a)

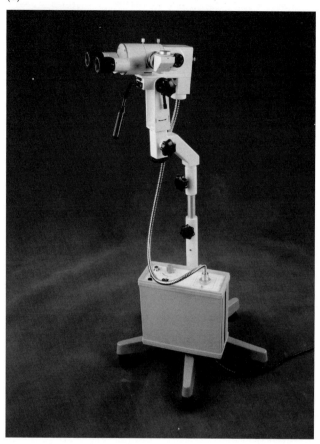

(b)

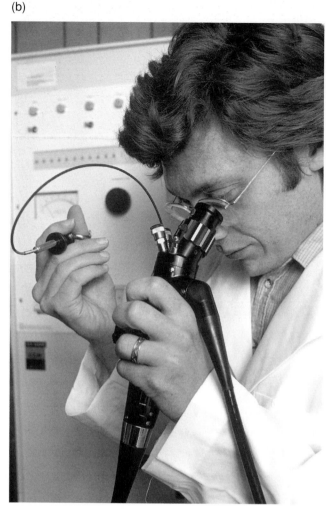

Figure 7.10. *(a) A colposcope is used to view the cervix. Copyright OperatingMicroscopes.com. Used with permission. (b) An endoscope. Source: NCI/Linda Bortlett.*

useful information and to guide minimally invasive surgical procedures (Figure 7.10b). Today, high resolution endoscopy is routinely used to screen patients at risk for esophageal cancer with a magnification of 400×.

More recently, a number of new optical microscopes have been developed in order to visualize cellular structure in living tissue. Optical techniques such as confocal microscopy (Figure 7.11) or optical coherence tomography use combinations of optical fibers and miniature optical components to obtain up to 1 micron resolution images at the subcellular level in near real time. Tandem advances in nanotechnology and molecular biology are under active development to enable real time, high resolution imaging of molecular and

genomic changes in living tissue for future point-of-care diagnostics.

Biosensors and bioinstrumentation

Messages in biological systems are often encoded as chemical or electrical signals. An important subdiscipline of bioengineering – the field of bioinstrumentation – is focused on developing sensors and instruments to quantitatively record and manipulate biological signals. A wide variety of biosensors and instruments are used clinically: an electroencephalogram (EEG) records the electrical activity of the brain (Figure 7.12) and is important in the diagnosis of seizure disorders and the evaluation of head injuries; a pulse oximeter monitors differential transmission of light at two wavelengths through

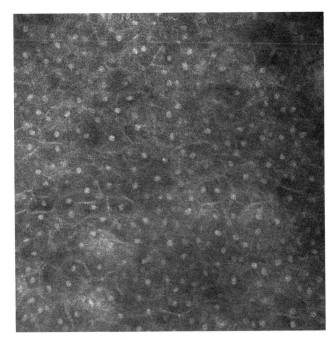

Figure 7.11. *Reflectance confocal microscopy image of normal human cervical tissue. The white dots represent cell nuclei within the tissue.*

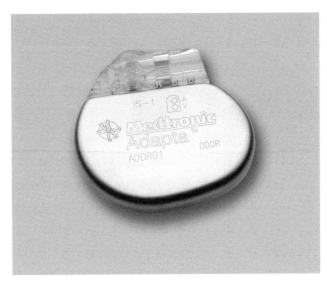

Figure 7.13. *A pacemaker, an implantable device for regulating a patient's heart rate. Source: Medtronic.*

a patient's finger or earlobe and is used to calculate the pulse rate and the percentage of hemoglobin which is saturated with oxygen; an artificial pacemaker is used to sense a patient's heart rate and trigger a timed electrical discharge to set the rate if it drops below a target value (Figure 7.13).

Biosensors serve as transducers, transforming biological signals into data that can be measured quantitatively, displayed, and, in combination with bioinstruments, potentially used to trigger and elicit a response if needed. Sensing signals *in vivo* presents special challenges, and the design of such systems must often integrate electrical instrumentation encapsulated within appropriate biomaterials.

The rapidly evolving field of brain–machine interfaces illustrates both the exciting promise and the great

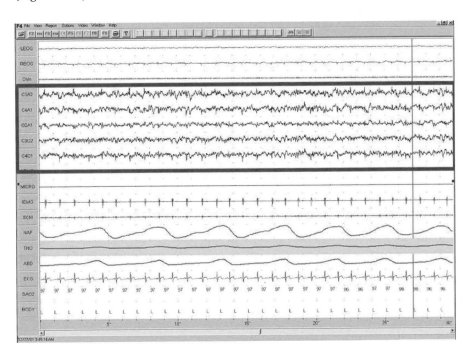

Figures 7.12. *A polysomnogram, the record of multiple tests performed while a patient sleeps. The outlined box includes the channels for the EEG. The other channels record oxygen saturation, eye movement, air flow, and many other measures. Source: Beaumont Hospital.*

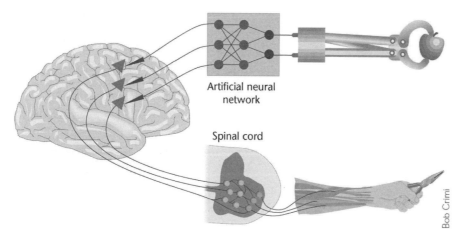

Artificial neural
network

Spinal cord

Figure 7.14. *Normally, neurons control voluntary arm movement. Microelectrodes could be used to record neural activity. These signals could be decoded using an artificial neural network and used to control a prosthetic arm. Used with permission from [12]. Reprinted with permission from Macmillan Publishers Ltd: Nature Neuroscience Jul; **2**(7): 583–4, © 1999.*

Bob Crimi

challenges associated with the field of bioinstrumentation. Brain–machine interfaces refer to devices that are designed to translate raw neuronal signals into motor commands; they have potential application to restore limb mobility in paralyzed patients and to provide a way to precisely control robotic prosthetic limbs with brain-derived signals. Several significant bioengineering challenges remain before such systems are a reality. A successful brain–machine interface requires a number of integrated components: (1) a fully implantable biocompatible recording device is needed to monitor electrical output of the appropriate groups of neurons; (2) real time computational algorithms are needed to interpret recorded neuronal output and trigger the appropriate response in the device to be controlled; (3) mechanisms to provide the brain with sensory feedback from the device are required; and (4) prostheses which can be controlled by brain derived signals must be developed [10].

Recent progress in this area has demonstrated the potential benefits which can result from solving these challenges. In 1999, researchers at the University of Pennsylvania and Duke University demonstrated that recordings made from ensembles of cortical neurons in rats could be used to control a robotic arm [11]. The animals were first trained to press a bar to move a robotic arm to get a drop of water. Electrodes placed in the animals' brains were used to record electrical signals from groups of 100–400 neurons; neuronal output was recorded as the animals moved the robotic arm [10, 12]. A computational program was developed to sum the electrical activity recorded from these neu-

rons; researchers found that the resulting neuronal signal could be used to drive the arm to retrieve the water. The researchers then switched control of the robotic arm from the bar that the animals pressed to the neuronal signal recorded from their brain. Most rats continued to operate the arm successfully; interestingly, the researchers found that the physical bar-pressing movement dropped off in these animals and that they used just the signals measured from their brain to control the arm [12]. The engineers hypothesized that this approach could eventually be used to restore limb movement in paralysis patients (Figure 7.14).

In order to achieve this clinical goal, several types of brain–machine interfaces have been developed. These approaches first differ in the way that signals are recorded from the brain. A major distinction is whether the system uses invasive intracranial recordings of brain electrical signals or relies on noninvasive recordings such as EEGs. EEG methods do not expose patients to the risks of brain surgery, but they provide limited bandwidth information, and can transmit only 5–25 bits per second [10, 13].

Better results can be obtained with electrodes that are implanted beneath the dura which covers the brain. Subdural electrodes can record the electrical activity of smaller groups of neurons. In order to achieve the full potential of such interfaces, current research is focused on using the ability of the brain to remodel neuronal connections (brain plasticity) to enable patients to incorporate a brain controlled prosthetic device into their own mental representation of their body (Figure 7.15). In this way, the prosthetic device can

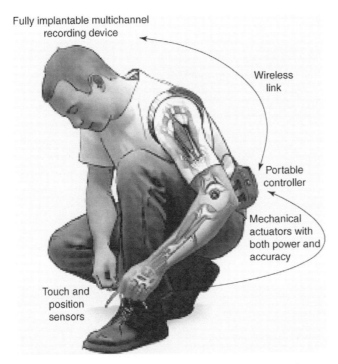

Fully implantable multichannel recording device

Wireless link

Portable controller

Mechanical actuators with both power and accuracy

Touch and position sensors

Figure 7.15. *A prosthetic system using a brain–machine interface. From [10]. Reprinted from* Trends in Neurosciences, *with permission from Elsevier.*

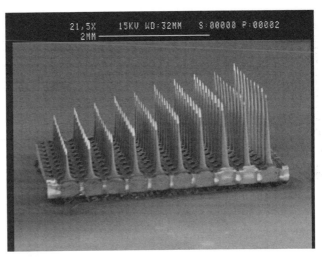

Figure 7.16. *Micrograph of implantable microelectrodes which are part of a brain–machine interface. Courtesy of Richard A. Normann, Ph.D., University of Utah.*

actually feel like the patient's own limb. A second key challenge is to develop microelectrodes which can be implanted for long periods of time. Current microelectrodes can make recordings for several months; however, the body frequently responds to the implant by making surrounding fibrous tissue [10].

Another important focus in the field of bioinstrumentation is to develop parallel arrays of sensors that can provide the opportunity to simultaneously monitor many different biological signals (Figure 7.16). Flow cytometry is an excellent example of a high throughput biosensor which is used both for patient care and biological and medical research. A flow cytometer is used to characterize the properties of each individual cell in a population of hundreds of millions of cells and to sort cells based on properties of interest. Cells to be analyzed and sorted are usually stained with a fluorescent dye that is targeted to an antibody which binds to a cell surface marker of interest. Labeled cells in suspension are loaded into the flow cytometer and made to pass single file in a line past a focused laser beam (Figure 7.17a,b). Some of the incident laser light is scattered by the cell in a way that depends on the cell size; if

the fluorescent dye binds to the cell, some of the incident laser light is also converted to fluorescence (Figure 7.18). Sensitive detectors record the intensity of scattered and fluorescent light at multiple wavelengths. If desired, a flow cytometer can also be used to sort cells based on the measured optical characteristics. After the cell passes through the laser beam it is given a net charge; cells which exhibit the desired characteristics are given a positive charge, and those which do not are given a net negative charge. An electrical field is then used to sort cells based on the desired characteristics. State-of-the-art flow cytometers can detect light at 18 different colors to simultaneously monitor the presence of a number of different markers of interest. Typically, flow cytometers can detect very minute amounts of fluorescent tagging and can sort cells with a purity of >99% [14].

Flow cytometry is an important tool for monitoring patients with HIV. HIV infects and kills certain types of white blood cells called CD4 lymphocytes. The number of CD4 lymphocytes (CD4 count) is critical to determine the clinical stage of HIV infection, to evaluate whether treatment is working and to determine when medications need to be changed. The CD4 count is measured in a flow cytometer. White blood cells are stained with a fluorescent antibody targeted against the CD4 surface marker in order to quantify the number of CD4 cells present in clinical samples. The cost of a flow

(a)

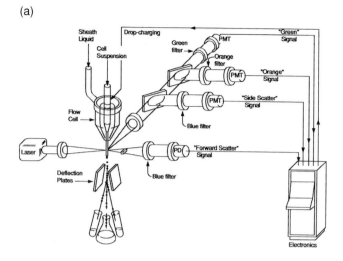

(b)

Figure 7.17. *(a) Schematic of how a flow cytometer works. Reprinted with permission of John Wiley & Sons, Inc. Flow Cytometry: First Principles; Alice L. Givan; Copyright © 1992 Wiley-Liss, Inc. (b) A flow cytometry system. Courtesy of the Center for Flow Cytometry, Czech Republic.*

(a)

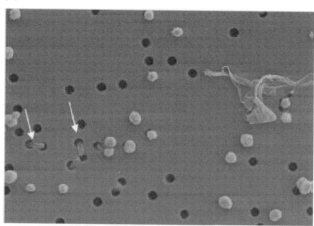

(b)

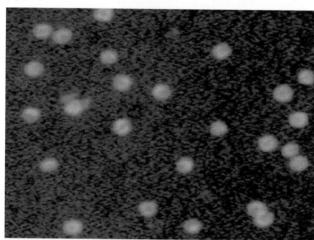

Figure 7.18. *(a) Whole blood sample processed through flow cell; (b) image of fluorescently stained CD4 lymphocytes. Source: Rodriguez et al. Plos Med. 2005. **Z** (7): el82.*

cytometer ranges from about $30,000 to $150,000; they are not available in many low-resource settings because of their high cost. As a result, many patients with HIV or AIDS do not currently have access to this important test [15].

Advances in microelectronics technology provide an exciting opportunity to reduce the cost of high throughput biosensors. Using microfabrication techniques, a number of lab-on-a-chip systems have been developed to carry out chemical analyses with pocket sized equipment. McDevitt and colleagues at the University of Texas have developed a microchip to rapidly quantify

CD4 lymphocytes at substantially reduced cost compared to flow cytometry [15]. Whole blood from the patient is introduced into a small flow cell; white blood cells are captured on a membrane which excludes red blood cells. A fluorescent antibody is used to stain CD4 lymphocytes, which are imaged using a color camera. Image processing algorithms are used to automatically identify and count the number of CD4 lymphocytes present.

Such sensors may have great applicability to improve healthcare in the developing world. In low resource settings, provision of laboratory services is frequently difficult because there may be limited access to running water or electricity and ambient temperature and humidity can fluctuate widely. In addition, consumable reagents required for diagnostic tests may frequently be unavailable. Even when laboratory facilities are available, there is often a lack of trained laboratory personnel in many developing countries [16]. Thus, there is an important need for simple, low cost techniques to perform diagnostic tests such as blood chemistries, immunoassays, and flow cytometry in low resource settings.

Disposable immunoassay tests, which use inexpensive components, can be mass produced, and are relatively affordable, have been successfully used in many developing countries. A disposable immunoassay test consists of a nitrocellulose membrane strip containing all of the dried reagents necessary to test for the presence of an antigen in a small amount of liquid sample solution (usually urine, blood or saliva). The strip test contains three different regions: (1) the sample pad contains tiny spheres made of gold or colored latex which are coated with antibodies that bind to the target antigen to be detected, (2) the test line contains a line of physically immobilized antibodies which will bind to the antigen to be tested, and (3) the control line contains a line of physically immobilized antibodies which will bind the antibody present on the surface of the spheres in the sample pad (Figure 7.19) [16].

In order to perform the test, the liquid sample is applied to the sample pad at the end of the test strip. This solubilizes the colored spheres coated with antibodies stored on the test strip. If the antigen is present in the sample, it binds to the solubilized antibody present on the surface of these colored spheres. The solution moves through the membrane by capillary action. As the solution passes the test line, a color change will occur only if the target antigen is present. If antigen is present, some of the complexes of antigen and labeled antibodies will bind to the immobilized antibody at the test line; essentially, the antigen acts as a "sandwich" to link the immobilized test antibody and the labeled antibody. The aggregation of spheres results in a color change at the test line location. If no antigen is present, then no color change occurs at the test line. As the solution passes the control line, some labeled

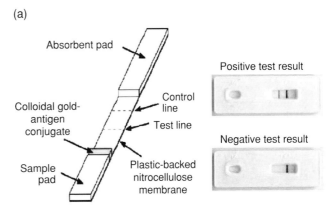

(a)

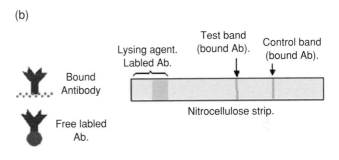

(b)

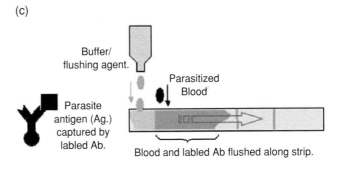

(c)

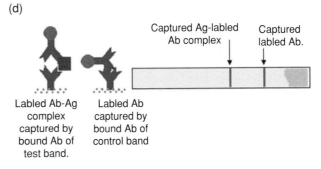

(d)

Figure 7.19. *(a) A disposable immunnoassay test, or strip test. Reprinted by permission from Macmillan Publishers Ltd: Nature* **442***: 412–18, © 2006. (b)–(d) Mechanism behind malaria rapid diagnostic test. Courtesy of WHO,* http://www.wpro.who.int/sites/rdt/whatis/mechanism.htm

**MCH Lab: June 26, 2007
Kim Malawi**

The district hospital lab was an interesting experience. I got several good ideas for possible design projects, and I learned a lot about the limitations that they're working under here. Tests that we take for granted in the States aren't possible here on a regular basis, and some are not possible at all.

The blood chemistry analyzer is out of order (and has been for months) so all of those kinds of tests are impossible. No cell counts. No enzyme levels. No basic diagnostic tests like those!

The automatic hemoglobin reader was also out of order. They were doing hemoglobin measurements by taking a hematocrit (they fill a capillary tube with blood, spin it in this special centrifuge, then use a device with an arm that you point at the division of plasma and blood and it gives you the packed cell percent. They then divide that number by three to get the Hb.)

They were using a glucometer (like the over-the-counter ones in the States for diabetics) to do blood and CSF glucose levels. For now, this is working. But when they run out of the proprietary test strips, they'll be out of luck on glucose tests.

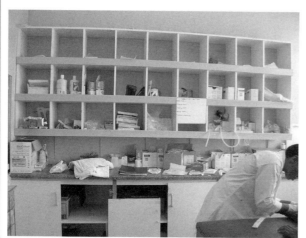

They don't have a histology department anymore because the pathologist left.

They do the "heat until the fluid begins to vaporize" step of the TB stain procedure by lighting a piece of cotton wool on fire and holding it with tongs. So not safe. (I watched the guy nearly light his sleeve.)

There is one person who spends his entire day reading malaria blood smears. (Most are negative, as it turns out.)

I'm interested now in seeing a rural health center (if I can) when we go out to Rhumpi or Chitipa, because those are where the bulk of health care in Malawi actually happens and Ellie says they are woefully underfunded and understaffed. She says it will be even less capable of conducting basic tests than the MCH lab. What a thought.

antibodies directly bind to the immobilized capture antibodies; the aggregation of spheres results in a color change. A color change at the control line indicates that reagent has passed this line and confirms that the antibodies present on the test strip are functional. Results are usually available in 5–15 minutes, and the test is read by looking for a change in color at the test line; quality control is ensured by looking for a change in color at the control line. Test sensitivity can be very high. For example, rapid diagnostic tests for the hepatitis B surface antigen can measure as little as 1.0 ng of antigen per ml of blood [17]. Typically, test strips are stable for months when properly protected from moisture and excessive heat. Such tests are routinely used to screen individuals for HIV infection in many developing countries. These disposable tests provide a yes/no answer and the test can be performed accurately by personnel with minimal training.

In some cases it is desirable to quantify the amount of antigen present (e.g. malaria) and in these cases a test which provides a quantitative or semi-quantitative result is necessary. To address this need, many researchers are developing point-of-care diagnostic systems that use a disposable card in conjunction with a low cost reader apparatus [16]. The sample is applied to the disposable card, which contains any necessary consumable reagents and calibration supplies, and retains waste materials; a quantitative result is obtained by inserting the card into the reader.

Figure 7.20 shows an example of a prototype device to measure small molecule analytes and drug metabolites in saliva developed at the University of Washington. Filtered saliva is placed at the sample port; the sample is transported through a series of microfluidic channels where it is separated, labeled with antibody, and transported to assay channels, which are then interpreted

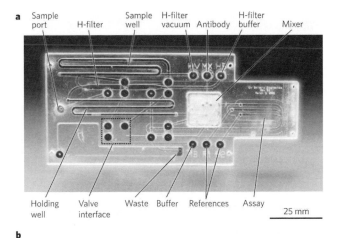

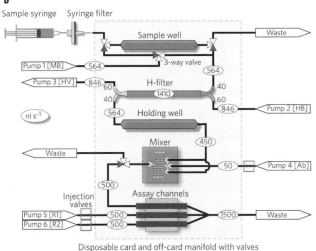

Disposable card and off-card manifold with valves

Figure 7.20. *Lab-on-a-chip technology using microfluidics to rapidly detect various small molecules and metabolites. Reprinted by permission from Macmillan Publishers Ltd: Nature,* **442**: 412–18, © 2006.

in the reader [16]. A similar approach has been used to develop a point-of-care diagnostic tool to identify pathogens responsible for enteric infections, which kill more than three million children each year, mostly in the developing world [16]. Tools used in developed countries (stool culture, enzyme immunoassay, and PCR), are not available in most laboratories in developing countries. As an alternative, a disposable card has been developed to identify the pathogens *Shigella dysenteriae* type 1, *Escherichia coli* (O157:H7), *Campylobacter jejuni* and *Salmonella*. A swab containing a stool specimen is inserted into the card. The card contains four microfluidic circuits which (1) capture and lyse the organism, (2) capture its nucleic acid, (3) amplify the nucleic acid, and (4) produce visual detection of the amplified prod-

ucts. Complete analysis takes less than 30 minutes and the cost is between $1 and $5 per disposable test [16].

The field of bioinstrumentation offers many opportunities to develop new medical sensors and implants, and devices to improve laboratory diagnostics in a wide variety of settings. Usually, new biomedical devices and sensors are developed by teams of bioengineers, electrical engineers, neuroscientists, materials scientists and chemists, working together with clinicians or specialists in laboratory medicine who understand the design requirements and clinical needs of their field. Such multi-disciplinary collaboration can rapidly lead to new prototype devices for clinical testing.

Biomechanics

The cells and tissues in our bodies are continuously exposed to a wide variety of mechanical forces. When we walk, muscles exert tensile forces which are transmitted through tendons to act on bones and move joints. Walking and running generate substantial compressive loads on cartilage and bones. As our heart pumps blood, changes in blood pressure generate hoop stress which causes blood vessels to dilate cyclically. The frictional forces associated with blood flowing past the vessel wall produce shear stress. The field of biomechanics is concerned with the study of mechanical forces in living systems and the use of engineering design to create prosthetic devices and tools for rehabilitation.

Understanding the complex interactions between the skeletal system, the muscular system and the nervous system required to produce coordinated movement is a challenging task. Experimental measurements during movement, such as the use of high speed cameras to track changes in the positions and orientations of body segments during motor tasks, coupled with surface electrodes to record the sequence and timing of muscle activity, have contributed greatly to our understanding of biomechanics. While these measurements can reveal data important to understand the kinematics and dynamics of body segment movement, they don't explain how muscles work together at each instant during motor tasks. Over the past decade, large scale computational models have been developed to produce realistic simulations of movement that include a model of

Profiles of translational innovation – Emil J. Freireich MD, Sc.D

Emil J. Freireich, a founding father in the field of clinical cancer research, started his career intending to become a family physician. As he tells the story, "I grew up in the depression and my role model was my doctor because he was the only male that wore a tie, looked dignified and was educated. Everyone else was just digging holes and working for the WPA." As fate would have it, every turning point in his career inadvertently led him away from family medicine toward the field of oncology.

In 1955, Dr. Freireich (image courtesy of Emil J. Freireich MD, Sc.D., The University of Texas MD Anderson Cancer Center) became a Public Health Service Officer at the newly opened National Institutes of Health (NIH) in Bethesda, Maryland. As he explains, the physical layout of the NIH facility was revolutionary for its time. "It was designed to expedite the interaction between the basic sciences and the clinic . . . the [patient] wards were separated from the laboratories by only a service corridor. There is no hospital like it in the world."

Upon the suggestion of Dr. Gordon Zubrod, the Medical Director of the cancer institute, Dr. Freireich dedicated his efforts to finding a cure for leukemia, a cancer of the blood cells. In the early 1950s a diagnosis of leukemia was essentially a death sentence, with patients succumbing to massive hemorrhage and infection. Dr. Freireich recounts the reaction of Dr. Zubrod upon making rounds of the leukemia service. "He said to me, you know Freireich, your ward is a big mess. These children are really suffering. There's blood everywhere! There's blood on the pillows, on the sheets, the nurses are covered in blood, you're covered in blood, it looks like a butcher shop. . . . You're a hematologist Freireich, why don't you do something about this bleeding?"

Accepting the challenge, Dr. Freireich set out to discover the cause of the bleeding. Examining the clinical records of his patients over a period of several years, he identified a quantitative relationship between platelets, a component of the blood, and a propensity to hemorrhage. The lower a patient's platelet count dropped, the greater the frequency and severity of hemorrhage. This groundbreaking work led him to hypothesize that transfusions of platelets could stop patients from hemorrhaging. Platelets were collected from donors using plasmapheresis, a technique whereby red blood cells are separated from the platelet rich plasma through centrifugation and returned to the donor. "We did a study where we matched up one donor with one child, [transfused] two units a week and could maintain them hemorrhage free for months." Following implementation of a prophylactic platelet transfusion program, hemorrhage was controlled. "After that, we didn't allow any blood on our ward. If there was blood on a pillow, we asked why didn't this patient get platelets."

Despite the success, patients were still at risk for developing life threatening infections. Inspired by their work with platelets, Dr. Freireich and his colleagues questioned whether they could take the same approach with transfusions of leukocytes (white blood cells). Whereas platelets could be easily harvested from a donor's blood, leukocytes proved more difficult for two reasons. First, normal healthy adults typically have very low numbers of leukocytes circulating in the blood at any given time. Second, leukocytes have density values very close to those of red blood cells, making it difficult to separate the cells from one another using simple centrifugation.

Initially these problems were overcome by using donors with a condition called chronic myelogenous leukemia or CML. As Dr. Freireich explains, "We had patients that had chronic myelogenous leukemia and that disease is characterized by having neutrophil [leukocyte] counts two hundred times normal. That's their disease. These neutrophils are not normal, they are leukemic neutrophils, but in the laboratory, functionally, they are about half as good as a granulocyte. . . . What if we collected granulocytes from donors with CML and transfused them into children?" Using this approach, Dr. Freireich demonstrated that infection in patients with acute leukemia could be controlled with daily transfusions of leukocytes.

While this approach proved successful, it was not sustainable on a large scale. They needed a device that could collect blood from normal, healthy donors, selectively harvest the leukocytes, and return the red cells and plasma. Dr. Freireich went back to the lab. "I tried to use capillary flow and I had tubes all around my lab. Then I tried electromagnetic things, I tried different charges, I tried electropheresis, but all these techniques are cumbersome and slow. The only thing that was fast was the centrifuge." While sitting in *his lab one day, Dr. Freireich was approached by George Judson, an IBM engineer whose son had come to the NIH for leukemia treatment. Judson wanted to help save his son's life and joined Dr. Freireich in his quest to create a device that would separate blood according to its components. "One day he [Judson] appeared in my office with a pile of junk on a cart. He went to the IBM storeroom and found rejected pieces of plastic and screws and bolts and he actually built a centrifuge. He couldn't test it, but it had all the ideas." The initial design was relatively simple, relying on centrifugal force to separate the blood cells according to density and gravity to collect the components into separate containers.*

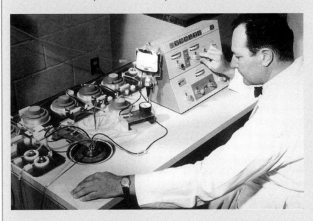

They tested the device in the lab using rejected blood, tweaking the device and adjusting the speed to achieve separation. Several iterations later, they were ready to test on a patient. "We wanted to do a CML patient because that would be easy. . . . We drew one unit of blood into the machine, separated it, and put nothing back, so it was perfectly safe and we showed it worked. The next step was to collect two units, separate it, and put one unit back." In this manner, they worked their way up to a fully functioning device that came to be known as the continuous flow blood cell separator. (The continuous flow blood separators found in blood banks around the world are based on the original device shown in the photograph.

Courtesy of NCI/G. Terry Sharrer, Ph.D., National Museum of History.) In 1965, the National Cancer Institute partnered with IBM to commercially produce the device to which Dr. Freireich holds the patent. It has since become a fixture in blood banks around the world.

In just ten years, Dr. Freireich developed strategies to control hemorrhage and infection, revolutionizing the treatment of acute leukemia. For the first time in history, children diagnosed with leukemia had a fighting chance of survival. In light of Dr. Freireich's commitment to educating the next generation of physicians and scientists, he offers a piece of advice. "Don't think small . . . think about big things. It's like hemorrhage or death or cancer or diabetes . . . the moon or space travel or energy. Young people have to think big, they have to tackle big problems. You have to be fearless."

the skeleton, the muscle paths, musculo-tendon actuation, and excitation contraction coupling between the nervous system and the muscular system (Figure 7.21). As more detailed models are developed and validated, they have the potential to evaluate planned surgical procedures designed to correct gait abnormalities in patients who have experienced stroke or have cerebral palsy [18].

In addition to understanding motion, the field of biomechanics is also concerned with the way in which tissues respond to mechanical forces. It has been known for many years that tissues respond to mechanical forces. Over the past decade, it has become increasingly clear that cells themselves are exquisitely sensitive to mechanical forces and that changes in tissue structure that occur in response to mechanical forces begin with cellular changes. Mechanical forces can initiate changes in gene expression which lead to protein synthesis, cell growth, death and differentiation. During normal growth and development, this mechanosensitivity is important to maintaining tissue homeostasis. For example, osteoblasts in bone are responsible for bone formation. They secrete proteins which make up the bone matrix. However, abnormal loading conditions can alter cell function and change the structure and composition of the extracellular matrix and produce pathologies such as osteoporosis, osteoarthritis, atherosclerosis, and fibrosis. The endothelial cells which line blood vessels secret matrix products as well as enzymes which break down structural proteins present in matrix. The endothelial cell response to abnormal forces in high blood pressure is an important component in the development of atherosclerosis, which can lead to heart attack [19].

Muscle Excitations

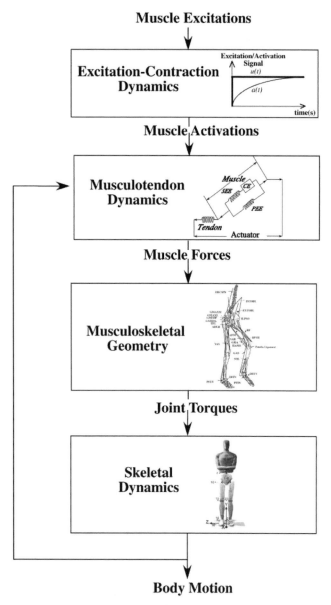

Figure 7.21. *Advances in modeling the movement of the human body have revolutionized the field of biomechanics. Used with permission from [18]. Reprinted, with permission, from the* Annual Review of Biomedical Engineering, *Volume* **3** *© 2001 by Annual Reviews www.annualreviews.org.*

Atherosclerosis illustrates the interplay between biomechanics and mechanobiology. As we will see in Chapter 12, atherosclerotic blockages in the coronary arteries that supply blood to the heart can lead to heart attack. Mechanical interactions play an important role in the development of these blockages – whether they produce symptoms and whether they rupture and lead to a heart attack. An important treatment of atherosclerosis is the use of a stent to restore blood flow through a

constricted artery. A stent is a tubular scaffold made of metal which is inserted into the blood vessel to dilate the artery and restore flow; the stent must provide sufficient radial strength to hold the artery open. However, a stent can subject the artery to abnormally high stresses; these can lead to undesirable biological responses that cause restenosis and treatment failure. Biomechanical simulations can be used to investigate the effects of changing the geometry of a stent on the resulting arterial stress. Stents consist of concentric rings of sinusoid like curves connected by straight struts of varying length. Figure 7.22a shows the main parameters of a stent which can be varied: the spacing between struts in the stent, the radius of curvature of the small bends in each strut, and the height of these small bends. Figure 7.22b shows several generic stents that were designed by varying these parameters. Results of finite element models of stented arteries (Figure 7.22c) show that stent designs which incorporate large axial strut spacing, large radius of curvature, and high amplitudes will expose the smallest arterial segment to high stress and will still maintain sufficient blood flow [20].

Biomechanics bridges the fields of biology and mechanical engineering; work in this area provides opportunities to integrate experimental studies that operate across many scales from the molecular to cellular to tissue level. Progress in this area is dependent on close interaction between multi-scale modeling efforts and experimentation. The results of such work promises to help understand basic physiologic processes, such as development, as well as important patho-physiologies.

Biomaterials and drug delivery

The successful ability to implant biomedical devices and artificial tissues has revolutionized the treatment of many diseases, ranging from replacement heart valves to treat children with congenital heart disease, to artificial hip replacements to treat patients suffering from osteoarthritis, to surgically implantable polymer wafers which slowly release chemotherapy drugs to treat patients suffering from inoperable brain tumors. Each year, more than 200,000 pacemakers, 100,000 heart valves, one million orthopedic devices, and five million intraocular lenses are implanted into patients

(a)

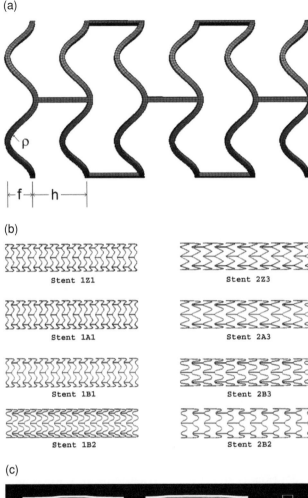

(b)

(c)

Figure 7.22. *(a) Stent parameters; (b) simulated stents; (c) hoop stress in stented arteries. Stress is lowest with design 2B3, which incorporates large strut length, large radius of curvature, and large amplitude [20].*

restore body function and come into contact with body fluids. In designing new biomaterials, it is important to understand both the chemical and mechanical requirements of the material. Because body chemistry is highly corrosive, and implanted materials may have to undergo many loading cycles per day, this is a particular challenge. In addition, the design of biomaterials must take into account the interactions which will occur between the implanted materials and the surrounding tissue. Host responses, such as immune reactions, inflammation, wound healing, infection and tumor generation, can all occur in response to an implanted device. Understanding and controlling these reactions are crucial to clinical success. Original attempts to develop biomaterials focused on the design of passive, inert materials. However, trials in animal models and human subjects showed that the vast majority of materials elicit some type of cellular response in the host. Instead of attempting to design biomaterials that will simply act as static implants, the interdisciplinary field of biomaterials design has today evolved to focus on the design of materials that interact with tissue in a predictable manner, with the goal of creating a controlled cellular response between the artificial material and the living tissue surrounding it [21].

The challenges associated with developing effective biomaterials depend on whether the material is to be implanted permanently or temporarily and whether it is to be implanted within the body or on the body surface. Challenges are often easier to address for extracorporeal biomaterials, such as catheters, tubing, wound dressings, and dialysis membranes, which are placed temporarily outside the body, but come into contact with body fluids and tissues. Challenges increase for temporarily implanted biomaterials, which are designed to be placed inside the body and degrade over time, e.g. degradable sutures, implantable drug delivery systems, or scaffolds for cell or tissue transplants. Some of the greatest challenges arise with permanent implants, such as cardiovascular devices, orthopedic devices, dental devices, and sensory devices, which are designed to be placed within the body and must function effectively over a period of years to decades.

worldwide; demand for implants of all kinds is growing at a rate of 5–15% each year [21].

The field of biomaterials engineering encompasses the design of any materials that are used to replace or

A number of different types of synthetic and naturally occurring materials have been used to address the challenges associated with this broad range of clinical applications. Ceramic materials, such as hydroxyapatite, calcium salts, and silicate ceramics, are often used to achieve hardness in implant surfaces such as those associated with joints or teeth. In addition, these materials can be easily bonded to bone surfaces to facilitate placement of an implant. They can also be used as the foundations of bone scaffolding materials in tissue engineering, where they can be manufactured to degrade at controlled rates. Metals, such as titanium and stainless steel, are frequently used in implants that function in load bearing applications such as walking or chewing. Finally, the use of polymers provides the ability to design implants that have both flexibility and stability, and can be used in the design of articulating surfaces which generate low friction. A number of synthetic biodegradable polymers are available (e.g. poly(glycolic acid), poly(ethylene glycol)), but biomaterials engineers have also derived polymers from natural sources, such as modified polysaccharides, or modified proteins [21].

Recently, a new class of biomedical implants – the drug eluting stent – was developed to improve the treatment for cardiovascular disease. Most stents are made of stainless steel or nitinol (an alloy of nickel and titanium) [22]. Computational models have been developed to predict the mechanical interactions between stent and artery wall, and we have seen how such models have been used to design stents which have a geometry that minimizes risk of clot formation and restenosis from the mechanical perspective [20]. However, this approach does not address the long-term biological response to the implanted stent. The biggest problem following placement of bare metal stents is restenosis of the vessel. Restenosis occurs as a result of trauma to the vessel wall induced by deployment of the stent; this trauma causes an aggressive healing response which overgrows the stent lumen [22].

Biomaterials engineers have developed special drug eluting stents which slowly release the drugs that inhibit this healing response. Drug eluting stents consist of a metal stent coated with a drug releasing polymer; together, this combination of materials provides both mechanical function as well as biological function. Early clinical results indicate that this new implant substantially reduces the rate of restenosis in patients treated for coronary artery disease.

The development of drug eluting stents illustrates the challenges that must be addressed in engineering new biomaterials. This design problem required an understanding of the biology of restenosis, development of drugs that target one or more pathways in the restenosis process, as well as the development of a stent as a controlled delivery platform for release of drug. Restenosis occurs due to a complex cascade of events, which may include blood clot formation, inflammatory response, vascular smooth muscle cell proliferation, and synthesis of extracellular matrix. Drugs which reduce the rate of restenosis have been identified and include compounds which suppress the patient's immune response, reduce cellular proliferation, and reduce inflammatory response. When these drugs are given systemically, typically they do not provide the desired effect. The challenge is that they are needed at high concentration only at the site of the stent. Biomaterials engineers focused on developing stents which release the drug where it is needed. Stents coated with drug-loaded polymers can be used to provide release of drug at the site and time of injury with minimal systemic toxicity. Drug eluting stents are coated with a drug-loaded polymer matrix which sustains drug release for up to four weeks following stent placement. The drug-releasing polymer coating is designed to be biologically inert and sterilizable and to be sufficiently flexible to follow changes in stent shape upon deployment [22].

The field of biomaterials and drug delivery integrates advances in materials science, polymer chemistry, pharmacology, and immunology. Implantable materials and drug delivery systems have the potential to impact a wide variety of disease ranging from improved therapy for advanced cancers to the development of systems to monitor and regulate blood glucose levels in diabetic patients. This field is characterized by multidisciplinary studies bridging engineering, chemistry and biology.

Profiles of translational innovation – Robert Langer, Sc.D.

Courtessy of Stu Rosner Photograhpy.

Robert Langer, Sc.D. *is a chemical engineer by training, but has taken a different road than most. In search of something more meaningful, Langer accepted a postdoctoral position with cancer researcher Judah Folkman, MD, at Children's Hospital, in Boston. It was during this time that he began his revolutionary work in drug delivery. Today, he is a professor of chemical and biomedical engineering at the Massachusetts Institute of Technology where, together with physicians and researchers, he pushes the frontiers of biotechnology and materials science.*

Langer initially faced great skepticism and a general negative reaction from the science community upon proposing the use of polymers for the slow release of large molecules in a controlled manner.

"I was very discouraged, but I just kept plugging. You write papers, you do talks, you do more experiments to convince the skeptics" [23].

However, three decades later Dr. Langer is distinguished among the most successful and renowned scientists in the biomedical engineering field. He has been one of the key pioneers in the field that paved the way for exciting new research in the areas of biomaterials and drug-delivery.

"When I started doing this in 1974 there were almost no engineers working in medicine" [23].

When Langer began his work, researchers used off-the-shelf polymers for medical purposes. For example, the plastic used in women's girdles was used in the first artificial heart because of their elastic properties and breast implants were initially filled with mattress stuffing.

"This type of approach has often led to a number of problems. For example, when blood hits the surface of the artificial heart, a clot may form and the patient may suffer a stroke. We were interested in creating biomaterials that would have the right properties from the engineering, chemistry, and biological standpoint, and then synthesize them from first principles" [24].

It took a novel approach and Dr. Langer's relentless effort to reach beyond this initial approach of biomaterials and ignite the field of biomedical engineering. Today, the fruits of Dr. Langer's work can be seen globally, from nicotine patches to the stents currently implanted in cardiac patients. The chemotherapy wafers he developed with Dr. Henry Brem of Johns Hopkins have lead to the first new brain cancer treatment approved by the FDA in over twenty years. Artificial skin based on his research has been approved by the FDA and his work in artificial cartilage, bone, corneas blood vessels, spines and vocal cords is under development and appears promising. Even though Dr. Langer has nearly 550 issued or pending patents and is one of 13 Institute Professors at MIT (MIT's highest honor) he continues to discover new treks to pursue in biomedical engineering.

"Some of the newer things we're looking at are solving problems with the delivery of genes, DNA, RNAi. Then there are remote control delivery, microchips, smart systems you can control- or maybe that you wouldn't have to control" [25].

And as for the future of biomedical research, "One of the most important contributions we've made is to train the next generation. We're bringing biomedical engineering to a different place and I think the contributions of those people will be ever greater as the years go on" [25].

Tissue engineering and regenerative medicine

Over the past 50 years, advances in surgical techniques and the discovery of new immunosuppressive drugs have resulted in the development of organ transplantation as a tool to successfully treat end-stage organ failure of the kidney, liver, heart, lung, pancreas and intestine [26]. While organ transplantation has reduced mortality associated with end-stage organ failure, access to this procedure varies widely throughout the world due to a combination of economic factors, availability of donor tissue, and access to specialist care. In the USA alone, more than 96,000 patients are waiting for transplant surgeries and 17 people die each day due to a shortage of donor organs [27, 28]. The field of tissue

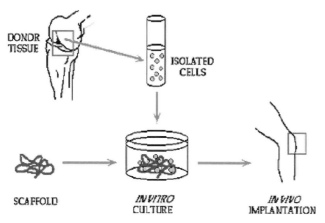

Figure 7.23. *Cells from donor tissue are extracted, isolated, and placed on a porous biodegradeable scaffold that directs the growth of new cells. Once implanted into the body, the scaffold eventually disintegrates. Used with permission of Christine Schmidt, Ph.D., University of Texas at Austin.*

engineering aims to integrate advances in engineering and life sciences to address this global need by developing living functional constructs that can be used to replace or regenerate damaged or diseased tissues in a less expensive and invasive manner.

The basic components used to engineer replacement tissues include: (1) cells to initiate development of new tissue (Figure 7.23a), (2) scaffolds to guide the three dimensional development of tissue (Figure 7.23b), and (3) signals to coordinate cell growth and differentiation in space and time [29]. The goal of tissue engineering is to integrate these three components in a manner that promotes the development and growth of functional, three dimensional tissues.

One strategy that has been successfully used to engineer tissues is to harvest cells from the intended recipient, expand these cells in culture outside the body, and then seed them onto a scaffold that drives formation of tissue until the cells can make their own supporting matrix [29]. Engineers have explored the use of scaffolds that are resorbed as new tissue grows as well as permanent scaffolds. In any case, scaffolds must be biocompatible, and must provide a template for tissue growth in three dimensions. Since the phenotype of cells is very dependent upon their microenvironment, frequently engineers strive to design scaffolds which replicate the microenvironment in which cells would

naturally grow [29]. Consideration must be given to both the biomechanical and biochemical components of the microenvironment – a structure which provides the appropriate temporal and spatial sequence of signals dictating cell growth and differentiation is needed to yield a tissue that can survive and provide the appropriate biological function after implantation [29]. Scaffolds can be made of biological materials, synthetic materials, or hybrid biomaterials, so long as the scaffold provides cells with the right cues so that they eventually synthesize and secrete their own matrix. New advances in tissue engineering are integrating advances in nano-structured composite materials with advances in drug delivery to provide targeted delivery of signaling molecules such as peptides, proteins and plasmid DNA and ensure repair of tissues in a timely manner.

Several clinical and commercial successes have been reported in the field of tissue engineering. A number of tissue engineered skin substitutes have been brought to market in the last decade to treat burns, chronic ulcers, surgical wounds, and other dermatologic conditions (Figure 7.24) [31]. Apligraf is a two-layer tissue engineered construct which mimics the structure of human skin. The bottom layer, the equivalent of the human dermis, is derived from neonatal foreskin fibroblasts in a contracted type I collagen matrix. The top layer, the equivalent of the human epidermis, is generated from keratinocytes seeded onto the bottom layer. The cells in the construct do not survive long term after implantation, but instead encourage ingrowth of the patient's own cells [31].

More recently, clinical trials have been carried out to transplant tissue engineered internal organs into patients. Anthony Atala at Wake Forest University School of Medicine led a team of engineers and clinicians who engineered a human bladder (Figure 7.25) and successfully transplanted it into patients. The team harvested about 1 million urothelial cells and muscle progenitor cells from bladder biopsies of seven children with malfunctioning bladders as a result of spina bifida. Muscle cells were seeded onto dome-shaped, biodegradable molds of synthetic polymer and collagen, while bladder urothelial cells were seeded onto the interior surface. The constructs were grown in culture

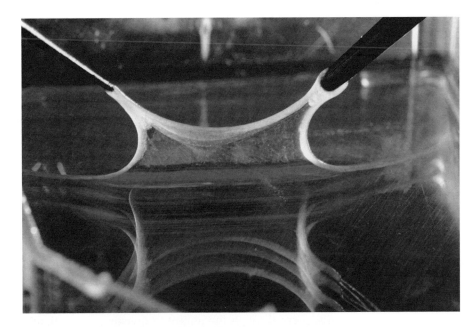

Figure 7.24. *Tissue engineered skin. Phototake © Jean Claude Revy, Phototake.*

Figure 7.25. *Tissue engineered bladder. Courtesy of Wake Forest Institute of Regenerative Medicine.*

for seven weeks, expanding the cells to about 1.5 billion in number, and then sewn to the patient's own bladder. After implantation, the increased bladder capacity reduced the risk of long term kidney damage in this group of patients [32].

The design of engineered tissues must also take into account the further remodeling which will take place following implantation. Once tissue thickness exceeds several hundred microns, it must be vascularized in order for cells to receive adequate oxygenation [33]. The development of appropriate vascularization and innervation are significant challenges currently faced by the field.

The recent development of techniques to isolate, culture, and differentiate embryonic and adult stem cells has provided considerable excitement in the field of tissue engineering. Stem cells are undifferentiated cells that can proliferate, self-renew, and differentiate to one or more types of specialized cells when grown under appropriate conditions. Adult stem cells are rare undifferentiated cells found among the more common differentiated cells in a tissue that can renew themselves [29]. Adult stem cells are the basis for natural pathways of tissue maintenance and repair, and targeted activation of adult stem cells can turn on the body's natural repair mechanisms. Embryonic stem cells are the most plastic cell source; they are totipotent – capable of differentiating to all cell lineages. Much research in the field of tissue engineering is now focused on how stem cells can be used as a source of cells in engineered tissues to reduce issues of immunogenicity and to increase the complexity of engineered tissues [29].

With current clinical successes and advances in tissue engineering and regenerative medicine, the possibilities for the future include the development of an insulin-secreting, glucose-responsive bioartifical pancreas, the development of heart valves that can be implanted into children with congenital heart defects and can grow with the infant or child, as well as the repair or regeneration of the central nervous system [30]. The field of tissue engineering offers opportunities to combine advances in developmental biology, materials science, and engineering design in order to advance medical science.

Systems biology and physiology

Bioengineers seek to develop quantitative models of physiologic systems to help understand normal function and disease and to guide the design of therapeutic interventions. Organ level models of the cardiovascular system have elucidated the quantitative relationships between intracardiac pressure and tissue perfusion and are used to help physicians assess patients with heart disease or valve disorders and to design appropriate interventions. Mathematical models which describe coordinated nerve conduction, muscle contraction and musculoskeletal forces and motions have helped to understand normal gait as well as gait disorders [18].

Understanding the physiology of a whole organism is a complex task – organs are made up of tissues and cells and the goal of systems physiology is to describe the behavior of the system as a whole, starting with the component parts. Advances in the field of molecular biology have made tremendous progress toward understanding the molecular origins of physiology and disease. In traditional biology, the focus has been to study individual genes or proteins one at a time; however, in systems biology, the focus is to investigate the behavior and relationships of all the elements in the system and to describe the interactions among them as part of one functioning system [34].

It is now possible to rapidly assess changes in gene expression, protein expression, and protein–protein interactions in cells and tissues. These new experimental techniques have generated extremely large and complex datasets, driving the need for models to pull them together in a way that helps us understand biology at a higher level, as a complex collection of networks and pathways.

This approach is important because the causes of many common diseases are multi-factorial in nature – they are not caused by changes in a single gene or protein, and are often not treated effectively with a single drug [35]. However, using systems biology and physiology to construct models of complex biological systems and diseases may aid in understanding complex pathophysiology and may guide the search for multi-target approaches to drug therapy [34].

At the cellular level, the focus is to develop models to understand how molecular pathways and networks relate to cellular functions such as metabolism, proliferation, differentiation and migration [34]. Developing computer models of cells (sometimes referred to as silicon cells) may enable one to test out potential drugs on the computer before they are tested in animals and used in clinical trials. This may have important global health implications; for example, Westerhoff and Bakker have developed a systems biology model of the parasite that causes African sleeping sickness, *Trypanosoma brucei* [37]. Through modeling, they discovered that a glucose transporter is the best predicted drug target rather than the target which was previously under intensive study. Thus, systems biology may lead to more efficient drug discovery through the ability to quickly and efficiently carry out simulations to predict the effect of many different proposed drugs. A scientist may think that she has developed an inhibitor that will block an enzyme pathway important in controlling disease. However, these pathways are highly connected into networks. Sometimes the effect of blocking one pathway results in unanticipated effects – systems biology provides a way to anticipate these interactions without having to do extensive experimentation [36, 37].

Advances in the field of systems biology and computational bioengineering over time will lead to the ability to model increasingly complex physiological systems. Some success has been achieved to develop integrative systems physiology models of the cardiovascular system. These cell models have been integrated into large scale, anatomically detailed models of electrical conduction to investigate the molecular basis of life threatening arrhythmias [34]. The field is characterized by close interplay between experts in the areas of genetics, molecular and cell biology, statistics, and computer science.

Molecular and cellular engineering

Molecular and cellular engineering uses engineering principles to understand and construct cellular and molecular systems with useful properties. At the molecular level, proteins can be engineered to modify the communication of cells with their environment; this approach can form the basis for rational design of targeted drug therapies to treat cancer. At the cellular level, metabolic engineering can be used to design cellular factories which manufacture pharmaceuticals or scaffolds for use in tissue engineering applications, or to be used as cellular biosensors to monitor the environment for toxic chemicals.

At the molecular level, much research in this field is focused on discovering the design principles that govern the behavior of macromolecular complexes and interacting networks of proteins found within cellular organelles, molecular motors and biological membranes. Complexes of biological macromolecules form the basis of many cellular processes, including signaling, motility, metabolism, and biomolecular transport. Bioengineers are developing new computational and experimental methodologies that provide unique insights into how biological macromolecules self-assemble, interact, and function collectively. Understanding the function of these supramolecular assemblies is essential to understanding complex biomolecular processes, and will create new avenues to predict, control, and manipulate biomolecular machinery.

As an example, cells contain a wide variety of biological nanomotors (Figure 7.26): flagellar motors propel bacteria; motor proteins, such as myosin, are responsible for muscle contraction; RNA based motors enable packaging of viral nucleic acids when viruses reproduce within host cells. Kinesin is a motor protein important in organelle transport and mitosis [38]. These motors are remarkable for their efficient conversion of chemical energy into mechanical work – they operate at greater than 50% efficiency, double that of the average engine used in cars. Combining tools of molecular biology with single molecule imaging and force measurements is providing a clearer picture of how these molecular motors operate within cells; this knowledge can then be translated to harness nanomotors to power nanodevices in analytical biosensors or molecular assembly platforms.

Biological nanomotors may provide a solution to the difficulties of moving biological fluids through nanofluidic devices. Moving solutions through nanodevices is particularly challenging because the ratio of surface area

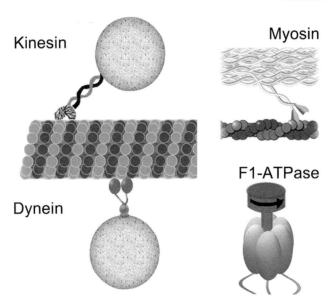

Figure 7.26. *Examples of biological nanomotors: kinesins move along microtubules to move cargo toward the periphery of a cell. Dynein moves in the opposite direction, to transport cargo to the center of the cell. Mysoin motors move along actin filaments. F1-ATPase is a rotary motor [38].*

to volume is high in these devices, dramatically increasing the effects of friction. An alternative approach is to bind the analyte of interest to a molecular shuttle (Figure 7.27) powered by a molecular motor; the motor can then be used to move the molecule of interest while leaving the bulk of the solution at rest [38]. Similarly, there is great interest in using molecular motors to direct and control macromolecular assembly. Exploiting the understanding of biological motors, efforts are underway to design networks of nanoscale conveyor belts which transport molecules and control their encounters with reaction partners in order to yield prescribed target products.

At the cellular level, engineers use quantitative tools to understand and manipulate the **network of metabolic reactions** within the cell. Using the techniques of molecular biology, it is now possible to modify specific enzyme controlled reactions within the metabolic network of a cell.

Cellular engineering refers to the improvement of cell properties through modification of specific biochemical reactions. An important goal of cellular engineering is to develop cell systems with desired properties, such

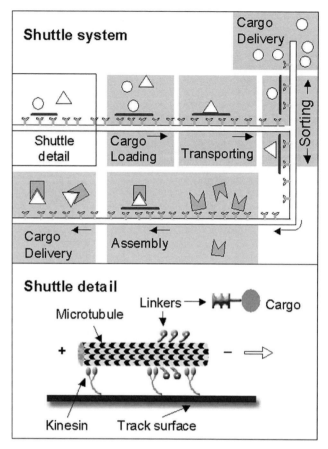

Figure 7.27. *A proposed molecular shuttle system [38].*

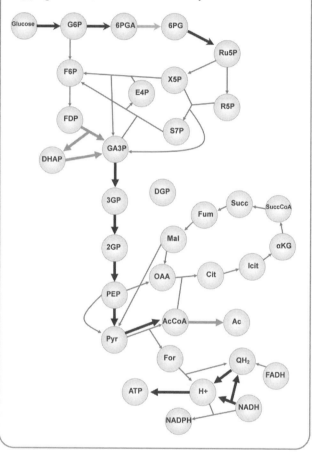

Example metabolic network model for ***Escherichia coli.*** The model incorporated data on 436 metabolic intermediates undergoing 720 possible enzyme-catalyzed reactions. Circles show abbreviated names of metabolic intermediates, and arrows represent enzymes. Heavy lines indicate links with high metabolic fluxes. From [39]. Figure copyright 2000, National Academy of Sciences USA.

as improved ability to synthesize natural products of interest, the ability to produce products that are new to the host cell, or improved ability to function in extreme environments such as hypoxic conditions [40]. The use of recombinant techniques, combined with cellular engineering, has improved the ability to engineer cells to produce protein pharmaceuticals such as insulin. There are currently more than 200 FDA approved peptide and protein pharmaceuticals. Natural sources of these compounds are often rare and expensive. Today, most of these are produced using recombinant methods in bacteria, yeast, animal cells, or plants [41].

The field of cellular engineering has also made contributions to understand and manipulate the interactions between a cell and its environment. The bi-directional communication between cells and their environment is referred to as cell signaling. Cell signaling controls many complex biological processes, such as development, tissue function, immune response, and wound healing.

In communicating with their environment, cells need to receive environmental signals, react to these signals by translating them into appropriate intracellular responses, and, if necessary, by sending an extracellular message back to the environment. Many diseases result from a breakdown in this communication: auto-immune diseases result from the failure of cells to correctly read signals (self vs foreign); in cancer, genetic mutations can hardwire a cell's signaling machinery in a pro-growth state, leading to uncontrolled growth even in absence of external signals to stimulate growth. Thus, understanding cellular signaling provides an opportunity to gain insight into a wide variety of disease processes, and the molecules important in cell signaling provide good targets for disease therapy [42].

One of the most studied cell signaling systems is the family of tyrosine kinase receptors, which includes the epidermal growth factor receptor (EGFR). These receptors mediate the signaling network that transmit extracellular signals into the cell, controlling cellular differentiation and proliferation [43]. Normally their activity is tightly regulated. Models of tyrosine kinase receptors have been instrumental in understanding both basic cell biology as well as important clinical features, such as the propensity for cancer cells to metastasize [42]. Studies of cell signaling and the tyrosine kinase receptors have led to new drugs which inhibit these receptors. Herceptin, Gleevec and Iressa are the first examples of targeted therapeutics for tyrosine kinase receptors and are used to treat advanced breast cancer, chronic myelogenous leukemia and gastro-intestinal stromal tumors, and lung cancer, respectively [43].

The field of cellular and molecular engineering is contributing both to our knowledge of how biological systems are organized and interact as well as to the development of new therapeutic molecules and ways to produce them efficiently and inexpensively. This field is characterized by collaboration between biochemists, biologists and engineers [44]; interaction between combinatorial experimental approaches and large scale computational modeling is particularly important to advances in this field.

Bioengineering and biotechnology to improve health in developing countries

Many of the technologies that we have just seen are available primarily in developed countries (recall that MRI systems cost millions of dollars and most bioengineering research requires expensive computational, instrumental, and material infrastructure). A recent panel of 28 scientific experts from around the world who are well acquainted with health problems of developing countries was asked "What do you think are the major biotechnologies that can help improve health in developing countries in the next 5–10 years?" [45]. The top ten technologies are profiled in Table 7.3. As we have

Table 7.3. *Top ten biotechnologies for improving health in developing countries [45].*

Rank	Technology
1	Modified molecular technologies for affordable, simple diagnosis of infectious diseases
2	Recombinant technologies to develop vaccines against infectious diseases
3	Technologies for more efficient drug and vaccine delivery systems
4	Technologies for environmental improvement (sanitation, clean water, bioremediation)
5	Sequencing pathogen genomes to understand their biology and to identify new anti-microbials
6	Female controlled protection against sexually transmitted disease, both with and without contraceptive effect
7	Bioinformatics to identify drug targets and to examine pathogen–host interactions
8	Genetically modified crops with increased nutrients to counter specific deficiencies
9	Recombinant technology to make therapeutic products (for example insulin, interferons) more affordable
10	Combinatorial chemistry for drug discovery

seen in this chapter, the field of bioengineering plays an important role in the development of many of these technologies, including the development of inexpensive tools for point-of-care detection of infectious disease, methods to more efficiently deliver drugs and vaccines, computational and combinatorial methods to develop new therapeutic products, and the use of recombinant techniques to produce drugs in a more cost effective manner.

Currently, most research efforts in the field of bioengineering and biotechnology are focused on health challenges faced by developed countries, and new devices are designed to meet the constraints present in laboratory and healthcare facilities in the developed world. It has been estimated that 90% of the health research dollars are spent on the health problems of 10% of the world's population [45]. Between 1975 and 1997, only 13 new chemical entities were developed for the treatment of tropical diseases [46]. Barriers which limit the development and dissemination of new technologies for the developing world include low profit margins in developing countries, lack of infrastructure, and regulatory constraints [47].

The importance of market forces increasingly plays a role in determining which potential new products receive private investment; the cost of bringing a new medicine to market in the USA has been estimated to be as high as $0.8–$1.7 billion [48]. This is a major disincentive to investment in drugs for rare diseases or those that predominantly affect the developing world. Recently, a number of new ideas have been proposed to address the failure of market forces to lead to technologies to address the health needs of developing countries. The USA has made a substantial increase in its support for biomedical research; from 1998 to 2005, the budget of the National Institutes of Health (NIH) doubled to nearly $28 billion annually; and as of 2004, the USA spent 0.25% of its GDP per year on health related research, more than double the average of other developed countries [49, 50]. However, the focus of the NIH is largely to address health priorities of importance in the USA. Some emerging economies, notably those of South Korea, China, and India, have benefited from strong increases in public investment in scientific and health-related research; for example, in South Korea, the number of health biotechnology related publications by South Korean researchers increased by tenfold from 1992 to 2002 [51]. A number of private organizations, including the Carter Center, and the Bill and Melinda Gates Foundation have made substantial contributions to research and development focused on meeting the health needs of developing countries.

Another approach to stimulate private investment in diseases that affect developing countries is for governments to guarantee markets for new technologies. In 2006, the G8 countries considered a plan to provide a guaranteed market for new vaccines that meet pre-determined safety and efficacy standards [52]. As a way to incentivize private investment in developing new vaccines, governments have agreed to provide subsidies ranging from $800 million to $6 billion depending on the disease to purchase the resulting vaccine. Once the initial subsidies have been spent, pharmaceutical companies would be required to provide vaccine to developing world customers at sharply discounted prices. A committee of experts advising the G8 recommended initially using guaranteed markets as a way to develop a vaccine for pneumococcal disease which kills more than 1.5 million people every year, many under the age of five. Guaranteed markets as a tool to encourage the development of a vaccine to prevent malaria are also of great interest [52].

In the next chapters, we will examine in detail the development of several classes of technologies designed to improve health. We will focus on new tools to treat, detect and prevent the leading causes of death throughout the world: infectious disease, heart disease and cancer. In Chapter 8, we will examine technologies to prevent infectious diseases, beginning with an overview of how our immune system protects us against disease, and then considering the steps involved in designing new vaccines to protect against disease. In Chapter 10, we will develop an understanding of the biology of cancer, and will examine the development of new technologies to detect cancers at a stage when they are still treatable, as well as technologies to prevent the development of cancer. Finally, in Chapter 12, we will examine cardiovascular physiology and pathophysiology and we will consider the engineering of technologies designed to treat heart disease as well as approaches to prevent heart disease. Along the way, we will find we need several additional tools to facilitate the translation of new technologies. In Chapter 9, we will consider the ethical guidelines which have been developed to

---

Bioengineering and Global Health Project
Project task 3: Evaluate current policy designed to develop or implement solutions to the problem
What investments is the health system in the region you have selected making to develop or implement new solutions? Are there other efforts from the private or public sectors to develop new solutions? Are there investments in basic or applied research? Are large clinical trials underway to test new solutions? What are the limitations of these approaches? Write a one page summary of current health policy regarding the health problem/region you have selected.

ensure that the rights of human subjects participating in clinical trials of new technology are adequately protected. In Chapter 11, we will learn how to assess the cost effectiveness of new interventions. In Chapter 13, we will learn how to design clinical trials and to choose a sample size to achieve statistically meaningful results.

Homework

1. Read the following abstract from an article recently published in *Nature*. Briefly explain how the steps the authors took correspond to the steps of the scientific method. All five steps are represented here.

   Letter: An unexpected cooling effect in Saturn's upper atmosphere C. G. A. Smith, A. D. Aylward, G. H. Millward, S. Miller and L. E. Moore http://www.nature.com/nature/journal/v445/ n7126/abs/nature05518.html. Reprinted with permission from Macmillan Publishers Ltd: *Nature*, Smith *et al.* An unexpected cooling effect in Saturn's upper atmosphere, **445**(7126): 399–401. Copyright 2007.

   The upper atmospheres of the four Solar System giant planets exhibit high temperatures that cannot be explained by the absorption of sunlight. In the case of Saturn the temperatures predicted by models of solar heating are 200 K, compared to temperatures of 400 K observed independently in the polar regions and at 30° latitude. This unexplained "energy crisis" represents a major gap in our understanding of these planets' atmospheres. An important candidate for the source of the missing energy is the magnetosphere, which injects energy mostly in the polar regions of the planet. This polar energy input is believed to be sufficient to explain the observed temperatures, provided that it is efficiently redistributed globally by winds, a process that is not well understood. Here we show, using a numerical model, that the net effect of the winds driven by the polar energy inputs is not to heat but to cool the low-latitude thermosphere. This surprising result allows us to rule out known polar energy inputs as the solution to the energy crisis at Saturn. There is either an unknown – and

   large – source of polar energy, or, more probably, some other process heats low latitudes directly.

2. Compare and contrast the first steps in the engineering and scientific methods. Why are these differences important?

3. Directions: After each description of health news just released in 2004, identify whether you think it would be better labeled as science or engineering, then briefly describe what characteristics of the example support your choice.

   (A) **Laboratory rat gene sequencing completed; humans share one-fourth of genes with rat, mouse**

   A large team of researchers, including a computer scientist at Washington University in St. Louis, has effectively completed the genome sequence of the common laboratory brown rat, *Rattus norvegicus*. This will make the third mammal to be sequenced, following the human and mouse.

   http://record.wustl.edu/news/page/normal/ 3222.html

   (B) **Chemists seek light-activated glue for vascular repair**

   Surgeons battle time and the body's defenses as they stitch together veins and arteries, whether after an injury or in the course of such treatments as transplants or bypasses. Loss of blood before a site is closed and too much clotting soon after challenge medical care. Virginia Tech researchers are creating biocompatible adhesives for use with vascular tissue that will speed the process of mending tissue.

   http://www.news-medical.net/?id=216

   (C) **New biomaterial may replace arteries, knee cartilage**

   A unique biomaterial developed by researchers at the Georgia Institute of Technology could be available in as few as five years for patients needing artery or knee cartilage replacement. It may also be used to speed repair of damaged nerves in patients with spinal cord injuries and as the basis for an implantable drug delivery system.

http://gtresearchnews.gatech.edu/
newsrelease/BIOMAT.html

(D) **New insight on cell growth could lead to method for stopping cancer**

West Lafayette, Ind. – Halting the development of certain pancreatic, ovarian, colon and lung cancers may be possible with therapy based on recent Purdue University research. By investigating a single molecule that influences cell growth, a research group in the Purdue Cancer Center, has gained new insight into the chain of events that make some cancer cells divide uncontrollably – insight that may eventually lead to a way to break that chain, stopping cancer in its tracks.

http://www.purdue.edu/UNS/html4ever/
2004/040328.Henriksen.ras.html

4. Identify a researcher involved in developing new health technologies at your institution. Arrange to interview them in person. Write a one page profile of this individual. Your profile should include at least the following.

- A summary of the researcher's educational background.

- A description of current technologies they are developing and the potential of these technologies to improve health.

- Your assessment of how or where their research and development efforts fit along the spectrum of translational research.

References

[1] Tredgold T. *ICE Council Minutes*. ICE Meeting of Council. 1827–1835 December 29;3.

[2] National Science Foundation. *Integrative Graduate Education and Research Traineeship*. 2007 [cited 2007 June 1]; Available from: http://www.igert.org/

[3] National Institutes of Health. *NIH Roadmap for Medical Research*. 2007 May 25 [cited 2007 June 1]; Available from: http://nihroadmap.nih.gov/

[4] Hopkin K. *Biography of Huda Y. Zoghbi, M.D.* Chevy Chase, M.D.: Howard Hughes Medical Institute; 2003.

[5] *Heart–Lung Machine*. Arlington, VA: The Whitaker Foundation; 2006.

[6] Stephenson LW. History of cardiac surgery. In: Edmonds LH, Cohn LH, eds. *Cardiac Surgery in the Adult*. New York: McGraw-Hill; 2003.

[7] Kasper D, Braunwald E, Fauci A, Longo D, Hauser S, Jameson JL. *Harrison's Principles of Internal Medicine*. 16th edn. New York: McGraw-Hill; 2005.

[8] American Heart Association. *Heart Disease and Stroke Statistics-2007 Update*. Dallas, TX: American Heart Association; 2007.

[9] *History of Medical Diagnosis and Diagnostic Imaging*. 2006 August [cited 2006 September 16]; Available from: http://www.imaginis.com/faq/history.asp

[10] Lebedev MA, Nicolelis MA. Brain–machine interfaces: past, present and future. *Trends in Neurosciences*. 2006 Sep; **29**(9): 536–46.

[11] Chapin JK, Moxon KA, Markowitz RS, Nicolelis MA. Real-time control of a robot arm using simultaneously recorded neurons in the motor cortex. *Nature Neuroscience*. 1999 Jul; **2**(7): 664–70.

[12] Fetz EE. Real-time control of a robotic arm by neuronal ensembles. *Nature Neuroscience*. 1999 Jul; **2**(7): 583–4.

[13] Brain-Computer Interfaces Come Home: National Institute of Biomedical Imaging and Bioengineering; 2006 November 28.

[14] Bonetta L. Flow cytometry smaller and better. *Nature Methods*. 2005; **2**(10): 785–95.

[15] Rodriguez WR, Christodoulides N, Floriano PN, Graham S, Mohanty S, Dixon M, *et al.* A microchip CD4 counting method for HIV monitoring in resource-poor settings. *PLoS Medicine*. 2005 Jul; **2**(7): e182.

[16] Yager P, Edwards T, Fu E, Helton K, Nelson K, Tam MR, *et al.* Microfluidic diagnostic technologies for global public health. *Nature*. 2006 Jul 27; **442**(7101): 412–18.

[17] PATH. *Rapid Diagnostic Test Technologies: Lateral-flow.* [cited 2007 April 10]; Available from: http://www.rapid-diagnostics.org/tech-lateral.htm

[18] Pandy MG. Computer modeling and simulation of human movement. *Annual Review of Biomedical Engineering*. 2001; **3**: 245–73.

[19] Wang JH, Thampatty BP. An introductory review of cell mechanobiology. *Biomechanics and Modeling in Mechanobiology*. 2006 Mar; **5**(1): 1–16.

[20] Bedoya J, Meyer CA, Timmins LH, Moreno MR, Moore JE. Effects of stent design parameters on normal artery wall mechanics. *Journal of Biomechanical Engineering*. 2006 Oct; **128**(5): 757–65.

[21] *Capturing the Full Power of Biomaterials for Military Medicine: Report of a Workshop*. Washington, D.C.: National Academy of Sciences; 2004.

[22] Burt HM, Hunter WL. Drug-eluting stents: a multidisciplinary success story. *Advanced Drug Delivery Reviews*. 2006 Jun 3; **58**(3): 350–7.

[23] Saulnier B. The Big Picture. *Cornell Alumni Magazine*. 2004 Sep/Oct; **107**(2).

[24] Langer RS. An interview with a distinguished pharmaceutical scientist: Robert S. Langer. *Pharmaceutical Research.* 1999 Apr; **16**(4): 475–7.

[25] Langer R. Robert Langer, ScD – Engineering medicine. Interview by M. J. Friedrich. *Jama.* 2005 Oct 5; **294**(13): 1609–10.

[26] *Timeline of Key Events in U.S. Transplantation and UNOS History.* 2007 [cited 2007 April 12]; Available from: http://www.unos.org/whoWeAre/history.asp

[27] *U.S. Transplantation Data. 2007* [cited 2007 April 12]; Available from: http://www.unos.org/data/default.asp?displayType=usData

[28] *Help Save a Life. 2007* [cited 2007 April 12]; Available from: http://www.unos.org/helpSaveALife/

[29] Polak JM, Bishop AE. Stem cells and tissue engineering: past, present, and future. *Annals of the New York Academy of Sciences.* 2006 Apr; **1068**: 352–66.

[30] Nerem RM. Tissue engineering: the hope, the hype, and the future. *Tissue Engineering.* 2006 May; **12**(5): 1143–50.

[31] Ehrenreich M, Ruszczak Z. Update on tissue-engineered biological dressings. *Tissue Engineering.* 2006 Sep; **12**(9): 2407–24.

[32] Atala A, Bauer SB, Soker S, Yoo JJ, Retik AB. Tissue-engineered autologous bladders for patients needing cystoplasty. *The Lancet.* 2006 Apr 15; **367**(9518): 1241–6.

[33] Aschheim K. Profile: Anthony Atala. *Nature Biotechnology.* 2006 Nov; **24**(11): 1311.

[34] Ideker T, Winslow LR, Lauffenburger AD. Bioengineering and systems biology. *Annals of Biomedical Engineering.* 2006 Feb; **34**(2): 257–64.

[35] Institute for Systems Biology. *Health Care in the 21st Century: Predictive, Preventive and Personalized.* 2006 [cited 2006 November 13]; Available from: http://www.systemsbiology.org/Intro_to_ISB_and_Systems_Biology

[36] Westerhoff HV, Palsson BO. The evolution of molecular biology into systems biology. *Nature Biotechnology.* 2004 Oct; **22**(10): 1249–52.

[37] Henry CM. Systems biology. *Chemical and Engineering News.* 2003 May 19: 45–55.

[38] Hess H, Bachand GD, Vogel V. Powering nanodevices with biomolecular motors. *Chemistry* (Weinheim an der Bergstrasse, Germany). 2004 May 3; **10**(9): 2110–16.

[39] Edwards JS, Palsson BO. The Escherichia coli MG1655 in silico metabolic genotype: its definition, characteristics, and capabilities. *Proceedings of the National Academy of Sciences of the United States of America.* 2000 May 9; **97**(10): 5528–33.

[40] Stephanopoulos G. Metabolic fluxes and metabolic engineering. *Metabolic Engineering.* 1999 Jan; **1**(1): 1–11.

[41] Wishart D. *Production of Protein Pharmaceuticals (Part 1)* [PowerPoint Slides]: University of Alberta; 2005.

[42] Asthagiri AR, Lauffenburger DA. Bioengineering models of cell signaling. *Annual Review of Biomedical Engineering.* 2000; **2**: 31–53.

[43] Bennasroune A, Gardin A, Aunis D, Cremel G, Hubert P. Tyrosine kinase receptors as attractive targets of cancer therapy. *Critical Reviews in Oncology/Hematology.* 2004 Apr; **50**(1): 23–38.

[44] Maynard J, Georgiou G. Antibody engineering. *Annual Review of Biomedical Engineering.* 2000; **2**: 339–76.

[45] Daar AS, Thorsteinsdottir H, Martin DK, Smith AC, Nast S, Singer PA. Top ten biotechnologies for improving health in developing countries. *Nature Genetics.* 2002 Oct; **32**(2): 229–32.

[46] Pecoul B, Chirac P, Trouiller P, Pinel J. Access to essential drugs in poor countries: a lost battle? *Jama.* 1999 Jan 27; **281**(4): 361–7.

[47] Free MJ. Achieving appropriate design and widespread use of health care technologies in the developing world. Overcoming obstacles that impede the adaptation and diffusion of priority technologies for primary health care. *International Journal of Gynaecology and Obstetrics: the Official Organ of the International Federation of Gynaecology and Obstetrics.* 2004 Jun; **85** Suppl. 1: S3–13.

[48] *Innovation or Stagnation?* Rockville, MD: US Department of Health and Human Services, Food and Drug Administration; 2004 March.

[49] Koizumi K. National Institutes of Health in the FY 2007 Budget. In: Intersociety Working Group, ed. *AAAS Report XXXI Research and Development FY 2007.* Annapolis Junction, MD: AAAS 2006.

[50] A.8. Health Related R&D. *OECD Science, Technology and Industry Scoreboard 2005 – Towards a knowledge-based economy* OECD Publishing 2005.

[51] Wong J, Quach U, Thorsteinsdottir H, Singer PA, Daar AS. South Korean biotechnology – a rising industrial and scientific powerhouse. *Nature Biotechnology.* 2004 Dec; **22** Suppl.: DC42–7.

[52] Phillips M. Global Vaccine Initiative Hits Snag. *Wall Street Journal.* 2006 July 7.

Prevention of infectious disease

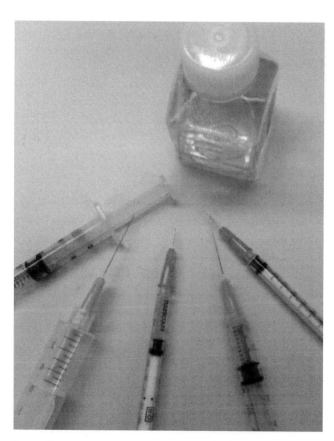

Figure 8.1. *Vaccines play a critical role in the prevention of infectious diseases.*

In Chapter 7, we examined the process of designing new technologies and the interdisciplinary translational research efforts needed to advance technologies from the laboratory to clinical practice. In the rest of this book, we will explore in detail the development of several types of new technologies which draw on advances in the different sub-disciplines of bioengineering. We begin by focusing on the development of vaccines to prevent infectious disease. We will see that scientific knowledge, such as an understanding of both the organisms that cause disease and the protective mechanisms of the immune system, is critical to enable the engineering of preventive vaccines. We will examine the development of vaccines from idea to product.

We have seen that infectious diseases are responsible for a large fraction of deaths, particularly in the developing world. In high-mortality developing countries, infectious disease is responsible for nearly half of all deaths, and the childhood cluster diseases (pertussis, poliomyelitis, diphtheria, measles, and tetanus) kill just over half a million children under the age of five each year [2]. In the developed world, the use of vaccines has dramatically reduced the incidence of infectious disease. As we explore how vaccines work, we will trace the tremendous improvements in world health that have resulted from mass childhood immunization and examine the global obstacles that remain. There are many diseases for which no vaccine is available. We conclude this chapter by considering the scientific and engineering challenges associated with developing new vaccines to prevent HIV infection.

The immune system

Vaccines manipulate the immune system of the recipient; thus to understand how vaccines work, we must first understand how the immune system prevents and fights infections. In Chapter 1, we examined the use of high dose chemotherapy for breast cancer. A side effect of that treatment was that it destroyed cells in the patient's bone marrow, leaving patients vulnerable to infection. Stem cells in the bone marrow are responsible for generating the cellular components of blood: the oxygen carrying red blood cells, the platelets which aid in blood clotting and the infection–fighting white blood cells. There are five main types of white blood cells: eosinophils, neutrophils, basophils, lymphocytes and monocytes (Figure 8.2). Eosinophils are important in fighting infections due to parasites, such as malaria. Neutrophils are important in fighting infections caused by bacteria, such as tuberculosis. Monocytes leave the bloodstream and mature into macrophages, which are also important in fighting bacterial infections. As we will see, lymphocytes are important in the fight against both bacterial and viral infections. The function of basophils is poorly understood, but they are believed to be important in allergic reactions. To understand how white blood cells work as part of the immune system to fight infection, we next examine how infectious agents, such as bacteria and viruses, cause disease [3].

How infectious agents cause disease

Bacteria and viruses cause disease in very different ways. Bacteria consist of cells with a cell membrane and, unlike human cells, usually also have a cell wall (Figure 8.3). Bacteria can survive outside of a host and are capable of reproduction without a host.

(a) (b) (c)

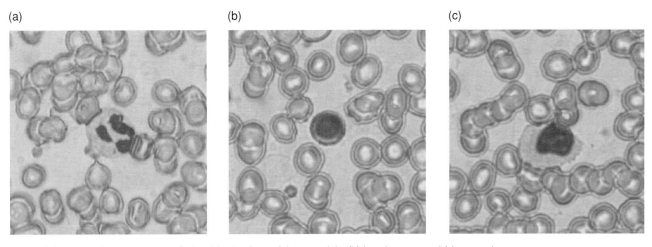

Figure 8.2. *Three of the main types of white blood cells are (a) neutrophils, (b) lymphocytes, and (c) macrophages.*

(a)

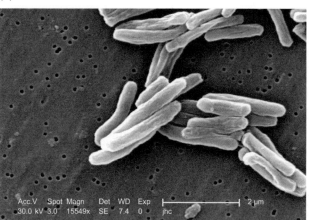

(b)

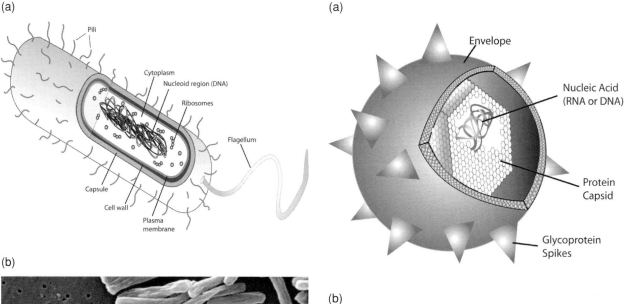

Figure 8.3. *(a) The structure of a bacterial cell and (b) an electron micrograph of bacterial cells. CDC/Dr. Ray Butler, Janice Carr.*

(a)

(b)

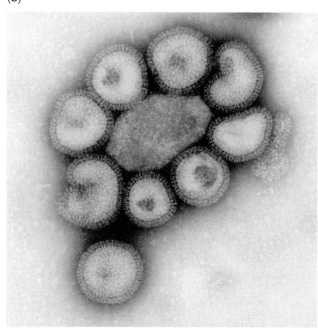

Figure 8.4. *(a) The structure of a virus and (b) an electron micrograph of a virus. CDC/ Dr. F. A. Murphy CDC.*

Bacteria can be killed or inhibited by antibiotics, which frequently destroy the bacterial cell wall. Viruses consist of a nucleic acid core surrounded by a protein envelope (Figure 8.4). Unlike bacteria, viruses must use the intracellular machinery of their host to reproduce, and they cannot be killed with antibiotics. There are more than 50 different viruses that can infect humans [4].

Whether virus or bacteria, there are three basic problems each pathogen must solve: (1) how to reproduce inside a human host, (2) how to spread from one person to another, and (3) how to evade the immune system. Because of their fundamental structural and functional differences, bacteria and viruses solve these problems and cause disease in very different ways. Bac-

teria invade a host, and then begin to reproduce. As they grow and reproduce, they produce toxins which disturb the function of normal cells. Viruses actually invade the cells of their host (Figure 8.5). They typically accomplish this invasion by binding to receptors on the membrane of the host cell, which then transport the virus into the cell in a process known as endocytosis. Once inside, the virus takes over the cell, using viral nucleic acid and

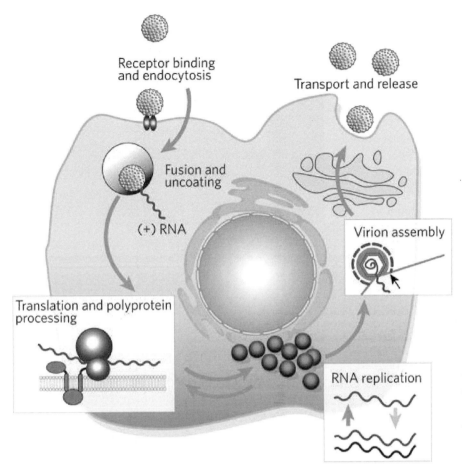

Receptor binding
and endocytosis

Transport and release

Fusion and
uncoating

(+) RNA

Virion assembly

Translation and polyprotein
processing

RNA replication

Figure 8.5. *The life cycle of a virus. Reprinted with permission from Macmillon Publishers, Ltd: Nature. Brett D. Lindenback: Charles M. Rice. Unravelling hepatitis C virus replication from genome to function. Nature Publishing Group. 2005. **436**(7053): 933–938.*

host cell resources to make new viral nucleic acid and proteins. As the virus directs the synthesis of new viral particles, more virus is released from the host cell. The new virus is disseminated when the virus either causes the host cell to lyse (break apart) or the newly formed viral particles bud from host cell surface [3].

Bacterial disease

Let's examine a once common bacterial infection – **pertussis**, also known as whooping cough – to see how bacteria typically cause disease. Pertussis is a highly infectious respiratory illness which is transmitted through contact with respiratory droplets from an infected person. Typically seven to ten days after exposure, patients develop cold-like symptoms, such as a runny nose, sneezing, low grade fever, and a mild cough. The cough gradually worsens, and after one to two weeks patients experience violent bursts of coughing. At the end of a coughing episode, a long gasp is accompanied by a high pitched sound (whoop). Pertussis is caused by the bacterium *Bordetella pertussis*. How does the *Bordetella pertussis* bacterium cause these symptoms? The respiratory tract is lined with ciliated epithelial cells; ordinarily these cilia play an important role in clearing the respiratory tract of mucus and secretions. The *Bordetella pertussis* bacterium binds to receptors on the surface of cells lining the respiratory tract (Figure 8.6); as the bacteria grow, they produce toxins which paralyze the respiratory cilia and interfere with the patient's ability to clear respiratory secretions. The violent coughing fits associated with pertussis are an attempt to clear secretions which build up due to impaired ciliary function. Ultimately, the immune system usually clears the bacterial invasion, and patients recover within several weeks. However, infants and young children are at particular risk of developing a secondary bacterial pneumonia which can lead to death [1].

[5]

At the end of a **pertussis** coughing episode, a long gasp is accompanied by a high pitched sound (whoop). You can listen to a typical "whooping" cough associated with pertussis at: http://www.whoopingcough.net/

Before the availability of the pertussis vaccine in the 1940s, more than 200,000 cases of whooping cough were reported each year in the USA. The disease is still an important cause of mortality in developing countries; in 2001, pertussis was estimated to cause more than 285,000 deaths in children in developing countries.

Figure courtesy of CDC http://www.cdc.gov/vaccines/pubs/pinkbook/downloads/pert-508.pdf

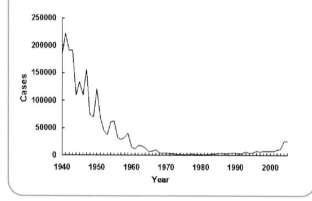

Pertussis—United States, 1940-2005

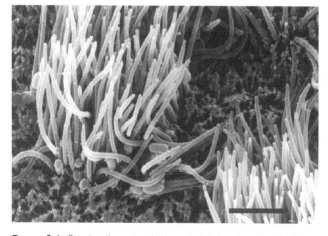

Figure 8.6. Bordetella pertussis *bacteria bind to the cilia of cells lining the respiratory tract. Reprinted from* Respiratory Medicine, *Vol. 94, Soane, M. C. et al.,* Interaction of Bordetella pertussis *with human respiratory mucosa in vitro, pp. 791–799, © 2000, with permission from Elsevier.*

Figure 8.7. *Influenza is spread from person to person through coughs or sneezing. Courtesy of Andrew Dandhazy, Rochester Institute of Technology.*

Viral disease

In contrast, viral pathogens have developed different ways to reproduce in a host, spread from one person to another, and evade the immune system. How does a common virus like influenza solve these problems? In order to reproduce, the influenza virus must get inside the human cell to use the cell's biosynthetic machinery. Influenza virus accomplishes this by binding to receptors on the host cell surface and then inducing receptor mediated endocytosis. Once it has been endocytosed, the influenza virus is trapped in a vesicle made of the cell membrane called an endosome. The virus acts to slowly reduce the pH in the endosome, creating a hole in the membrane through which the virus releases its single stranded RNA and polymerase proteins. These RNA segments and polymerase proteins enter the nucleus of the infected cell and direct the cell to begin making many copies of the viral RNA and viral coat proteins. These new viral particles then exit the nucleus and bud from the cell. During this reproduction, the viral polymerase proteins don't proofread reproduction, and as a result nearly every virus produced in an influenza-infected cell is a mutant, differing slightly from the original infecting virus [4].

How does the influenza virus spread from one person to another? Generally, this happens when an infected person sneezes or coughs (Figure 8.7), and micro-droplets containing viral particles are inhaled by

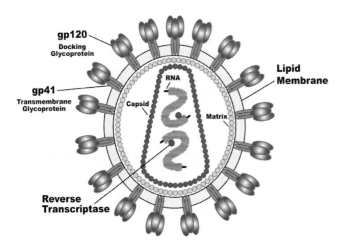

gp120
Docking
Glycoprotein

gp41
Transmembrane
Glycoprotein

RNA

Capsid

Matrix

Lipid
Membrane

Reverse
Transcriptase

Figure 8.8. *The HIV virus. Courtesy of NIAID.*

another person. The influenza virus is particularly adept at penetrating epithelial cells lining the respiratory tract, killing cells that it infects. The resulting inflammation triggers a cough reflex to clear airways of foreign invaders. During influenza infection, the immune system produces large quantities of a substance called interferon. Interferon leads to the common symptoms of the flu: fever, muscle aches, headaches and fatigue [4].

Let's look at a more deadly pathogen – the HIV virus. HIV consists of a central core of RNA and an enzyme

called reverse transcriptase surrounded by a protein core (Figure 8.8). The entire virus is surrounded with a lipid membrane; this membrane is studded with special proteins called gp120 envelope proteins that enable it to bind to the surface of host cells. In particular, this envelope protein is recognized by receptors on the surface of a special type of lymphocyte [3].

Figure 8.9 illustrates what happens when the HIV virus binds to the surface of a host lymphocyte. The membrane of the viral particle fuses with the host cell membrane, allowing the contents of the viral particle to enter the cytoplasm of the lymphocyte. The viral enzyme reverse transcriptase uses the host machinery to copy the viral RNA into viral DNA. The viral DNA then directs the lymphocyte to produce new copies of viral protein and RNA which are assembled into protein coated viral particles within the lymphocyte. These mature particles can then bud out from the cell to release new copies of the virus into the host. Thus, HIV infection destroys an important component of the immune system; without treatment, patients develop AIDS and severe immunodeficiency. Patients with severely compromised immune function become susceptible to a variety of opportunistic infections which can result in death [3].

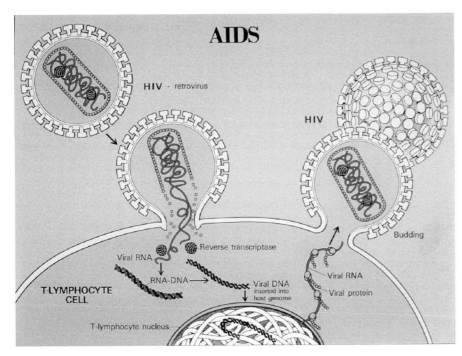

AIDS

HIV - retrovirus

HIV

Budding

Reverse transcriptase

Viral RNA

RNA-DNA

Viral DNA
inserted into
host genome

Viral RNA

Viral protein

T-LYMPHOCYTE
CELL

T-lymphocyte nucleus

Figure 8.9. *Life cycle of the HIV virus. NCI/Trudy Nicholson.*

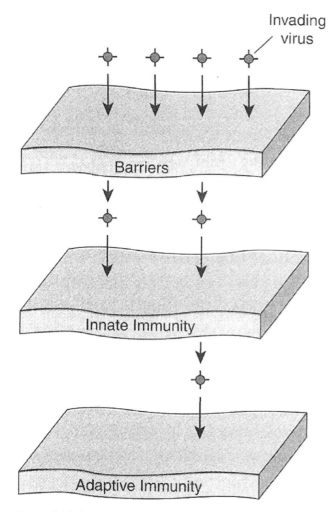

Figure 8.10. *Layers of immunity.*

Figure 8.11. *A microscopic view of the mucous membrane of the uterine cervix.*

How the immune system fights pathogens

How are we protected from bacterial and viral attack? Evolution has provided two simple protective strategies: (1) keep pathogens out, and (2) kill them if they get in. To accomplish these goals, we are protected by three layers of immunity (Figure 8.10). First, physical barriers act to keep pathogens out. The most important physical barriers are the skin and mucous membranes. These barriers must defend an enormous area – humans have over two square meters of skin and 400 square meters of mucous membranes [6]. Second, all animals possess an innate immune system to fight pathogens that make it past these physical barriers. The innate immune system recognizes molecular patterns typically associated with pathogens and responds with a general inflammatory response to fight those pathogens which penetrate physical barriers. Third, vertebrates possess an adaptive immune system, capable of recognizing and adapting itself to defend against any invader. The adaptive immune system becomes important when the innate immune system cannot defend against attack. The adaptive system also provides the immune system with "memory" [3].

Figure 8.11 shows a microscopic view of a protective physical barrier – in this case, the mucous membrane lining of the uterine cervix. This lining is about 250 microns thick (about the diameter of a human hair) and consists of multiple layers of specialized epithelial cells. The epithelial cells at the bottom of the layer are rapidly dividing and are responsible for regenerating the epithelial tissue as it dies. As we move toward the surface of the epithelium, the cells become more mature. Cells at the very top layer are dead, but the tight junctions between these cells provide an important barrier which is difficult for many pathogens to cross. In addition, substances present on the surface of many mucus membranes and the skin provide a chemical barrier, which functions to trap pathogens and may contain enzymes or other molecules with anti-bacterial activity.

What happens when a pathogen is able to cross this barrier? You have no doubt experienced this if you have ever gotten a splinter (Figure 8.12). In this case, a sharp piece of wood crosses the skin, enabling bacteria to penetrate beneath the epithelial lining. What are the symptoms you experience? Frequently the area can become red, swollen, and warm. Sometimes the area will ooze pus. These symptoms are all signs that the second line of defense – the innate immune

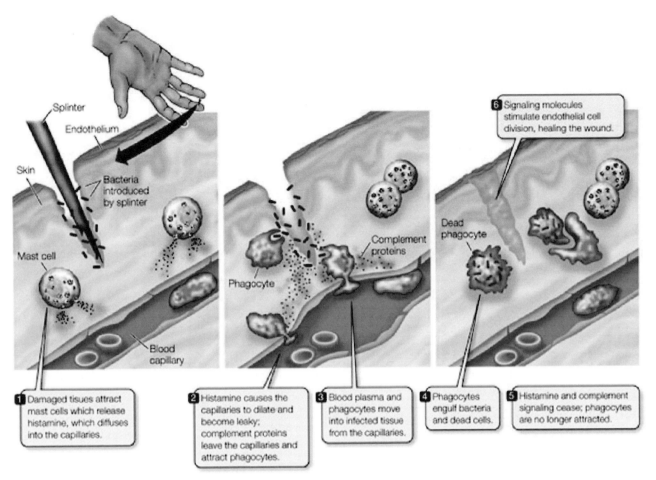

Splinter
Endothelium
Skin
Bacteria introduced by splinter
Mast cell
Phagocyte
Blood capillary
Complement proteins
Dead phagocyte

6 Signaling molecules stimulate endothelial cell division, healing the wound.

1 Damaged tisues attract mast cells which release histamine, which diffuses into the capilaries.

2 Histamine causes the capilaries to dilate and become leaky; complement proteins leave the capilaries and attract phagocytes.

3 Blood plasma and phagocytes move into infected tissue from the capilaries.

4 Phagocytes engulf bacteria and dead cells.

5 Histamine and complement signaling cease; phagocytes are no longer attracted.

Figure 8.12. *When a splinter carrying a bacterial pathogen crosses the physical barrier provided by epidermis of the skin, the innate immune response provides a second line of defense. Tissue dwelling macrophages recognize the bacteria and begin to phagocytose the pathogen. Activated macrophages secrete chemical messengers which increase blood flow, increase the permeability of blood vessels, and recruit other white blood cells to the site of the invasion. Neutrophils recruited in this manner aid in phagocytosing bacteria. This response is known as inflammation, and results in the redness, swelling, heat and pus that are sometimes present at the site of an infection. Sadava et al. Life: The Science of Biology, Eighth Edition. Sinauer Associates, Inc. and W. H. Freeman and Co. © 2008.*

system – is kicking into gear. In most cases, the innate immune system can respond to the pathogen. A specialized kind of cell called a macrophage continually patrols beneath the epithelium, to detect foreign invaders and signal the immune system to respond. Macrophages are derived from monocytes, one of the five types of white blood cells. You can think of the macrophages as guards that patrol just beneath the physical barriers (skin and mucous membranes). When macrophages encounter bacteria on a splinter they ingest the bacteria in a process known as phagocytosis (Figure 8.13). When macrophages are activated in this manner, they produce chemicals which increase local blood flow. It is this increase in blood flow that makes the area around the splinter appear red and warm. These chemicals also cause the blood vessels in the area to become leaky. Fluid leaking out of the blood vessels produces the swelling around the area with the splinter. The chemicals released by the macrophages also recruit other white blood cells, such as neutrophils, to the site of the infection. Neutrophils aid in phagocytosing bacteria, and these cells make up the pus which can sometimes be present in the area [6].

The innate immune system is primarily effective against pathogens outside of cells. How do macrophages recognize extracellular invaders as foreign? Macrophages recognize foreign invaders in two ways. There are particular molecular patterns found on the

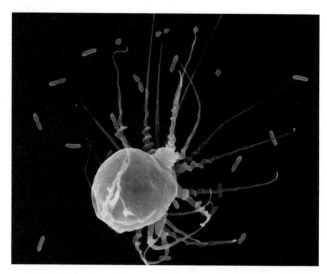

Figure 8.13. *A macrophage "eats" a bacterium. Copyright © 2004 Dennis Kunkel Microscopy, Inc.*

Video microscopy showing a macrophage phagocytosing cells of the fungus *Candida albicans* can be seen at:

http://www.cellsalive.com/mac.htm

When activated, macrophages recruit circulating neutrophils to the site of inflammation. The video shows neutrophils rolling along the surface of a venule; macrophages secrete chemicals which cause neutrophils to exit vessels and enter tissue. Follow the link below for video microscopy of neutrophils rolling along a blood vessel.

http://www.cbrinstitute.org/labs/springer/lab_goodies/lab_goodies.html

surface of many pathogenic microorganisms. These signatures are recognized by receptors on the surface of the macrophage. In order to evade this route of detection by the immune system, many bacteria have evolved to hide their surface markers behind a polysaccharide capsule [3]. In some cases, proteins within the blood of the host can bind to pathogens and mark them for destruction by macrophages as a way to guide the innate immune system. In any case, if macrophages identify an invader, they become activated. Once activated they send signals to recruit other immune system cells (neutrophils), they become killers, and they activate the third line of defense, the adaptive immune system [6].

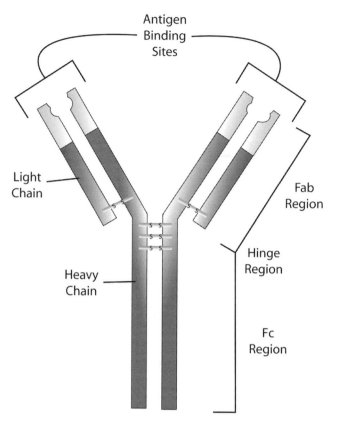

Figure 8.14. *Structure of an antibody.*

The adaptive immune system has two main components. The first component, called humoral immunity, relies on large proteins called antibodies (Figure 8.14) to recognize and fight pathogens outside of cells. The second component, called cell-mediated immunity, relies on several types of white blood cells to kill pathogens inside of cells. The innate immune system recognizes general molecular signatures of pathogens and provides a generalized response. In contrast, the adaptive immune system recognizes specific molecular signatures called antigens, associated with individual pathogens. The two components of the adaptive immune system accomplish the recognition using different strategies. The humoral component of the adaptive immune system relies on the chemical specificity of antibodies to recognize different pathogens. Recognition by the cell-mediated component of the adaptive immune system is facilitated by specific receptors on the surface of lymphocytes [3].

Let's first examine how antibodies help to recognize and kill pathogens. Antibodies are Y-shaped proteins, about 12 nm in length, which are made by the immune

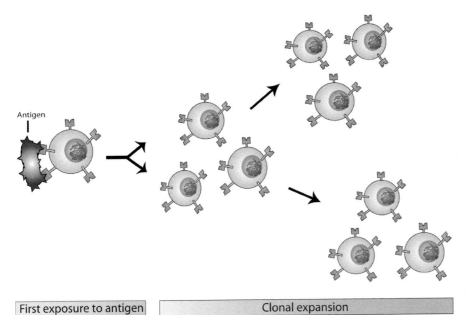

Antigen

| First exposure to antigen | Clonal expansion |

Figure 8.15. *When a B cell binds an antigen it rapidly proliferates to generate thousands of clones.*

system. The bottom of the Y is known as the Fc region, while the top of the Y has two antigen binding sites (Fab region) which can bind either to the surface of a free bacteria or virus or to the surface of a virus-infected cell [4, 7]. The free Fc region hanging off the pathogen then binds to macrophages and neutrophils and induces them to phagocytose the tagged pathogen. The Fc region can also bind to a special kind of lymphocyte known as a natural killer cell to induce destruction of the invader [4]. Essentially, you can think of an antibody as a bridge between a pathogen and the tool to kill it. The Fab portion of the antibody recognizes the antigen while the Fc region interacts with other components of the immune system to initiate destruction of the pathogen.

Antibodies are made by a type of lymphocyte called B cells. These B cells have special receptors on their surface that are designed to recognize foreign pathogens. These receptors are antibodies with their Fc region inserted in the lymphocyte cell membrane. We all have 100 million different types of B cells, each with different surface receptors. These B cell receptors are so diverse they can recognize every organic molecule; thus, they provide the ability to recognize specific pathogens [4]. Lymphocytes respond only to foreign antigens because self-reactive lymphocytes are eliminated during their development. When a B cell binds an antigen it begins to rapidly divide and proliferate – in one week, a clone of 20,000 identical B cells can be created [6]. This process

is known as *clonal expansion* (Figure 8.15). Effector B cells secrete antibodies which will recognize the specific pathogen targeted by that B cell.

Antibodies are helpful to recognize and target pathogens outside of cells. However, many viruses hide inside our cells. How do we kill viruses once they are inside the cell, where antibodies cannot reach them? This process is carried out by a special class of white blood cells called T lymphocytes. T cells recognize protein antigens. Again, we require a way to let T cells know which cells have been invaded by viruses. Nucleated cells in your body have special molecules on their surface known as major histocompatibility (MHC) molecules. These molecules help your immune system recognize the cells of your body as "self" so that they do not come under attack by the immune system [7]. When a virus invades a cell, fragments of viral proteins are loaded onto MHC proteins (Figure 8.16). T cells inspect the MHC proteins on the surface of your cells and use this as a signal to identify infected cells (Figure 8.16) and target them for destruction. Like B cells, when T cells bind antigen, they undergo clonal expansion [4].

The process of clonal expansion enables the adaptive immune system to have memory (Figure 8.17). The first time the adaptive immune system is activated by an antigen, your body builds up a clone of B cells and T cells. This process takes about a week [6]. After the infection is over, most of these cells die off; however, some cells,

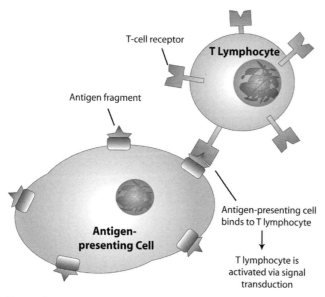

Figure 8.16. *An infected cell presents antigen fragments on its surface, provoking a response from T cells.*

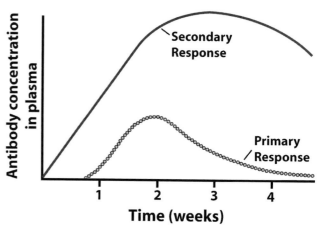

Figure 8.18. *After the first exposure to an antigen, the immune system can respond much more quickly to future exposures.*

known as memory cells, remain. If the immune system is activated a second time by the same antigen, these memory cells are much easier to activate (Figure 8.18). The response of the immune system is much faster, generally so much more rapid that you don't experience any symptoms associated with the infection [6]. This memory explains why you are "immune" to most diseases after a first exposure (think chicken pox) and is also what allows us to develop vaccines to establish this immunity in a safer way.

Why then do we get the flu more than one time? Influenza virus particles are usually 80–120 nm in diam-eter, and consist of an outer lipid envelope, an inter-mediate protein capsid, and a central core of RNA. There are three major types of influenza virus: A, B, C. Most serious cases of the flu in humans are caused by type A influenza. Figure 8.20 shows a schematic drawing of an influenza virus; there are two kinds of proteins found in the lipid envelope of the influenza virus. Hemaglutinnin mediates attachment of viral parti-cles to the host cell membrane; neuraminidase mediates release of newly formed viral particles from the host cell. Type A virus contains 8 single stranded pieces of RNA [3].

The influenza virus can evade immune extinction in two ways: antigenic drift and antigenic shift. As influenza reproduces, reproduction errors occur that

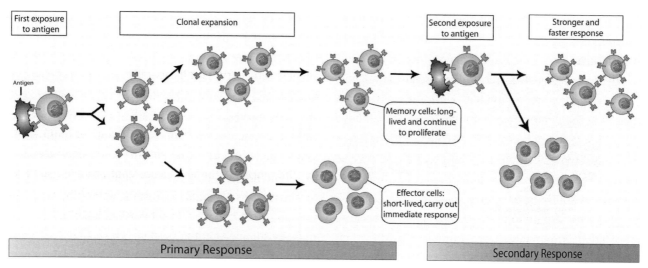

Figure 8.17. *After the initial exposure to an antigen, memory cells "remember" the antigen, allowing for a faster response to future exposures.*

Seasonal influenza and pandemics

Antiviral drugs are effective against the influenza virus if taken within two days of the onset of symptoms. Many of these drugs work by hindering the change in pH necessary for influenza virus to escape the endosome following endocytosis. Other antiviral drugs block the effect of neuraminidase, to inhibit release of new viral particles. H5N1 is resistant to some antiviral drugs, but responds to others.

Mortality due to influenza fluctuates seasonally. The CDC tracks mortality, to monitor for epidemics of influenza, where the mortality rate rises above the seasonally anticipated baseline level. In 2004, an influenza epidemic occurred in the USA.

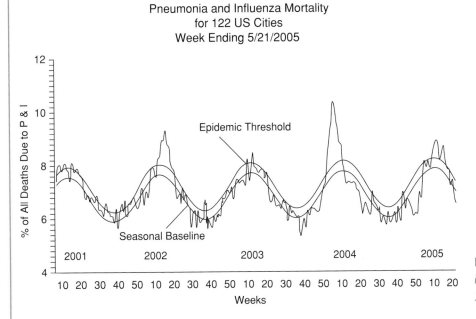

Figure 8.19. *Pneumonia and influenza mortality rates for years 2001–2005. Source: CDC.*

http://www.cdc.gov/flu/weekly/weeklyarchives2004–2005/images/bigpicurvesumary04–05.gif

Three influenza pandemics have been recorded; the most deadly occurred in 1918. New strains of influenza continue to emerge, and may result in future pandemics.

http://www.news.cornell.edu/stories/Oct05/Avian_Torres.kr.html

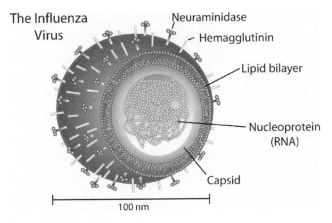

Figure 8.20. *The influenza virus.*

change the structure of the virus. The changes are so slight that the virus is still capable of infection and reproduction but significant enough that the immune system does not possess memory for the changed virus. Mutations in the virus have resulted in a number of different strains of influenza virus, and these are characterized by differences in the two major coat proteins, such as influenza A (H2N1) [3]. This antigenic drift is why the structure of the flu vaccine changes annually.

Antigenic shift refers to much larger changes in the structure of the virus. Sometimes animals can be co-infected by different strains of the influenza virus.

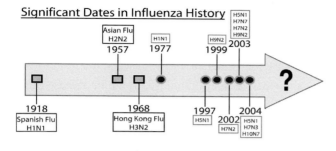

Significant Dates in Influenza History

Major Pandemic

The appearance of a new influenza strain in the human population

Figure 8.21. *Significant influenza appearances since 1918. Adapted from http://www.news.cornell.edn/stories/Oct05/Avian_Torres.kr.html*

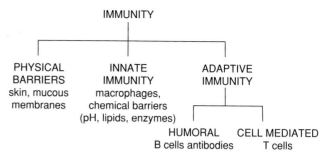

Figure 8.22. *Overview of the components of the immune system.*

During viral replication, viral gene segments from different influenza strains are present inside the same cell; occasionally, these gene segments can randomly reassociate to create a completely new strain of the virus. Reassortment can lead to new strains of influenza that are capable of infecting humans. If a new strain can be easily transmitted from person to person, an influenza pandemic (global outbreak) can occur because people do not have any immunity to this new strain [3].

As a result of genetic shift, there have been three pandemics of flu in the past century (Figure 8.21). The largest occurred in 1918, when the Spanish flu (H1N1) affected between 20–40% of the world's population, killing more than 50 million people worldwide and 675,000 in the USA [9, 10]. The virus had a particularly high mortality rate in young adults. The impact of this pandemic on the USA was so significant that life expectancy dropped by ten years [10]. Why did the Spanish flu kill so many? To gain insight into why this virus was so lethal, scientists recently reconstructed the H1N1 influenza virus and used it to infect macaque monkeys; the experiment was carried out in a biosecure facility [11]. They found that the virus triggered a severe immune reaction in the animals; the immune reaction was so severe that it provoked the body to begin killing its own cells and could rapidly destroy large amounts of lung tissue, leading to death.

Today, there is particular concern regarding the strain of the flu known as avian flu, or influenza A(H5N1). This strain of flu is ordinarily carried by wild birds and does not grow well in human cells. The virus is spread from bird to bird through fecal contact, and is very contagious among birds. While it usually does not cause illness in wild birds, it can quickly devastate populations of domesticated birds – with a mortality rate approaching 100% [9]. H5N1 does not normally infect humans, but in rare cases it can be transmitted from birds to people. Over the last few years, several hundred cases of H5N1 have been reported in humans, mostly in Asia. Fatality rates have been reported to be approximately 50% [12]. The majority of these cases were acquired via bird to human transmission; however, a few cases of human to human transmission have been reported [13]. Today, there is considerable concern that the H5N1 will mutate into a highly infectious strain that can spread rapidly from person to person. This could lead to a pandemic with potentially very high mortality. Later we will examine the challenges associated with preventing such a pandemic.

Summary

The genetic simplicity of many infectious pathogens allows them to undergo rapid evolution; our immune system must constantly cope with pathogens that develop selective advantages to avoid detection and destruction by the immune system. The complexity of the immune system is remarkable, and we have seen only an overview of how this system functions to protect us from disease (Figure 8.22). Physical barriers prevent many pathogens from entering tissue. The innate immune system provides a rapid response system to detect and attack many extracellular pathogens that make it past these physical barriers. The adaptive immune system provides the diversity to recognize over 100 million antigens and to remember which have been encountered previously. Antibody-mediated immunity can fight pathogens outside of cells, and cell-mediated

Table 8.1. *Advantages and disadvantages of three types of vaccines [5].*

Type of vaccine	Advantages	Disadvantages
Noninfectious vaccines: Vaccinate with killed pathogen	Stimulates humoral immunity Minimal danger of infection	Does not stimulate cell-mediated immunity Usually need booster vaccines
Live attenuated bacterial or viral vaccines: Vaccinate with weakened pathogen	Stimulates both humoral and cell-mediated immunity Usually provides lifelong immunity	Poses some risk of disease, particularly in immuno-compromised host
Subunit vaccines: Vaccinate with pathogen subunit	Stimulates humoral immunity Minimal danger of infection	Does not stimulate cell-mediated immunity Usually need booster vaccines

immunity can fight pathogens within cells. In the next section, we examine how we can exploit the properties of the adaptive immune system to provide immunity to dangerous pathogens in a process called vaccination.

How vaccines work

Vaccines are the most cost effective medical intervention known to prevent death or disease. Vaccination is the practice of manipulating the adaptive immune system to artificially induce immunity. The goal of vaccination is to stimulate the adaptive immune system to make memory cells that will protect the vaccinated person against future exposure to a pathogen, without causing the symptoms of disease.

How can we stimulate the adaptive immune system to make memory cells? It is easiest to stimulate humoral immunity because this component of the immune system responds to extracellular pathogens. In order to create memory B cells, we simply need to expose B cell receptors to the pathogen or some part of the pathogen. It is more difficult to stimulate cell-mediated immunity because this component of the immune system responds to intracellular pathogens. Memory killer T cells are created only when antigen presenting cells are infected with a pathogen. Our goal in making a vaccine is to provide this exposure in a safe way.

There are several types of vaccines which can stimulate the adaptive immune system to provide memory and protect against future exposure to a pathogen. Here, we will examine three types of vaccines including: (1) inactivated organism vaccines, (2) live attenuated vaccines, and (3) pathogen subunit vaccines. Table 8.1 compares the advantages and disadvantages of each of these strategies.

Inactivated organism vaccine

Generally considered to be the safest type of vaccine, noninfectious vaccines are based on a pathogen which has been killed or inactivated so that it can still elicit immunity, but it can not cause disease. For example, pathogens can be treated with chemicals (like formaldehyde) to kill them. When a patient is exposed to the dead pathogen, they mount an immune response without becoming infected. The Salk (inactivated) Polio vaccine, the hepatitis A vaccine, and the rabies vaccine are examples of killed pathogen vaccines [5]. The immune system encounters the pathogen outside of cells, so that this approach stimulates humoral immunity. Because the pathogen has been killed it does not infect cells, so this approach does not stimulate cell-mediated immunity. As a result, booster vaccines are usually required to maintain lifelong immunity to these diseases. When using vaccines based on an inactivated organism alone, the vaccine may not stimulate the immune system sufficiently. In order to increase the response of the immune system, the vaccine is sometimes formulated with an adjuvant – a substance that increases the response of the immune system. Aluminum salts are frequently used as an adjuvant in vaccines [14].

Live attenuated vaccines

A stronger immune response can be produced by vaccinating with a live pathogen which has been weakened so that it does not cause disease, but still elicits immunity. In this approach, the pathogen is grown in host cells in the cell culture laboratory in order to produce mutations which weaken the pathogen so it cannot produce disease in healthy people, but can still produce a sufficiently strong immune response that protects against future infection. The Sabin Polio vaccine (oral Polio), measles, mumps, rubella (MMR), and varicella vaccines are examples of vaccines made using this approach [5]. This approach has a number of advantages. Live, attenuated organism vaccines contain the target antigens for humoral antibody, the pathogen molecular patterns for stimulating innate immunity, and, because of the invasiveness of the organism, it can deliver antigens effectively [14]. The immune system treats live, attenuated vaccines in just the same way as it would an infectious pathogen. Thus, these vaccines stimulate both humoral and cell-mediated immunity. As a result, they usually provide lifelong immunity. However, such vaccinations can produce disease in an immunocompromised host [5]. A technical challenge of this approach is that it is not always feasible to produce strains of a pathogen that have been attenuated sufficiently. In addition with pathogens that undergo antigenic drift, there is a finite risk of reversion to a virulent form [14]. Vaccines based on live attenuated pathogens account for approximately 10% of total vaccine sales [15].

Subunit vaccines

Our immune system generally recognizes and responds to a portion of an infectious organism known as an antigen. If we can identify the antigen that will produce an immune response, we can purify that antigen and use it as the basis for a vaccine. This type of vaccine is very safe because there is no risk that it can lead to infection, even in an immunocompromised host [5]. Several strategies have been developed to produce subunit vaccines. For example, many bacteria produce disease by secreting toxins that interfere with normal cell function. We can create an immune response to these toxins by vaccinating with purified bacterial toxins that have been chemically treated to make them harmless. This type of vaccine is known as a toxoid vaccine and it produces an immune response without symptoms of disease [14]. The diphtheria and tetanus vaccines are examples of toxoid vaccines [5]. Alternatively, a subunit vaccine can use part of a pathogen to induce an immune response but not disease. Our immune system responds to the polysaccharides found on the surface of certain bacteria. We can grow bacteria in culture and extract

these polysaccharides from the culture media the bacteria are grown in to develop a vaccine. The haemophilus influenza type b (HiB), and pneumonoccocal vaccines are examples of vaccines made using this approach [5].

In a related approach, we can use the tools of genetic engineering to manufacture a pathogen protein; again exposure to the pathogen protein provides immunity without causing disease. The vaccine for Hepatitis B is based on a protein found on the surface of the virus. We can insert the genes that encode this surface protein into yeast, and use the yeast as a factory to produce the protein for the vaccine [5].

History of vaccines

Throughout history vaccination has protected individuals against disease. As early as the seventh century, there are records of Indian Buddhists drinking snake venom to induce immunity (likely through a toxoid effect) [17]. In the second millennium, vaccination against smallpox was carried out in central Asia, China and Turkey. People recognized that those who had been exposed to variola (also called cowpox, which produces only mild symptoms) usually did not contract smallpox, an often fatal disease. These two antigens are sufficiently similar that exposure to one protects against the other. In 1721, the idea of variolation against smallpox moved from Turkey to England [18].

Independently in 1796, Edward Jenner, a country doctor living in England, noted the relationship between smallpox and cowpox. He observed that dairy milkmaids frequently contracted cowpox, which caused lesions similar to that of smallpox. The milkmaids who had cowpox almost never got smallpox. Jenner carried out an experiment where he injected cowpox pus into a young boy named James Phipps. He later injected Phipps with pus from smallpox sores and noted that Phipps did not contract smallpox. (We will return later to discuss the ethical problems associated with Jenner's experiment.) Despite the ethical flaws of the experiment, it was a scientific success and Jenner was the first to introduce large scale, systematic immunization against smallpox. While his work ultimately led to the elimination of smallpox from the world, it was not immediately embraced by all. Many people were deeply suspicious of the practice of introducing animal products into

Edible vaccines

Subunit vaccines can also be produced in plants; genes are introduced which induce the plant to make the protein that will stimulate immunity.

Usually subunit vaccines are expensive because you must purify the protein grown in culture. With plants, you don't have to do this. Plants can be grown locally, avoiding problems with vaccine transport.

Usually subunit vaccines are injected, because the digestive system will destroy the protein in the stomach before it can be presented to the immune system. However, in plants, the protein is protected by the cell walls of the plant cells. As a result, the protein survives until it reaches the intestine, where immune cells in the intestinal wall are activated.

Studies have shown that tomatoes and potatoes can synthesize antigens from major causes of diarrhea: Norwalk virus, enterotoxigenic *E. coli*, and *Vibrio cholera*. Feeding these plants to animals can evoke immune response, and provide partial exposure to the real toxin. Small trials in human volunteers have produced immune reactivity in people.

Scientists are grappling with the problem of getting the plants to produce a sufficient amount of antigen. When plants produce large quantities of antigen, they tend to grow poorly. Also, some plants require cooking in order to be palatable. Plants containing edible vaccines cannot be cooked because heating denatures the proteins, thereby preventing an appropriate immune response [16].

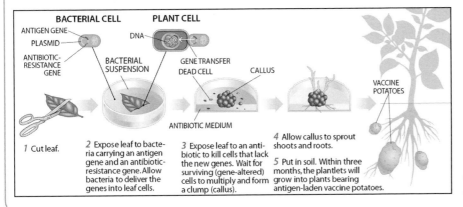

Figure courtesy of Jared Schneidman Design.

their own bodies. During the 1800s, cartoons appeared mocking Jenner and depicting the transformation of the recently vaccinated into sickly cows and fantastic beasts (Figure 8.23) [1].

Despite these concerns, rapid progress followed. In 1885, Louis Pasteur developed the concept of an attenuated vaccine and produced the first vaccine against rabies [1]. In the early 1900s, the concept of toxoid vaccines was developed, leading to vaccines for diphtheria and tetanus [5]. In the 1950s, the tools to maintain cells alive in tissue culture were developed [19]. This scientific advance led to a live attenuated vaccine for polio, a discovery for which Enders, Robbins, and Weller won the Nobel prize. Vaccines for measles, mumps, and rubella were developed in the 1960s [17].

Figure 8.23. *Cartoon mocking Edward Jenner and vaccination. Wellcome Library, London.*

Table 8.2. *The incidence of many infectious diseases in the United States was dramatically reduced by vaccines [20].*

Disease	Peak # of Cases	# Cases in 2000	% Change
Diphtheria	206,929 (1921)	2	−99.99
Measles	894,134 (1941)	63	−99.99
Mumps	152,209 (1968)	315	−99.80
Pertussis	265,269 (1952)	6755	−97.73
Polio	21,269 (1952)	0	−100
Rubella	57,686 (1969)	152	−99.84
Tetanus	1560 (1923)	26	−98.44
HiB	~20,000 (1984)	1212	−93.14
Hep B	26,611 (1985)	6646	−75.03

Table 8.2 shows the dramatic reduction in the incidence of infectious disease in the USA following routine vaccination. For example, there were more than 200,000 cases of diphtheria in the USA in 1921, the year of peak incidence. In 2000, only two cases of diptheria were reported in the USA; the dramatic reduction in incidence is due to routine childhood immunization. Polio has been eliminated in the USA due to vaccination. In 2005, the CDC announced that rubella is no longer a health threat in the USA [21]. In 1965, there were 12.5 million cases of rubella in the USA. As a result, more than 12,000 babies were born deaf, blind or both and 6200 children were stillborn. In 2004, there were only nine rubella infections in the USA [22].

Smallpox is one of world's deadliest diseases, having caused more deaths in history than any other disease.

However, smallpox is also the first human disease to be eradicated from the face of the Earth by a global immunization campaign. As a result, we no longer routinely immunize against smallpox. The Jenner vaccine was first available in the early 1800s [1]. However, it was difficult to keep the vaccine viable enough to deliver in the developing world. In the 1950s a much more stable, freeze dried vaccine was developed, making it practical to deliver vaccine worldwide. In 1959, the Twelfth World Health Assembly set a goal to eradicate smallpox from the globe. However, little progress was made until 1967 when sufficient economic resources were dedicated to vaccinate at least 80% of all populations and to survey for and contain outbreaks. At that time, approximately ten million cases of smallpox occurred per year. On May 8, 1980, the Certification of Smallpox Eradication declared the world to be smallpox free, and we no longer routinely vaccinate for smallpox [18].

In the developed world, routine childhood immunization has dramatically reduced the incidence and mortality associated with many diseases. However, the situation is drastically different in the developing world. In 1974, only 5% of the world's children received six vaccines recommended by WHO [23]. At that time, the WHO set a goal to immunize at least 80% of the world's children against these six diseases by 1990 [17]. The program has been a tremendous success, and as of 2004, vaccine coverage has reached nearly 80%. As a result of vaccination, 20 million lives have been saved over the last two decades. Figure 8.24 shows the dramatic reduction in the incidence of reported cases of measles and pertussis during this time period. While these achievements have dramatically reduced child

Measles and Pertussis Cases Reported to the WHO

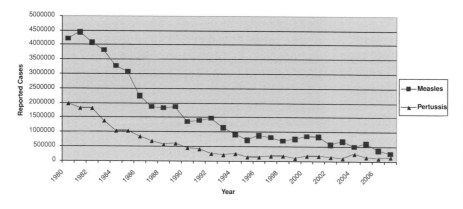

Figure 8.24. *The worldwide incidence of measles and pertussis have dropped drastically in response to efforts to increase childhood vaccinations. Source: WHO, Immunization surveillance, assessment and monitoring.*

mortality, the 20% of children who do not receive these vaccines account for nearly 1.4 million preventable deaths each year due to pertussis, diphtheria, polio, measles, tetanus and tuberculosis [24].

Listen to the story of the first volunteer to test a vaccine designed to prevent Ebola in a phase I clinical trial [25].

NPR Story – Nurse Takes Plunge in Ebola Test

With new technologies, it is likely that nearly a dozen vaccines will soon be available. Recently licensed new vaccines include Gardasil (Merck) to prevent HPV infection and Prevnar (Wyeth Pharmaceuticals) for pneumococcal disease. Half of all vaccines have been developed in the past 25 years (about one per year, compared to about one every five years before this) [14].

How do we test the effectiveness of new vaccines?

Vaccines are first tested in the laboratory to see if they initiate a response in cell culture or tissue culture systems. If successful, the vaccine is then tested in animal model systems. The animal must be susceptible to infection by the agent against which the vaccine is directed and should develop the same symptoms as humans. Vaccines which are successful at preventing disease in animals can then enter human trials.

We will learn more about the process of testing drugs and devices in patients later in Chapter 9, but briefly, these trials have three phases. In phase I trials, the vaccine is tested in a small number of volunteers (20–80 persons) [26]. Usually phase I trials are carried out in healthy adults and last a few months. The goal of phase I trials is to determine the vaccine dosages necessary to produce protective levels of immunity, as well as to evaluate side effects at these dosages. Before phase I trials can be carried out, the Food and Drug Administration (FDA) must approve the vaccine as an Investigational New Drug (IND) [27].

If successful results are obtained in phase I trials (immune protection with minimal side effects), then the vaccine goes into phase II clinical trials. In a phase II trial, a larger number of volunteers (several hundred) are tested, over a period lasting a few months to a few years [28]. Generally, a phase II trial is a controlled study, with some volunteers receiving the vaccine and some receiving a placebo (or existing vaccine). The volunteers are monitored to see if they mount an immune response or contract the disease (vaccine effectiveness) or to see if they develop side effects (vaccine safety).

Finally, vaccines enter phase III clinical trials involving large numbers of volunteers (several hundred to several thousand) [28]. These trials last years and are usually carried out as controlled double blind studies, with some volunteers receiving vaccine and some receiving placebo (or existing vaccine) [26]. The trial is referred to as double blind because neither patients nor physicians know which was given.

If the vaccine is proven to be safe and effective in phase III clinical trials, the manufacturer can apply to the FDA to sell the vaccine. Licensure by FDA is required before a company can market the vaccine. Generally this requires about a decade [28]. Vaccines must be made following strict manufacturing guidelines and quality control procedures known as current Good Manufacturing Practices (cGMP) [26]. These regulations ensure that the plant, equipment and procedures are designed to make a vaccine that is as safe as possible. Each batch of vaccine must be tested for safety, potency, purity, and a sample lot must be sent to the FDA [28].

After approval to market a vaccine is given, the FDA continues to monitor vaccine safety in a process known as post-licensure surveillance. Doctors must report adverse reactions after vaccination to the FDA and the Centers for Disease Control and Prevention (CDC). The reporting system is known as the Vaccine Adverse Events Reporting System (VAERS); it receives as many as 12,000 reports per year, of which 2000 are serious. Most are unrelated to the vaccine, but some can indicate rare but serious side effects that were not observed in phase III clinical trials [29].

How effective are vaccines?

In general, about one to two of every 20 people immunized will not have an adequate immune response to a vaccine [30]. Yet, vaccination has largely eliminated many diseases. This occurs because of a phenomenon

Global Diphtheria Cases Reported to WHO

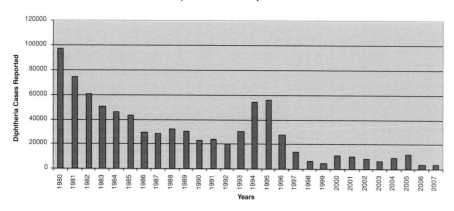

Figure 8.25. *The global incidence of diphtheria, 1980–2007. Source: WHO, Immunization surveillance, assessment and monitoring.*

known as herd immunity. Vaccinated people have antibodies against a pathogen, and as such they are much less likely to transmit that germ to other people. As a result, even people that have not been vaccinated are protected. About 85–95% of the community must be vaccinated to achieve herd immunity [1].

When herd immunity is lost, outbreaks of once uncommon diseases can occur. In the early 1990s, eastern Europe experienced an outbreak of diphtheria. Universal childhood immunization against diphtheria was introduced throughout the Soviet Union in 1958, and diphtheria incidence dropped and remained at very low levels for more than thirty years. However in the early 1990s, childhood immunization rates fell in the newly independent states of the former Soviet Union. Declining economic conditions, large population migrations and low immunization rates led to a resurgence of the incidence of diphtheria (Figure 8.25) [31]. Massive efforts to vaccinate children and adults were required to bring the epidemic back under control.

As vaccines become more widely available, they have often lost their allure in many countries [1]. With the success of widespread immunization, the threat of illness that initially led to the support of vaccines has diminished. Instead, attention has become increasingly focused on the risks of vaccination. One needs only to Google the terms "vaccine safety" to see the results of this shift in public perception in the United States [32]. In 1954, more Americans knew about the field trial of the Salk Polio vaccine than knew the full name of US President Dwight David Eisenhower. At the time, the March of Dimes carried out extensive media campaigns,

which increased awareness of the risks of polio and efforts to develop a vaccine. As a result, most Americans understood the risks of polio and were anxious to be vaccinated [33]. As vaccination has reduced the incidence of infectious disease, groups opposing vaccines have proliferated. An increasingly cynical public sometimes regards information campaigns about the benefits of new vaccines as hype to increase the revenue of pharmaceutical companies. The Internet has facilitated the spread of dissenting views about the risks of vaccination [32].

For example, some watchdog groups have questioned the link between a rise in autism and the use of the preservative thimerosal in vaccines. A careful series of scientific studies have shown there is no link between thimerosal in vaccines and austism. Even so, the FDA ceased to license thimerosal-containing vaccines [1].

Similar claims linking autism and the MMR vaccine have arisen [1]. In 1998, the journal *The Lancet* published a paper that investigated the link between chronic gastro-intestinal disease and severe developmental regression and autism in a small group of children [34]. The paper noted that most instances occurred after MMR immunization, but the researchers noted they had not yet proved a causal link. Though the paper was appropriately cautious, the lead author of the study, London-based researcher Andrew Wakefield, held a press conference and warned parents that it would be safer if their children received individual vaccinations for measles, mumps and rubella, rather than the combined shot [32]. Fearing a decline in immunization rates, the UK Department of Health urged parents

Table 8.3. *Immunization schedule for children from birth to six years of age (courtesy of CDC).*

Recommended Immunization Schedule for Persons Aged 0–6 Years—UNITED STATES • 2007

Vaccine ▼ Age ▶	Birth	1 month	2 months	4 months	6 months	12 months	15 months	18 months	19–23 months	2–3 years	4–6 years
Hepatitis B[1]	HepB	HepB		see footnote 1	HepB					HepB Series	
Rotavirus[2]			Rota	Rota	Rota						
Diphtheria, Tetanus, Pertussis[3]			DTaP	DTaP	DTaP		DTaP				DTaP
Haemophilus influenzae type b[4]			Hib	Hib	*Hib[4]*	Hib		Hib			
Pneumococcal[5]			PCV	PCV	PCV	PCV				PCV PPV	
Inactivated Poliovirus			IPV	IPV		IPV					IPV
Influenza[6]						Influenza (Yearly)					
Measles, Mumps, Rubella[7]						MMR					MMR
Varicella[8]						Varicella					Varicella
Hepatitis A[9]						HepA (2 doses)				HepA Series	
Meningococcal[10]										MPSV4	

Legend: Range of recommended ages · Catch-up immunization · Certain high-risk groups

This schedule indicates the recommended ages for routine administration of currently licensed childhood vaccines, as of December 1, 2006, for children aged 0–6 years. Additional information is available at http://www.cdc.gov/nip/recs/child-schedule.htm. Any dose not administered at the recommended age should be administered at any subsequent visit, when indicated and feasible. Additional vaccines may be licensed and recommended during the year. Licensed combination vaccines may be used whenever any components of the combination are indicated and other components of the vaccine are not contraindicated and if approved by the Food and Drug Administration for that dose of the series. Providers should consult the respective Advisory Committee on Immunization Practices statement for detailed recommendations. Clinically significant adverse events that follow immunization should be reported to the Vaccine Adverse Event Reporting System (VAERS). Guidance about how to obtain and complete a VAERS form is available at http://www.vaers.hhs.gov or by telephone, 800-822-7967.

The Recommended Immunization Schedules for Persons Aged 0–18 Years are approved by the Advisory Committee on Immunization Practices (http://www.cdc.gov/nip/acip), the American Academy of Pediatrics (http://www.aap.org), and the American Academy of Family Physicians (http://www.aafp.org).

DEPARTMENT OF HEALTH AND HUMAN SERVICES • CENTERS FOR DISEASE CONTROL AND PREVENTION
SAFER • HEALTHIER • PEOPLE™

not to reject MMR vaccinations. Between 1998 and 2004, fueled in part by inflammatory medial coverage, MMR immunization rates declined in Britain to only 80%, falling to just 62% in some areas of London. Two subsequent, larger studies in 1999 showed there was no link between the MMR vaccine and autism. In 2004, collaborators on Wakefield's paper publicly rejected the link between autism and the MMR vaccine. That same year, Wakefield was accused of having misled the editors of *The Lancet* by concealing the fact that his research was partially funded by the legal team seeking compensation for parents who believed their children were injured by the MMR vaccine [32].

Childhood illness and vaccines

Who receives vaccines that have been licensed by the FDA? Generally, recommendations are made by the Centers for Disease Control and Prevention (CDC) working in conjunction with expert physician groups regarding when the vaccine should be used and who should receive it. In making recommendations, these experts weigh the risks and benefits of the vaccine, as well as the costs of vaccination [28]. In addition, some vaccinations are required by law. In 1905, the US Supreme Court ruled that the need to protect public health by requiring smallpox vaccination outweighed an individual's right to privacy [1]. All 50 states have school immunization laws.

These laws provide for exemptions based on medical reasons (50 states), religious reasons (48 states), and philosophical reasons (15 states) [28]. Tables 8.3 and 8.4 indicate the CDC recommended childhood and adolescent vaccination schedules in the USA, respectively. Children are now routinely immunized against 16 diseases (Figure 8.26) [35].

How vaccines are made

There are substantial scientific and engineering challenges associated with developing new vaccines. For new vaccines to impact public health, we must be able to manufacture hundreds of millions of doses of vaccine. Each and every dose must be safe and effective and equivalent. Because vaccines are given to healthy children and adults, the burden of ensuring that each dose of vaccine is safe is particularly high [36]. Large scale manufacturing of vaccines involves substantial engineering challenges. It requires the ability to take a candidate vaccine developed in a basic research lab and scale up the manufacturing process to make millions of doses. As we will see, to do this effectively

Table 8.4. *Immunization schedule for persons from seven to 18 years of age (courtesy of CDC).*

Recommended Immunization Schedule for Persons Aged 7–18 Years—UNITED STATES • 2007

Vaccine ▼ Age▶	7–10 years	11–12 YEARS	13–14 years	15 years	16–18 years
Tetanus, Diphtheria, Pertussis[1]	see footnote 1	Tdap	Tdap		
Human Papillomavirus[2]	see footnote 2	HPV (3 doses)	HPV Series		
Meningococcal[3]	MPSV4	MCV4	MCV4[3] MCV4		
Pneumococcal[4]		PPV			
Influenza[5]		Influenza (Yearly)			
Hepatitis A[6]		HepA Series			
Hepatitis B[7]		HepB Series			
Inactivated Poliovirus[8]		IPV Series			
Measles, Mumps, Rubella[9]		MMR Series			
Varicella[10]		Varicella Series			

Range of recommended ages

Catch-up immunization

Certain high-risk groups

This schedule indicates the recommended ages for routine administration of currently licensed childhood vaccines, as of December 1, 2006, for children aged 7–18 years. Additional information is available at http://www.cdc.gov/nip/recs/child-schedule.htm. Any dose not administered at the recommended age should be administered at any subsequent visit, when indicated and feasible. Additional vaccines may be licensed and recommended during the year. Licensed combination vaccines may be used whenever any components of the combination are indicated and other components of the vaccine are not contraindicated and if approved by the Food and Drug Administration for that dose of the series. Providers should consult the respective Advisory Committee on Immunization Practices statement for detailed recommendations. Clinically significant adverse events that follow immunization should be reported to the Vaccine Adverse Event Reporting System (VAERS). Guidance about how to obtain and complete a VAERS form is available at http://www.vaers.hhs.gov or by telephone, 800-822-7967.

The Recommended Immunization Schedules for Persons Aged 0–18 Years are approved by the Advisory Committee on Immunization Practices (http://www.cdc.gov/nip/acip), the American Academy of Pediatrics (http://www.aap.org), and the American Academy of Family Physicians (http://www.aafp.org).

DEPARTMENT OF HEALTH AND HUMAN SERVICES • CENTERS FOR DISEASE CONTROL AND PREVENTION

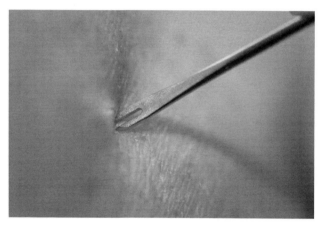

Figure 8.26. *By age two children must receive more than 20 shots. Sometimes as many as five shots are required in a single visit to the pediatrician. Courtesy of CDC/James Gathany.*

requires that scientists and bioprocess engineers work closely together; typically a team of at least ten people is required to lead and coordinate such a complex project [36]. As an example of these challenges, we next consider how the seasonal influenza vaccine is made, and the hurdles that must be overcome in order to produce sufficient vaccine to prevent future influenza pandemics.

Seasonal influenza vaccine

Influenza is the seventh leading cause of death in the USA. It is the leading cause of death in children aged from one to four years old, and pneumonia associated with influenza causes 90% of deaths in people over the age of 65 [15]. The Advisory Committee on Immunization Practices, a branch of the CDC, recommends an annual influenza vaccine for children between the ages of six months and five years, pregnant women, people 50 years of age and older, people of any age with certain chronic medical conditions and people who live in nursing homes and other long term care facilities [37].

The antigenic drift of the influenza virus presents special challenges for developing an influenza vaccine. The number and type of influenza strains circulating among the population varies dramatically from year to year. Generally, it takes two to four weeks following vaccination for people to develop protective immunity [15]. Thus, to be effective the influenza vaccine must be available in advance of the peak influenza season. Because of the long lead time to produce millions of doses of the flu vaccine and the time required for vaccinated people to develop immunity, the choice of strain to be used in the vaccine must be made months in advance of flu

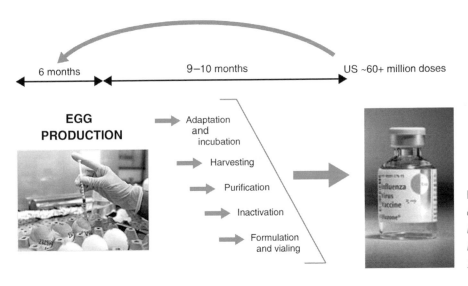

6 months   9–10 months   US ~60+ million doses

EGG PRODUCTION

Adaptation and incubation

Harvesting

Purification

Inactivation

Formulation and vialing

Figure 8.27. *Manufacturing process of current flu vaccine using chicken eggs. Reprinted with permission from the National Academies Press, Copyright 2000, National Academy of Sciences.*

season, increasing the chances of selecting the wrong strain.

When the influenza vaccine was first developed in the 1940s, it provided protection against only one strain of the influenza virus (monovalent vaccine). In the 1960s and 1970s, vaccines were developed that protected against two strains (bivalent vaccine), increasing the chances that more people would be protected against circulating strains. Starting in 1978 and continuing to today, three strains are included (trivalent vaccine): two A strains and one B strain. Between 1970 and 2004, the formulation has changed 40 times for one of the strains. On eight occasions, changes have been made in two strains, and on one occasion, all three strains were changed [15].

Recent shortages in the availability of the influenza vaccine highlight the engineering challenges associated with producing vaccines. In 2003–4, two companies manufactured 83 million doses of flu vaccine for the US market: 48 million doses were made by Aventis in the USA and an additional 35 million doses were made by Chiron in Liverpool, England. That year, the influenza epidemic started early and the media broadcast many stories describing patients who were hospitalized and died from influenza and subsequent pneumonia. The demand for vaccine exceeded supply and many people could not obtain the vaccine. The shortage of vaccine was even more dramatic the following year. Although more doses were manufactured (Aventis made 55 million doses and Chiron made 48 million doses), there was

a manufacturing error at the Chiron plant, and those doses could not be sold [33].

As a result of these shortages, scientists have begun to examine alternative, more rapid methods of manufacturing the influenza vaccine. The most common influenza vaccine used today is based on an inactivated form of the virus; chicken eggs are used as small bioreactors to grow sufficient quantities of the virus, which is then harvested and inactivated. The current manufacturing process relies largely on technology developed more than 60 years ago [14].

Once a year, for each hemisphere, experts gather to decide upon the vaccine composition. This is a time consuming step, requiring about seven weeks [15, 38]. Figures 8.27 and 8.28 show the steps in the process of manufacturing the current flu vaccine. In order to produce enough vaccine, approximately 300 million chicken eggs are required; egg production occurs in parallel with strain selection [38]. After the strain of virus has been selected, a form of the actual virus to be grown in eggs must be developed through a process called reassortant preparation. In this process, cells in culture are co-infected with the wild type strain and a strain which has been adapted to grow very efficiently in eggs. The goal is to create a new strain of the virus – one which is capable of producing immunity in people but will grow well in eggs. Eggs are then inoculated with the reassortant preparation in order to produce large amounts of the virus. Fluid containing the virus is then harvested from the chicken eggs, purified using

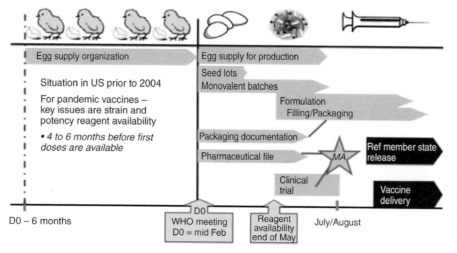

Figure 8.28. *Timeline of yearly flu vaccine manufacturing process. Reprinted with permission from the National Academic Press, Copyright 2000, National Academy of Sciences.*

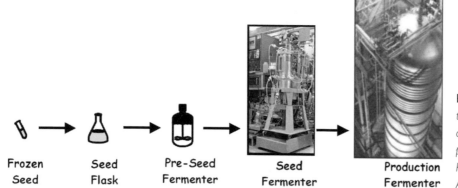

Figure 8.29. *Manufacturing process used to produce virus in cultured mammalian cells as an alternate form of vaccine production [40]. From Estell, 2006. Reprinted with permission of the National Academy of Engineering.*

centrifugation and filtration, and inactivated using formalin. This process occurs for each of the three strains. Purified, inactivated virus from each strain is then combined, and packaged into doses of the trivalent influenza vaccine [15]. Inactivated influenza vaccine is stored in the refrigerator and loses its immunogenicity if frozen [39]. After the vaccine is made, the manufacturer must still carry out phase I, phase II and phase III clinical trials to test the vaccine efficacy and safety.

This entire process (including licensing and safety testing) must be repeated each year. All unused vaccine is discarded. The monovalent concentrates cannot be reused after 12 months [15]. Because the yields of new strains are not known in advance, it is difficult to ensure that the manufacturing process will give sufficient quantities of a new strain.

There are some important advantages of this manufacturing process. Because it has been used for many years it is well tested and understood. The only part that changes from year to year is the structure of the virus to

be produced – all other elements in the process can stay the same from year to year. However, because the process is cumbersome and involves long lead times, there are concerns that it will not be possible to produce sufficient vaccine to prevent an influenza pandemic should a virulent new strain of flu emerge. Until recently, egg production was seasonal raising the likelihood that a pandemic might occur at a time when no eggs are ready. Because of this concern, industry has recently changed to a cycle of continuous egg production. Furthermore, because of concerns that an avian influenza virus could infect populations of chickens that produce eggs to make vaccine, flocks associated with egg production are now under strict biosafety control so they cannot be wiped out [15].

Updating the manufacturing process may increase the speed with which new vaccine could be produced. Instead of using eggs as bioreactors to grow the influenza virus, it is possible to grow the virus in mammalian cells in culture (Figure 8.29). The process of developing the vaccine is similar, except it is made in the mammalian cells rather than in eggs. The mammalian cells are kept in a bioreactor. The cell line to be used must be able to grow the influenza virus in large numbers, and be suitable for a wide variety of flu strains. Several cell lines which meet these requirements are available and they can be grown in chemically defined, synthetic growth media. The purification and inactivation procedures are similar to those used in egg based systems. An important advantage of mammalian cell culture based systems is that it completely avoids the need to grow eggs from biosecure flocks. In addition, growing virus in this manner may lead to higher initial purity [41].

Can we switch to systems that use mammalian cell culture? In the event of pandemic influenza, cell culture based manufacturing of vaccines could provide an advantageous alternative to traditional egg-based systems. At present, there are approximately 1.5 million liters of cell culture capacity in the USA. However, this capacity is currently dedicated to the production of other essential drugs and on a very limited basis, vaccines. Given the high cost for construction and validation of a biological production facility (upwards of a billion dollars), it is not economically feasible to have production facilities sitting idle. In the event of an emergency, such as pandemic influenza, cell culture facilities would need to halt production of other essential drugs in order to accommodate the demand for vaccines [38].

Another approach to vaccine production is to use recombinant methods to produce a subunit vaccine for influenza. We have seen that the influenza virus has two major antigens on its surface: the hemagglutinin and neuraminidase proteins. The recombinant process involves taking the DNA for HA and inserting it into another type of cell which can then be used as a bioreactor to produce large amounts of the HA protein (Figure 8.30) for use in a vaccine. Yeasts are currently used to make the recombinant proteins in the hepatitis B vaccine and in the HPV vaccine [42].

Pandemic influenza vaccine

Given the existing egg based production capacity, let's examine how long it would take to produce sufficient flu vaccine for global coverage in the event of a flu

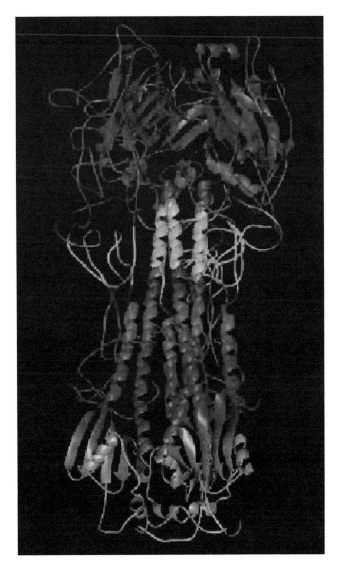

Figure 8.30. *Hemagglutinin protein produced through recombinant methods for use in a subunit vaccine Isin, Doruker, Bahar. Functional Motions of Influenza Virus Heamagglutinin (HA): A Structure-Based Analytical Approach.* Biophysical Journal, 2002;**82**(2):569–81.

pandemic. The current trivalent vaccine contains 15 µg of the HA antigen for each strain of the virus. Each year, approximately 300 million doses of the trivalent seasonal vaccine are produced. At current production capacity then, approximately 1 billion doses of monovalent vaccine could be produced. This is enough to immunize only about 1/6 of the world's population [14]. Table 8.5 shows that it would take almost five years to produce sufficient vaccine to provide global coverage.

Unfortunately, it is unlikely that we will have five years to produce vaccine in the event of an influenza

Table 8.5. *Current influenza vaccine production capability [14]. Reprinted with permission from Macmillan Publishers Ltd: Nature Publishing Group [14], copyright 2006.*

Scale-up of influenza vaccine production		
Production time (including lead time)	Worldwide capacity (monovalent doses of 15 μg)	Worldwide coverage (%)
1 year	~1,000,000,000	~17
2 years	~2,500,000,000	~40
4 years and 9 months	~6,500,000,000	100

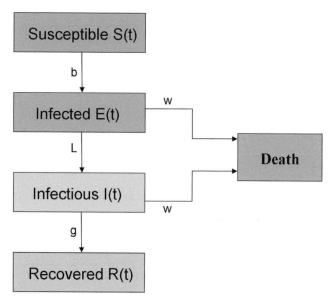

Figure 8.31. *Progression of infectious diseases [44].*

pandemic. Because antigenic shift is a random occurrence, it is difficult to predict in advance the structure of vaccines that might be protective. To get some idea of how rapidly we might need to manufacture vaccine in the event of a pandemic, we can use simple modeling techniques to predict how rapidly a pandemic might spread throughout the world once a new strain emerges.

In the simplest modeling approach, we divide the population into groups, based on their disease status, and track how the number of people in each group changes over time (Figure 8.31) [43]. Consider a population of N people, divided into the following groups:

S = the number of people susceptible to infection

E = the number of infected people who are not yet contagious

I = the number of infectious people

R = the number of people who have been infected and recovered.

When a new strain of flu emerges, the population initially is all susceptible. Over time, people acquire disease, transmit it from person to person, and either die or recover. The decrease in the number of susceptible people as a function of time is given by Equation (8.1).

$$\frac{dS}{dt} = -bSI \tag{8.1}$$

where b represents the person to person transmission rate.

The time between the point of infection and the point where a person becomes infectious is known as the incubation period, $1/L$. The incubation period is approxi-

mately 1.2 days for the influenza virus [43]. The mortality rate due to influenza, w, is estimated to be 0.0005 per day for a virulent pandemic. Thus, the change in the number of infected people versus time is given by Equation (8.2)

$$\frac{dE}{dt} = bSI - LE - wE. \tag{8.2}$$

The infectious period, $1/g$, is approximately 4.1 days for influenza [43]. We can express the change in the number of people who are infectious versus time with Equation (8.3)

$$\frac{dI}{dt} = LE - gI - wL. \tag{8.3}$$

Finally, the change in the number of people who have recovered versus time is given by Equation (8.4)

$$\frac{dR}{dt} = gI. \tag{8.4}$$

During this period we assume that the birth rate and the death rate due to other causes are the same. We can solve this simple system of differential equations to predict the duration of an influenza pandemic. In order to solve the equations, we need to specify the initial conditions. We assume that the population is at some initial number and begins with one initial infected person; everyone else is uninfected.

(a)

(b)

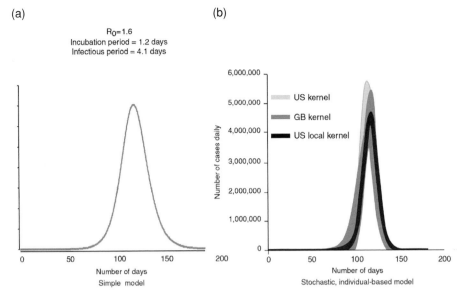

Figure 8.32. *(a) Model demonstrating spread of virus over first 200 days. (b) Model demonstrating spread of virus taking into account international patterns of population migration. Reprinted with permission from the National Academies Press © 2000, National Academy of Sciences.*

The predictions of this simple model indicate that the vast majority of cases would occur in the first 200 days of an epidemic [43]. Figure 8.32 shows the predicted number of cases per day following the initial case. The predictions of this very simple model (Figure 8.32a) agree quite well with predictions of much more sophisticated models that take into account factors such as international patterns of population migration (Figure 8.32b) [43]. In addition, these predictions agree well with epidemiologic data from the 1918 Spanish influenza pandemic, and illustrate the scary prospect that in the event of a flu pandemic, vaccine will be primarily available to survivors unless improvements are made in the manufacturing process [45].

Interestingly, the models predict that it will be far more difficult to control an influenza pandemic than SARS, because a large part of the infectious period associated with influenza occurs before the onset of symptoms. A number of researchers have developed computational tools to understand the most effective approach to prevent an influenza pandemic. Efforts such as border restrictions are unlikely to be effective. Anti-virals must be given within one or two days of symptoms to have an effect. Models indicate that the best way to prevent a pandemic is to pre-vaccinate the population. These models show that you need to immunize at least 1/3 of the population if you have a vaccine of perfect efficacy in order to prevent a pandemic [43]. However, with current manufacturing capabilities, making this amount of vaccine will take more than one year [14].

Economic challenges

Are there economic incentives in place to encourage the type of investment in vaccine research and development and manufacturing processes that is needed to meet global health needs? Vaccines are made by pharmaceutical companies, and pharmaceutical companies are businesses. The current economic outlook for vaccine products is not encouraging. In 1967, there were 26 companies that made vaccines used in the USA; in 2004 there were only five. Since 1998, nine of twelve vaccines recommended for children in the USA have been in short supply [33]. As a result, children were delayed in receiving vaccines that they needed, and some children never caught up. It is particularly worrisome that vaccines for seven childhood diseases have only a single manufacturer. What happens if this company experiences a business problem or production failure [1]?

Today the vaccine industry faces major hurdles. The research and development process for a new vaccine is increasingly expensive, lengthy and risky. Companies must build expensive manufacturing facilities, operate within a complex regulatory environment, and deal with a growing anti-immunization movement and a surge in liability litigation. As a result of these factors, it now costs between $110–$800 million to bring a new drug to market, and typically takes more than a decade to bring a vaccine from early development to finished product launch [26]. In 2003, a report from the Institute of Medicine noted with concern the lack of financial incentives for vaccine manufacturers and called for reforms that would encourage investment [46]. In 1998, Warner Lambert stopped making Fluogen vaccine for influenza because of economic considerations and regulatory challenges [1].

Over the past 50 years, there have been many mergers in the pharmaceutics industry. Companies which previously only made vaccines now make both drugs and vaccines. In these companies vaccine products compete with potential drug products for limited research and development dollars [33]. In general, the market of a drug is much larger – vaccines are only given once, and are often purchased by the federal government. Today, 43% of childhood vaccines are purchased by the private sector. The remaining 57% are purchased through a federal contract which covers children on Medicaid or without health insurance. In 2005, the price for vaccines recommended for children before entering elementary school was $474 if purchased on the federal contract and $782 if purchased privately. Yet, every dollar spent on vaccines, saves $5.80 in direct medical costs [26]!

Litigation has also played a role in increasing the costs of vaccines in the USA. In the 1970s and 1980s a series of personal injury lawsuits claiming that the pertussis vaccine resulted in complications such as sudden infant death syndrome and mental retardation were filed. Although scientific studies showed there was no link between the vaccine and these adverse events, due to litigation the cost of the pertussis vaccine increased from 17 cents per dose to $11 per dose [33]. To stem this trend, in 1986, the National Vaccine Injury Compensation Program (VICP) was established in the USA. The program is funded by a 75 cent tax on each dose of vaccine. Litigants claiming to have been injured by a vaccine must first file a claim through this program. Some injuries are automatically eligible for compensation with no need to prove that the vaccine caused the injury. If the injury is outside the rule, the claimant must prove that the vaccine was the cause of injury.

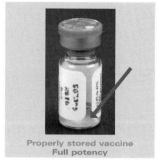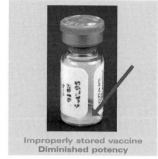

Figure 8.33. *Vaccines with full potency (left) and diminished potency (right). For most vaccines, potency cannot be determined by simply looking at the vial [39].*

Claimants are free to either accept or reject the decision and award of the VICP. As a condition of accepting an award from the VICP, claimants agree not to pursue further legal action against the vaccine manufacturer. The VICP has paid about $600 million for injuries caused by vaccines administered between 1988 and 2004. Only a small fraction of these funds (2%) were used to cover lawyers' fees; the rest went to the claimant [26].

Challenges of vaccination in developing countries

Developing countries now wait an average of 20 years between when a vaccine is licensed in industrialized countries and when it is available for their own populations [47]. Economic, infrastructural, and scientific hurdles all contribute to this long delay.

Vaccines are complex biological substances; they can lose their potency over time. They are more likely to lose potency if exposed to temperatures which are too cold or too warm; some vaccines are also sensitive to exposure to ultraviolet light. This loss of potency is permanent and irreversible. For most vaccines, it is impossible to tell if they have lost potency simply by looking at them (Figure 8.33) [39]. If individuals are vaccinated with vaccine that has been damaged in this way, they will not have the desired immune response.

The cold chain is a system that has been developed to ensure that vaccines remain potent as they make the trip from manufacturer to the patient being immunized. The cold chain has three main components – transport and

The Cold Chain

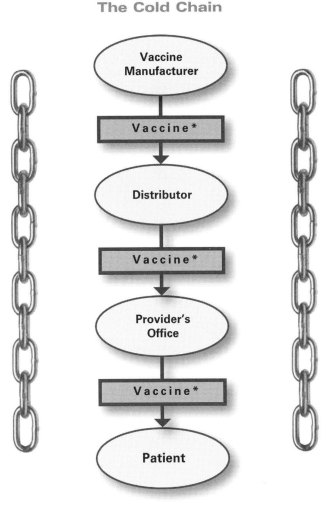

Figure 8.34. *Cold chain system. This is designed to ensure vaccines remain potent during transport from the manufacturer to the patient [39].*

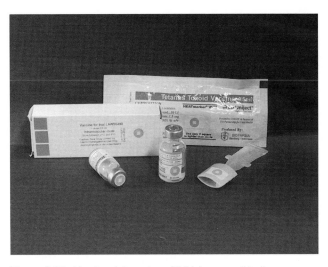

Figure 8.35. *Vaccine vial monitors (VVMs) are small indicators designed to irreversibly change color upon exposure to a specific heat level. Reprinted with permission from PATH and photographer Jennifer Fox. Vaccine vial monitors by TEMPTIME Corporation.*

storage equipment, trained personnel, and management procedures. As shown in Figure 8.34, the cold chain begins with the refrigerator at the manufacturing plant, extends as the vaccine is transfered to the distributor, delivered to the provider's office and ends with vaccine administration. It is essential to maintain proper temperature at each step in the process. It has been estimated that 17–37% of providers expose vaccines to improper storage temperatures [39].

Maintaining the cold chain is a challenge in developing countries, where a lack of infrastructure can make it difficult to maintain proper storage temperatures. When workers suspect that a container of vaccine has not been properly transported or stored, they must throw it away rather than risk using inactive vaccine. A number of tools have been developed to monitor the temperature history of vaccines during the transport and storage processes. In 1996, vaccine vial monitors (VVM) were developed based on technology originally devised for use in the food industry (Figure 8.35). VVMs are small indicators adhered to the tops of vaccine vials; the inner square of the VVM is chemically active and changes color irreversibly with exposure to heat. VVMs can be manufactured for a variety of heat-exposure specifications. Since March 1996, all oral polio vaccine supplied through UNICEF carry VVMs, adding only pennies to the cost of a vial. As of January 2001, all vaccines supplied by UNICEF are required to have VVMs. More than 1 billion VVMs have been delivered to developing countries, and 16 of 25 UN pre-qualified vaccine suppliers include VVMs on their products. The use of VVMs has led to a significant reduction in vaccine wastage because there is an accurate record of their temperature history [48]. Freeze watch indicators (Figure 8.36) have been developed to monitor whether vaccines have been exposed to temperatures below 0 °C. The freeze watch indicator consists of a small vial of red liquid contained in a plastic casing. If exposed to temperatures below 0 °C for more than one hour, expansion of the liquid causes the vial to burst, releasing the red liquid [49].

(a)   (b)

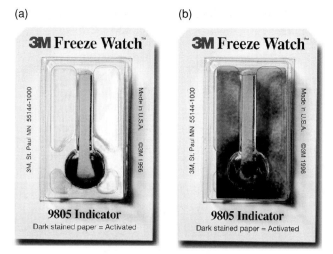

Figure 8.36. *Freeze watch indicators monitor whether a vaccine has been exposed to temperatures below 0 °C. Courtesy of 3M.*

Most vaccines must be given by injection. This is a particular challenge in developing countries, where healthcare workers may not have access to an adequate supply of sterile needles. As a result, disposable syringes are often saved and reused. It has been estimated that over 50% of injections given in developing countries follow unsafe injection practices, which can lead to the spread of blood borne diseases [50]. A simple technologic solution is now in place to address this challenge. The BD SoloShot(TM) auto-disable syringe (Figure 8.37) is designed so that when the syringe is filled to a preset level, the plunger stops and can't be pulled back, further ensuring that the correct amount of vaccine is delivered. After one use, the plunger automatically locks so that it can't be reused. The BD SoloShot(tm) syringe is manufactured and marketed by Becton Dickinson and Company [51]. The price of the auto-disable syringe is rapidly dropping and is within 1 cent of disposable syringes. Since its commercial introduction in 1992, more than 2.5 billion immunizations have been delivered using BD SoloShot(TM) syringes in more than 40 countries in Africa, Asia, Eastern Europe, and Latin America. UNICEF provides only auto-disable syringes to countries that request disposable syringes [50].

An alternative approach is to develop needle free methods to deliver vaccines. In developed countries, jet injector guns are used to deliver vaccines without needles. These devices rely on a liquid stream at high

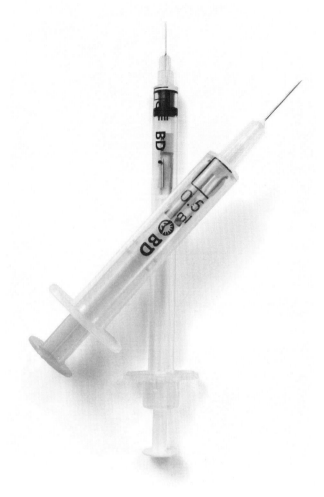

Figure 8.37. *BD SoloShot auto-disable syringes automatically lock after a single use. Courtesy and © Becton Dickenson and Company.*

pressure that is used to penetrate the skin [49]. The jet injector was initially developed for use in mass injection campaigns and could immunize between 600 and 1000 people per hour. From the 1950s to the 1980s they were widely used in US school immunization campaigns and throughout the developing world. However, their use was discontinued in the 1980s, when it was recognized that the multiuse jet injectors had a small risk of transmitting blood borne pathogens from one person to another. Recently, single-use jet injectors, such as Biojector2000 (Figure 8.38) have been developed. However, they are currently too expensive for developing countries [52].

The Global Alliance for Vaccines and Immunization (GAVI) is a partnership between many public and private organizations – including UNICEF,

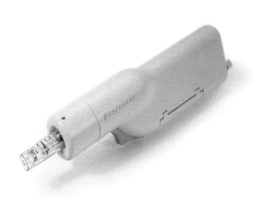

Figure 8.38. *Jet injector guns, like this Biojector2000, deliver vaccines without needles, using a sterile, single-use syringe. Courtesy of Bioject, Inc.*

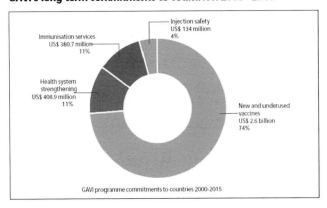

Figure 8.39. *By the end of 2007, US $3.5 billion had been approved for spending in countries up to 2015. Source: GAVI Alliance Secretariat, 2008.*

the WHO, the Bill and Melinda Gates Foundation, members of the vaccine industry, and NGOs. GAVI was formed in 1999 to address the long delay between vaccine availability in industrialized countries and developing countries. Between 2000 and 2015, GAVI has committed more than $2.6 billion for new and underused vaccines around the world (Figure 8.39) [53]. Scientific advances that would help make more vaccines available in developing countries include the development of temperature stable vaccines, development of vaccines that required less than three doses to immunize, and the development of needle free methods to administer vaccines [52].

Despite the truly remarkable advances in public health as a result of vaccines, there are still many infec-tious diseases for which no vaccine exists. The big three challenges of most importance to developing countries include vaccines to prevent HIV, malaria and tuberculosis. We now examine the obstacles which stand in the way of developing a vaccine for HIV.

Designing a new vaccine: HIV

To understand the challenges involved in developing an HIV vaccine, we must first consider the pathophysiology of HIV/AIDS in more detail. An HIV infection begins when virus is deposited on a mucosal surface. An initial acute infection can sometimes produce mono-like symptoms. Viral dissemination follows, and the patient will exhibit an HIV specific immune response. As the virus replicates, it destroys an important component of the immune system – a special kind of T lymphocyte called the CD4+ lymphocytes. The rate of progression of the disease is strongly correlated with viral load. Following initial infection, patients typically experience a long latent period with no clinical symptoms. Eventually, as more and more lymphocytes are destroyed, patients develop Acquired Immunodeficiency Syndrome (AIDS) (Figure 8.40). AIDS is characterized by immuno-logic dysregulation, accompanied by many opportunistic infections and cancers. The risk of opportunistic infection is correlated inversely with the number of CD4+ lymphocytes. Left untreated, the average patient with AIDS dies in one to three years [54]. HIV infection is identified by measuring whether a person is producing antibodies against the HIV virus.

The virus that causes HIV and AIDS was first discovered by Robert Gallo in 1984. At that time, Margaret Heckler, then US Secretary of Health Education and Welfare, predicted that an HIV vaccine would be developed within two years. Thirteen years later, in 1997, President Clinton declared that, "an HIV vaccine will be developed in a decade's time." In 2003, President Bush asked congress to appropriate $15 billion to combat the spread of HIV in Africa and the Caribbean, yet there is still no vaccine available to prevent HIV [57]. See Table 8.6 for an overview of HIV vaccines undergoing clinical trials.

There are many reasons that a vaccine has proven so difficult to develop. HIV represents a unique challenge; our bodies can eliminate most acute viral infections. In

Table 8.6. *Overview of HIV vaccines currently undergoing clinical trials (June, 2006). Data reproduced by permission of IAVI Report and VAX, International AIDS Vaccine Initiative.*

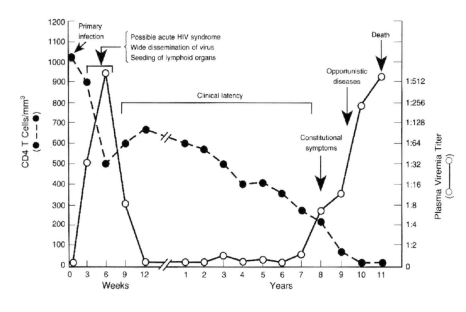

Figure 8.40. *The time course of the HIV/AIDS disease progression. Source: G. Pantaleo et al. Mechanisms of Disease: the Immunopathogenesis of HIV Infection. NEJM.* **328** *(327–35) © 1993. Massachusetts Medical Society. All rights reserved.*

contrast, our natural immune response does not destroy HIV. In fact, HIV infection results in the production of large amounts of virus, even in the presence of killer T cells and antibody. In developing a vaccine, we are faced with the challenge of trying to elicit an immune response that does not exist in nature. Therefore, we don't know exactly what type of immune response a vaccine should develop [58].

How does HIV outwit the immune system so successfully? As HIV replicates inside host cells, it frequently

HIV testing

We test for the presence of antibodies against HIV using an ELISA (enzyme linked immunosorbent assay). In this procedure, blood is taken from a person who may be infected with HIV. In the lab blood is added to laboratory HIV virus. HIV antibodies from the blood, if present, attach to HIV antigens. Next a chemical that attaches only to antibody/antigen complexes is added. If the solution changes color, the person is making antibodies against HIV. The advantage of ELISA tests is that they are very sensitive, meaning they don't often miss disease if it is present; the disadvantage of ELISA tests is that they are not very specific, meaning that they sometimes generate a falsely positive result in someone who does not have disease. Thus, a positive ELISA tests requires another test to confirm the presence of disease. The second test is called a western blot. Western blots are not as sensitive, but are more specific. Together, these two tests reduce the rate of false positives to 1/250,000. If this combination of HIV tests is positive, it indicates that the person is infected with the HIV virus, though he or she may not have AIDS yet. A negative ELISA test indicates either that a person is not infected with HIV, or that the person is infected with HIV but not yet making detectable level of antibodies. In general, it may take one to three months (and in rare cases six to twelve months) before the body makes enough antibodies to be detected by an ELISA [55, 56].

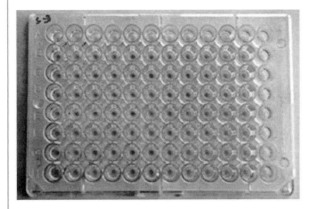

Courtesy of the Nevada Department of Agriculture.

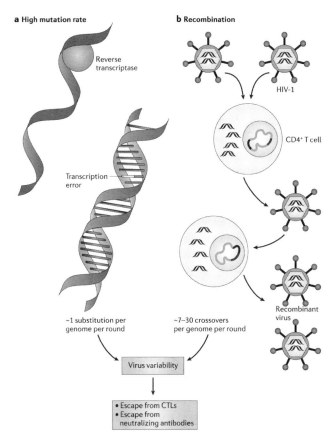

Figure 8.41. *Ways in which HIV can undergo mutation. From [59]. Reprinted with permission from Macmillan Publishers Ltd:* Nature Reviews Immunology, *copyright 2006.*

undergoes mutation. Many researchers believe that it is this continuous mutation of the HIV virus that enables it to escape destruction by the immune system. A high mutation rate increases the probability that a new form of the virus will emerge with a genetic advantage that enables it to survive. Figure 8.41 illustrates the ways in which the HIV virus can undergo mutation. As HIV replicates, it uses an enzyme called reverse transcriptase to copy its RNA into double stranded DNA. This DNA is inserted into the host chromosome, where it then directs production of more viral proteins that ultimately assemble into new viral particles. The HIV reverse transcriptase does not proofread this reproduction process, and on average each time the enzyme copies RNA into DNA, the new DNA differs at one base site [59]. HIV is the most variable virus known [60]. Additionally, if

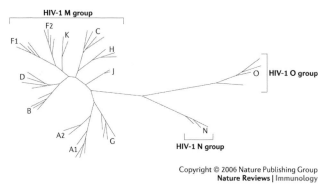

HIV-1 M group
F2
K
C
F1
H
D
J
B
A2
G
A1
N

HIV-1 O group
O

HIV-1 N group

Figure 8.42. *Individual strains or clades of HIV. From [59]. Reprinted with permission from Macmillan Publishers Ltd.* Nature Reviews Immunology, *copyright 2006.*

two genetically distinct forms of the HIV virus with different genetic sequences infect the same cell, DNA from both can integrate into the host genome and produce viral RNA. When new viral particles are packaged, RNA from the different parent viruses can combine to give rise to forms of HIV with entirely new genomes. HIV replicates at a very high rate, so the odds are high that useful mutations will occur over time. Over a ten year period, thousands of generations of viral reproduction have occurred; during that ten year period, the virus can undergo as much genetic change as humans would undergo in millions of years [60]!

As a result of this high rate of mutation, there are many forms of HIV which a vaccine must provide protection against. The two major types of HIV are HIV-1 and HIV-2. HIV-1 causes a more serious form of the disease and is responsible for the majority of HIV disease throughout the world. There are three main groups of HIV-1; more than 90% of HIV-1 disease is caused by group M HIV-1 [61]. Within this group, there are genetically distinct individual strains, called clades (Figure 8.42). It is thought that each strain may require a different vaccine [62].

Finally, development of a vaccine is complicated because there are many routes of transmission for HIV, including sexual contact and contact with contaminated blood. HIV can be transmitted by contact with virus alone or by contact with cells infected with the virus. Recall that cell-free virus is recognized and eliminated by antibodies, while cells infected with virus are recognized and eliminated by cell-mediated immunity. Thus

an HIV vaccine must generate both cell-mediated immunity and antibody-mediated immunity [58].

What are the design goals for an HIV vaccine? A successful vaccine must produce both antibody-mediated immunity and cell-mediated immunity against multiple forms of the virus. To develop antibody-mediated immunity, the immune system must see virus or viral debris. To produce cell-mediated immunity, HIV viral proteins must be presented to immune system on MHC receptors. We have seen three strategies for developing vaccines thus far: inactivated organism vaccines, subunit vaccines and vaccines based on live, attenuated pathogens.

Noninfectious HIV vaccine strategies

A noninfectious vaccine, made using killed virus or a viral subunit, will only stimulate antibody-mediated immunity, and thus will not meet the design goal. Animal trials with inactivated whole virus have shown only antibody-mediated immunity to a small number of HIV viral subtypes [63]. Similarly, trials of viral subunit HIV vaccines have shown modest antibody-mediated immunity effective against a limited number of HIV strains. One type of subunit vaccine has advanced to phase III clinical trials; this vaccine is based on the gp120 protein found in the envelope of the HIV virus. The gp120 protein is needed for HIV to enter cells. Researchers theorized that if patients could make antibody against this protein it could prevent free virus from infecting the patient's cells. In animal models and phase I clinical trials, it has been shown that this vaccine does elicit production of antibodies against gp120. The antibodies produced neutralized HIV in a test tube. But they only recognized strains of HIV similar to those used to generate the vaccine [58]. As we have seen, the HIV virus is notoriously susceptible to mutation, and these mutations change the structure of the gp120 protein over time so that the antibody may no longer be effective against it.

Despite poor results in animal trials, gp120 subunit vaccines have progressed to human clinical trials. The company Vaxgen developed a subunit vaccine based on the gp120 protein from two clades of HIV-1; the vaccine (called AIDSVAX) entered phase III clinical trials

Rural village outreach:
June 15, 2007
Christina Lesotho

We went to a rural village where our mentor, Dr. Dudley, had promised the family of a little girl that had passed away that she would return and give back their records, etc. We got there and were greeted by the grandmother who was taking care of the little girl and seemed to be the one in charge around that small corner of the village. She had gathered other children to be tested for HIV, and soon, her little single room Basotho (what they call people or things that are from Lesotho) hut became a testing center for a few different families that came by. A social worker with us performed his first pre-testing counseling, HIV tests, and post-testing counseling for each person tested, and it was interesting to see the reactions to his explanations and what the people being tested were and weren't comfortable discussing related to HIV and its transmission. I could not believe this was my first time seeing an HIV test kit.

Sophie and I stepped outside for a moment to get some fresh air and see all the children that had gathered. They were all friendly and as soon as the camera came out, the requests for their picture to be taken did not end. They were especially interested in seeing the shots on the screen after they were taken. A few of the older girls spoke English well and one of them was telling me about her interest in school, science, social studies, and traveling. She said she wants to travel to so many countries in Africa and beyond and she hopes to become a nurse and treat others around the world. I encouraged her to return to Lesotho, of course, and kept asking her about her future schooling. She then surprised me with the fact that she cannot pay for high school, so next year may be her last year of school. I could not believe the fees for high school were so high and I really got aggravated by this. I am almost certain it is to keep too many kids from qualifying for university since the government provides college scholarships for most students who make it to that point.

The ride to and from the village was really beautiful and it has been so nice not being in a busy city or hectic area. Back at the clinic, we were having lunch and some of the women on the staff at the reception and in social work/counseling started talking about handouts and food assistance and were adamantly against it. They were explaining to us that a mother came in earlier today and started crying when her child tested negative for HIV because that meant she would not get the food supplements

the clinic gives to patients on medication. I was so amazed and terrified by this thought. The staff member went on to explain the extreme economic and social problems created by WFP, they call it (World Food Program, I think) and how it has decreased productivity since people have this food source to turn to. She talked about the amount of excess WFP food in villages that she has seen at funerals and other village events, and how terrible it is that the country is being destroyed by foreign aid. I was completely frustrated at the thought.

I feel settled and like I am learning so much each day. I enjoy the people around me and have had an interesting time learning Sesotho words/phrases today. Hopefully I can try out some of my introductions tomorrow.

in 1998; results were announced in 2003. More than 5000 volunteers participated in the randomized, double blind, placebo controlled trial. At the beginning of the trial all were HIV negative; 3330 volunteers received AIDSVAX, and the remainder received the placebo. After three years, researchers compared the number of HIV infections for those receiving the vaccine and the placebo. 5.7% of those receiving the vaccine developed HIV, whereas 5.8% receiving placebo developed HIV. The difference was not statistically different. Most researchers believe that the effectiveness of this vaccine in preventing HIV infection was limited because it does not induce cellular immunity and provides antibodies against a limited number of HIV strains.

Researchers and public health workers have expressed concern that a vaccine with limited efficacy could actually increase the rate of new HIV infections. They fear that people who receive the vaccine will engage in riskier behaviors if they believe they are protected by a vaccine [62]. Such behavioral changes could possibly negate the benefit of the vaccine with limited efficacy. It is currently not known whether vaccines that do not prevent infection will delay disease progression in infected individuals. This could be an important benefit, especially in parts of the world where access to HAART is limited [59]. Furthermore, it is not known whether such vaccines could reduce viral loads in infected individuals, and this could help curb the spread of disease [64–66].

Live attenuated HIV vaccine strategies

As we saw earlier, an advantage of vaccines that are based on live attenuated forms of a pathogen is that they stimulate both antibody- and cell-mediated immunity. Vaccines made using this approach stimulate both B cells and killer T cells and this approach is the most likely to stimulate the necessary immune response. However, because the HIV virus mutates so rapidly, this approach presents unique potential dangers. Because the HIV virus mutates constantly, there is a chance that the attenuated form of the virus used in a vaccine could undergo a mutation that restores its strength. If this occurs, the consequences would obviously be devastating for the person receiving the vaccine. Vaccines based on a live attenuated form of the simian immunodeficiency virus (SIV) have been tested in macaque monkeys. SIV infects monkeys and is closely related to HIV. The vaccine successfully protected the animals when they were exposed to SIV. However, many of the animals vaccinated progressed to AIDS like symptoms, even when not exposed to the wild type virus, although more slowly than those infected with unaltered virus [58].

Thus, new vaccine strategies are required to develop vaccines to prevent HIV infection. In the remainder of this chapter, we consider several new approaches under investigation.

DNA vaccines

An interesting new approach to develop vaccines that stimulate both antibody- and cell-mediated immunity is to directly inject DNA that codes for viral protein into a patient. This leads the host cells to produce the protein that the DNA codes for; it is processed and loaded onto the MHC receptors and stimulates cell-mediated immunity, without any danger of causing infection. DNA vaccination approaches have shown very successful results in animal trials, generating a strong cell-mediated immune response [57]. Somewhat less successful results have been reported in human trials, where much larger quantities of DNA must be injected to generate immune response. While, many DNA based vaccines are currently in clinical trials, there are also concerns that scientists will not be able to identify a single protein that will elicit immune response against many HIV strains [62].

Carrier vaccines

The strength of the immune response elicited by a DNA vaccine can be strengthened by using a virus or bacterium that does not cause disease to carry the viral genes of interest to the host cell. Again, protein is produced by the host cells and loaded onto MHC receptors where it stimulates cell-mediated immunity. This approach stimulates both humoral and cell-mediated immunity without the danger of real infection. However, immunocompromised individuals can become ill from the carrier. A limitation of the approach is that the carrier must be one that individuals are not already immune to [62]. For the same reasons, booster vaccines cannot be made with the same carrier.

Promising results have been observed using what is known as a prime/boost strategy (Figure 8.43). In this approach, a prime vaccine is first given using a carrier to stimulate cell-mediated immunity. This is followed by a boost vaccine using a subunit vaccine to further stimulate antibody-mediated immunity. Phase III clinical trials of such a strategy began in 2003 using a canarypox vector to deliver HIV-1 genes that code for several proteins, followed by a boost with the AIDSVAX gp120 subunit vaccine [59].

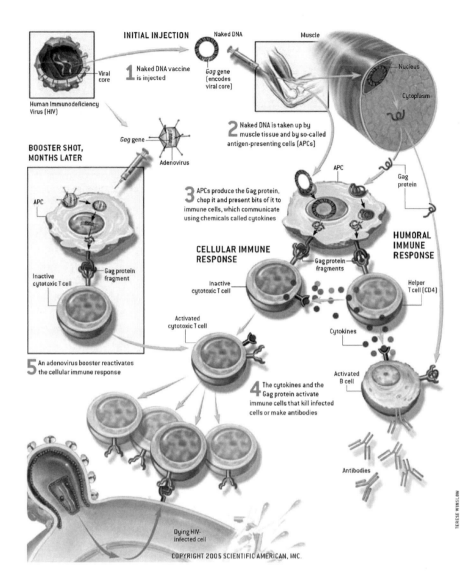

Figure 8.43. *Prime/boost vaccine strategy designed to initiate both antibody- and cell-mediated immune responses against HIV [62]. © 2002 Terese Winslow.*

Clearly, developing a new HIV vaccine will rely on the contributions of thousands of volunteers who are willing to participate in clinical trials. Scientific and medical progress rely on the sacrifices that healthy people are willing to make for the sake of research – and these risks are real. For example, one risk of participating in an HIV vaccine trial is that it may cause future HIV tests to be positive. This is because rapid tests measure antibodies against HIV and the vaccine may cause people to produce antibodies. Thus, a person participating in a vaccine trial may test positive for HIV, even though they do not have HIV. False positive HIV test results can lead to social stigma. People with positive HIV tests are not allowed to donate blood. Such results can also interfere with the ability of people to

get health insurance, obtain employment, or to travel to some foreign countries. Society has an obligation to provide protection for these volunteers. For example, we can pass laws to require insurance companies to change their screening procedures. If an applicant tests positive on an HIV rapid test, they can be required to administer a more sensitive and specific Western blot to discern between false positive rapid test and true HIV infections [47].

While many volunteers willingly participate in clinical trials, public mistrust has made it more difficult to carry out large trials of candidate HIV vaccines. There is limited public knowledge about HIV vaccine research, and many people harbor suspicion that HIV vaccine research is not being carried out for the greater good.

**Rabid dogs and AIDS parades:
June 6, 2007
Tessa Swaziland**

I've been meaning to blog about the impressive awareness about HIV/AIDS. I have no idea the extent of HIV/AIDS education (i.e., how much Swazis know about how HIV/AIDS is spread and why adherence to treatment is important) but everyone knows it's a problem. Last weekend, when Dave and I were picking up some fruit from the market (where grandmothers, also called "gogos," sell produce), we were briefly impeded by a parade of young adolescents holding up signs with various slogans: "No balloon; No party!" and other similar phrases. Two of my favorite campaigns are "I love you, positive or negative" and "I'm over it." There is a huge stigma against people with HIV here,

and the "positive or negative" campaign is geared toward that problem. Even the condoms say, "I love you, positive or negative." (I only know that because there are condoms in random places around the clinic.) The "I'm over it campaign," is pretty entertaining. There are billboards in Mbabane that have virtual text message conversations that go something like this: "My wife's at work. Wanna cum work on me?" "No. Thanks. I'm over it." There are several different ones. That's the one that really stuck in my mind though. They're pretty funny. I think the awareness is restricted mostly to the cities though. My guess is that in rural areas, people probably know a lot less. That reminds me . . . education here is not free. Many people can't afford to send their children to school. I was actually talking to my friend Treasure (she works in administration here at the Baylor clinic) about her education experience. She was very lucky and had been sponsored by a woman in California since she was three. The mystery woman paid for all the school fees, uniforms, books, etc. Then Treasure did well enough on her exams to get a scholarship for a university.

The National Institutes of Health conducted a survey of public knowledge and attitudes regarding HIV vaccine research in 2001. Nearly half (46%) of the population surveyed strongly agreed, somewhat agreed or did not know when asked if there was already a vaccine for HIV that was being kept a secret. An even higher proportion of minorities (72% of African American and 49% of Latinos) answered this way. When asked if HIV vaccines being tested could give a person HIV, 69% of the general population strongly agreed, somewhat agreed or did not know. In Chapter 9, we consider the ethical dilemmas that arise in research involving human subjects and we outline the framework of requirements that we have put in place to ensure that such research provides an appropriate balance between the risks and benefits of research.

Bioengineering and Global Health Project
Project task 4: Define the problem that your design will address

In this task, you will need to be much more specific about the particular problem you are trying to solve. For example, you may have identified the need for better treatments for tuberculosis in Project tasks 1–3. In this task, you may want to consider the particular problem of developing a treatment for tuberculosis that increases patient compliance. Turn in a one-page summary of the specific health need that your design will address.

Homework

1. The immune system.
   a. What is an antibody? Describe its structure and its function in the immune system.
   b. Explain the term "immunologic memory."
   c. Describe the cellular-level processes that enable the adaptive immune system to have immunologic memory.
2. When you get a splinter in your toe, the area can become red, hot, swollen and ooze pus. Describe the specific causes of each of these symptoms.
3. Name the three general types of immunity and give an example of each.
4. Most common anti-HIV drugs work by inhibiting key steps in viral uptake and reproduction.
   a. Make a drawing which shows the major steps that occur when HIV infects a CD4+ lymphocyte. Indicate on your figure where in the viral life cycle the following classes of drugs act: (1) fusion inhibitors, (2) reverse transcriptase inhibitors and (3) protease inhibitors.
   b. Beginning in the mid 1990s, an increasing number of HIV-infected individuals began a drug regime called highly active anti-retroviral therapy (HAART), a combination of three or more anti-HIV drugs taken at the same time. Why is taking a combination of drugs, each targeted against a different aspect of the viral life cycle, so much more effective than taking a single drug?
5. When a TB skin test is performed, a small amount of harmless TB antigen is injected under the skin. The patient monitors for redness and swelling at the site of injection. If a patient has been previously exposed to TB, but does not currently have an active TB infection will redness and swelling be observed? Why or why not?
6. Oh no! You return to Student Health two days after receiving a routine PPD skin test. You have a red bump on your forearm that measures 12 mm in diameter. Every year up until now, your test had been negative.
   a. How does the PPD skin test work, and why does a red bump form for individuals infected with TB?
   b. Assuming you have no significant health problems, what are the odds that the bacterium will remain in a latent, inactive state for the rest of your life?
   c. The PPD skin test is imperfect. Describe one instance in which the PPD skin test fails by giving a false-negative result, and describe another instance in which the test fails by giving a false-positive result. Why does the test fail in each circumstance?
7. If you are exposed to the varicella virus as a child and have not been vaccinated, you will likely develop chicken pox. If you are exposed again as an adult, you probably will not develop the disease again.
   a. At first exposure, what type of immunity fights off the varicella virus?
   b. The varicella vaccine contains a live virus. Is this safe? Why or why not? What is the advantage of this type of vaccine over a vaccine made of a dead virus?
8. A 24-year-old HIV-positive man is hospitalized because he developed pneumonia. The doctor starts the patient on antibiotics and measures the number of CD4 helper T cells in the patient's blood. The patient has a low CD4 count.
   a. What are two of the three major transmission routes by which this man might have become HIV-positive?
   b. The doctor then performs a test and finds that the man's serum is positive for antibodies to gp41 and gp120, the HIV envelope glycoproteins. Name and briefly describe this test the doctor ordered.
9. Answer the following questions about pathogens, the immune system, and vaccines.
   a. Check to indicate pathogen type(s) for which each statement applies.

Trait	Bacteria	Virus
Uses host cellular machinery to reproduce		
Can be killed or inhibited by antibiotics		
Short pathogen peptide sequences are displayed in MHC surface receptors.		
Living cells, usually having both a membrane and cell wall		
Protein capsid houses nucleic acid core		
Can reproduce without a host		
Tens of nanometers in size		

b. How can T cells identify cells infected with viruses?

c. Antigen binding to B-cell surface receptors and interaction with activated helper T cells activates B-cells to produce and secrete antibodies. Compare the onset and magnitude of the B-cell response for primary (initial) and secondary (subsequent) exposure to a particular antigen.

d. Identify the following vaccine types from the descriptions provided. Which one is likely to confer lifelong immunity?

   _____: The pathogen is treated with chemicals or irradiated. The early version of the polio vaccine and the rabies vaccine are examples.

   _____: Mutations have been introduced to the pathogen. This form is used to prevent measles, mumps and rubella.

10. The incidence of many diseases has been reduced by widespread vaccination. However, vaccines are not available for some diseases.

   a. Name three diseases for which vaccines are most critically needed to improve world health.

   b. For one of the diseases you listed in part a, explain the major scientific and economic challenges associated with developing a vaccine.

11. Technologies for vaccine development and delivery are considered among the top ten biotechnologies that may improve health in developing countries. Imagine that you are a member of GAVI evaluating new vaccination strategies for adoption by the organization. You are asked to choose between an oral live attenuated (Sabin) and an injectable inactivated (Salk) polio vaccine for use in sub-Saharan Africa. Polio is a viral disease that can produce paralysis. It is passed through fecal–oral transmission. Assume that the two vaccines have equal efficacy in preventing polio infection.

   a. What does GAVI stand for?

   b. Place the following stages of viral infection and replication in the correct order.
      — Synthesis of viral proteins
      — Viral budding or cell lysis
      — Endocytosis/injection of viral contents
      — Binding to cell membrane

   c. How does a live attenuated vaccine differ from an inactivated vaccine?

   d. Which of the following components of the immune system must a vaccine stimulate?
      i. macrophages
      ii. B cells and T cells
      iii. neutrophils
      iv. innate immunity
      v. complement

12. It has been shown that unvaccinated contacts of babies who receive the Sabin Polio vaccine will develop antibodies to the virus, while unvaccinated contacts of babies who receive the Salk Polio vaccine will not.

   a. Explain why this might be the case.

   b. List two reasons why the Sabin vaccine might be preferable to the Salk for use in sub-Saharan Africa.

   c. What is the main risk of using the Sabin vaccine in an immunocompromised population?

   d. There has been much interest in eliminating poliovirus worldwide. What is the only infectious disease that has been eradicated to date?

e. Discuss why the use of a vaccine led to the eradication of this disease while other diseases for which vaccines exist have not been eradicated.

f. List two properties that are necessary in order for a disease to be eradicable.

g. Which vaccine would you recommend that the GAVI adopt? Name two reasons why.

h. Why has there been so much focus on and investment in vaccination as a strategy in world health?

13. Portions of the following article appeared in the *Austin American Statesman* on May 10, 2005. Please read the article and answer the following questions.

Questions about pertussis article:

a. The article states that the pertussis vaccine does not protect 15–20% of children who receive it. Discuss how the concept of "herd immunity" will protect these children. What fraction of the population must be vaccinated to achieve 'herd immunity'?

b. The article describes a new booster vaccine called Boostrix. It states that the new vaccine may be available commercially next month. What process will the FDA use to ensure that the vaccine is safe after it is approved for general use? Why is this process necessary?

c. Pertussis is generally a mild disease in adults and older children. What arguments would you make in support of widespread distribution of the booster vaccine?

Travis County investigating outbreak of whooping cough

County leads state in number of cases; State also could have another bad year for pertussis
By Mary Ann Roser
AMERICAN-STATESMAN STAFF
Tuesday, May 10, 2005
Local health officials are investigating an outbreak of whooping cough as Travis County copes with the bleak distinction of having the state's most reported infections of the highly contagious disease, also known as pertussis, so far this year.

The Austin/Travis County Health and Human Services Department reported 58 confirmed cases of pertussis since Jan. 1, an unusually high number. The county has not had a whooping cough death since 2003. That year, infant Serena King died of the illness, which causes a violent cough followed by a whooping sound.

King was younger than 2 months, the age at which babies get their first pertussis vaccination, when she died. State health officials are awaiting confirmation of a suspected pertussis death this year, but the patient was not from Central Texas, said Rita Espinoza, an epidemiologist at the Department of State Health Services.

Pertussis is on the upswing nationally, and if current trends continue in Texas, 2005 could be one of the worst years since vaccines have been available.

As of April 30, the state had a preliminary count of 269 pertussis cases, compared with 192 during the same period a year earlier, according to the state health department. The worst year for whooping cough since the introduction of vaccines in the 1940s was 2002, when the state reported 1,240 pertussis cases, Espinoza said.

Health officials are worried.

In 2004, the preliminary count was 1,174 whooping cough cases statewide, compared with 670 in 2003. Travis County reported 97 cases in 2004 (the state's count for Travis County was higher, at 125; the two will reconcile the numbers later), and the county had 62 whooping cough cases in 2003, said Dr. Adolfo Valadez, the health authority for the Austin/Travis County department.

"It's a concern all over the state," Espinoza said. "I was just down in the Valley last week, and I was informed of 20 to 25 cases in an area where we usually don't hear of that many. We need to find a way to curb the cycle."

The outbreak is a warning to parents to keep their children's immunizations up-to-date, Valadez said.

It has picked up steam in the past five to six weeks, and most of the cases are in babies younger than a year old and children from ages 10 to 15, Valadez said. Schools in Austin and Pflugerville are seeing sporadic cases, but "no schools have had to be closed," he said. "Quite, honestly, we're looking forward to school ending. That's how it spreads."

Espinoza and Valadez said other factors could be contributing to the uptick in cases in recent years: growing awareness of pertussis, a quicker test to diagnose the illness and waning immunity from the whooping cough vaccine.

The vaccine has been changed to reduce some side effects, which could cause immunity to wear off in less than five to 10 years, Espinoza said. Also, the vaccine is far from foolproof. It does not protect 15 percent to 20 percent of the children who get it, which means adolescents can get pertussis and spread it to young children and babies who are at greatest risk of serious illness.

A week ago, the Food and Drug Administration approved the use of a pertussis booster vaccine, Boostrix, for children from ages of 10 and 18. Valadez said it is expected to be available commercially as early as next month, and he was encouraged that the tool was coming to the public health arsenal.

Now, children are vaccinated for pertussis at 2 months, 4 months, 6 months and between 15 months and 18 months, with a booster between ages 4 and 6, Espinoza said.

Pertussis bacteria live in the nose, mouth and throat and escape into the air when people sneeze, cough and talk. The disease is usually mild in older children and adults but can cause breathing problems, pneumonia and swelling of the brain. It begins like a cold, with a mild fever and cough, which slowly worsens and leads to coughing fits that sometimes end in vomiting.

14. Google the terms:
    a. Vaccine and safety
    b. Vaccine and dangers
       Do you think the sites that pop up on the two searches contain accurate health information? Why or why not? If you were a pediatrician, what would you tell the parents of your patients who had performed similar searches? A short paragraph is sufficient.

15. You have been asked to write a 500–525 word column on the Avian influenza situation for the *BIOE Tribune*. Your editor informs you that you must write a critique of the US plan in case of an Avian influenza pandemic. Your critique should include the scientific, economic, and public health aspects of this plan. Other topics, including potential vaccine strategies, may be addressed as well. The CDC website http://www.cdc.gov/flu/avian/ may provide information which will be helpful in completing this assignment. REMEMBER this is for a newspaper so make it compelling and make it interesting, but also make it TRUE!

16. Who sings "**The Avian Flu . . . a three minute summary**"? (Hint: she also sings "King of the Rollerama" and "The Great Metric Threat of 79.")

References

[1] Stern AM, Markel H. The history of vaccines and immunization: familiar patterns, new challenges. *Health Affairs (Project Hope)*. 2005 May–Jun; **24**(3): 611–21.

[2] WHO. *Mortality: Revised Global Burden of Disease (2002) Estimates*. Geneva: World Health Organization; 2002.

[3] Kasper DL, Braunwald E, Fauci AS, Hauser SL, Longo DL, Jameson JL, eds. *Harrison's Principles of Internal Medicine*. 16th edn. New York: McGraw-Hill; 2005.

[4] Sompayrac L. *How Pathogenic Viruses Work*. Sudbury, MA: Jones and Bartlett Publishers 2002.

[5] Centers for Disease Control and Prevention. *Epidemiology and Prevention of Vaccine-Preventable Diseases*. 10th edn. Washington DC: Public Health Foundation; 2007.

[6] Sompayrac L. *How the Immune System Works*. 2nd edn. Boulder Blackwell Publishing; 2003.

[7] Silverthorn DU. *Human Physiology : an Integrated Approach*. 3rd edn. San Francisco: Pearson/Benjamin Cummings; 2004.

[8] Centers for Disease Control and Prevention. *2004–05 U.S. Influenza Season Summary*. Atlanta, GA; 2005 July 5.

[9] WHO Regional Office for the Eastern Mediterranean. *Avian Influenza (Bird Flu): an Introduction*. Division of Communicable Disease Control Newsletter. 2005 November (7).

[10] Tumpey TM, Basler CF, Aguilar PV, Zeng H, Solorzano A, Swayne DE, *et al.* Characterization of the reconstructed 1918 Spanish influenza pandemic virus. *Science* (New York, NY). 2005 Oct 7; **310**(5745): 77–80.

[11] Kobasa D, Jones SM, Shinya K, Kash JC, Copps J, Ebihara H, *et al.* Aberrant innate immune response in lethal infection of macaques with the 1918 influenza virus. *Nature*. 2007 Jan 18; **445**(7125): 319–23.

[12] *Cumulative Number of Confirmed Human Cases of Avian Influenza A/(H5N1) Reported to WHO*. 2007 April 11 [cited 2007 April 23]; Available from: http://www.who.int/csr/disease/avian_influenza/ country/cases_table_2007_04_11/en/print.html

[13] World Health Organization. *Avian Influenza Frequently Asked Questions*. 2005 December 5 [cited 2007 April 23]; Available from: http://www.who.int/csr/disease/ avian_influenza/avian_faqs/en/index.html

[14] Ulmer JB, Valley U, Rappuoli R. Vaccine manufacturing: challenges and solutions. *Nature Biotechnology*. 2006 Nov; **24**(11): 1377–83.

[15] Matthews JT. Egg-based production of influenza vaccine: 30 years of commerical experience. *The Bridge*. 2006; **36**(3): 17–24.

[16] Langridge WHR. Edible vaccines. *Scientific American*. 2000; **283**(3): 66–71.

[17] Bloom BR, Lambert PH, eds. *The Vaccine Book*. San Diego: Academic Press; 2003.

[18] Fenner F, Henderson DA, Arita I, Jezek Z, Ladnyi ID. *Smallpox and its Eradication*. Geneva: World Health Organization; 1988.

[19] Smith KA. Medical immunology: a new journal for a new subspecialty. *Medical Immunology*. 2002 Sep 30; **1**(1): 1.

[20] Abbas AK, Lichtman AH. *Basic Immunology: Functions and Disorders of the Immune System*. 2nd edn. Philadelphia: W.B. Saunders; 2004.

[21] Mahmoud A. The vaccine enterprise: time to act. *Health Affairs*. 2005; **24**(3): 596–7.

[22] The infrastructure for vaccine development. *Health Affairs*. 2005; **24**(3): 598.

[23] Levine R. *Millions Saved: Proven Successes in Global Health*. Washington D.C.: Center for Global Development; 2004.

[24] Unicef. *Immunization Summary 2006*. New York: Unicef and WHO; 2006 January.

[25] Nurse Takes Plunge in Ebola Test. *All Things Considered*: National Public Radio; 2003.

[26] Orenstein WA, Douglas RG, Rodewald LE, Hinman AR. Immunizations in the United States: success, structure, and stress. *Health Affairs (Project Hope)*. 2005 May–Jun; **24**(3): 599–610.

[27] Vaccine Education Center – Children's Hospital of Philadelphia. *Frequently Asked Questions*. 2003 [cited 2007 April 25]; Available from: http://www.chop.edu/ consumer/jsp/division/generic.jsp?id=75743

[28] National Network for Immunization Information. *Why Immunize? How Childhood Vaccines are Selected for Routine Use*. 2004 [cited 2007 April 25]; Available from: http://www.immunizationinfo.org/parents/ howVaccines_selected.cfm

[29] National Network for Immunization Information. *Why Immunize? Monitoring Vaccine Safety*. 2007 [cited 2007 April 25]; Available from: http://www.immunizationinfo.org/parents/ monitoringSafety.cfm

[30] National Network for Immunization Information. *Why Immunize? How Vaccines Work*. 2007 [cited 2007 April 25]; Available from: http://www.immunizationinfo.org/ parents/howVaccines_work.cfm

[31] Centers for Disease Control and Prevention. Diptheria epidemic – new independent states of the former Soviet Union, 1990-1994. *Morbidity and Mortality Weekly Report*. 1995; **44**(10).

[32] Colgrove J, Bayer R. Could it happen here? Vaccine risk controversies and the specter of derailment. *Health Affairs (Project Hope)*. 2005 May–Jun; **24**(3): 729–39.

[33] Offit PA. Why are pharmaceutical companies gradually abandoning vaccines? *Health Affairs (Project Hope)*. 2005 May–Jun; **24**(3): 622–30.

[34] Wakefield AJ, Murch SH, Anthony A, Linnell J, Casson DM, Malik M, *et al.* Ileal-lymphoid-nodular hyperplasia, non-specific colitis, and pervasive developmental disorder in children. *The Lancet*. 1998 Feb 28; **351**(9103): 637–41.

[35] Centers for Disease Control and Prevention. *Childhood & Adolescent Immunization Schedules 2007* [cited 2007 April 27]; Available from: http://www.cdc.gov/nip/ recs/child-schedule.htm#presentation

[36] Buckland BC. The process development challenge for a new vaccine. *Nature Medicine.* 2005 Apr; **11**(4 Suppl.): S16–19.

[37] Smith NM, Bresee JS, Shay DK, Uyeki TM, Cox NJ, Strikas RA. Prevention and control of influenza recommendations of the Advisory Committee on Immunization Practices (ACIP). *Morbidity and Mortality Weekly Report.* 2006 July 28; **55**(RR-10).

[38] Scannon PJ. Pharmaceutical preparedness for an epidemic. *The Bridge.* 2006; **36**(3): 10–16.

[39] Centers for Disease Control and Prevention. *Vaccine Storage and Handling Toolkit.* 2005 June [cited 2007 April 30]; Available from: http://www2a.cdc.gov/nip/isd/shtoolkit/splash.html

[40] Estell D. Adapting industry practices for the rapid, large scale manufacture of pharmaceutical proteins. *The Bridge.* 2006; **36**(3): 39–44.

[41] Rappuoli R. Cell culture based vaccine production: technological options. *The Bridge.* 2006; **36**(3): 25–30.

[42] Shaw A. Alternative methods of making influenza vaccines. *The Bridge.* 2006; **36**(3): 31–8.

[43] Anderson RM. Planning for pandemics of infectious diseases. *The Bridge.* 2006; **36**(3): 5–9.

[44] Clancy C, Callaghan M, Kelly T. A multi-scale problem arising in a model of avian flu virus in a seabird colony. *Journal of Physics: Conference Series.* 2006; **55**: 45–54.

[45] Heuer AH. Engineering and vaccine production for an influenza pandemic. *The Bridge.* 2006; **36**(3): 3–4.

[46] Institute of Medicine. *Financing Vaccines in the 21st Century: Assuring Access and Availability.* Washington D.C.: National Academies Press; 2003.

[47] McCluskey MM, Alexander SB, Larkin BD, Murguia M, Wakefield S. An HIV vaccine: as we build it, will they come? *Health Affairs (Project Hope).* 2005 May–Jun; **24**(3): 643–51.

[48] *Vaccine Vial Monitors (VVMs).* Seattle: PATH; 2005.

[49] *Safe Vaccine Handling, Cold Chain and Immunizations.* Geneva: World Health Organization; 1998.

[50] *HealthTech Historical Profile: Technologies for Injection Safety.* Seattle: PATH; 2006 January.

[51] *BD Immunization.* 2007 [cited 2007 May 1]; Available from: http://www.bd.com/aboutbd/global/immunization.asp

[52] Levine MM. Can needle-free administration of vaccines become the norm in global immunization? *Nature Medicine.* 2003 Jan; **9**(1): 99–103.

[53] *GAVI Alliance Progress Report.* 2007; Available from: http://www.gavialliance.org/media_centre/publications/index.php

[54] Beal J, Orrick J, Alfonso K, Rathore M, eds. *HIV/AIDS Primary Care Guide*: Florida/Caribbean AIDS Education Training Center; 2006.

[55] Centers for Disease Control and Prevention. *Frequently Asked Questions About HIV and HIV Testing* [cited 2007 May 3]; Available from: http://www.hivtest.org/subindex.cfm?FuseAction=FAQ#1

[56] Chou R, Huffman LH, Fu R, Smits AK, Korthuis PT. Screening for HIV: a review of the evidence for the U.S. Preventive Services Task Force. *Annals of Internal Medicine.* 2005 Jul 5; **143**(1): 55–73.

[57] Smith KA. The HIV vaccine saga. *Medical Immunology.* 2003 Feb 14; **2**(1): 1.

[58] Baltimore D, Heilman C. HIV vaccines: prospects and challenges. *Scientific American.* 1998 July; **279**(1): 98–103.

[59] Letvin NL. Progress and obstacles in the development of an AIDS vaccine. *Nature Reviews.* 2006 Dec; **6**(12): 930–9.

[60] Nowak MA, McMichael AJ. How HIV defeats the immune system. *HIV: 20 Years of Research*: Scientific American; 2003.

[61] Carmichael M. How it began: HIV before the age of AIDS. *PBS Frontline.* 2006 May 30.

[62] Ezzel C. Hope in a Vial. *HIV: 20 Years of Research: Scientific American* 2003: 38–43.

[63] Singh M. No vaccine against HIV yet – are we not perfectly equipped? *Virology Journal.* 2006; **3**: 60.

[64] Eaton L. AIDS vaccine may offer hope only for some ethnic groups. *BMJ* (Clinical research edn. 2003 Mar 1; **326**(7387): 463.

[65] AIDS vaccine only limited success. *BBC News.* 2003 February 24.

[66] Kresge KJ. VRC starts Phase II vaccine trial. *VAX.* 2005 October; **3**(10).

Ethics of clinical research

The practice of medicine cannot improve in the absence of medical research. Advancing clinical medicine requires controlled experiments to compare the performance of a new intervention to the current standard of care. In many cases, initial experiments can be carried out in the laboratory using cell cultures or animal models, but eventually new techniques must be tested in humans to ensure that they are safe and effective. Unfortunately as we will see, people have not always treated each other humanely in the pursuit of medical research. How do we ensure that medical research involving human subjects is carried out in a fair and ethical manner? In Chapter 9, we will examine the ethical principles that guide research involving human subjects, and how we ensure that researchers adhere to these principles.

For centuries, the actions of physicians have been guided by the Hippocratic principle of "first do no harm." This principle guides clinical practice to improve an individual patient's health. Often the goal of medical research is to improve the health of future patients, and a subject participating in a research project may receive absolutely no benefit. In fact, participating in a research study may involve risks not fully understood at the beginning of a study. In the 1800s, scientists began to formally articulate ethical principles to guide medical research. In his 1865 book, *An Introduction to the Study*

of Experimental Medicine, Claude Bernard stated that one could never perform an experiment on man "which might be harmful to him in any extent, even though the result might be highly advantageous to science" [1].

For many years, ensuring that scientists and physicians adhered to these ethical principles was largely left to the discretion of individual researchers, not always with success. Table 9.1 chronicles some historical examples of ethically questionable research involving human subjects.

The atrocities committed by the Nazis and by Japanese forces in World War II in the name of medical research shocked the world, and led to a new era in the regulation of medical research [15]. As a result, several codes governing the ethical conduct of research have been developed to provide guidelines for patients, practitioners and scientists. Later in Chapter 9, we will examine the ethical principles laid out in The Nuremberg Code of 1949, The Helsinki Declaration of 1964, and The Belmont Report of 1979. But we begin Chapter 9 by examining several case studies of research carried out in the United Sates which further motivated the development of these codes of conduct.

Tuskegee Syphilis Study

The Tuskegee Syphilis Study was begun in 1932 in Macon County, Alabama. The goal of this study was

Table 9.1. *Historical examples of ethically questionable research [2–14].*

Year	Example
1796	Edward Jenner injects healthy eight year old James Phipps with cowpox, then six weeks later with smallpox. Ultimately Jenner's experiments gave rise to the first smallpox vaccine [2].
1845–1849	J. Marion Sims performs experimental surgeries on enslaved African women in an attempt to repair vesicovaginal fistulas – a severe complication of prolonged childbirth. While historical record suggests that the women voluntarily participated, Sims has been criticized for experimenting on a vulnerable population [3].
1896	Dr. Arthur Wentworth performs spinal taps on 29 infants and children at Children's Hospital in Boston to determine if the procedure is harmful. Upon reporting results, Wentworth is criticized by peers for failing to obtain parental consent and for performing non-therapeutic procedures [4].
1897	Italian bacteriologist Giuseppe Sanarelli injects five subjects with what he believes to be a filtered, inactivated solution of the yellow fever bacillus, producing yellow fever like symptoms in several of the patients [5]. The experiment was carried out without the subjects' permission or consent [6]. Walter Reed and James Carroll later disprove Sanarelli, demonstrating that the injected bacillus was actually a member of the hog cholera family [7].
1906	Richard Strong, head of the Philippine Biological Laboratory innoculates 24 inmates of a Manila prison with a cholera vaccine that is contaminated with plague. Thirteen of the inmates die [8]. It is unclear whether or not contamination was accidental [9].
1939–1945	Dr. Shiro Ishii, a physician and officer in the Japanese army, directs programs throughout China dedicated to biological warfare research, including the infamous Unit 731. Prisoners of Chinese and Russian nationality were innoculated with a variety of diseases including plague, typhoid, cholera, smallpox, and hemorrhagic fever. Additional experiments were carried out on local populations by contaminating wells and food sources. Precise estimates of casualties are not possible, but number likely in the thousands [10].
1941–1945	Nazi physicians conduct sterilization experiments on prisoners at Auschwitz and Ravensbrueck concentration camps in an effort to identify a means of carrying out mass sterilization campaigns [11].
1941–1945	Nazi physicians conduct typhus experiments on prisoners of Buchenwald and Natzweiler concentration camps. Prisoners were given experimental vaccines and chemical substances and infected with typhus leading to hundreds of deaths [11].
1942–1943	Nazi physicians conduct hypothermia experiments on approximately 300 male prisoners in Dachau concentration camp by immersing the prisoners in tanks of ice water [12].
1942–1945	US Chemical warfare service conducts mustard gas experiments on approximately 4000 servicemen. Soldiers were placed in gas chambers and field testing situations in order to test experimental protective clothing and collect data on exposure levels that produce injury [10].
1944–1946	Four hundred prisoners in the Illinois Statesville Penitentiary volunteer to participate in malaria experiments headed by Dr. Alf Alving through the University of Chicago Medical School. At the conclusion of the two year program a considerable portion of the prisoners received parole in return for their participation [9].
1950–1953	The US Atomic Energy Commission and Quaker Oats Company sponsor researchers at Harvard and MIT to conduct a study of nutrient absorption at the Fernald School – a residential institution for mentally disabled children. Children were fed cereals containing radioactive tracers and received calcium tracer injections. Parents, while told of a study, were not informed of the details [10].
1994–1996	In a series of studies, 100 young, predominantly minority boys with a personal or family history of aggression are administered fenfluramine in an effort to test whether aggression can be predicted by chemical changes in the brain. Fenfluramine has since been taken off the market due to evidence that long term use may give rise to heart valve defects in adults [13,14].

to examine the natural history of untreated syphilis. At the time the study began, the standard medical therapy for syphilis was to give patients heavy metals, like bismuth and arsenic. The cure rate for this treatment was less than 30%, and the side effects were sometimes fatal [16]. While this treatment did appear to reduce mortality, it was unclear whether some of the complications of syphilis were associated with the disease itself or were side effects associated with the heavy metals [17]. Because these side effects were so debilitating, the investigators felt that the treatment was potentially as toxic as the disease. In an attempt to separate the side effects of treatment from the natural progression of disease, researchers recruited a group of 600 low income black men, 399 with syphilis and 201 without syphilis [16]. The researchers withheld treatment from the group with disease; they felt they could justify withholding treatment because the side effects of the treatment were potentially as serious as the symptoms of syphilis. However, the participants did not voluntarily consent to participate in a research study. In fact, they were lured to participate in the study when researchers offered free treatment for "bad blood" – a generic term then used to describe a range of symptoms. The men were misinformed that some study procedures, like spinal taps, were free "extra treatment" (Figure 9.1) [17].

Ten years after the study began, the investigators noted that the death rate of non-treated patients was

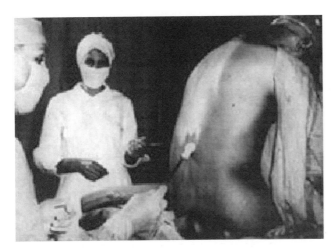

Figure 9.1. *A subject in the Tuskegee Syphilis Study undergoes a lumbar spinal tap. National Archives.*

twice as high as for treated patients, yet treatment was still withheld. In the 1940s even when penicillin became the clear drug of choice to treat syphilis, the study was still not interrupted and the men were not informed that penicillin was available. The study continued until 1972, when a researcher voiced concern to a reporter and the study was widely reported in the media [17]. As a result of the publicity, the study ended in 1972, and participants were offered monetary reparations. In 1973, Congressional investigations into the study commenced, and the NAACP won a $9 million settlement on behalf of the participants [18]. On May 16, 1997, US President Bill Clinton apologized to the surviving participants of the Tuskegee Syphilis Study [19].

Willowbrook School Study

Another ethically questionable study was the Willowbrook School Study, which was carried out from 1963 to 1966 and sought to examine the natural history of infectious hepatitis A. The study subjects were children at the Willowbrook State School, an institution for "mentally defective persons." Subjects in the study were deliberately infected with hepatitis A by feeding them stool from infected persons. Later in the study, as the virus became better defined, subjects were injected with the virus. The investigators justified their actions because the vast majority of children admitted to the Willowbrook State School acquired hepatitis anyway. Parents of children participating in the study gave consent for their children to participate. However, during the time of this study the Willowbrook State School was at times closed to new patients due to crowding. Because the hepatitis project had its own space, in some cases the only way to gain admission to the school was to agree to participate in the study [17].

Jewish Chronic Disease Hospital Study

In 1963, in the Jewish Chronic Disease Hospital Study live cancer cells were injected into debilitated patients in a hospital for the elderly. The purpose of the study was to develop information about the transplant rejection process and to study rejection of cancer cells. Patients hospitalized with various chronic debilitating diseases were injected with live cancer cells. Consent to participate in the study was negotiated orally, but not documented. Patients were not told that cancer cells would be injected because researchers felt that this might scare them unnecessarily. The investigators justified this because they were reasonably certain the cancer cells would be rejected. Researchers knew that healthy patients reject cancer cell implants quickly, while cancer patients reject the same cancer cell implants much more slowly. They wanted to understand whether this was due to impaired immunity because of the cancer or a more general manifestation of debility in cancer patients [17].

San Antonio Contraceptive Study

The goal of the San Antonio Contraceptive Study was to understand which side effects of oral contraceptive pills (OCPs) are due to the drug and which are simply by-products of everyday life. The study, carried out in the 1970s, was a randomized trial comparing a placebo and OCPs. Study subjects were 76 impoverished Mexican-American women with previous multiple pregnancies who had come to a public clinic seeking contraceptive assistance. The experiment was designed as a randomized, double-blind, placebo controlled trial – meaning that a fraction of the participants received placebo while the remainder received OCPs. The study utilized a cross-over design – during the middle of the trial, the placebo group was given OCPs and the OCP group was given placebo. All women were instructed to use vaginal cream as contraceptive during the study, but none of the women were told that the study involved a placebo. During the study, 11 women became pregnant, 10 while using placebo [17].

Codes of conduct for human subjects research

As a result of these and other examples, scientists and policy makers have developed codes to govern research involving human subjects. As a result of atrocities discovered in German concentration camps, The Nurem-berg Code was adopted in 1949. The Nuremberg Code states that in research, voluntary consent of the human subject is absolutely essential, and the subject should be at liberty to end the experiment at any time. All research involving human subjects should yield fruitful results for the good of society, which are obtainable in no other way. Experiments involving human subjects should avoid all unnecessary mental and physical suffering, and no experiment should be performed if it is believed that death or disabling injury may occur. The degree of risk to human subjects should never exceed the humanitarian importance of the problem to be solved. Finally, research involving human subjects should be conducted only by scientifically qualified persons [11].

In an international move to establish common ethical principles to guide medical research, the World Medical Association worked to develop and adopt the Declaration of Helsinki in 1964. The primary principle established in this document is to place the interests of the individual patient before those of society, stating that the primary goal of a physician is to "protect the life, health, privacy and dignity of the human subject [20]." The Helsinki Declaration affirms many of the principles of the Nuremberg code: that research subjects must be informed of the risks of a study and must voluntarily consent to participate, even if they are minors; and that studies should be designed and conducted by scientifically qualified personnel; and that risks of a study should not outweigh possible benefits. The Helsinki Declaration calls for formal review of research protocols by independent committees.

Despite these guidelines, abuses continued. Largely as a result of publicity associated with the Tuskegee trials, the US Department of Health, Education and Welfare issued *The Belmont Report* (Figure 9.2), a statement of basic ethical principles and guidelines to resolve ethical problems associated with conduct of research with human subjects, in 1979 [21]. *The Belmont Report* drew distinctions between clinical practice and research. Clinical practice includes interventions designed solely to enhance well being of an individual patient that have a reasonable expectation of

Figure 9.2. The Belmont Report, *published in 1979, is a statement of basic ethical principles that must be followed in research on human subjects.*

success. In contrast, research involves an activity to test a hypothesis that will permit conclusions to be drawn, and will contribute to generalizable knowledge. Research should be described in a formal protocol that sets forth an objective and procedures to reach that objective.

The Belmont Report established three basic ethical principles which must be followed in all research involving human subjects [21].

Respect for persons. Respect for persons demands that subjects enter into research voluntarily with enough information to make a decision about whether to participate. Further, persons with diminished autonomy (e.g. prisoners, children) are entitled to special protection.

Beneficence. Beneficence requires that researchers design experiments which do not harm study participants. Experiments which will injure one person are not allowed regardless of benefits that may come to others. Instead, researchers must make every effort to secure the well being of study participants, by maximizing all possible benefits and minimizing all possible harms.

Justice. This principle addresses who should receive benefits of research and who should bear its burdens. Justice requires that all individuals should

be treated as autonomous agents, and that the selection of research subjects must be scrutinized to determine whether some participants are being selected because of easy availability, compromised position or manipulability.

The Belmont Report provided guidelines for researchers to follow in order to ensure that these three principles were applied. First, researchers must obtain voluntary informed consent from all study participants. In order for a participant to give informed consent, they must fully understand the research procedure, the purpose of study, the potential risks and anticipated benefits, any alternative procedures that are available to them and they must be told that they may withdraw from the study at any time. Researchers must present this information in a way the subject can understand. It cannot be disorganized, presented too rapidly, or be above the subject's educational level. This consent must be given voluntarily, and persons in positions of authority cannot urge a particular course of action [21].

Second, research must be justified based on a favorable risk/benefit ratio for the participants, and researchers must select subjects fairly. Here, risk is defined as the possibility that harm may occur and benefit is defined as a positive outcome related to the health or welfare of a participant. Brutal or inhumane treatment of subjects is never justified. Instead, studies should be designed to reduce risks to only those necessary to achieve the research objective. Researchers must also select subjects fairly. They must not select only "undesirable" persons for risky research. Distinctions should be drawn between groups that should and should not be asked to participate in research based on ability of that group to bear burdens. For example, adults should be asked to bear burdens of research before children, when possible. Methods used to avoid exploiting vulnerable patients include: choosing subjects who are not vulnerable, distributing benefits so that those who participate benefit, getting community consultation to hear many points of view from those being studied, and using lottery systems when there are insufficient pools of new therapy [21].

A summary of the history of regulations

Fifth Century BC: Hippocratic Oath

The medical ethics standard "first do no harm" is attributed to Hippocrates. The oath became mandatory for physicians prior to practicing medicine in the fourth century AD [22].

1949: Nuremberg Code

Nazi physicians were charged with war crimes for research atrocities performed on prisoners of war. An American military war crimes tribunal conducted the proceedings against 23 Nazi physicians and administrators who willingly participated in war crimes. The judgment, known as the Nuremberg Code, was the first internationally recognized code of research ethics. It set forth ten standards for human subject research [11].

Volunteers must freely consent to participate in research.

Researchers must fully inform volunteers concerning the study.

Risks associated with the study must be reduced where possible.

Researchers are responsible for protecting participants against harms.

Participants can withdraw from the study at any time.

Research must be carried out by qualified researchers.

If adverse effects emerge, research must be stopped.

Society should benefit from study findings.

Research on humans should be based on previous animal or other work.

No research study should begin if there is a reason to believe that death or injury may result.

1964: Helsinki Declaration

The 18th World Medical Assembly met in Helsinki, Finland, and issued recommendations to guide biomedical research involving human subjects. The primary principle of the Declaration of Helsinki was to place individual patient interests before those of society. The basic principles of the Declaration of Helsinki are as follows [20].

The physician's duty is to protect the life, health, privacy and dignity of the human subject.

Research involving humans must conform to scientific principles and methods.

Research protocols should be reviewed by an independent committee.

Research protocols should be carried out by scientifically and medically qualified individuals.

The risks and burden to human subjects should not outweigh the benefits.

Research should be stopped if risks are found to outweigh potential benefits.

Research is justified only if there is a reasonable likelihood that the population under study will benefit from the results.

Participants must be volunteers and informed about the research study.

Every precaution must be taken to respect privacy, confidentiality, and participants' integrity.

Consent must be obtained from minors if they are able to do so.

Investigators are obliged to preserve the accuracy of results; negative and positive results should be publicly available.

1979: *The Belmont Report*

National Commission for the Protection of Human Subjects of Biomedical and Behavioral Research published *The Belmont Report* which set forth three basic ethical principles to guide research involving human subjects [21].

Respect for persons: participants must give voluntary consent; participants with diminished autonomy (e.g. children, prisoners) are entitled to special protection.

Beneficence: research must maximize possible benefits and minimize possible harms.

Justice: the benefits and risks of research must be distributed fairly.

Reexamining the Tuskegee study in light of the three principles of *The Belmont Report* illustrates its many ethical failures. Participants did not give consent to participate, and they were not informed of the study. Risks to participants were not minimized; indeed, participation increased risks. Participants were limited to disadvantaged, rural black men, but the disease under study is not limited to this population. A much broader population benefited from the findings of the research.

How do institutions work to ensure that studies conform to these guidelines? Today, US institutions carrying out research involving human subjects have a special, independent committee called the Institutional Review Board (IRB). The role of the IRB is to work with investigators to be sure that the rights of subjects are protected, to educate the research community and public about ethical conduct of research, and to be a resource center for information about Federal guidelines. Research involving human subjects cannot begin until the IRB has approved the research protocol and the informed consent document, a written document that subjects sign indicating their willingness to participate. An IRB approved informed consent document can be found in the appendix to this chapter. The research protocol is written for review by the physicians and scientists who are members of the IRB, while the informed consent document is written for potential participants.

Informed consent is a critically important part of research. The Nuremberg Code speaks to the voluntary consent of human subjects being essential: "This means the person involved should have 'legal capacity' to give consent; should be situated to exercise 'free power of choice', without the intervention of any element of force, fraud, deceit, duress, over-reaching or other ulterior motives; over-reaching or other ulterior form of constraint or coercion, and should have sufficient 'knowledge', and 'comprehension' of the elements of the subject matter involved as to enable him to make an understanding and enlightened decision [11]." Therefore, for informed consent to be valid: the subject must be competent, the consent voluntary, their participation informed, and their understanding complete.

Conceptual framework for the process of obtaining informed consent

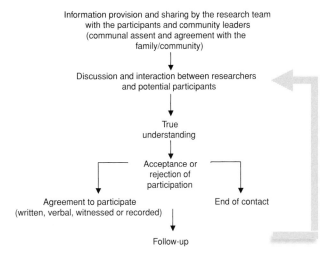

Figure 9.3. *The steps involved in obtaining informed consent. Used with permission from [15].*

Figure 9.3 provides an overview of the process of obtaining informed consent. The research team must provide full and understandable information about the proposed research. The participant must understand what is being asked of him or her and must freely agree to participate. Comprehension is a key element in the informed consent process; the investigator must ensure that the subject understands both the risks and benefits involved in participation. Technical procedures must be explained in lay terms at the appropriate educational level and using interpreters and translators as necessary.

Researchers must document that participants have given informed consent. Most frequently, consent is documented by having participants sign a written informed consent document. Table 9.2 shows the elements typically included in such a document. The appendix to this chapter provides a sample informed consent document; as you will see, informed consent documents often use complex language and seem to be written to provide legal protection to researchers rather than to provide information for participants [15]. Unfortunately, there is currently little emphasis on assessing a participant's understanding of a project before they sign an informed consent document. Researchers are not required to test or document participant understanding,

Table 9.2. *The components of an informed consent document for human research subjects.*

Invitation	Clear invitation to participate
Statement of overall purpose	Explanation of the purpose of the research in laymen's language
Basis for selections	Why have you, the individual patient, been asked to participate in this study
Explanations of procedures	A description of procedures to be followed, with identification of any procedures that are experimental. A statement of where and when the research will be done, and how much time will be involved in participating in the research
Description of the discomforts and risks	Description of foreseeable risks, discomforts, and inconveniences to the subject, the likelihood that they may occur, and steps taken to minimize risk
In case of injury	Description of the availability of medical therapy as well as the compensation for disability that may result from participating
Description of benefits	Description of benefits are hoped for but not guaranteed. If participants will not benefit, this must be indicated, e.g. "the purpose of _____ is to develop knowledge useful in developing improved therapies for your disease. Thus we hope to provide benefits in the future for persons like you"
Disclosure of alternatives	Description of alternative and routine therapies
Confidentiality assurances	Disclosure of who may review the chart; this usually involves discussion of who is supporting the study, who monitors trials for that group, and any other state or federal authorities likely to review the research work
Financial considerations	Description of any economic advantages in participating in a clinical trial, such as any financial inducements for participation, and explains that patients are usually not eligible for patent or royalty rights of invention
Offer to answer questions	Information about how to contact scientific, medical and administrative personnel in case the participant has questions regarding the study.
Continuing disclosure	Statement that the PI will notify subjects of any new findings obtained during the course of the study that may impact their decision to continue to participate in the research

although it has been suggested that simple questionnaires or interviews could be used to document understanding prior to informed consent (Figure 9.4) [15].

Continuing controversies

Despite explicit ethical guidelines, recently, a number of high profile ethical dilemmas have arisen in research projects involving human subjects. We conclude Chapter 9 by reviewing the debate surrounding some of these dilemmas.

Blinded seroprevalence studies

In 1988, the CDC and state health departments carried out studies to determine the prevalence of HIV in the population. They tested blood samples for HIV to determine the portion of the population infected with HIV. The study was blinded, so that researchers did not

Figure 9.4 *The use of street theater can improve community knowledge to facilitate informed consent. ©Giacomo Pirozzi/Panos Pictures.*

have access to any patient identifiers. Before proceeding with the research, it was reviewed for ethical concerns.

Informed consent was considered unnecessary because the data had been anonymized, and the researchers did not have access to information which could identify the subjects. However, this prevented the researchers from notifying infected individuals. As treatments evolved for HIV, and the importance of early clinical intervention with anti-retroviral drugs was revealed, the studies came under attack. Several legislators argued that the studies should be unblinded. Nettie Mayersohn, a democratic representative in the New York State Assembly expressed concern that infected babies who were identified through the study had a right to treatment if their test results were positive [23]. US Congressman Gary Ackerman introduced legislation to unblind the study. Ackerman warned, "There was one point in our society, a very dark day when people were allowed to walk around after being tested with a dread disease just so the medical establishment could . . . see what happens . . . " [24]. Because of these concerns, the CDC suspended the study in 1995 [23].

Did this study adequately protect the rights of human subjects? Most experts agree that the study conformed to the ethical guidelines of *The Belmont Report*. These guidelines permit experiments to be carried out using patient specimens which will normally be discarded without consent, so long as patient identities are not released to investigators and the study has been reviewed by an IRB. The purpose of the study was to identify populations at risk for HIV so that effective interventions could be designed for these groups. None of the study participants were prevented or discouraged from seeking voluntary HIV testing [23].

Study of HIV transmission in Uganda

From 1994 to 1998 a team led by researchers at Columbia University tested 15,000 adults in ten rural Ugandan communities for HIV and other sexually transmitted diseases (STDs) [25]. The goal of the study was to determine whether treatment of STDs like syphilis and chlamydia could reduce the transmission of HIV. All participants in five villages were treated for STDs, while participants in five control villages were simply told of their results and were referred to free clinics for treatment. Results showed that treatment with antibiotics lowered the rate of other STDs, but did not affect the rate of HIV transmission. When the study was ended all participants were given antibiotics. After the study was concluded, the researchers analyzed their data to see what other factors might affect HIV transmission. They matched sexual partners and identified 415 partners where one partner was infected and the other was not at the beginning of the study. They found that the most significant factor likely to increase transmission from the infected to the uninfected partner was the amount of virus in the infected person's blood [26].

The study was criticized by Marcia Angell, editor of the *New England Journal of Medicine*, who was troubled that the researchers did not inform the at-risk partners. The researchers did not identify the discordant couples until after the study had been concluded, and argued that even if they had known, they could not have informed at risk partners, because Uganda has a national policy that prevents health workers from telling a third party about an individual's HIV status [26]. Angell was also troubled that HIV positive participants were not offered treatment with anti-retroviral drugs. Angell believed that the Helsinki Declaration requires that researchers provide the best available treatment to their subjects. She argued that it did not matter that such care is not usually available in the setting where the research was conducted; the researchers had an ethical obligation to provide the same treatment that would be available in a developed country [26]. Edward Mbidde, a medical oncologist in Uganda, pointed out that if studies in the developing world were held to the same standards of care available in developed countries, research to develop new treatments affordable for use in developing countries would be impractical [26].

Developing country HIV prevention trials

In the USA, a study was carried out to determine whether treatment could interrupt transmission of HIV from mothers to babies. The trial was called the AIDS Clinical Trial Group (ACTG) Study 076 [27]. It showed a dramatic reduction in transmission for women who received the intervention compared to women who received placebo. The effect was so dramatic that the study was stopped early, so that no additional women received placebo. In this trial, drug was administered during the last 26 weeks of pregnancy. Drug was also given intravenously during delivery and to the baby for six weeks after delivery. While successful, the intervention cost $800 for drug alone [28]. Because of this high cost and the long duration over which drug must be given, many people believed its use would be impractical in many developing countries, where women don't deliver in hospitals, don't seek care until later in their pregnancies, and can't afford an $800 drug [28].

Studies were started in nine developing countries to determine whether much cheaper alternatives could also reduce maternal to child transmission of HIV. The goal of these studies was to evaluate the effectiveness of a regimen which provided drug only during the last three to four weeks of pregnancy, reducing the cost of the intervention to just $80 [29]. This could be afforded by two of the countries, and international agencies made a commitment to provide drug to other resource poor countries participating in the trials [28]. The trial was designed as a randomized trial in which some mothers got the new regime and others received a placebo. These trials were sponsored by the CDC and the NIH and all were subject to careful ethical review [28].

The study led to a bitter ethical debate regarding the appropriate standard of care to be used in the control arm. Marica Angell, editor of the *New England Journal of Medicine*, criticized the trials on September 18, 1997, saying "The justifications are reminiscent of the Tuskegee study: Women in the Third World would not receive antiretroviral treatment anyway, so the investigators are simply observing what would happen to the subject's infants if there were no study [30]." She cited the Declaration of Helsinki as preventing the trials. Angell argued that the new intervention should have been compared to the full ACTG 076 protocol, which was the standard of care in the developed world.

Researchers argued that investigators would learn more in a shorter time if they did a placebo controlled trial. The placebo control was necessary to establish the baseline rates of maternal to child HIV transmission, because these vary throughout the world. Rates

Bouncing PORECO Babies!:
June 19, 2007
Dave Swaziland

Today is the one-year anniversary of the initiation of these babies into the clinic's PORECO program. [PORECO stands for Pilot Operational Research and Community Based Project.] The aim of the PORECO program is to prevent the transmission of HIV from mothers to their babies. To celebrate, we have a huge, bouncy, inflatable castle . . . and an enormous, enormous cake (like four feet by five feet)!

And, boy, are they happy today!

of transmission can be influenced by the health state of the mothers and babies. Mothers in developing countries are often anemic and malnourished, so researchers wanted to measure the baseline transmission rate in order to know whether the new treatment reduces the rate of transmission below the baseline rate. Also, the drug itself causes anemia, so researchers believed that a placebo control was needed to determine whether the drug increased anemia. Other ethicists argued that the trial was ethical only if it was accompanied by a plan to make the treatment available to the local population if it proved to be effective [29].

Amidst the controversy, the CDC sponsored study in Thailand took place and showed that the reduced course of therapy did dramatically reduce maternal to child HIV transmission rates – although not as much as

Figure 9.5. *A billboard in Gabarone, Botswana advocates participation in programs to Prevent Maternal to Child Transmission (PMTCT).*

Most experts agree that a new definition of the standard of care is needed, which permits different standards for research in developing countries. However, these discrepancies should be subject to approval by ethical review committees in the host country. Rather than requiring that patients have access to the highest attainable standard of care, it has been suggested that researchers provide access to highest attainable and *sustainable* therapeutic method. The level of therapy generally available in a host country is the least that is ethically acceptable. Researchers must commit to provide a level of treatment that can continue in a host country after the research has been completed. If therapy is not

the ACTG 076 protocol. Within weeks after the study findings were made public, agencies started supplying drug to women in studies around the world who were previously on placebo. Glaxo Wellcome, the drug manufacturer announced it would cut prices of drug for sale in developing countries. Thus, the study enabled worldwide programs designed to prevent maternal to child transmission of HIV (Figure 9.5) [29].

Standard of care: a new definition?

Many of the current controversies center on debate over what should be the appropriate standard of care for research involving human subjects. How do we decide what is a reasonable standard of care for research subjects in developing countries? Should we automatically use the standard set by developed countries?

What is the danger of simply imposing the highest attainable standard of care for all research throughout the world? If we require that subjects in the control group receive the same treatment that would be available to them in a developed country, we may never develop sustainable techniques to improve health in developing countries [28]. What is the danger of accepting less than the highest attainable standard of care? We may find that researchers choose to carry out phase I drug studies in Africa because it is cheaper and less regulated.

> ## Suggested guidelines for research involving human subjects in developing countries
>
> Carry out research on a health problem of the developing country population.
>
> Research objectives, not vulnerability of the population, should be used to justify conduct of the research in a developing country.
>
> Ensure that benefits of participating in trial outweigh the risks.
>
> Only undertake research that benefits the community participating.
>
> Translate research findings into accessible care in the community participating.
>
> Involve members of the host community in design and conduct of trial; they must decide if benefits outweigh risks.
>
> Provide subjects with care they would not ordinarily get in the country where the trial is carried out.
>
> Ensure that trial does not widen disparities by taking resources away from the healthcare system of the host country.
>
> Interventions proven safe and effective through research should be made available in those countries [28, 31, 32].

sustainable, then results can never be made available to the inhabitants of the country [28].

Bioengineering and Global Health Project
Project task 5: Define the constraints that a solution must satisfy

These should be quantitative measures that include both technical performance and economic constraints that your solution must satisfy. If there are existing solutions, you should identify the performance capabilities and cost of these solutions. Your solution should provide an advantage compared to existing solutions. You should carefully justify trade-offs made between expected performance and cost. Examples of constraints that you might consider include necessary educational level of primary user, detection limits of new diagnostic methods, efficacy rates of new therapies, power requirements, cost and size. Turn in a one page table summarizing the design constraints for your problem. Each row in the table should indicate a specific constraint (e.g. unit cost of device). The table should include at least two columns – one or more which represents the current performance of available technologies and one which represents the constraint that your design must satisfy.

Homework

1. *The Belmont Report* establishes the three fundamental ethical principles that guide the ethical conduct of research involving human participants: (1) Respect for persons; (2) Justice; and (3) Beneficence. These principles require that all subjects participating in medical research give informed consent.

   a. Define informed consent.

   b. The following story appeared in *The Oregonian*. Read it and answer the following question. Suppose you are a member of the OHSU IRB. Would you have voted to approve this trial? Why or why not? Support your answer using the principles of *The Belmont Report*.

Blood trial could omit consent form
Doctors seek community consensus to test a blood substitute on trauma patients who may not be conscious
ANDY DWORKIN

How would you feel knowing that a doctor could experiment on you, without your permission, while you were unconscious? What if that experiment could help save your life and test a possible treatment for wounded soldiers or car crash victims? Doctors want Portland-area residents to ponder those questions as they move toward joining a study of a blood substitute called PolyHeme. Trauma medics with Legacy Health System, Oregon Health & Science University and local ambulance companies would take part in a national trial comparing PolyHeme with the salt-water solution now carried on ambulances.

This is no ordinary research project. In most trials, scientists must tell each potential participant about the possible risks and rewards before getting their agreement to participate, a process called "informed consent." But PolyHeme would go to people unconscious from blood loss when treatment starts. A seldom-used and ethically controversial 1996 Food and Drug Administration regulation lets researchers waive informed consent to test potential life-saving treatments when there is no other way to conduct the research. Instead of individual consent, the FDA says researchers must teach local residents about the trial and gauge their feelings. So Legacy and OHSU workers are mailing letters to local officials and holding three public meetings to explain the trial and ask for feedback. "This is not a sure thing that the study will happen," said Lise Harwin, a Legacy communications coordinator who helped plan the public education. "What we're trying to do now is get feedback to determine if it will." Portland researchers have spent more than a year planning the trial, and both hospitals' researchreview boards have approved the idea.

But those boards won't give their final approval until they consider public reaction. Scientists have spent decades searching for a blood substitute, which trauma doctors say is desperately needed. Donated blood is too delicate and has too short a shelf life to carry on ambulances. Instead, paramedics use durable saline solution. But saline can't carry oxygen through the body; PolyHeme does. PolyHeme, which is made from expired blood donations, has a longer shelf life than blood and can be administered to a person of any blood type. Local research boards "haven't established a particular percent or number" of negative responses from the community that would cause them to stop the trial, Allee said. One reason is that researchers assume people worried about the process are more likely to comment than those who support it.

2. The following text contains a portion of an article which appeared in the *Austin American Statesman*. Read the text and answer the following questions.

Federal researchers tested AIDS drugs on foster children without advocate protections
At least seven states, including Texas, participated in studies, which are now under investigation.
By John Solomon ASSOCIATED PRESS Thursday, May 05, 2005. Used with permission of the Associated Press Copyright © 2009. All rights reserved.
WASHINGTON – Government-funded researchers tested AIDS drugs on hundreds of foster children over the past two decades in at least seven states, including Texas, often without providing them a basic protection afforded in federal law and required by some states, an Associated Press review has found. The research funded by the National Institutes of Health was most widespread in the 1990s as foster care agencies sought treatments for their HIV-infected children that weren't yet available in the marketplace. The practice ensured that foster children – mostly poor or minority – received care from world-class researchers at government expense, slowing their rate of death and extending their lives. But it also exposed a vulnerable population to the risks of medical research and drugs that were known to have serious side effects in adults and for which the safety for children was unknown.

Several studies that enlisted foster children reported that patients suffered side effects such as rashes, vomiting and sharp drops in infection-fighting blood cells as they tested antiretroviral drugs to suppress AIDS or other medicines to treat secondary infections. In one study, researchers reported a "disturbing" higher death rate among children who took higher doses of a drug. That study was unable to determine a safe and effective dosage. Research and foster agencies declined to make foster parents or children in the drug trials available for interviews, or to provide information about individual drug dosages, side effects or deaths, citing medical privacy laws. Some foster children died during studies, but state or city agencies said they could find no records that any deaths were directly caused by experimental treatments.

The government provided special protections for child wards in 1983. They required researchers and their oversight boards to appoint independent advocates for any foster child enrolled in a narrow class of studies that involved greater than minimal risk and lacked the promise of direct benefit. Some foster agencies required the protection regardless of risks and benefits. Advocates must be independent of the foster care and research agencies, have some understanding of medical issues and "act in the best interests of the child" for the entirety of the research, the law states.

However, researchers and foster agencies said foster children in AIDS drug trials often weren't given such advocates even though research institutions many times promised to do so to gain access to the children. Illinois officials say they think none of their nearly 200 foster children in AIDS studies got independent monitors even though researchers signed a document guaranteeing "the appointment of an advocate for each individual ward participating in the respective medical research." New York City could find records showing 142 – less than a third – of the 465 foster children in AIDS drug trials got such monitors even though city policy required them. The city has asked an outside firm to investigate.

Researchers typically secured permission to enroll foster children through city or state agencies. They frequently exempted themselves from appointing advocates by concluding the research carried minimal risk and the child would directly benefit because the drugs had already been tried in adults. If they decline to appoint advocates under the federal law, researchers and their oversight boards must conclude that the experimental treatment affords the same or better risk-benefit possibilities than alternate treatments already in the market-place. They also must abide by any additional protections required by state and local authorities.

Many of the studies that enrolled foster children occurred after 1990 when the government approved using the drug AZT – an effective AIDS treatment – for children. Those studies often involved early Phase I and Phase II research – the riskiest – to determine side effects and safe dosages so children could begin taking adult "cocktails," the powerful drug combinations that suppress AIDS but can cause bad reactions like rashes and organ damage. Some of those drugs were approved ultimately for children, such as stavudine and zidovudine. Others were not.

Arthur Caplan, head of medical ethics at the University of Pennsylvania, said advocates should have been appointed for all foster children because researchers felt the pressure of a medical crisis and knew there was great uncertainty as to how children would react to AIDS medications that were often toxic for adults. "It is exactly that set of circumstances that made it absolutely mandatory to get those kids those advocates," Caplan said. "It is inexcusable that they wouldn't have an advocate for each one of those children."

Those who made the decisions say the research gave foster kids access to drugs they otherwise couldn't get. And they say they protected children's interest by explaining risks and benefits to state guardians, foster parents and the children themselves. "I understand the ethical dilemma surrounding the introduction of foster children into trials," said Dr. Mark Kline, a pediatric AIDS expert at Baylor College of Medicine. He enrolled some Texas foster kids in his studies, and said he doesn't recall appointing advocates for them. "To say as a group that foster children should be excluded from clinical trials would have meant excluding these children from the best available therapies at the time," he said. "From an ethical perspective, I never thought that was a stand I could take."

Illinois officials directly credit the decision to enroll HIV-positive foster kids with bringing about a decline in deaths – from 40 between 1989 and 1995 to only 19 since.

NIH did not track researchers to determine whether they appointed advocates. Instead, the decision was left to medical review boards made up of volunteers at each study site. A recent Institute of Medicine study concluded those Institutional Review Boards were often overwhelmed, dominated by scientists and not focused enough on patient protections.

a. What are the three basic ethical principles of *The Belmont Report*? Define each principle.
b. Discuss the ethical and legal issues that arise when new medical technologies are tested in vulnerable populations, such as foster children. Do you think that the studies described

adequately protected the rights of this population? Give the reasons for your position in terms of the principles outlined in *The Belmont Report*.

c. The article states that the studies ensured that foster children received care from world-class researchers at government expense, extending their lives. Illinois officials credit the decision to enroll HIV-positive foster children with bringing about a decline in deaths. Describe how these outcomes influence your reasoning in part b above.

3. Briefly describe the Willowbrook study to investigate the natural history of infectious hepatitis. List the principles of *The Belmont Report* which were violated in this study. Support your answer with evidence.

4. A clinical trial recently carried out at Johns Hopkins University tested the effects of a chemical irritant to understand why some people get asthma. Three healthy volunteers with normal respiratory systems inhaled the chemical. Two days after inhaling the chemical, Ellen Roche, 24, a technician at the Johns Hopkins Asthma and Allergy Center, developed a cough, fever and muscle pain. She quickly developed respiratory distress, and within a month she was dead. The chemical she inhaled turned out to be far more toxic than the researchers realized. In fact, the lead investigator's literature search of the most common databases (which date back only to 1960), did not turn up earlier studies hinting at the chemical's potential dangers, but after-the-fact searches using different search engines and databases did turn up references to the potential risks to humans. In a review of the study, the FDA raised questions about the informed-consent forms that Roche and two other subjects had signed. On them, hexamethonium is referred to as a "medication" and as "(having) been used as an anesthetic" – giving subjects a sense that it was an FDA-approved medicine and therefore safe. Another criticism: Togias failed to report that his first subject (Roche was the third) had developed a cough. It went away, and Togias assumed it had to do with a

viral infection making the rounds at Bayview at the time. Source: "At Your Own Risk: Some Patients Join Clinical Trials Out of Desperation, Others to Help Medicine Advance," *Time*, April 22, 2002. Discuss any problems associated with the protection of human subjects using the principles of *The Belmont Report*.

5. Use the following link to read the article "Placebos break taboo in cancer drug tests: Study seeks hope for desperately ill" that was first printed in the *Boston Globe*: http://www.irbforum.com/forum/read/2/78/78. You have just been named the Director of the National Cancer Institute. You control an annual budget of $6 billion. You must decide whether any of these funds can be used to support placebo controlled research studies for terminally ill cancer patients. Your decision will determine whether any studies of this type will receive any funding. Using the article as a reference point, prepare an argument in favor of or against such studies. Your argument should be no more than one typed page. Limit your argument to either the pro or con stance and prepare a convincing case as to why you ruled the way you did.

6. Discuss the ethical and legal issues that arise when new medical technologies are tested in developing countries. In what ways can this benefit the population of the developing country? In what ways can the population be harmed? If the researchers are based in the United States, what legal and ethical responsibilities do they have?

References

[1] Bernard C. *An Introduction to the Study of Experimental Medicine*. New York: Macmillan; 1927.

[2] Stern AM, Markel H. The history of vaccines and immunization: familiar patterns, new challenges. *Health Affairs (Project Hope)*. 2005 May–Jun; **24**(3): 611–21.

[3] Wall LL. The medical ethics of Dr J Marion Sims: a fresh look at the historical record. *Journal of Medical Ethics*. 2006 Jun; **32**(6): 346–50.

[4] Grodin MA, Glantz LH. *Children as Research Subjects : Science, Ethics, and Law*. New York: Oxford University Press; 1994.

[5] Lederer SE. *Subjected to Science : Human Experimentation in America before the Second World War*. Baltimore: Johns Hopkins University Press; 1995.

[6] Pierce JR, Writer J. *Yellow Jack : How Yellow Fever Ravaged America and Walter Reed Discovered its Deadly Secrets*. Hoboken, N.J.: John Wiley; 2005.

[7] Phillip S. *Hench Walter Reed Yellow Fever Collection*. The United States Army Yellow Fever Commission (1900–1901). 2001 August 8 [cited 2007 May 7]; Available from: http://yellowfever.lib.virginia.edu/reed/commission.html#_edn1

[8] Chernin E. Richard Pearson Strong and the iatrogenic plague disaster in Bilibid Prison, Manila, 1906. *Reviews of Infectious Diseases*. 1989 Nov–Dec; **11**(6): 996–1004.

[9] Hornblum AM. *Acres of Skin : Human Experiments at Holmesburg Prison : a Story of Abuse and Exploitation in the Name of Medical Science*. New York: Routledge; 1998.

[10] Moreno JD. *Undue Risk : Secret State Experiments on Humans*. New York: W.H. Freeman; 2000.

[11] *Trials of War Criminals before the Nuremberg Military Tribunals under Control Council Law No. 10. Nuremberg, October 1946–April 1949*. Washington D.C.: U.S. G.P.O 1949–1953.

[12] Michalczyk JJ. *Medicine, Ethics, and the Third Reich : Historical and Contemporary Issues*. Kansas City, MO: Sheed & Ward; 1994.

[13] Hilts PJ. Experiments on children are reviewed. *The New York Times*. 1998 April 15.

[14] Bernstein N. 2 Institutions faulted for tests on children. *The New York Times*. 1999 June 12.

[15] Bhutta ZA. Beyond informed consent. *Bulletin of the World Health Organization*. 2004 Oct; **82**(10): 771–7.

[16] Centers for Disease Control. *US Public Health Service Syphilis Study at Tuskegee*. 2007 March 27 [cited 2007 June 1]; Available from: http://www.cdc.gov/nchstp/od/tuskegee/time.htm

[17] Levine RJ. *Ethics and Regulation of Clinical Research*. 2nd edn. New Haven: Yale University Press; 1986.

[18] Around the Nation; judge upholds deadline on syphilis case settlement. *The New York Times*. 1981 June 18.

[19] Mitchell A. Clinton regrets 'clearly racist' U.S. study. *The New York Times*. 1997 May 17.

[20] World Medical Association. *Declaration of Helsinki*. 52nd World Medical Association General Assembly; 2000; Edinburgh, Scotland; 2000.

[21] United States. National Commission for the Protection of Human Subjects of Biomedical and Behavioral Research. *The Belmont Report : Ethical Principles and Guidelines for the Protection of Human Subjects of Research*. Washington: Dept. of Health, Education, and Welfare, National Commission for the Protection of Human Subjects of Biomedical and Behavioral Research: for sale by the Supt. of Docs., US Govt. Print. Off. 1978.

[22] National Library of Medicine. *Greek Medicine: The Hippocratic Oath*. [cited 2007 May 9]; Available from: http://www.nlm.nih.gov/hmd/greek/greek_oath.html

[23] Fairchild AL, Bayer R. Uses and abuses of Tuskegee. *Science (New York, NY)*. 1999 May 7; **284**(5416): 919–21.

[24] Testimony of Gary Ackerman. *Hearing Before the Subcommittee on Health and Environment of the Committee on Commerce, House of Representatives*. 1st Session, 11 May ed 1995:Serial No. 104–22:8.

[25] Quinn TC, Wawer MJ, Sewankambo N, Serwadda D, Li C, Wabwire-Mangen F, *et al*. Viral load and heterosexual transmission of human immunodeficiency virus type 1. Rakai Project Study Group. *The New England Journal of Medicine*. 2000 Mar 30; **342**(13): 921–9.

[26] Vogel G. Study of HIV transmission sparks ethics debate. *Science (New York, NY)*. 2000 Apr 7; **288**(5463): 22–3.

[27] Connor EM, Sperling RS, Gelber R, Kiselev P, Scott G, O'Sullivan MJ, *et al*. Reduction of maternal-infant transmission of human immunodeficiency virus type 1 with zidovudine treatment. Pediatric AIDS Clinical Trials Group Protocol 076 Study Group. *The New England Journal of Medicine*. 1994 Nov 3; **331**(18): 1173–80.

[28] Levine RJ. Some recent developments in the international guidelines on the ethics of research involving human subjects. *Annals of the New York Academy of Sciences*. 2000 Nov; **918**: 170–8.

[29] Beardsley T. Coping with HIV's ethical dilemmas. *Scientific American*. 1998 July; **279**(1): 106–7.

[30] Angell M. The ethics of clinical research in the Third World. *The New England Journal of Medicine*. 1997 Sep 18; **337**(12): 847–9.

[31] Benatar SR, Singer PA. A new look at international research ethics. *BMJ (Clinical research edn)*. 2000 Sep 30; **321**(7264): 824–6.

[32] Lo B, Bayer R. Establishing ethical trials for treatment and prevention of AIDS in developing countries. *BMJ (Clinical research edn)*. 2003 Aug 9; **327**(7410): 337–9.

Appendix: Informed Consent Document

Informed Consent to Participate in Research
The University of Texas at Austin

You are being asked to participate in a research study. This form provides you with information about the study. The Principal Investigator (the person in charge of this research) or his/her representative will also describe this study to you and answer all of your questions. Please read the information below and ask questions about anything you don't understand before deciding whether or not to take part. Your participation is entirely voluntary and you can refuse to participate without penalty or loss of benefits to which you are otherwise entitled.

Title of research study
Evaluating the Effectiveness of Evidence-Based Teaching Strategies in BME 301: Biotechnology and World Health

Principal Investigator(s) (include faculty sponsor), UT affiliation, and Telephone Number(s)
Rebecca Richards-Kortum, Ph.D. Professor of Biomedical Engineering 512–471-2104

Funding source
Howard Hughes Medical Institute

What is the purpose of this study?
The purpose of the study is to investigate the effectiveness of learner-centered, open-ended problem solving and cooperative learning strategies in BME 301. The total number of students registered for BME 301 was approximately 60. Thirty seven students participated from BME 301. Your participation will serve as a control group.

What will be done if you take part in this research study?
From a pool of undergraduate students, we are requesting volunteers for an interview protocol. We will ask participants to volunteer to take part in an activity in which they are given a newspaper article related to the BME 301 course material and asked to critically discuss it in groups of three to five. Participant responses will be videotaped. Students will be compensated for their time with a $20.00 gift certificate from Barnes and Noble.

What are the possible discomforts and risks?
There are no physical risks or discomforts that apply with this research. However, if you wish to discuss the information above or any other risks you may experience, you may ask questions now or call the Principal Investigator listed on the front page of this form. Some participants may be uncomfortable with being videotaped initially. If you feel uncomfortable, you may discontinue participation at any time. The only treatment for participating in this research that differs from discussing a newspaper article casually with a peer is your agreement to being videotaped.

What are the possible benefits to you or to others?
Your participation in this study will provide data which will permit researchers to identify learning behaviors that are positively associated with content mastery in this course when comparing various instructional techniques. This study will result in a deeper understanding of learner centered environments and the effect of this on learning. The study may result in the development of reliable and valid instruments which can measure learning in more effective ways than are currently used. This will improve the knowledge base for the

science of learning and ultimately the knowledge disseminated from the study could improve the teaching of undergraduate curricula beyond the University of Texas at Austin.

If you choose to take part in this study, will it cost you anything?
There are no costs associated with participating in this study.

Will you receive compensation for your participation in this study? What if you are injured because of the study?
Students who complete the interview will be compensated for their time with a $20.00 gift certificate from Barnes and Noble. There are no physical risks associated with this research. University students may be treated at the usual level of care with the usual cost for services at the Student Health Center for any injury, related or not, but no payment can be provided in the event of a medical problem.

If you do not want to take part in this study, what other options are available to you?
Participation in this study is entirely voluntary. You are free to refuse to be in the study, and your refusal will not influence current or future relationships with The University of Texas at Austin or your grades in any course. Your decision to participate will not bestow any competitive academic or occupational advantage over any other University of Texas at Austin students who do not volunteer, and the researchers will not impose any academic or occupational penalty on those University of Texas at Austin students who do not volunteer.

How can you withdraw from this research study and who should you call if you have questions?
If you wish to stop your participation in this research study for any reason, you should contact: Deanna Buckley at (512)-471–3068. You are free to withdraw your consent and stop participation in this research study at any time without penalty or loss of benefits for which you may be entitled. Throughout the study, the researchers will notify you of new information that may become available and that might affect your decision to remain in the study.

In addition, if you have questions about your rights as a research participant, please contact Clarke A. Burnham, Ph.D., Chair, and The University of Texas at Austin Institutional Review Board for the Protection of Human Subjects, 512–232-4383.

How will your privacy and the confidentiality of your research records be protected?
Authorized persons from The University of Texas at Austin and the Institutional Review Board have the legal right to review your research records and will protect the confidentiality of those records to the extent permitted by law. If the research project is sponsored, then the sponsor also has the legal right to review your research records. Otherwise, your research records will not be released without your consent unless required by law or a court order.

If the results of this research are published or presented at scientific meetings, your identity will not be disclosed.

Because these studies will use video recordings, you should know that the CDs will be: (a) coded so that no personally identifying information is visible on them; (b) kept in a secure locked location in the co-investigator's office (Deanna Buckley); (c) heard or viewed only for research purposes by the investigator and his or her associates; (d) possibly retained for future research analysis.

Will the researchers benefit from your participation in this study?
Your participation will allow researchers to collect objective data to be analyzed for publications in educational and scientific research journals and presentations to other scientific researchers and educators. No other benefits are expected at this time.

Signatures

As a representative of this study, I have explained the purpose, the procedures, the benefits, and the risks that are involved in this research study.

_____

Signature and printed name of person obtaining consent   Date

You have been informed about this study's purpose, procedures, possible benefits and risks, and you have received a copy of this Form. You have been given the opportunity to ask questions before you sign, and you have been told that you can ask other questions at any time. You voluntarily agree to participate in this study. By signing this form, you are not waiving any of your legal rights.

_____

Printed Name of Subject   Date

_____

Signature of Subject   Date

_____

Signature of Principal Investigator   Date

We may wish to present some of the tapes from this study at scientific conventions or as demonstrations in classrooms. Please sign below if you are willing to allow us to do so with the tape of your performance.

_____

Signature of Subject   Date

I hereby give permission for the video tape made for this research study to be also used for educational purposes.

_____

Signature of Subject   Date

Technologies for early detection and prevention of cancer

10

In Chapter 8, we saw how technology could be used to prevent infectious diseases, one of the leading killers in the developing world. In this chapter, we examine how technology can be used to diagnose disease. Our focus is the detection of cancer, where early detection (Figure 10.1) can mean the difference between life and death. We begin by examining the global burden of cancer. Next, we examine how cancers develop and why early detection is so crucial. Finally, we examine three cancers in detail – cervical cancer, ovarian cancer and prostate cancer – and look at existing and new technologies to aid in the early detection and prevention of each disease.

The burden of cancer in the United States

Cancer is the second leading cause of death in the United States, responsible for nearly one out of every four deaths (Table 10.1). Only 66% of all cancer patients live more than five years past their initial diagnosis, a statistic known as the five year survival rate. Cancer is

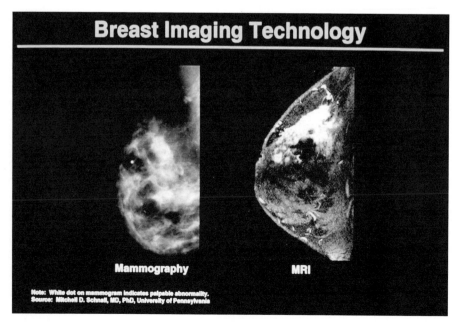

Breast Imaging Technology

Mammography   MRI

Note: White dot on mammogram indicates palpable abnormality.
Source: Mitchell D. Schnall, MD, PhD, University of Pennsylvania

Figure 10.1. *Mammography is one method used to screen for breast cancer. Source: Mitchell D. Schnall, M.D., Ph.D. University Of Pennsylvania, National Cancer Institute.*

Table 10.1. *Cancer is the second leading cause of death overall in the USA, and the leading cause of death for people under 85 years of age [2].*

Rank	Cause of Death	No. of Deaths	% of Deaths
	U.S. Mortality, 2004		
1	Heart Disease	654,092	**27.2**
2	Cancer	550,270	**22.9**
3	Cerebrovascular diseases	150,147	**6.2**
4	Chronic lower respiratory diseases	123,884	**5.2**
5	Accidents (Unintentional injuries)	108,694	**4.5**
6	Diabetes mellitus	72,815	**3.0**
7	Alzheimer's disease	65,829	**2.7**
8	Influenza and pneumonia	61,472	**2.6**
9	Nephritis	42,762	**1.7**

A comprehensive set of statistics regarding cancer incidence and mortality in the United States is compiled each year by the National Cancer Institute; the report, called *Cancer Facts and Figures*, can be found at www.cancer.org. It was predicted that 1,444,920 new cases of cancer would be detected in the USA in 2007, and that 559,650 people would die as a result of cancer [1].

Table 10.2 ranks the most commonly occurring cancers in men and women in the United States (excluding basal cell and squamous skin cancers). Prostate cancer is the most common cancer in US men, accounting for 1/3 of cancer incidence. Breast cancer is the most common cancer in USA women, accounting for nearly 1/3 of cancer incidence; ovarian cancer, which we will study in detail later, accounts for 3% of cancer incidence in US women [3].

important from an economic perspective as well. In the USA alone, cancer cost approximately $206 billion in 2006. Of this, $78.2 billion was spent on direct medical costs; $17.9 billion represents lost productivity to illness and $110.2 billion represents lost productivity due to premature death [1].

Table 10.3 ranks the most common causes of cancer mortality in men and women in the United States. In both sexes, lung cancer is the leading cause of cancer

Table 10.2. *The most commonly occurring cancers in men and women in the USA in 2005.*

Estimated US Cancer Cases in 2009*

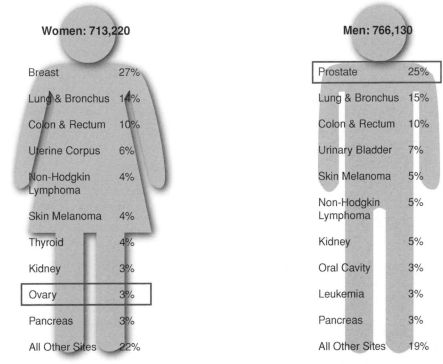

Women: 713,220		Men: 766,130	
Breast	27%	Prostate	25%
Lung & Bronchus	14%	Lung & Bronchus	15%
Colon & Rectum	10%	Colon & Rectum	10%
Uterine Corpus	6%	Urinary Bladder	7%
Non-Hodgkin Lymphoma	4%	Skin Melanoma	5%
Skin Melanoma	4%	Non-Hodgkin Lymphoma	5%
Thyroid	4%	Kidney	5%
Kidney	3%	Oral Cavity	3%
Ovary	3%	Leukemia	3%
Pancreas	3%	Pancreas	3%
All Other Sites	22%	All Other Sites	19%

*Excludes basal and squamous cell skin cancers and in situ carcinomas except urinary bladder.

Adapted and reprinted with permission from American Cancer Society. Cancer Facts and Figures 2009. Atlanta: American Cancer Society, Inc.

Table 10.3. *The most common causes of cancer mortality in men and women in the USA in 2005.*

Estimated US Cancer Deaths in 2009*

Women: 269,800

Lung & Bronchus	26%
Breast	15%
Colon & Rectum	9%
Pancreas	6%
Ovary	5%
Non-Hodgkin Lymphoma	4%
Leukemia	3%
Uterine Corpus	3%
Liver & Intrahepatic Bile Duct	2%
Brain/ONS	2%
All Other Sites	25%

Men: 292,540

Lung & Bronchus	30%
Prostate	9%
Colon & Rectum	9%
Pancreas	6%
Leukemia	4%
Liver & Intrahepatic Bile Duct	4%
Esophogus	4%
Urinary Bladder	3%
Non-Hodgkin Lymphoma	3%
Kidney	3%
All Other Sites	25%

*Excludes basal and squamous cell skin cancers and in situ carcinomas except urinary bladder.

Adapted and reprinted with the permission from the American Cancer Society. Cancer Facts and Figures 2009. Atlanta: American Cancer Society, Inc.

Trends in cancer incidence and mortality in the USA

From 1993 to 2002, cancer death rates in the United States dropped by 1.1% per year. The decrease in cancer mortality is attributed to a combination of better treatment, better early detection and cancer prevention. Death rates dropped more for men (1.5%/year than for women 0.8%/year). Cancer incidence rates in the United States have been stable since 1992 [4].

---

death, even though it is only the second most common cancer in men and women separately. As we will see later in this chapter, routine tests are available to screen older men and women for prostate cancer and breast cancer, so they tend to be diagnosed at an earlier, more curable stage. Because we do not have good screening tests for lung cancer, it tends to be diagnosed at a much later stage with worse prognosis. The situation is similar for ovarian cancer. Although it is responsible for only 3% of cancer incidence in women, it accounts for 6% of cancer mortality in women [3].

The global burden of cancer

Globally, cancer is an important cause of mortality, accounting for 12% of all deaths worldwide (Figure 10.2). Cardiovascular disease is the leading cause of mortality worldwide, followed by infectious disease and cancer. Today, more than 11 million new cases of cancer are detected worldwide every year, and 6.7 million deaths can be attributed to cancer [6]. Table 10.4 shows the leading causes of cancer mortality worldwide. In men, lung cancer is the most common cause of cancer death, while in women, breast cancer is the most common cause of cancer death. The third most common cause of cancer death in women worldwide is

Table 10.4. *The number of estimated cancer deaths worldwide in 2002.*

Estimated Worldwide Cancer Deaths in 2002*

Women: 2,927,896

Breast	14%
Lung	11%
Cervix uteri	9%
Stomach	9%
Colon & Rectum	8%
Liver	6%
Ovary	4%
Esophagus	4%
Pancreas	4%
Leukemia	3%
All Other Sites	28%

Men: 3,795,991

Lung	22%
Stomach	12%
Liver	11%
Colon & Rectum	7%
Esophagus	7%
Prostate	6%
Leukemia	3%
Pancreas	3%
Urinary Bladder	3%
Non-Hodgkin Lymphoma	3%
All Other Sites	23%

*Excludes basal and squamous cell skin cancers

Source: International Agency for Research on Cancer (IARC). GLOBOCAN 2002 estimates.

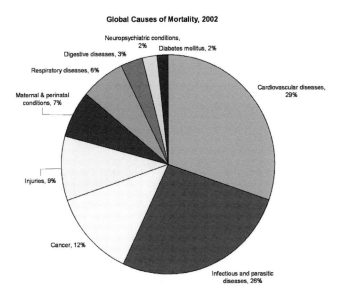

Figure 10.2. *The most common causes of death worldwide in 2002. Cancer is the third leading cause of mortality, worldwide [5].*

be attributed to the use of screening tools to detect cervical cancer at an early stage. In the developed world, the Papanicoloau (Pap) smear is used to screen the general female population for cervical cancer and its precursors. The early detection and treatment of these conditions prevents the development of invasive cervical cancer. Unfortunately, owing to limited resources, cervical cancer screening is not implemented in many developing countries; as a result, cervical cancer is the leading cause of cancer death for women in developing countries [6].

The maps in Figure 10.3 illustrate global variations in the mortality of cancer today, and the changes predicted in cancer mortality throughout the world in the year 2020. Both the global incidence and mortality of cancer are predicted to increase. In the next 20 years, it is estimated that global cancer incidence will increase by nearly 50% and global cancer mortality will double. The largest rates of increase are predicted to occur in developing and newly industrialized countries. In 2020,

cervical cancer; note that cervical cancer was not among the top ten causes of cancer incidence or mortality in the USA. Again, as we will see later, this large difference can

(a)

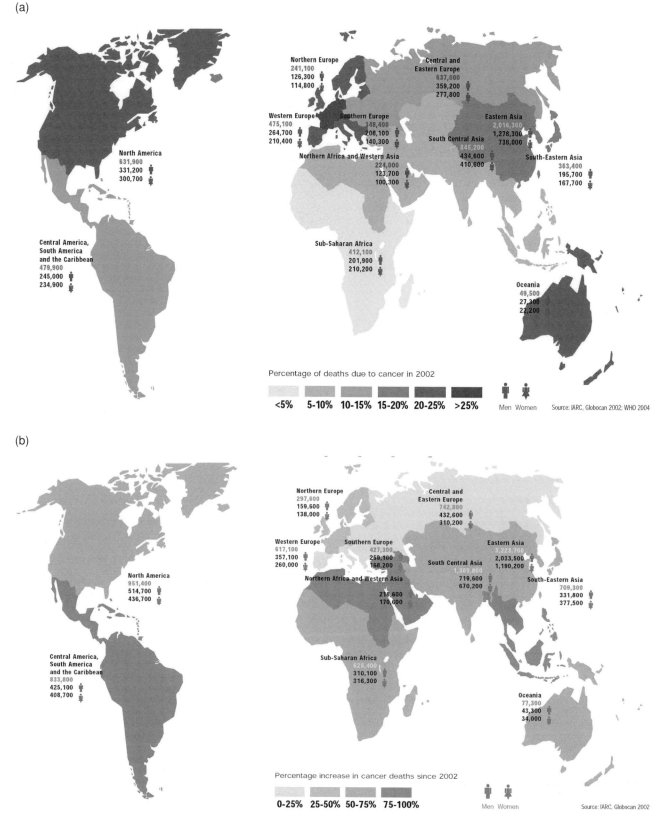

Percentage of deaths due to cancer in 2002

<5%   5-10%   10-15%   15-20%   20-25%   >25%   Men Women   Source: IARC, Globocan 2002; WHO 2004

(b)

Percentage increase in cancer deaths since 2002

0-25%   25-50%   50-75%   75-100%   Men Women   Source: IARC, Globocan 2002

Figure 10.3. *(a) Global distribution of cancer deaths in 2002. (b) Predicted global distribution of cancer deaths in 2020. Cancer is predicted to kill more than 10 million people worldwide each year.*

Table 10.5. *Risk for US women of developing cancer over the course of a lifetime.*

Lifetime Probability of Developing Cancer, Women, US, 2003-2005*

Site	Risk
All sites[†]	1 in 3
Breast	1 in 8
Lung & bronchus	1 in 16
Colon & rectum	1 in 20
Uterine corpus	1 in 40
Non-Hodgkin lymphoma	1 in 53
Urinary bladder[‡]	1 in 84
Melanoma[§]	1 in 58
Ovary	1 in 72
Pancreas	1 in 75
Uterine cervix	1 in 145

* For those free of cancer at beginning of age interval.
† All Sites exclude basal and squamous cell skin cancers and in situ cancers except urinary bladder.
‡ Includes invasive and in situ cancer cases
§ Statistic for white women.
Source: DevCan: Probability of Developing or Dying of Cancer Software, Version 6.3.0 Statistical Research and Applications Branch, NCI, 2008. http://srab.cancer.gov/devcan

Table 10.6. *Risk for US men of developing cancer over the course of a lifetime.*

Lifetime Probability of Developing Cancer, Men, 2003-2005*

Site	Risk
All sites[†]	1 in 2
Prostate	1 in 6
Lung and bronchus	1 in 13
Colon and rectum	1 in 18
Urinary bladder[‡]	1 in 27
Melanoma[§]	1 in 39
Non-Hodgkin lymphoma	1 in 45
Kidney	1 in 57
Leukemia	1 in 67
Oral Cavity	1 in 72
Stomach	1 in 90

* For those free of cancer at beginning of age interval.
† All Sites exclude basal and squamous cell skin cancers and in situ cancers except urinary bladder.
‡ Includes invasive and in situ cancer cases
§ Statistic for white men.
Source: DevCan: Probability of Developing or Dying of Cancer Software, Version 6.3.0 Statistical Research and Applications Branch, NCI, 2008. http://srab.cancer.gov/devcan

more than 16 million new cancer cases are predicted, and 10.3 million people are expected to die of cancer [7]. Although the probability of being diagnosed with cancer is twice as high in developed countries, cancer survival rates are much lower in developing countries. In developed countries, about 50% of cancer patients die as a result of their cancer, while in developing countries more than 80% of cancer patients already have late-stage incurable tumors at the time of their diagnosis [8]. It is estimated that by 2020, at least 70% of cancer deaths will occur in developing countries, where resources for early detection and treatment are least available [9].

If you live in the USA, what is your lifetime risk of developing cancer? If you are female, you have a 33% chance of developing cancer at some time in your life, with a 14% chance of developing breast cancer at some point in your life (Table 10.5). If you are male, you have a 50% chance of developing cancer at sometime in your life, with nearly a 17% chance of developing prostate cancer at some point (Table 10.6) [10].

How can you reduce your cancer risk? More than 1/3 of cancers are preventable, through three approaches: (1) reducing tobacco use, (2) implementing existing screening techniques worldwide, and (3) adopting a healthier lifestyle and diet. Globally, 43% of cancer deaths are due to tobacco use, inappropriate diet or infection [7]. In developing countries, infectious agents

are responsible for nearly 25% of cancers, while only 9% of cancers are due to infectious agents in developed countries. These include hepatitis B and C (can lead to liver cancer), the sexually transmitted human papillomavirus (HPV) (can lead to cervical cancer), and *Helicobacter pylori* (can lead to stomach cancer). As we will see later, vaccination may be the key to preventing these cancers [11]. Figure 10.4 shows the relationship between per capita cigarette consumption and lung cancer rates in men and women [12]. There is a 20–25 year delay between the peak in cigarette consumption and the peak in lung cancer incidence, reflecting the long period of exposure and resulting biological changes which occur in lung cancer. Rates of lung cancer incidence in women peak about a decade later than in men, reflecting the delay in when women began to smoke. While tobacco use has declined in many developed countries, it is rising in many developing countries. Worldwide about 35% of men in developed countries smoke, while the fraction of men who smoke is 50% in developing countries. China represents a particular concern; with more than 300 million male smokers today, future increases in lung cancer incidence and mortality are a likely consequence [13].

Changes in diet can also reduce cancer risk. The American Cancer Society recommends that persons consume five servings of fruits and vegetables daily to reduce their cancer risk. Unfortunately, less than 1/4

Tobacco Use in the US, 1900–2004

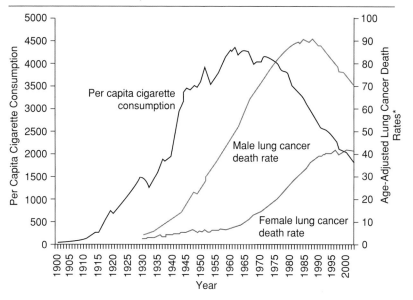

*Age-adjusted to 2000 US standard population.
Source: Death rates: US Mortality Data, 1960–2004, US Mortality Volumes, 1930–1959, National Center for Health Statistics, Centers for Disease Control and Prevention, 2006. Cigarette consumption: US Department of Agriculture, 1900–2004.

Figure 10.4. *An increase in the mortality due to lung cancer in the USA followed an increase in cigarette consumption that began in the 1930s. Reprinted by the permission of the American Cancer Society, Inc. from www.cancer.org. All rights reserved.*

of Americans follow this recommendation; only 24.3% eat five or more servings of fruit and vegetables daily, a figure that has changed little over the past decade [14].

The pathophysiology of cancer

Now that we have examined the global burden of cancer, let's turn to how cancers develop. While the growth and differentiation of normal cells is carefully coordinated by growth signals, cancers are characterized by uncontrolled growth and spread of abnormal cells. Unlike normal cells, cancer cells continue to grow in the absence of external growth signals, and they ignore signals to stop dividing, to specialize or to die [15]. Cancer cells can not only sustain themselves, they can expand and migrate. Normal cells can be transformed into cancer cells by a number of factors, including exposure to external carcinogens such as tobacco smoke, certain chemicals, or ionizing radiation, exposure to certain infectious agents, and exposure to certain hormones.

At the cellular level, the process of cancer development is remarkably similar in many different tissues. More than 85% of cancers arise in the epithelial tissues that line our organs, such as the skin, the digestive

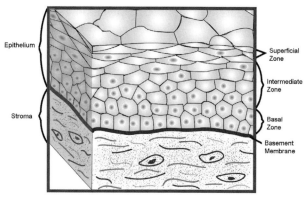

Figure 10.5. *Cartoon of squamous epithelium. Multiple layers of epithelial cells sit atop a basement membrane. Cells become progressively more specialized as you move from the basement membrane toward the epithelial surface.*

tract, the respiratory tract, and the genitourinary tract. Figure 10.5 shows a cartoon of one type of epithelial lining. In many tissues, the epithelial surface consists of multiple layers of epithelial cells. These cells sit on top of a special membrane called the basement membrane. Beneath the basement membrane there are layers of muscle and connective tissue that give the organ its structural stability. The epithelium is exposed to

Part I: Peace Corps and rice visits: July 2–July 4, 2007
Tessa Swaziland

I got up around the same time as I usually do for clinic (6:45-ish) to shower and finish packing for my long awaited visit to a Peace Corps volunteer's site. Carrie had suggested I stay with a volunteer as soon as I got here, but once WFP stuff started, I didn't have any time to escape the clinic. We both thought it would be a good opportunity to see where the COE's patients come from – not just physically, but culturally, emotionally, etc . . . Like, what kind of customs and beliefs exist in their communities? How are decisions made? What is the family environment like? How are orphans and abandoned children cared for? What is the system for governing? And also, this would give me an opportunity to talk with someone whose work and goals were similar to mine but who had been here much longer and worked on many more projects.

Carrie gave me a volunteer's number a couple weeks ago, and I had called her and set up a time to come. Tandi (that's the volunteer's Seswati name) and I met downtown by an Internet café. She was very nice and happy to answer all of my questions. In what ended up being a very rushed morning (there had been some confusion with the new WFP system, and Dave called me to the clinic mid-shower) I hadn't managed to squeeze in breakfast, so we sat down for omelets and coffee before heading out to her homestead. I was certainly glad for her company. I'd been on combies before (the vans used for public transport here), but only for short trips between the clinic and Mbabane. Finding my way from Mbabane to Manzini, switching to another combie, riding from Manzini to Siphoneni, transferring to yet another combie, and riding out to her village (name was hard to pronounce, and now I've forgotten it completely) would've been quite an adventure (possibly an unsuccessful one).

When we got off the last combie, we walked by a row of gogos (old women selling fruit or other items on the side of the road) and she greeted all of them with "Sanibonani" (the "hello" you use to address a group of people). A chorus of "Yebo"s echoed back, and the exchange continued for a minute. Everyone we passed, Tandi greeted and waved to. In that particular community, they used a two hand wave, which proved difficult since we were both carrying quite a bit of stuff. Discovering I had no Seswati name, Tandi enlisted the help of the gogos in naming me. I am now officially Zandile Dlamini. "Zandile" means "too many girls," and Dlamini is the most common last name in Swaziland. I'd say about 40% of our COE patients are Dlamini's.

We hiked for about 20 minutes to reach the homestead where she had been living for almost a year. The paths were dirt trails randomly winding and crisscrossing through brush and occasionally along pastures and homesteads. It reminded me quite a bit of the landscape and random layout of the community I lived in five years ago in Nicaragua.

At her home, I met her gogo (literally translate as "grandmother") and the children who lived there. Her babe (father of the house) wasn't home. In fact, he rarely is there, she said. Two of his three wives reside at that particular homestead, but neither was there at the time. One was shucking corn on the other side of the nearby mountain, and I'm not sure where the other one was. The one shucking was Tandi's "mage" (mother . . . pronounced ma-ge with a soft g sound), and I got to meet her later. She was very well educated and easy to talk to. In fact, she had met her husband when they were both studying at a university in the UK. At that time he already had two wives back in Swaziland. She had hosted Peace Corps volunteers several times before and seemed like a very good host mom. She kept trying to make Tandi stop translating. (She thought I was

another PC volunteer and thought I should know the language by now.) She said all the children there were her own, but Tandi told me later that in fact, they were all children of her husband, but none actually belonged to her. They were the children of all of his girlfriends. (Keep in mind; this is a completely normal arrangement for a family. Other than the fact that they were wealthy for a rural Swazi family, this was very representative of many of the homesteads all over Swaziland. A homestead is basically a collection of homes that belong to one extended family – a gogo, a babe, several mages, and many children).

Anyway, after I met the gogo (and before I met the mage) Tandi took me over to the hospital and VCT (voluntary counseling and testing facility) to see what they were like. It looked much like the Vuvulane clinic except bigger and cleaner (still nothing close to the COE standards). We talked to the VCT employees for a bit. They basically serve as a site where community members can come in and get tested for HIV. If they test positive, there are support groups and counseling available. Also, if they've been diagnosed and prescribed ART, they can pick up the meds there. One nice thing about this particular VCT was the fact that there was a woman who worked there whose sole responsibility was to deal with adherence. She was Swazi, and I believe she actually grew up there or nearby. She had obtained a grant for her project and was now trying to improve adherence in the community. Unfortunately, she had left for the day, so I didn't get to speak with her.

On the way back to the homestead, Tandi told me more about her experiences living in Swaziland. She said adjusting to the culture wasn't too difficult. She didn't really get homesick, and the community welcomed her. Apparently the last volunteer was kind of angry. Whenever Tandi had events, people would always come up to her afterwards and say, "Thank you for not yelling at me, Sisi." (Sisi means sister and is the word everyone uses to address a young woman who isn't married). Evidently, the last volunteer had been a yeller. The one thing Tandi said was most difficult to adjust to was the number of deaths. Every week, there are about three vigils to mourn the deceased. They last all night and end at about 7 a.m. with a funeral.

the external world; it serves as an important protective barrier, and is constantly regenerating itself. The cells adjacent to the basement membrane are responsible for this regeneration, therefore they are the most metabolically active cells in the epithelium. Epithelial cells become more specialized and mature as we move closer to the surface of the epithelium. Those cells at the top of the epithelium are dead, and will eventually be sloughed off.

Thus, there is a gradient of cell differentiation throughout the epithelium with the most differentiated cells at the epithelial surface and the least differentiated cells at the basement membrane, and the morphology of normal cells differs along this gradient. As we move from the bottom to the surface of the normal epithelium, the nucleus of each cell occupies progressively less and less of the cell volume. This ratio is called the nuclear to cytoplasmic ratio (N/C ratio). A large N/C ratio is characteristic of a rapidly dividing immature cell, while a small N/C ratio is characteristic of a mature, terminally differentiated cell. Figure 10.6 shows a photograph of a biopsy from the oral mucosa that has been sectioned

transversely and stained with hematoxylin and eosin dyes; the hematoxylin colors the nucleus purple. The cells adjacent to the basement membrane have a large N/C ratio and this ratio becomes progressively smaller as we move up through the normal epithelium.

An epithelial cancer begins with transformation of a single epithelial cell. As this transformed cell grows, it fails to differentiate, continuing to actively divide. When the lower 1/3 of the epithelium is filled with transformed cells, the condition is known as low grade pre-cancer. When the lower 2/3 of the epithelium is occupied by transformed cells, the condition is known as high grade pre-cancer. Figure 10.7 shows a photograph of an oral mucosa biopsy with a high grade pre-cancer on the right. Note the cells with increased N/C ratio in the lower 2/3 of the epithelium. These lesions are called pre-cancerous lesions because they are not yet cancers, but they have the potential to develop into a cancer. As shown on the left of Figure 10.7, in some cases transformed cells can break through the basement membrane, entering the stroma beneath. In this case, we no longer have an organized epithelium and stroma; instead the tissue is

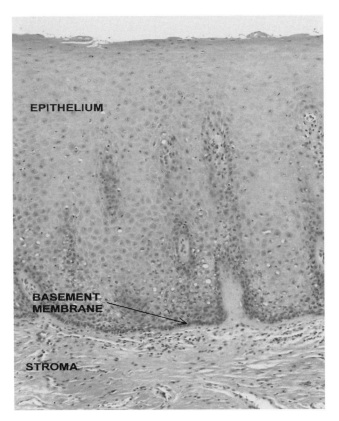

Figure 10.6. *Photo of stained biopsy of the normal oral mucosa. Multiple layers of epithelial cells sit atop a basement membrane. Cells become progressively more specialized as you move from the basement membrane toward the epithelial surface.*

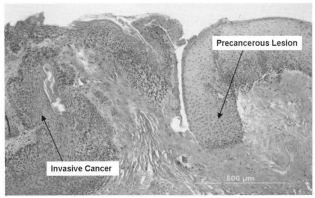

Figure 10.7. *Photo of stained biopsy of the oral mucosa. On the right side of the biopsy, a high grade pre-cancerous lesion is present. Note the cells with large N/C ratio that occupy most of the epithelium. The left side of the biopsy shows an invasive cancer of the oral mucosa. Nests of cancer cells are intermixed with stromal tissue.*

a mix of nests of cancer cells and surrounding stroma. This is a significant phenomenon – it is at this point that we go from pre-cancer to invasive cancer, and as we will later see, the prognosis and treatment of pre-cancer and cancer are radically different.

How do epithelial cells become transformed to initiate the development of a cancer? A cell is transformed through a series of mutations that affect its DNA; DNA mutation leads to production of mutant proteins. For example, a mutated gene can produce a defective protein that causes the growth factor receptors on the cell's surface to be constantly on. This type of 'gain of function' mutation can produce a transformed cell which continues to grow in the absence of external signals. The DNA in cancer cells can also undergo mutations which result in a loss of function; for example, ignoring signals to stop growth, or losing the function to repair or destroy defective cells.

Thus in cancer, oncogenic mutations disrupt the cell cycle and the careful coordination of growth, differentiation and death that characterizes normal cells and tissues. Cancer cells don't respond to signals that regulate cell growth and division. They can grow in the absence of signals to grow, and they ignore signals to stop growth. Changes in the gene expression profile allow cancer cells to replicate indefinitely. For example, normal cells can divide a finite number of times. Cancer cells overcome this limitation to become immortalized. Normal cell division is regulated by telomeric DNA at chromosome ends (Figure 10.8). This DNA functions to prevent end to end fusion of chromosomes. Normal cells shorten telomeric DNA with each division. After a certain point telomeres fuse and cells die. Most cancer cells activate an enzyme called telomerase. This enzyme extends the telomeres so that the cell can go through unlimited cycles of cell division [16].

The mutations that lead to cancer start in one cell, and as this cell divides, further mutations can occur in daughter cells. It is the accumulation of mutations that irreversibly transforms a normal cell into a cancer cell. Usually five to seven mutations are required to transform a cell. These mutations accumulate over time, which is why cancer is more common with increasing age. Fewer than 10% of mutations that lead to cancer are inherited; most are due to environmental factors [16].

Learn more about it

A comprehensive overview of cancer biology can be found at www.insidecancer.org. The section "Pathways to Cancer" contains 3D animation that illustrate the abnormalities in signaling pathways associated with cancer cells. Courtesy of Dolan DNA Learning Center, Cold Spring Harbor Laboratory.

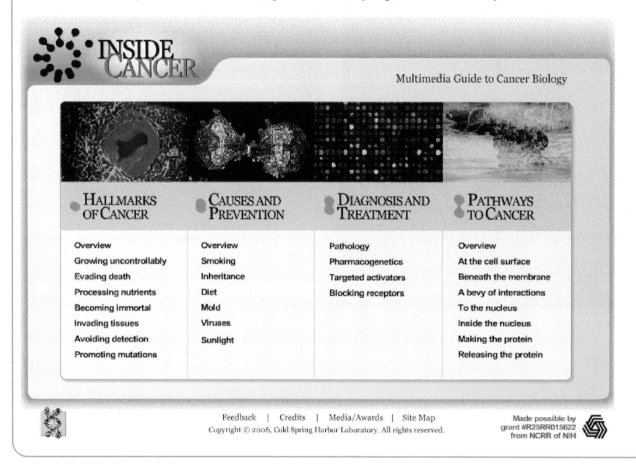

We can think of tumor development as analogous to Darwinian evolution. The transformation of a normal cell into a cancer cell occurs via a succession of genetic changes; together, these changes lead to the progressive conversion of normal cells to cancer cells [15]. Because these changes confer a growth advantage relative to normal tissue, the unchecked growth of cancer cells results in a mass of tumor cells. Once a nest of transformed cells begins to grow, the energy demands of these cells rapidly outstrip the capacity of the normal vasculature to supply nutrients [16]. As a result, transformed cells can induce the formation of new blood vessels, in a process called angiogenesis (Figure 10.9). Angiogenesis can occur in the early pre-cancerous stages. Frequently, the vessels formed in a tumor are abnormal – they are tortuous and leaky; we will see later that we can exploit these properties to aid in the early detection and treatment of tumors.

When cancer cells are confined to the organ in which they originated, we refer to the lesion as a primary tumor. Cancer cells have the ability to spread beyond the primary organ site (Figure 10.10). As cancer cells invade through the connective tissue in the primary organ site, they can intravasate into blood vessels and lymph vessels in that organ. From there, they can travel to distant organs and extravasate out of the blood vessels to form metastatic nests of tumor cells in distant organs [17]. Some of these nests of tumor cells will survive, grow and expand to form metastatic tumors. Figure 10.10 illustrates how a single transformed cell can

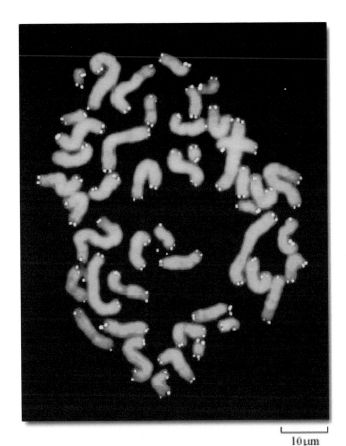

Figure 10.8. *A fluorescent stain indicates telomeric DNA (lighter areas) at the tips of chromosomes. Courtesy of Peter M. Landsdrop, BC Cancer Research Centre.*

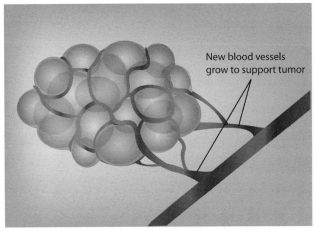

New blood vessels grow to support tumor

Figure 10.9. *During angiogenesis, tumor cells induce formation of new blood vessels.*

lead to the development of a metastatic tumor. Metastasis is responsible for a large fraction (90%) of deaths due to cancer [16].

In summary, there are more than 100 different types of cancer, yet the process of tumor development is remarkably similar across different organ sites. The formation of tumors is a multi-step process, during which six essential alterations occur in cell physiology: (1) cells develop self-sufficiency in growth signals, (2) they become insensitive to signals of growth inhibition, (3) they evade programmed cell death, (4) develop limitless replicative potential (5) they can sustain angiogenesis, and (6) acquire the ability to invade tissue and metastasize [15]. As we will see later in this chapter, the development of new diagnostic and therapeutic techniques for cancer increasingly focuses on these six common elements of cancer cells.

Why is early detection so important?

In his 1971 State of Union address, President Nixon declared "war" on cancer and requested $100 million for cancer research. On December 23, 1971, Nixon signed the National Cancer Act into law and said, "I hope in years ahead we will look back on this action today as the most significant action taken during my Administration [18]." Today, the US government still makes a substantial investment in cancer research; the National Cancer Institute (NCI) will spend over $4.6 billion for cancer research in 2007 [19]. The mission of the NCI is to eliminate suffering and death due to cancer by 2015 [20].

How have cancer incidence and mortality rates changed as a result? In 1950, heart disease was the leading cause of death, followed by cancer, cerebrovascular disease, and infectious disease (Figure 10.11). More than 50 years later, in 2004, the age adjusted mortality due to heart disease, cerebrovascular disease, and infectious disease have all dropped by more than half; that due to cancer has decreased only slightly [21]. Table 10.7 shows the five year survival rates for patients diagnosed with different cancers during three different time periods: 1975–1977, 1984–1986 and 1996–2004. In the mid 1970s, the overall five year survival rate was 50% and this did not change appreciably over the next decade. In the late 1990s to early 2000s, the five year survival rates had risen to 66%. Prostate cancer

Progression of Malignant Cancer

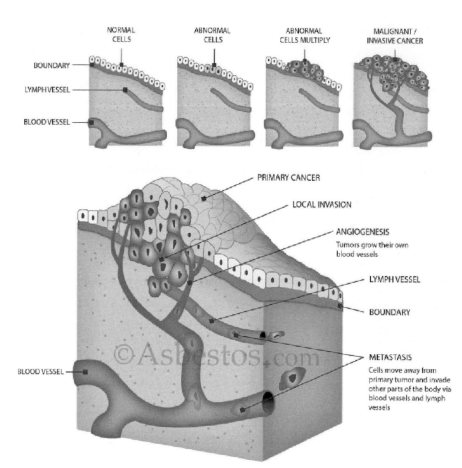

Figure 10.10. *Development of a metastatic tumor. Courtesy of Asbestos.com.*

showed great improvement between the 1980s and 1990s, reflecting the introduction of a screening test designed to detect early disease [1]. Lung cancer and pancreatic cancer survival rates have not changed substantially during this period; these cancers tend to be diagnosed at a relatively advanced stage when available therapies are not particularly effective.

Why is early detection so important to reducing cancer mortality? Figure 10.12 compares the five year survival rates for three different cancers as a function of the stage at which they are diagnosed. The five year survival rates exceed 90% for patients whose cancer is diagnosed when it is still confined to the local organ site [1]. The five year survival rate drops when cancer is first detected after metastasis to a regional location. For those patients whose cancer has metastasized to

a distant location prior to diagnosis, the five year survival rates are dismally low. Thus, an important strategy to reduce the mortality associated with cancer is to develop improved detection technologies designed to identify cancer at the earliest possible stages, when the available therapies are more likely to result in cure.

Strategies for early detection

Typically, cancers do not produce symptoms until a fairly late stage. How can we identify disease in asymptomatic patients? This is the goal of a process called "cancer screening." Screening refers to the use of simple tests in a healthy population. The goal of screening is to identify individuals who have disease, but do not yet have symptoms. The goal of screening is not to

Table 10.7. *Changes over time in five year survival rates [1].*

Trends in 5-year Relative Survival Rates* (%) by Race and Year of Diagnosis, US, 1975-2004

Site	All races 1975-77	All races 1984-86	All races 1996-2004	White 1975-77	White 1984-86	White 1996-2004	African American 1975-77	African American 1984-86	African American 1996-2004
All sites	50	54	66†	51	55	68†	40	41	58†
Brain	24	29	35†	23	28	34†	27	33	39†
Breast (female)	75	79	89†	76	80	91†	62	65	78†
Colon	52	59	65†	52	60	66†	46	50	55†
Esophagus	5	10	17†	6	11	18†	3	8	11†
Hodgkin lymphoma	74	79	86†	74	80	87†	71	75	80†
Kidney	51	56	67†	51	56	67	50	54	66†
Larynx	67	66	64†	67	68	66	59	53	50
Leukemia	35	42	51†	36	43	52†	34	34	42
Liver#	4	6	11†	4	6	10†	2	5	8†
Lung & bronchus	13	13	16†	13	14	16†	11	11	13†
Melanoma of the skin	82	87	92†	82	87	92†	60§	70§	78
Myeloma	26	29	35†	25	27	35†	31	32	33
Non-Hodgkin lymphoma	48	53	65†	48	54	66†	49	48	58
Oral cavity	53	55	60†	55	57	62†	36	36	42†
Ovary	37	40	46†	37	39	45†	43	41	38
Pancreas	3	3	5†	3	3	5†	2	5	5†
Prostate	69	76	99†	70	77	99†	61	66	96†
Rectum	49	57	67†	49	58	67†	45	46	59†
Stomach	16	18	25†	15	18	23†	16	20	25†
Testis	83	93	96†	83	93	96†	82‡	87‡	87
Thyroid	93	94	97†	93	94	97†	91	90	95
Urinary bladder	74	78	81†	75	79	82†	51	61	66†
Uterine cervix	70	68	73†	71	70	74†	65	58	65
Uterine corpus	88	84	84†	89	85	86†	61	58	61

* Survival rates are adjusted for normal life expectancy and are based on cases diagnosed in the SEER 9 areas from 1975-1977, 1984-1986, and 1996-2004, and followed through 2005. † The difference in rates between 1975-1977 and 1996-2004 is statistically significant (p <0.05). ‡ The standard error of the survival rate is between 5 and 10 percentage points. § The standard error of the survival rate is greater than 10 percentage points. # Includes intrahepatic bile duct.
Source: Ries LAG, Melbert D, Krapcho M, et al (eds.). *SEER Cancer Statistics Review, 1975-2005*, National Cancer Institute, Bethesda, MD, seer.cancer.gov/csr/1975_2005/, 2008.

American Cancer Society, Surveillance and Health Policy Research, 2009

American Cancer Society. Cancer Facts and Figures 2009. Atlanta: American Cancer Society, Inc.

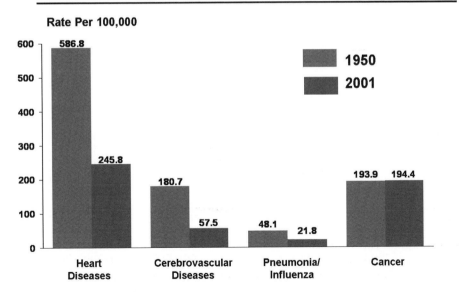

Change in the US Death Rates* by Cause, 1950 & 2001

* Age-adjusted to 2000 US standard population.
Sources: 1950 Mortality Data - CDC/NCHS, NVSS, Mortality Revised.
2001 Mortality Data–NVSR-Death Final Data 2001–Volume 52, No. 3.
http://www.cdc.gov/nchs/data/nvsr/nvsr52/nvsr52_03.pdf

Figure 10.11. *Changes in US death rates since 1950. Source: CDC.*

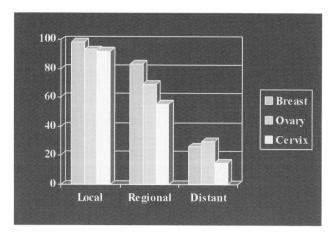

Figure 10.12. *Five year relative survival rates due to several different cancers [3].*

diagnose disease, but rather to use an inexpensive and simple test to identify those individuals who should have a more expensive and accurate test to confirm a diagnosis of disease.

In the USA, we routinely screen for four cancers, including female breast cancer with clinical breast examination and screening mammography, cervical cancer with HPV testing and the Pap smear, prostate cancer using the serum PSA test and digital rectal examination, and colon and rectal cancer, using a combination of the fecal occult blood test, flexible sigmoidoscopy, and colonoscopy [24].

Table 10.8 summarizes the recommendations of the American Cancer Society regarding screening for breast cancer; yearly mammograms to screen for breast cancer are recommended for women over the age of 40. While 69% of women over age 40 report having received a mammogram, women with no health insurance are significantly less likely to have received a mammogram. Table 10.9 summarizes the recommendations of the American Cancer Society regarding screening for cancers of the colon and rectum. While the percentage of people age 50 or more reporting a recent flexible sigmoidoscopy has increased in the past few years, only 45% of patients have had this test. The percentage of patients with no health insurance reporting this screening test is only 17% [14].

Table 10.8. *Recommendations for breast cancer screening from the American Cancer Society. Reprinted with permission from the American Cancer Society, Inc. from www.cancer.org. All rights reserved.*

Screening Guidelines for the Early Detection of Breast Cancer, American Cancer Society
• Yearly mammograms are recommended starting at age 40 and continuing for as long as a woman is in good health.
• A clinical breast exam should be part of a periodic health exam, about every three years for women in their 20s and 30s, and every year for women 40 and older.
• Women should know how their breast normally feel and report any breast changes   promptly to their health care providers. Breast self-exam is an option for women starting in their 20s.
• Women at increased risk (e.g., family history, genetic tendency, past breast cancer) should talk with their doctors about the benefits and limitations of starting mammography screening earlier, having additional tests (i.e., breast ultrasound and MRI), or having more frequent exams.

Table 10.9. *Recommendations for screening for cancers of the colon and rectum from the American Cancer Society. Reprinted with permission from the American Cancer Society, Inc., from www.cancer.org. All rights reserved.*

Screening Guidelines for the Early Detection of Colorectal Cancer, American Cancer Society 2003

• Beginning at age 50, men and women should follow one of the following examination schedules:
   • A fecal occult blood test (FOBT) every year.
   • A flexible sigmoidoscopy (FSIG) every five years.
   • Annual fecal occult blood test and flexible sigmoidoscopy every five years*.
   • A double-contrast barium enema every five years.
   • A colonoscopy every ten years.

*Combined testing is preferred over either annual FOBT, or FSIG every 5 years alone. People who are at moderate or high risk for colorectal cancer should talk with a doctor about a different testing schedule.

Cancer pain management: China

Cancer pain: it is a component of the disease that many patients fear more than death itself. Its severity has been described as intolerable and excruciating, and it only increases with the progression of the cancer. While cancer pain is treatable with the use of standard analgesics, including opioids such as morphine, this cheap and effective analgesic is not the standard of care everywhere.

"It was clear that many places didn't even have aspirin for cancer pain relief. And in many countries, the idea of having more potent drugs, even codeine let alone morphine, which is what we use to manage severe cancer pain, was nonexistent," recalls Dr. Charles Cleeland, Chair of the Department of Symptom Research at the University of Texas M.D. Anderson Cancer Center (image courtesy of Dr. Cleeland).

In 1993 the morphine consumption in China was estimated at 0.01 mg per capita, a surprisingly low number compared to the estimated 66.53 mg per capita consumption for Denmark [22]. Morphine was strictly managed. Suffering Chinese patients needed a certificate to receive morphine, and could only receive one ampoule of short-lasting morphine per day for four days as long as they returned an empty ampoule every time [22]. Patients had to choose the hour of their pain relief. Such telling statistics urged the World Health Organization (WHO) to initiate a plan to improve cancer pain management globally.

Up to this point, Dr. Cleeland, whose work focuses on pain assessment and treatment, had worked in countries such as Mexico and the Dominican Republic, instructing small groups in the usage of opioids for cancer pain relief. "At that time I had a postdoctoral fellow from China who said, 'Why don't you take on something really big?' So she got me connected with a friend of hers who was the minister of the health for the district of Beijing. So he came over and we tried to think of things to do to start."

Thus began a collaboration between the WHO, the government of China and Dr. Cleeland's pain research group to address the inadequate cancer pain management in China. The initial step was to study the epidemiology of the problem. To do this Dr. Cleeland launched a study led by Dr. Shelley Wang in 1992. The study helped elucidate the current state of pain control in the greater area of Beijing by examining 200 different cases. "And what we found was not a surprise; morphine was to be used rarely if at all. These patients had very high levels of pain compared to more developed countries."

To try and explain the kind of pain the group encountered, Dr. Cleeland says, "On a 0 to 10 scale, if we break that scale up you and I probably experience from time to time pain maybe up to 3 or 4. Many cancer patients after their disease has metastasized have a continual 10 pain."

The lack of pain control was evident, and the reasons behind this were many. "There were regulatory problems. There were issues of concern with addiction and a lack of any kind of distribution of controlled substances. It was a multifaceted problem."

The next step was to set up large meetings in Beijing and other major cities to introduce the epidemiology of the problem and to begin instruction. Over 500 individuals selected by government officials attended these three-day meetings. "They chose very well. They picked people who were administrators for major hospitals. They picked drug regulators. They picked nursing people as well as physicians."

Once the problem was discussed, a series of "trainer training" programs began. Groups composed of a nurse, a pharmacist and a physician from different hospitals would attend the program and learn about pharmacology of opioids and pain assessment. "The impressive thing, always, was when we'd train the students and then send them into the wards. They would ask the physicians how many patients have pain and the physicians would say "Well, of course, none of my patients have pain." Then they would go around and ask and find it was quite a different story."

Gradually, the training was handed over to Chinese professionals. To support their work, an evidence-based text developed by the WHO in the subject of cancer pain treatment was translated to Chinese. Concomitantly, the government of China adopted a new cancer pain relief policy, adjusting the inhibiting narcotics control policy, approving new opioid analgesics for sale and distribution, and increasing opioid manufacturing through joint ventures and other means. Gradually the pain alleviation for suffering cancer patients spread throughout the country.

Five years later, a comparison study was done. "It was a tremendous change from the majority of patients being under-managed to a majority of patients being managed very well according to the WHO standards. It was a 50% drop, from 60% under-managed in 1992 to 30% under-managed in 1997," Dr. Shelley Wang recalls.

Today opioids such as immediate and sustained-release morphine, fentanyl patch, meperidine, methadone and codeine are available to cancer patients in China. Physicians can prescribe these analgesics for up to 5 days, and in the case of more severe pain a physician can prescribe 15-day relief in the form of the fentanyl patch [23]. And, although the work is still not complete, a network of committed policy makers, oncologists and pain experts have been able to tremendously change the excruciating pain experience for a nation.

Unfortunately, the beginning of this story is not uncommon. "You have about 10 million cancer deaths a year worldwide, and probably 2/3 of them will experience significant pain. And, unfortunately, at least more than half will be inadequately managed. So they'll have a period of 3 or 4 months at the end of life which will be just miserable," states Dr. Cleeland.

Presently, Dr. Cleeland, Dr. Wang and their group continue to address the cross-cultural issue of pain and agree that its poor treatment is due to the lack of appreciation and assessment of pain. They hope to reproduce the successful model established in China and are involved in joint efforts to help other nations such as Russia, Korea and Japan, bring relief to cancer patients [22, 23].

Effectiveness of screening

How do we judge the effectiveness of a screening test? Let's take the example of screening for breast cancer. Imagine that you are a patient being screened with mammography. We can envision four possible results. If you have breast cancer and the mammogram is positive, the result is a "true positive." However, if you have breast cancer but the mammogram is negative, the result is a "false negative." If you do not have breast cancer, and the test is negative, the result is a "true negative." Finally, if you do not have breast cancer, but the test is positive, the result is a "false positive." We can arrange these possible outcomes in a 2 × 2 table as shown in Figure 10.13.

We define the sensitivity of a test as the probability that given DISEASE, the patient tests POSITIVE. The sensitivity is a measure of the ability of the test to cor-

rectly detect disease when it is present, or the ability to find true positives. Sensitivity can range from a low of 0% to a high of 100%. We define the specificity of a test as the probability that given NO DISEASE, the patient

	Mammogram Positive	Mammogram Negative
Patient Has Breast cancer	True Positive (TP)	False Negative (FN)
Patient Does Not Have Breast Cancer	False Positive (FP)	True Negative (TN)

Figure 10.13. *Possible outcomes of a screening mammogram for breast cancer.*

Table 10.10. *Comparing the effectiveness of a breast exam versus a mammogram in screening for breast cancer [25].*

Accuracy of Breast Cancer Screening		
	Sensitivity	Specificity
Clinical breast exam	54%	94%
Mammography	75%	92%

	Test Positive	Test Negative	
Disease Present	TP	FN	Number with Disease = TP + FN
Disease Absent	FP	TN	Number without Disease = FP + TN
	Number who Test Positive = TP + FP	Number who Test Negative = FN + TN	Total Number Tested = TP + FN + FP + TN

Figure 10.14. *Possible outcomes of a diagnostic or screening test.*

tests NEGATIVE. Specificity characterizes the ability of a test to avoid calling normal things disease, or the ability to avoid false positives. Specificity can also range from a low of 0% to a high of 100%. A perfect test has a sensitivity of 100% and a specificity of 100%. If a test performs better than chance alone (or better than the toss of a coin), the sum of the sensitivity and specificity is greater than 100%.

Table 10.10 lists the average reported sensitivity and specificity of two different screening tests for breast cancer – clinical breast exam, and mammography. The average sensitivity of mammography is 75%. This is higher than the 54% sensitivity of clinical breast exam. The specificity of mammography is 92%, slightly lower than that of clinical breast exam [25]. How do we measure the sensitivity and specificity of a screening test? If we screen a population of patients, some of whom are known to have disease and some who are known to be disease free, we can calculate the sensitivity and specificity of the test. Figure 10.14 shows the possible outcomes of the testing.

The sensitivity can be calculated as:

$$Se = \frac{TP}{(\text{number with disease})} = \frac{TP}{(TP + FN)}. \quad (10.1)$$

	Test Positive	Test Negative	
Disease Present	TP = 11	FN = 2	Number with Disease = TP + FN = 13
Disease Absent	FP = 15	TN = 208	Number without Disease = FP + TN = 223
	Number who Test Positive = TP + FP = 26	Number who Test Negative = FN + TN = 210	Total Number Tested = TP + FN + FP + TN = 236

Figure 10.15. *Data to calculate the sensitivity and specific of MRI screening for breast cancer. Used with permission from [26].*

The specificity can be calculated as:

$$Sp = \frac{TN}{(\text{number without disease})} = \frac{TN}{(TN + FP)}. \quad (10.2)$$

As an example, let's calculate the sensitivity and specificity of a new test suggested for breast cancer screening – magnetic resonance imaging or MRI. In 2004, results from a clinical trial of 236 women were reported to assess the performance of MRI for screening for breast cancer in women at high risk of developing the disease [26]. During the first year of the study, each woman had an MRI exam; 26 women had an abnormal MRI. To confirm the presence of breast cancer, additional testing was performed in women with an abnormal breast MRI; these tests confirmed the presence of breast cancer in 11 women. An additional two women who had a normal MRI exam were found to have breast cancer based on other tests. Figure 10.15 shows the 2×2 table filled out for this example. We can calculate the sensitivity and specificity of MRI in this clinical trial as follows.

$$Se = \frac{TP}{(TP + FN)} = \frac{(11)}{(11 + 2)} = 84.6\% \quad (10.3)$$

$$Sp = \frac{TN}{(TN + FP)} = \frac{(208)}{(208 + 15)} = 93.3\% \quad (10.4)$$

In order to calculate the sensitivity and specificity of a new test, we must develop criteria to determine whether the test result is normal or abnormal. As these criteria change, our estimate of the test's sensitivity and specificity change. We can characterize the performance of a test by plotting the test sensitivity and specificity as we vary these criteria. The resulting plot of sensitivity vs. specificity is known as a receiver-operator characteristic curve (ROC curve). Figure 10.16 shows the ROC curve for MRI used to screen for breast cancer in high risk women calculated from the study above [26]. As the

Individual Modalities

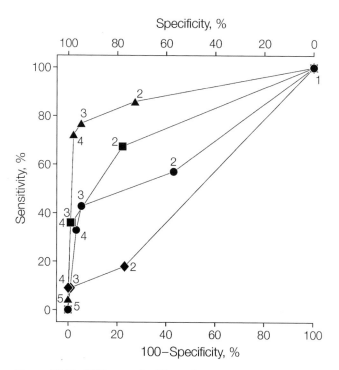

Screening Modalities
▲ Magnetic Resonance Imaging
■ Mammography
● Ultrasound
◆ Clinical Breast Examination
1-5 Minimum BI-RADS Grade

Figure 10.16. *ROC curves for different breast cancer screening modalities. The area under the curve is highest for MRI. Used with permission from [26]. JAMA, 2004, **292**; 1317–25. Copyright © 2004 American Medical Association. All rights reserved.*

Table 10.11. *MRI is significantly more expensive than other clinical methods of breast cancer screening [25].*

Screening Method	Medicare Reimbursement
Clinical Breast Exam	$39
Mammography	$90
Ultrasound	$74
MRI	$1108

sensitivity increases, the specificity decreases. The area under the ROC curve is often used to provide a measure of test accuracy. A perfect test has an ROC curve with area 1; a test that performs no better than chance has an area under the curve of 0.5. Figure 10.16 also compares the ROC curves for several screening methodologies in the same group of patients; the area under the ROC curve is highest for MRI. Unfortunately, the cost of MRI is substantially higher than the cost of clinically accepted technologies (Table 10.11) [25]. In Chapter 11, we will examine how to decide whether the additional resources required to implement a new technology represent a good investment.

As a patient, you wish to be screened with a test that has both a high sensitivity and specificity. But how high do these values need to be for the test to be useful to you? If you receive a positive screening test result, what is the likelihood that the result is a true positive or a false positive? Similarly, if you receive a negative screening test result what is the likelihood that the result is a true negative or a false negative? The sensitivity and specificity of the test don't provide enough information to answer these questions. Instead, we must calculate the positive and negative predictive value of the test, which give these probabilities.

The positive predictive value (PPV) is the probability that, given a POSITIVE test result, you have DISEASE. PPV ranges from 0% to 100%. The negative predictive value (NPV) is the probability that given a NEGATIVE test result, you do NOT HAVE DISEASE. Again, NPV ranges from 0% to 100%. We can use our 2 × 2 table to calculate PPV and NPV.

$$PPV = \frac{TP}{(\# \text{testing positive})} = \frac{TP}{(TP + FP)} \quad (10.5)$$

$$NPV = \frac{TN}{(\# \text{testing negative})} = \frac{TN}{(FN + TN)} \quad (10.6)$$

Again, we can use our MRI example of Table 10.11 to illustrate how to calculate positive and negative predictive value.

$$PPV = \frac{(11)}{(26)} = 42\% \quad (10.7)$$

$$NPV = \frac{(208)}{(210)} = 99\% \quad (10.8)$$

These statistics tell a patient that, given an abnormal screening MRI, there is only a 42% chance that she actually has breast cancer. Further testing is required to confirm whether cancer is actually present. However,

Breast cancer screening in China: is breast self-examination an alternative to mammography?

While screening mammography can reduce the mortality associated with breast cancer, it is not available in many developing countries due to limited resources. In such settings, breast self-examination may provide a less expensive alternative. A recent randomized clinical trial of breast self-examination was conducted in Shanghai, China to determine whether breast self-examination could reduce breast cancer mortality.

Beginning in 1989, more than 266,000 women in Shanghai were randomized into two groups. One group of 132,979 women received initial instruction in breast self-examination, with reinforcement sessions one and three years later. These women practiced breast self-examination every six months for five years; and 133,085 women were assigned to a control group. All women were followed through December 2000 for mortality from breast cancer.

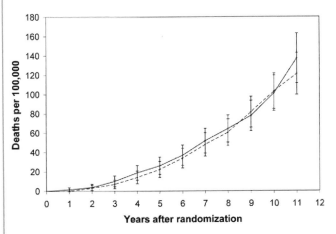

From [27] with permission from Oxford University Press.

The graph shows the cumulative breast cancer mortality per 100,000 women in the breast self-examination group (solid line) and the control group (dashed line). There were a total of 135 breast cancer deaths in the group who received instruction in breast self-examination, compared to 131 breast cancer deaths in the control group. In addition, more benign breast lesions were discovered in the group of women who performed breast self-examination. Unfortunately based on this study, it does not appear that breast self-examination can reduce mortality from breast cancer in this setting [27].

given a normal screening MRI, there is a 99% chance that the patient truly does not have breast cancer.

Clearly, the NPV and the PPV of a test depend on the sensitivity and specificity of the test. But they also depend on the prevalence of the disease that we are screening for. Prevalence is a measure of whether a disease is common or rare. Recall that prevalence of disease in a population, p, is defined as:

$$p = \frac{(\# \text{ in population with disease})}{(\text{total } \# \text{ in population})}. \tag{10.9}$$

In terms of our 2×2 table, prevalence can be calculated as:

$$p = \frac{(TP + FN)}{(TP + FP + TN + FN)}. \tag{10.10}$$

If we know the prevalence of a disease and the sensitivity and specificity of a test, we can calculate the positive and negative predictive values of the test as follows.

$$PPV = \frac{(p)(Se)}{[(p)(Se) + (1-p)(1-Sp)]} \tag{10.11}$$

$$NPV = \frac{(1-p)(Sp)}{[(1-p)(Sp) + (p)(1-Se)]} \tag{10.12}$$

Using our example of MRI to screen for breast cancer again, the prevalence of disease is:

$$p = \frac{(13)}{(236)} = 0.055 \text{ or } 5.5\%. \tag{10.13}$$

We can use the formulas above to calculate positive and negative predictive value.

$$PPV = \frac{(0.055)(0.846)}{[(0.055)(0.846) + (0.945)(0.067)]} = 42\% \tag{10.14}$$

$$NPV = \frac{(0.945)(0.933)}{[(0.945)(0.933) + (0.055)(0.154)]} = 99\% \tag{10.15}$$

We obtain exactly the same results as calculated previously. As the prevalence of disease decreases, the PPV of a test decreases. In our example of MRI to screen for breast cancer, the study was designed to screen women who were at high risk for breast cancer. As a result, 5.5% of the population studied had breast cancer, a prevalence that is much higher than that in the general population. Let's consider what would happen to the predictive value of the test if we were to use the same test to screen for breast cancer in all women. In the United States approximately 131 new cases of breast cancer are identified per 100,000 women [28]. Let's calculate the prevalence, positive and negative predictive value under these conditions. The prevalence is:

$$p = \frac{(131)}{(100,000)} = 0.0013 = 0.13\%. \qquad (10.16)$$

The sensitivity and specificity are independent of disease prevalence, and as before are:

$$Se = 84.6\% \qquad (10.17)$$
$$Sp = 93.3\%. \qquad (10.18)$$

However, the positive and negative predictive values differ.

$$PPV = \frac{(0.0013)(0.846)}{[(0.0013)(0.846) + (0.99987)(0.67)]} = 1.6\% \qquad (10.19)$$

$$NPV = \frac{(0.99987)(0.933)}{[(0.99987)(0.933) + (0.0013)(0.154)]} = 99.98\% \qquad (10.20)$$

The PPV deceases substantially, while the NPV increases slightly. Suppose a woman in our study has a positive MRI. What is the likelihood that she has breast cancer? This is the same as the positive predictive value and is only 1.6%! The low PPV illustrates the challenge of screening for a rare disease. For this reason, we generally screen for breast cancer in older women, because the prevalence of breast cancer increases with age.

While screening for breast cancer using mammography clearly reduces breast cancer mortality, it has been estimated that we must screen 1224 women for more than 14 years in order to prevent one death from breast cancer. Among women between the ages of 40–49 years, we must screen 1792 women for 14 years to prevent one death from breast cancer [29].

With this introduction, we now examine three cancers in detail – cervical cancer, prostate cancer, and ovarian cancer. In each case, we will examine the efficacy of existing screening technologies. We will also examine the new technologies in development to improve early detection.

Early detection of cervical cancer

In 2007, there were predicted to be 11,150 new cases of cervical cancer in the USA, and 3670 deaths due to cervical cancer [1].

Worldwide, cervical cancer is an important problem. In 2002, 493,000 new cases of cervical cancer were reported globally. Some 83% of cervical cancers occur in the developing world, with the highest incidence in central and South America, southern Africa and Asia (Figure 10.17). Cervical cancer caused 274,000 deaths in 2002 worldwide, and was the **leading cause of female cancer mortality in developing countries** [6]. Cervical cancer affects relatively young women, and is the single largest cause of years lost to life due to cancer in the developing world [30].

The cervix is located between the vagina and the uterus (Figure 10.18). The cervical os is the opening into the uterus; during conception, sperm travel from the vagina through the os to fertilize an egg. Throughout a pregnancy, the cervix provides the structural stability to hold the fetus inside the womb. During labor and delivery, the cervix thins and stretches to enable the baby to travel through the birth canal. Thus, the wall of the cervix contains both collagen and elastin fibers to provide both strength and elasticity. The outer surface of the cervix comes into contact with both semen and potentially dangerous bacteria and viruses. The cervix is lined with multiple layers of epithelial cells that play an important role both in preventing infection and facilitating conception. These epithelial cells produce cervical mucus; the mucus changes consistency throughout the menstrual cycle in order to facilitate travel of sperm during ovulation and to prevent travel and growth of pathogens.

Global incidence of cervical cancer, projections for 2005

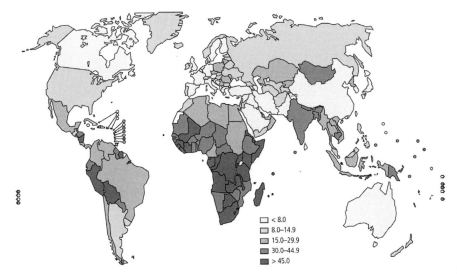

☐ < 8.0
☐ 8.0–14.9
☐ 15.0–29.9
☐ 30.0–44.9
☐ > 45.0

Source: This map is based on mortality, incidence and burden of disease projections produced for **Preventing chronic diseases: a vital investment.**
Geneva: World Health Organization: 2005. http://www.who.int/chp/chronic_disease_report/en.

Figure 10.17. *Global predictions for cervical cancer incidence in 2005.*

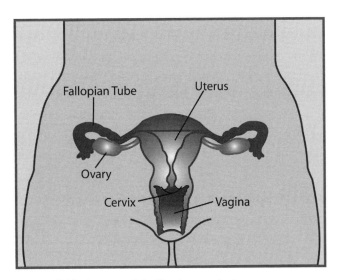

Figure 10.18. *Anatomy of the female reproductive tract.*

Cervical cancers begin in the epithelial lining of the cervix. Two types of epithelial tissue line the cervix (Figure 10.19); in both cases the epithelial cells are separated from the supporting stromal tissue below by a thin basement membrane. Surrounding the os, the surface of the cervix has small finger-like projections and is lined by a single layer of columnar epithelial cells. The outer edges of the cervix are flat and lined by multiple layers of squamous epithelial cells. The squamous epithelium is typically 200–300 microns thick. The junction between the columnar and squamous epithelium is known as the squamo-columnar junction (SQ junction). Most cervical cancers begin when an epithelial cell in the squamous epithelium becomes transformed and begins to proliferate. When an epithelial cell near the basement membrane becomes transformed, it loses the capacity to terminally differentiate, and the epithelium gradually fills with actively dividing cells that have large nuclei. When the bottom 1/3 of the epithelium is filled with transformed cells, the condition is referred to as a low grade squamous intraepithelial lesion (LGSIL). When the bottom 2/3 of the epithelium is filled with transformed cells, the condition is referred to as a high grade squamous intraepithelial lesion (HGSIL), and when the complete epithelium is transformed, it is called a carcinoma *in situ* (CIS). At this stage, the lesions are considered to be pre-cancerous. However, if the transformed cells break through the basement membrane and migrate into the stroma, the condition is known as a micro-invasive cancer. This cycle is known as the pre-cancer to cancer sequence. The prognosis of micro-invasive cancer is much more serious and the treatment is much more invasive than that of pre-cancers. Thus, the focus of cervical cancer screening programs is to identify cervical pre-cancers (when they can be easily treated) before they become cervical cancers (when they are difficult, painful and expensive to treat). Most low grade pre-cancers regress on their own, while 20–45% of high

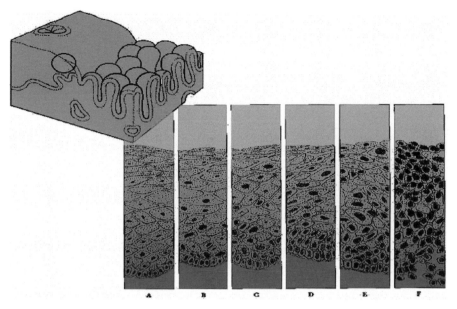

Figure 10.19. *A cross section (left) showing the two types of tissues lining the cervix; in normal cervical tissue (right) squamous cells change in appearance from top to bottom (A), but in pre-cancer this gradient becomes increasingly absent. Reprinted from Journal of Nurse-Midwifery, **3b**(5), McGraw R.K., Gynecology: A Clinical Atlas © 1991, with permission from Elsevier.*

grade lesions progress to cervical cancer if untreated. The progression from pre-cancer to cancer has been estimated to take about a decade [31].

What causes transformation of cervical epithelial cells?

In the 1990s, researchers demonstrated that infection with human papillomavirus (HPV) is the central causative factor in squamous cell carcinoma of the cervix. HPV infection is the most common sexually transmitted disease; asymptomatic HPV infections can be detected in 5–40% of women of reproductive age [32]. The majority of women with HPV infection do not develop invasive cervical cancer. In most young women, HPV infections are transient; the immune system clears them with no ill effects. However, if HPV infection persists past age 30, there is a much greater risk of developing cervical cancer [33].

There are more than 100 different types of the human papillomavirus; not all of them are carcinogenic. Fifteen types of HPV are commonly linked to cervical cancer, with HPV types 16 and 18 the most commonly found high risk types of virus (Figure 10.20) [34]. Human papillomaviruses have double stranded circular DNA chromosomes with about 8000 nucleotide pairs [35]. In an HPV infection, the HPV genetic material is transported to the nucleus of infected cervical epithelial cells.

In a wart or benign infection, the HPV chromosomes are stably maintained in the basal epithelium as plasmids whose replication keeps step with the chromosomes of the host (Figure 10.21, left). A cell becomes transformed when the viral DNA is integrated into a host chromosome. This alteration of the viral gene environment can disrupt control of their expression. The unregulated production of viral proteins tends to increase the

Global Prevalence of HPV Types in Cervical Cancer

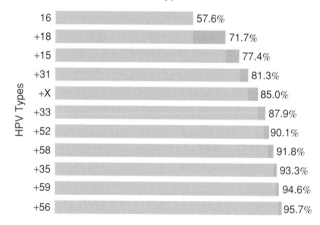

Figure 10.20. *Fraction of cervical cancer accounted for by different types of HPV. The two vaccines under development and testing protect against HPV types 16 and 18, which together account for about 70% of cervical cancers [34]. Credit: X. Bosch And N. Munoz/Iarc, Ibsccs, And Multicentric Studies (N = 3045). From Science 29 April 2005: Vol. **308**. no. 5722, pp. 618–621. Reprinted with permission from AAAS.*

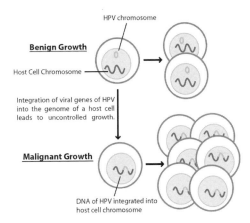

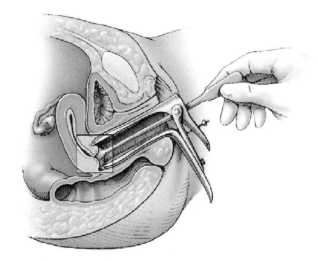

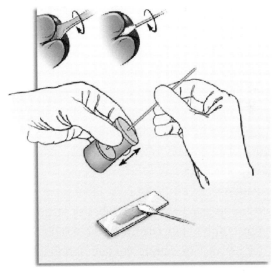

Figure 10.21. *When HPV DNA is integrated into the host cell DNA, growth is no longer regulated and a malignant tumor can form.*

rate of cell division, thereby helping to generate a cancer (Figure 10.21, right). For example, the HPV E6 protein appears to alter cell growth through effects on p53, an endogenous tumor-suppressor protein. E6 binds to p53, targeting it for destruction [35].

The signs and symptoms of cervical cancer include abnormal vaginal bleeding, in between periods or especially related to intercourse, and pelvic pain. Advanced cervical cancer is treated using surgery, and a combination of radiation therapy and chemotherapy. In the USA, the five year survival rate for localized cervical cancer is excellent at 92%. Slightly more than half of cervical cancers in the USA are diagnosed at this stage [1]. Such a large fraction of cervical cancers are detected early because we have a good screening test. In fact, most lesions are caught at the stage where they are still pre-cancers, and can be treated easily before they progress to cancer.

How do we detect cervical cancer and its precursors?

We screen for cervical cancer and its precursors using a test called the Papanicoloau (Pap) smear. The use of the Pap smear to screen has resulted in dramatic decreases in the incidence and mortality of cervical cancer, and is largely viewed as the most successful cancer screening test in medical history. The diagnosis of cervical cancer and its precursors is made using a confirmatory follow-up test, colposcopy and biopsy.

Figure 10.22. *A wooden spatula is used to obtain a Pap smear sample. By permission of the Mayo Foundation for Medical Education & Research. All rights reserved.*

In a Papanicoloau smear, a speculum is inserted into the vagina enabling the healthcare provider to visualize the cervix. A small wooden spatula is scraped against the cervix; the spatula is placed at the squamo-columnar junction and rotated to scrape off epithelial cells all around the junction (Figure 10.22). The cells collected on the spatula are then smeared on to a glass slide and allowed to dry. In obtaining a successful Pap smear, more than 50,000–300,000 cells, including both columnar and squamous epithelial cells, will be placed on the slide. The cells are then stained and examined by a trained cyto-technologist. Any abnormal appearing cells (cells with large nuclei or abnormal chromatin)

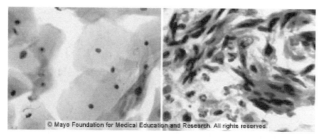

Figure 10.23. Normal appearing cells (left) and abnormal appearing cells (right) in a Pap smear. Copyrighted and used with permission of Mayo Foundation for Medical Education and Research, all rights reserved.

are noted (Figure 10.23). Based on these changes, Pap smears are classified into several categories: normal, infection/repair, atypical cells of uncertain significance, low grade pre-cancer, high grade pre-cancer, and cancer. Interpretation of the Pap smear is subjective, and the reproducibility of this interpretation has been found to be poor. An individual clinician agrees with their own prior diagnosis about 78% of the time and agrees with the diagnosis of others only between 28–72% of the time [36].

While Pap smears are helpful in identifying cervical cancer and its precursors, a number of factors can lead to false positive and false negative results. Only a small fraction of Pap smears contain abnormalities, and in those cases only a small percentage of cells may show cancerous changes, so positive cases are sometimes missed due to human error. Since the Pap smear samples only a fraction of cells from the cervix, abnormal cells present on the cervix may not be exfoliated when the sample is collected. Finally, benign changes such as infection or inflammation and tissue repair can cause cells to have the appearance of pre-cancerous cells.

It is difficult to measure the accuracy of the Pap smear for several reasons. For example, in many studies of Pap test accuracy, only patients with an abnormal Pap smear receive further testing to confirm the presence of disease. Studies of this type suffer from what is known as verification bias; only enough data are collected to allow one to calculate the specificity of the test, but not the sensitivity. Recent studies designed to verify both positive and negative results indicate that the sensitivity

US regulations to ensure quality in cytology laboratories

In the mid 1980s, a number of cases were brought to light in which women had developed cervical cancer despite having routine Pap smears with normal results. Media coverage of the cases implied that the false negatives resulted from laboratory errors due to carelessness. At the time, many commercial clinical laboratories set very high target rates for cytotechnologists to screen a certain number of slides per day or risk being fined part of their salary. Several such cases received extensive media coverage. As a result, the US Congress passed the Clinical Laboratory and Improvement Amendments of 1988 (CLIA). CLIA limited the number of slides that cytotechnologists in the USA can review to no more than 100 slides per day. CLIA also mandated that 10% of slides with a "normal" diagnosis be re-screened in order to limit the number of false negative diagnoses [37–39].

of the Pap test ranges from 30% to 87% (average 47%), while the specificity ranges from 86% to 100% (average 95%) when low grade Pap smears and worse are considered to be abnormal [40]. Figure 10.24 shows sensitivity and specificity of 62 different studies comparing Pap test results with biopsy [41]. The sensitivity and specificity vary widely from one study to another, and it is clear that it is difficult to achieve simultaneously high sensitivity and specificity using the Pap test.

Because the Pap smear is a screening test, an abnormal Pap smear is usually followed by a diagnostic procedure called colposcopy. In colposcopy, a speculum is inserted and a low power microscope (called a colposcope) is used to view the cervix (Figure 10.25a,b). A solution of weak acetic acid (vinegar) is applied to the cervix. The vinegar washes away any cervical mucus and also causes any pre-cancerous areas of tissue to turn white. Any suspicious areas on the cervix are then biopsied, using a metal biopsy forceps to remove a pea-sized portion of tissue. The biopsy is cut, stained and examined under the microscope by a pathologist. The

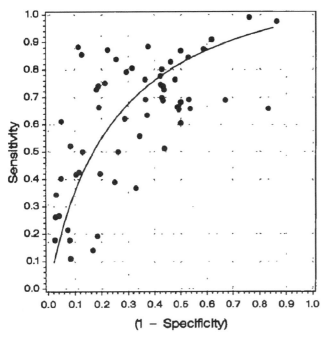

Figure 10.24. *The sensitivity and specificity of the Pap smear from 62 different studies can be used to estimate the ROC curve of the Pap test. From [41]. Used with permission from Oxford University Press.*

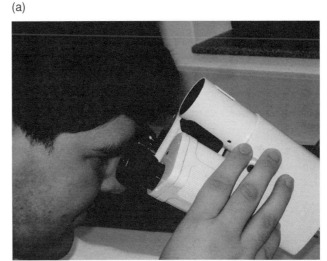

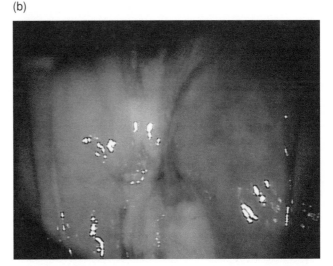

Figure 10.25. *(a) The use of a colposcope to view the uterine cervix. (b) Colposcopic photo of cervix.*

sensitivity of visual examination using the colposcope is excellent, at 96%; however, the specificity is quite low, only 48% [42]. The low specificity of colposcopy is the reason that a confirmatory biopsy must be obtained; but due to the low specificity, more than half of all biopsies obtained at colposcopy show only benign changes.

Figure 10.26a shows a histologic section of normal cervix from the squamous epithelium prepared from a biopsy obtained under colposcopic guidance; the cervix is lined by about 10–15 layers of epithelial cells. In the normal cervix, the basal layer of cells has the largest N/C ratio. The cells at the top of the epithelium have the smallest N/C ratio. Figure 10.26b shows a histologic section of a high grade pre-cancer. Clearly, the N/C ratio is increased throughout the entire epithelium. In this specimen, the cells have not yet invaded the basement membrane to form a micro-invasive cancer.

In summary, we screen for cervical cancer and its precursors using the Pap smear, and we confirm the diagnosis using colposcopy and biopsy. Pre-cancerous cervical lesions can be removed using a simple outpatient electro-surgical procedure to remove the transformed epithelium; this treatment preserves fertility. Because

we have a good screening test, most lesions are caught at the stage where they are still pre-cancers, and can be treated easily before they progress to cancer. As a result, by screening for pre-cancer, we can actually reduce the incidence of cervical cancer.

Before the introduction of screening programs, the incidence of cervical cancer in North America and Europe was similar to that seen in developing countries today. In every country in which organized screening programs based on the Pap smear have been introduced, rates of cervical cancer incidence and mortality have decreased [30]. While cervical cancer screening

(a)

(b)

Figure 10.26. *(a) Photograph of a normal cervical biopsy. (b) Photograph of high grade pre-cancer.*

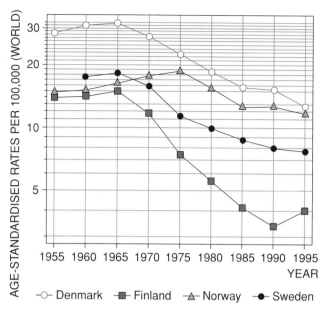

Figure 10.27. *Incidence rates of cervical cancer in four Nordic countries. Decreases in the incidence rate parallel the introduction and extent of screening programs. The article was published in Vaccine, Vol.* **24S3***, D. Maxwell Parkin and Freddie Bray, The burden of HPV-related cancers, pp. S3/11–S3/25, © Elsevier (2006).*

has never been tested in randomized clinical trials, reductions in cervical cancer mortality and incidence in countries where screening is practiced provide evidence that screening is effective. Figure 10.27 shows the declines in cervical cancer incidence in Nordic countries where screening programs were introduced in the 1960s to 1970s. Figure 10.28 compares changes in incidence rates over time in four countries. Decreases in Shanghai, China, reflect the introduction of an intensive screening program; in contrast, incidence rates have been stable in Bombay, India, where screening is largely unavailable [30].

The Pap smear is viewed as one of the most successful public health measures ever introduced. Given the relatively low sensitivity and specificity of the Pap test, it is sometimes surprising that screening has been so successful in reducing the incidence and mortality of cervical cancer. In large part, this is because, on average, to go from cervical pre-cancer to invasive cervical cancer requires a decade [44]. Even if a woman has a falsely negative Pap smear one year, chances are it will be detected when she has her next Pap smear, before it has progressed to cancer. However, the Pap smear does miss some cervical cancers. It has been estimated that

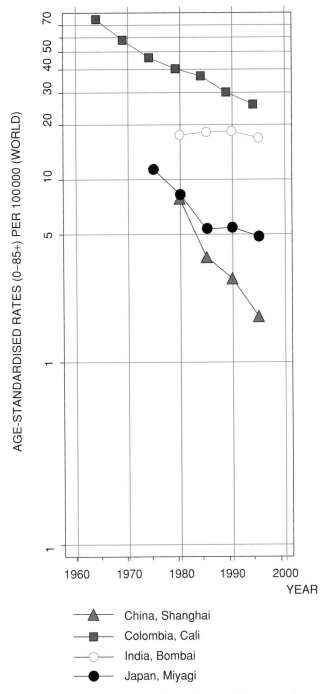

Figure 10.28. *Incidence rates of cervical cancer in four countries. Where screening programs are not available, incidence rates have been stable. The article was published in Vaccine, Vol. **24S3**, D. Maxwell Parkin and Freddie Bray, The burden of HPV-related cancers, pp. S3/11–S3/25, © Elsevier (2006).*

3% of preventable cervical cancer deaths can be traced to false negative readings; an estimated 50% of these are due to sampling error that would not be detected with re-screening [45]. We spend more than $6B annually in

the USA following up low grade Pap smears that likely will not yield any health benefits [46].

Because of the costs and infrastructure requirements associated with the test, the Pap smear is not available to a large segment of the world's population and cervical cancer continues to kill many young women. There are many barriers to cervical cancer screening in developing countries; developing countries face a lack of trained cytotechnologists and cytology labs. A further complication is the lack of facilities to follow up abnormal Pap smears and treat pre-cancerous lesions. It is difficult for women who live in rural areas to come for multiple visits required to screen, diagnose and treat cervical cancer and its precursors. Finally, the costs of screening in many developing countries exceeds a family's daily income, putting the test out of reach for most [47].

A number of new technologies have been developed to address the limitations of the Pap smear. In large part, these technologies have three goals: (1) reduce the false positive and false negative rates of the test, (2) develop tests that can give instantaneous results so that women could be treated at the initial visit if the test were positive, and (3) reduce the costs of the test so that it can be implemented in the developing world. Here, we will examine four new technologies to screen for cervical cancer: liquid based cytology, automated Pap smears, HPV testing, and optical testing.

Liquid cytology

One of the primary limitations of the conventional Pap smear is that only a fraction (estimated to be 20%) of the cells collected are transferred onto the slide which is later stained and examined for transformed cells [44]. A new technique called thin layer, liquid based cytology has been developed to improve the conventional Pap smear. In this technique, the brush used to collect the Pap smear is dipped into a vial containing a cell preservative, ensuring that all cells which are collected are available for analysis. A robotic preparation device is then used to remove blood and inflammatory cells and transfer a thin layer of representative cells in a circular area onto a slide. In one liquid cytology procedure (ThinPrep®), a spinning cylinder is lowered into the specimen vial and used to break up any clumps of cells.

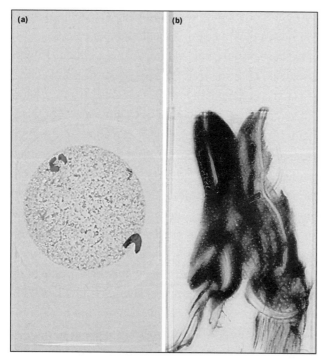

Figure 10.29. *(a) ThinPrep® Pap and (b) conventional Pap smear. Used with permission from [44]. Reprinted with permission from Elsevier (The Lancet Oncology, 2001, Vol. 2 No. 1, pp. 27–32).*

The cell suspension is then drawn upward through a polycarbonate filter until an approximate single layer of cells covers the filter. The filter is then briefly adhered to a glass slide in order to transfer the cells from the filter to the slide. Several slides, each containing a representative population of the exfoliated cells can be prepared from a single suspension [48].

Figure 10.29 shows a photograph of slides prepared using the conventional manner (b) and using the ThinPrep® device (a). Trials comparing conventional Pap to a thin layer Pap showed that the thin layer method results in an overall 18% higher detection rate of abnormalities than conventional cytology. Based on these results, this new technology was approved by the US FDA in 1996. Further studies indicate that the use of the thin layer Pap decreases the proportion of inadequate specimens, improves the sensitivity, and reduces the specimen interpretation time compared to the conventional Pap smear [44]. There is additional cost associated with the preparation of the liquid based cytology specimen. A conventional Pap costs around $15; ThinPrep adds about $15–25 to this cost [49].

How often should women be screened using the Pap smear?

The American Cancer Society issued Screening Guidelines for the Early Detection of Cervical Cancer in 2002. These guidelines recommend that screening should begin approximately three years after women begin having vaginal intercourse, but no later than 21 years of age. Screening should be done every year with regular Pap tests or every two years using liquid based cytology. If a woman has had three normal tests in a row and has reached 30 years of age, she may reduce the frequency of screening to every two to three years. However, doctors may suggest a woman get screened more often if she has certain risk factors, such as HIV infection or a weakened immune system. Women 70 and older who have had three or more consecutive Pap tests in the last ten years may choose to stop cervical cancer screening.

Do women follow these recommendations? Around 79% of women in the USA report having had a Pap smear in the last three years. Adherence to screening is slightly lower for women with no health insurance and for women with less than a high school education [14, 50].

Automated Pap smears

Currently, Pap smears are examined by highly trained cytotechnologists. A significant amount of training is required to be able to accurately identify smears containing pre-cancerous or cancerous cells. The lack of trained personnel is a barrier to screening in many developing countries [47]. Automated cytology devices use a microscope with autofocus and a motorized, computer controlled stage coupled to a high resolution video camera (Figure 10.30). Digital images are captured and sent to a computer, where image processing algorithms are applied to interpret the images and classify the slides. Images are first segmented, to separate cells from background objects, like debris or inflammatory cells. Morphologic parameters are then calculated such as the cell size, nuclear size, the nuclear to cytoplasmic ratio and

Figure 10.30. *An automated Pap smear machine. Courtesy and © Becton, Dickinson and Company.*

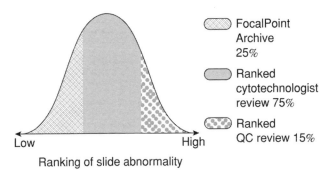

Figure 10.31. *Slides obtained from a Pap smear are ranked according to abnormality before being reviewed by a cytotechnologist [44]. Reprinted with permission from Elsevier (The Lancet Oncology, 2001, Vol. **2** No. 1, pp. 27–32).*

the texture of chromatin within the nucleus. In addition to features that cytologists normally use, more advanced morphologic parameters can be calculated. Abnormalities can then be detected by comparing the distributions of measured cells to those of known normal and abnormal reference cases. Classification algorithms are used to combine measured parameters and make a determination of whether the specimen is normal or abnormal. Statistically based algorithms, hierarchical decision trees or neural networks are examples of types of classifiers, each of which consists of a set of rules to classify the data [48].

In 1998, the FDA approved the use of such a device called the AutoPap® Primary Screening System to sort out 25% of smears that do not require human review because they are negative. The device can scan about 200 slides per day [44]. Slides containing potential abnormalities are ranked in order of abnormality. Slides with the lowest probability of abnormality are not ranked and reported as requiring "no further review." This approach can reduce the workload of the cytotechnologist, allowing him or her to focus on those slides most likely to contain abnormalities. In addition, the device is used to rank the 15% of slides with the greatest likelihood of abnormalities for re-review (Figure 10.31). These slides may be used instead of the 10% random selection of slides for quality control as mandated by CLIA [51].

Clinical studies of this technology have shown that the AutoPap® device outperformed human review by a factor of five to seven times when used to rescreen the 10% of negative Pap smears which must be examined for quality control purposes [44]. The use of AutoPap® adds between \$3-\$10 to the cost of the Pap test [52].

HPV DNA testing

Cervical cancer is caused by infection with the human papillomavirus (HPV). A new test has been developed to determine whether a patient is infected with HPV. Following a Pap smear, the remaining material on the spatula can be tested to determine if HPV DNA is present. The DNAwithPap™ Test is FDA-approved for routine adjunctive screening with a Pap test for women age 30 and older [53]. HPV is found so frequently in women under the age of 30 that it is not useful to indicate risk of cervical cancer and its precursors. However, if viral infection persists after age 30, there is an association with increased risk of cervical cancer and its precursors. Clinical studies have shown that the sensitivity of DNAwithPap™ is greater than that of the Pap smear alone or a liquid Pap. The sensitivity of DNAwithPap™ is 80–90%, while the specificity is 57–89% [54]. In Europe, as of 2006, the use of HPV tests was not currently included in basic screening [55].

VIA

The use of Visual Inspection with Acetic Acid (VIA) is being explored as an alternative to Pap smear and colposcopic examination in many developing countries. VIA consists of simple visual examination of the cervix with the naked eye by a trained healthcare provider before and after application of acetic acid. VIA relies on the acetowhitening of pre-cancerous lesions. It requires only

HPV DNA testing: how does it work?

Hybrid Capture System

As of 2007, Digene's hc2 High-Risk HPV DNA Test(tm) (DNAwithPap™), based on Hybrid Capture® 2 technology, is the only FDA approved method for HPV DNA testing. When used for the purpose of cervical cancer screening, the FDA requires that it only be used in women over the age of 30, in conjunction with a Pap test. In most cases, HPV DNA testing may be performed on the same sample of cells collected for the Pap test. Following collection, DNA is extracted from the cell sample and denatured and single stranded RNA probes for the 13 highly oncogenic types of HPV are added. If HPV DNA is present, it hybridizes to the probe RNA. Antibodies specific to DNA–RNA hybrids capture the hybrids and bind them to the wells of a microtiter plate. The complex is then enzymatically digested resulting in the emission of light produced by a chemiluminescent substrate conjugated to the enzyme. The intensity of light indicates the presence or absence of HPV DNA in the patient's sample [44, 53, 56].

The figure in this box is reproduced with kind permission of QIAGEN GmbH, Germany; QIAGEN copyright.

low technology equipment, and results are available in a few minutes. A recent review of the performance of VIA in nine studies involving more than 40,000 women in South Africa, India, Zimbabwe, China, and the Philippines found that VIA has similar sensitivity to that of Pap smear screening, but lower specificity, although some studies suffered from verification bias [57].

In a study of 18,675 women in India, Sankaranarayanan found that the sensitivity of VIA for detection of high grade squamous intraepithelial lesions (HSIL) was 60.3% and the specificity was 86.8% relative to the gold standard of colposcopic directed biopsy of colposcopically abnormal lesions [61]. The advantage of VIA is that it is an inexpensive test that does not require lab infrastructure. Providers can be trained to perform the test in five to ten days. Consumables required are cheap and universally available. Because results are available immediately, patients can be treated at the same visit. However, there are concerns that the low specificity of VIA may lead to over-diagnosis and treatment [62].

Because VIA relies on visual interpretation, defining objective criteria for a positive lesion and training operators to correctly implement these criteria are crucial. In a series of 1921 women screened in Peru, Jeronimo found that the VIA positivity rate dropped from 13.5% in the first months to 4% during subsequent months of a two year study; the drop in positivity rate was hypothesized to be due to a learning curve for the evaluator [47].

DIA

The use of digital image analysis (DIA) may provide a simple solution to reduce the subjectivity and improve the specificity of VIA. Advances in consumer electronics have led to inexpensive, high dynamic range CCD cameras with excellent low light sensitivity. At the same time, advances in vision chip technology allow high quality image processing in real time. These advances may enable acquisition of digital images of the cervix in a relatively inexpensive way, with or without magnification. Moreover, automated image diagnosis algorithms based on modern image processing techniques has the potential to replace clinical expertise, which may reduce a considerable amount of the system cost. A recent pilot study showed that digital images of the cervix can be obtained using a simple and inexpensive device, and that automated image analysis

Vaccines to prevent HPV infection and cervical cancer

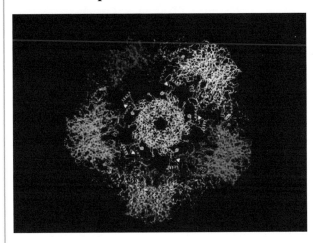

Figure courtesy of CDC.

HPV is the most common sexually transmitted disease in the USA. Today, more than 20 million people in the US harbor HPV. 80% of women will test positive for HPV by age 50. As we have seen, HPV infection usually does not cause any symptoms, but in some cases it can lead to cervical cancer.

In 2006, a new vaccine to prevent HPV infection was licensed for use in girls and women aged from nine to 26 years in the USA. The vaccine, Gardasil, protects against four strains of HPV. Two of these HPV types (16, 18) are responsible for 70% of cervical cancers; combined, the 4 HPV strains covered by Gardasil account for about 90% of genital warts. At the end of 2006, Gardasil had been approved in 49 countries.

Gardasil is made by inserting the gene for a protein found in the HPV capsid (L1) into a different virus or yeast. Recombinantly produced HPV capsid protein then self-assembles into virus like particles (VLPs). While these empty shells do not contain the cancer causing DNA of HPV, their shape is sufficiently similar to that of the HPV virus so that the immune system triggers a protective response against future HPV infection (the figure shows the L1 protein of HPV16 assembling into a virus like particle that does not contain HPV DNA).

While Gardasil can protect against new HPV infections, it is not effective for women who have already been exposed to HPV. The length of time that patients will be protected following vaccination is currently not known. As of 2006, the vaccine had been tested in more than 3000 women who had been followed for five years. The vaccine was protective throughout this period, but it is not known whether booster shots will be required over longer periods of time. Trials of Gardasil and a second promising HPV vaccine made by Glaxo SmithKline (GSK) are currently underway in more than 50,000 subjects.

HPV Vaccine Efficacy Trials				
Manufacturer	Vaccine	Location	Participants	Projected End
Merck	VLPs of L1 protein from HPV 6/11/16/18, made in yeast, aluminum adjuvant	U.S., S. America, Europe	17,800 women, 16 to 26 years old	2007
		U.S., S. America, Europe, Asia	3800 women, 24 to 45 years old	2008
		U.S., S. America, Europe, Asia, Africa	3700 men, 16 to 24 years old	2008
GSK	VLPs of L1 protein from HPV 16/18, made in baculovirus, AS04 adjuvant	U.S., S. America, Europe, Asia Pacific	18,000 women, 15 to 25 years old	2010
		Costa Rica (run by NCI)	12,000 women, 18 to 25 years old	2010

Currently, Gardasil is given as a series of three shots over a six month period; the cost of the vaccine is $360. This cost is a barrier even in developed countries, and is likely to limit its immediate impact in developing countries. For example, the HBV vaccine was licensed in 1981 in industrialized countries, but it took 10–15 years for it to be used in wealthier developing countries and over 20 years before children in the poorest countries had access to the vaccine [60]. Developing countries may also face difficulties in providing widespread access to a vaccine that is targeted towards girls and young women. Vaccines for adolescents are often given through school programs, but girls in developing countries are less likely to be in school than boys. Gender specific immunization may be culturally unacceptable in some settings. Many vaccination programs have been damaged due to rumors that vaccination is a plot to sterilize girls [60]. The stigma associated with a vaccine targeted against an STD may exacerbate such rumors. Such rumors derailed polio eradication campaigns in Nigeria and India, resulting in global consequences.

Will the HPV vaccine eliminate the need for cervical cancer screening? Currently available vaccines do not protect against all types of HPV that cause cervical cancer, so women who receive the HPV vaccine will still need to be screened for cervical cancer. Additionally, if women don't get all three doses of the vaccine or if they have already been exposed to HPV prior to being vaccinated they may not be protected [58–60].

Figure credit: X. Bosch and N. Munoz/IARC, IBSCCS, and Multicentric Studies (N = 3045). From *Science* 29 April 2005: Vol. **308**. no. 5722, pp. 618–621. Reprinted with permission from AAAS.

The next day, Tandi took me to the NCP (Neighborhood Care Point). There were several in the community, but this one in particular was also the kagogo, which is the central meeting place for the community. I met the secretary of the kagogo, and he asked me lots of questions about Baylor. No one in the community knew about the clinic, and he was curious as to who was eligible to go there, if it was free, and how they could become an outreach site. Most of the people wouldn't be able to afford the 42 rand (US $6) it would take to go to and from the clinic, so ideally, he would get Baylor to come to the community.

The NCPs are where orphans and vulnerable children can come for the day to receive meals and a bit of education. Tandi said that the community was pretty good about taking in the children but struggled to support them. Then NCPs filled this gap and provided as much support as possible, although often, it isn't enough either. Talia (a Canadian who visited with Rachel and Lindsay) worked a lot with orphans in Botswana. One of the services her NGO provided was gift baskets for the orphans. Once families learned about this, they started taking in as many orphans as they could in order to receive the baskets, which they would then sell. The orphans remained just as abandoned and starving as before. Tandi said that this wasn't really a problem in her community, but there were many others. For example, an orphan could go to school if they could prove (with death certificates) that they were indeed orphans. This is virtually impossible for many reasons. Many of them never knew their fathers, who left the mother when she was pregnant. Even if they knew both of their parents, no one gets a death certificate unless they go to a city far away and deal with some complicated legal procedure. So, unless there is someone who cares enough for the child to deal with the hassle and who is wealthy enough to afford it, there is no way for the orphan to prove that they lack parents. Thus, all they are left with are the NCPs.

I listened for a bit as Tandi and the secretary discussed some of their projects – a community garden, fundraising for NCP renovations (most of them were dirty, stick-in-the-mud structures), education campaigns, and other events. During this conversation, I discovered that children become sexually active as early as twelve. I also learned that men fear the HIV stigma

more than women (probably because it might limit the number of girlfriends they could have), while women were much more open and willing to address the problem. He told us that there was an article in the paper about a doctor who was telling many of his patients that they weren't actually HIV-positive even though they'd been told at a VCT (volunteer counseling and testing) clinic that they were. He thought they were lying about it because they were afraid that they would lose their jobs if the HIV rate dropped and funding for HIV/AIDS programs dropped. Tandi responded that it is much more likely that the one doctor was lying than everyone at the VCTs, and in addition, many people try to place blame elsewhere in order to avoid taking responsibility for their actions (which caused them to get the disease).

After that, I visited the school and nearby clinic. They were pretty much what I expected – about the same as the Vuvulane clinic, and the school was much like the school I worked at in Nicaragua. We looked at the picture of the map Tandi was painting with her class and I took photos of some of the HIV/AIDS awareness signs. Realizing that her watch had stopped, we rushed off to catch a combie and make our way back to Manzini. At that point, we split up. She headed over to her friend's community to help with a workshop, and I headed back to the COE in Mbabane.

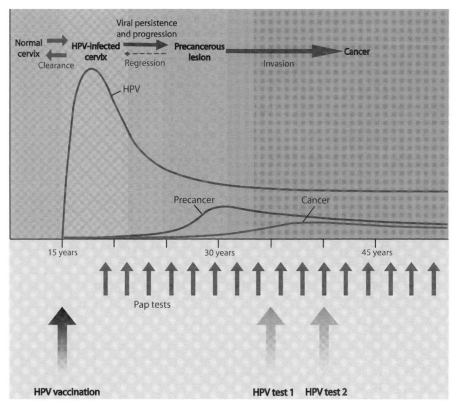

Figure 10.32. *Many young women develop an HPV infection during adolescence or young adulthood. In some women, HPV infection leads to pre-cancerous changes in the cervix. Regular Pap tests can identify these pre-cancerous changes, allowing treatment before cervical cancer develops. In the future, the availability of an HPV vaccine may reduce the incidence of HPV infection and reduce the frequency with which screening is needed. Alternative methods of screening, such as the HPV DNA test, may improve the sensitivity of screening. Schiffman, Castle. The promise of Global Cervical Cancer Prevention. 2005 NEJM. **353**: 2101–04. Copyright © 2005 Massachusetts Medical Society. All rights reserved.*

algorithms correctly identify histologically neoplastic tissue areas with a sensitivity of 79% and a specificity of 88% [63].

In summary, although cervical cancer is a completely preventable disease, it is the third leading cause of cancer death in women in the world [6]. Cervical cancer is caused by infection with HPV. HPV infection can initiate a transformation that results in a pre-cancerous lesion. If we detect and treat these common pre-cancerous lesions, we can prevent the development of cervical cancer. Current screening and detection using the Pap smear followed by colposcopy and biopsy has been proven to reduce both the incidence and mortality of cervical cancer. However, we have insufficient resources to screen using these technologies in developing countries. New technologies, such as automated reading of Pap smears, HPV testing, visual inspection, and digital image analysis (VIA and DIA) technologies may provide the improvements in performance at a sufficiently low cost to enable screening in resource poor settings where the vast majority of cervical cancer occur. Coupled with vaccines to prevent HPV infection, these tech-

nologies have the potential to reduce both the incidence and mortality of cervical cancer (Figure 10.32).

Prostate cancer

As we have seen, prostate cancer is the most common cancer diagnosed in men in the USA, with 218,890 new cases diagnosed annually. Prostate cancer is the second leading cause of cancer death in men, causing more than 27,050 deaths each year in the United States [1]. Worldwide, more than 679,023 new cases of prostate cancer are detected annually, making prostate cancer the second most common cancer in men [6]. Figure 10.33 shows the incidence and mortality rates of prostate cancer throughout the world. Risk factors for development of prostate cancer include advanced age, race (incidence rates are one and a half times higher in African Americans), and a family history of prostate cancer [1].

Figure 10.39 shows the location of the prostate gland. The prostate gland contributes enzymes, nutrients and other secretions to semen. Figure 10.34a shows a photograph of the normal prostate, while Figure 10.34b shows

Incidence   Mortality

	Incidence	Mortality
Northern America	119.9	15.8
Australia/New Zealand	79.9	18.1
Western Europe	61.6	17.5
Northern Europe	57.4	19.7
Caribbean	52.4	28.0
South America	47.0	18.0
Southern Africa	40.5	22.4
Southern Europe	35.5	13.2
Central America	30.6	15.5
Middle Africa	24.5	21.1
Micro/Polynesia	20.3	10.8
Western Africa	19.3	16.0
Eastern Europe	17.3	9.7
Eastern Africa	13.8	11.8
Japan	12.6	5.7
Western Asia	10.9	6.0
South-Eastern Asia	7.0	4.5
Northern Africa	5.8	4.9
Melanesia	5.7	3.1
South Central Asia	4.4	2.8
China	1.6	1.0

140 120 100 80 60 40 20 0 20 40 60 80 100 120 140

Age standardized per 100,000

Figure 10.33. *The incidence of prostate cancer is highest in North America. Mortality rates of prostate cancer are substantially lower than incidence rates. Parkin, Bray, Ferlay, Pisani. Global Cancer Statistics. 2002. CA Cancer Journal for Clinicians, **55**: 74–108.*

histologically stained sections of normal prostate tissue. The normal prostate consists of several branched glands leading to the urethra. These glands are covered by a single layer of columnar epithelial cells. In the normal prostate, the nuclei of these cells occupy approximately ¼ of the cell area (Figure 10.35). However, in pre-cancerous lesions, the nuclei of these epithelial cells become substantially enlarged (Figure 10.36) and multiple layers of cells stack atop one another. As these cells invade beneath the basement membrane lining the ducts, invasive prostate cancer develops. Initially, the lesion is localized to the prostate; at the microscopic level, the cancerous epithelial cells are found throughout the entire prostate (Figures 10.37a,b).

Prostate cancer is a slow but continuously growing cancer. Generally, pre-clinical asymptomatic forms of the disease can develop as early as age 30. This disease can remain latent for up to 20 years. In some patients, pre-cancerous lesions can progress to aggressive, malignant cancer. The peak incidence of prostate cancer occurs in the seventh decade of life [11]. Figure 10.38 shows the risk of developing prostate cancer in the next five years as a function of a patient's current age; the risk rises dramatically with increasing age. Prostate cancer is often asymptomatic in the early stages. When present, the signs and symptoms of prostate cancer include weak or interrupted urine flow or the inability to urinate. These symptoms are the same

Serum PSA test

Prostate specific antigen is a glycoprotein with a molecular weight of 34,000 Dalton. It is responsible for liquefaction of semen. PSA is highly specific for prostate tissue; it was first discovered when scientists were searching for a potential marker that could be used in investigation of rape crimes.

The PSA test is a blood test to measure levels of PSA in the serum. In the test, a serum sample is added to a tube containing two types of anti-PSA antibodies that recognize different antigenic sites on the PSA. One type of anti-PSA antibody is conjugated to an enzyme called alkaline phosphatase; the second type of anti-PSA antibody is conjugated to paramagnetic particles. If PSA is present in the sample it binds to both antibodies forming a sandwich complex. A magnetic field is applied to separate the magnetic particles. The sample is washed to remove unbound alkaline phosphatase conjugate; thus, the remaining alkaline phosphatase is proportional to the amount of PSA present in the sample. A chemiluminescent substrate called Lumigen PPD is added. Alkaline phosphatase causes cleavage of the phosphate group on the Lumigen PPD producing an intermediate product. The intermediate product decomposes and generates chemiluminescence; this signal decays with a half life of several minutes. The light intensity produced is a direct measure of enzyme present.

Courtesy Lumigen, Inc.

The serum PSA test was first approved by the FDA in 1986 to monitor patients who had been treated for prostate cancer to determine whether they had a recurrence of disease. In the early 1990s physicians began to use the test to screen patients who were at risk for developing prostate cancer. Two large studies have been carried out to study the accuracy of the PSA test. When the cut-off value for an abnormal PSA test is set at 4 ng/l, its sensitivity has been reported to be 44–46%, with a specificity of 91–94%. The cost of a PSA test is approximately $50 [65–70].

A number of approaches have been suggested to improve the sensitivity and the specificity of PSA based screening.

Adjust cut-offs with age since PSA levels increase with age.

Adjust cut-offs with ethnicity since African American males tend to have higher PSA values.

Monitor annual increases in PSA levels rather than absolute values.

Adjust PSA levels by the size of the prostate (PSA density).

Measure the fraction of free PSA relative to that bound to plasma proteins.

as those of prostate enlargement, thus are not diagnostic [1].

Prostate cancer is treated with a combination of surgery, radiation therapy, hormone therapy, and chemotherapy [64]. In the United States, the five year survival rate for all stages of prostate cancer combined is quite high at 99.9%. This is due to the effectiveness of treatments for cancer which is localized to the prostate, where the five year survival rate is 100%. When disease has metastasized to distant organs, the five year survival rate of prostate cancer is only 33.3% [1]. Thus, early detection of prostate cancer is important.

There are two tests which have been widely used to screen the general male population for prostate cancer, although there is considerable controversy regarding the most appropriate use of these tests. The first test is a

(a)

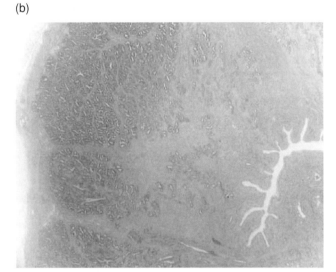

©2004 Bostwick Laboratories, Inc

(b)

Figure 10.34. *(a) Photograph of the normal prostate; 2004 Bostwick Laboratories, Inc. (b) Histologically stained section of a normal prostate. Reprinted with permission from Andrei Gunin, the Department of Obstetrics and Gynecology Medical School Chuvash State University.*

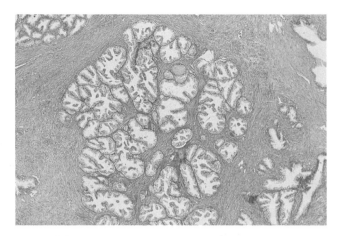

Figure 10.35. *A slide showing a normal prostate; note that the nuclei of normal cells take up roughly a quarter of the cell area. Photo courtesy of Laura P. Hale, M.D. PhD., Duke University Medical Center.*

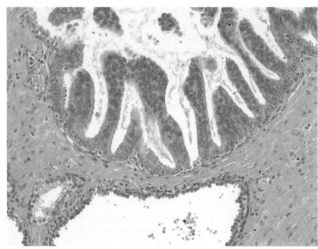

Figure 10.36. *A slide showing a pre-cancerous prostate gland, in which the nuclei of the cells have become enlarged. Dharam M. Ramnani, M.D.; WebPathology.com.*

simple blood test to measure levels of a protein called prostate specific antigen (PSA). PSA is a protein found on the surface of epithelial cells in the prostate. When prostate cancer develops, the number of epithelial cells increases and the amount of PSA found in the blood increases. A blood test can measure the levels of serum PSA quantitatively. However, other conditions which cause an increase in the number of prostate epithelial cells, such as benign enlargement of the prostate, can also cause PSA levels to be elevated [64]. The second test is to palpate the size of the prostate gland in a procedure called a digital rectal exam (DRE). The prostate

gland lies close to the rectum, and its size can be felt by placing a gloved finger inside the rectum. Prostate enlargement can be a sign of either prostate cancer or benign prostate enlargement [64]. Screening using the PSA and DRE tests has become one of the most commonly used cancer screening tests. More than a half of men over age 50 report having a recent serum PSA test and a digital rectal examination, although these figures drop to less than 33% for men without health insurance [14].

If screening tests for prostate cancer are positive, further diagnostic tests to confirm or exclude the presence

(a)

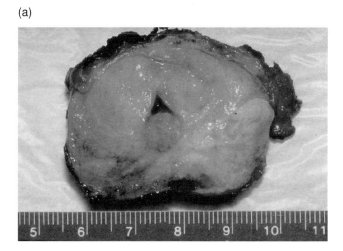

(b)

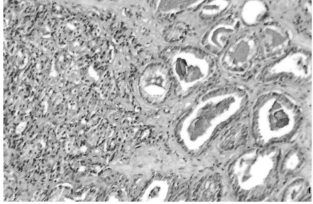

Figure 10.37. *Invasive prostate cancer at the (a) macroscopic and (b) microscopic levels. Courtesy of the University of Washington, Department of Pathology, http://www.pathology.washington.edu. NCI/Otis Brawley.*

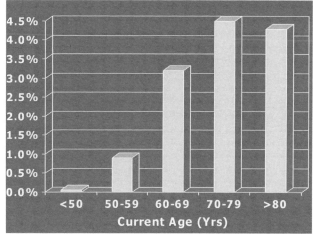

Figure 10.38. *As the graph above shows, the risk of developing prostate cancer in the next five years increases dramatically with age. BC Cancer Agency. PSA Screening information for patients, May 2009. Available at http://www.bccancer.bc.ca/NR/rdonlyres/375628D8-AB6F-4523-BFD4-C854FDA705F8/4510/PSAwebBrochure.pdf.*

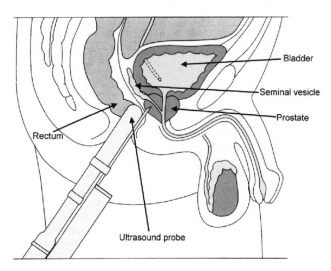

Figure 10.39. *A diagram showing the prostate biopsy procedure.*

of prostate cancer are required. To confirm the presence of cancer, physicians obtain small pieces of prostate tissue called core needle biopsies. Biopsies are then sectioned, stained and observed under a light microscope to examine the epithelial cells of the prostate. A biopsy of prostate tissue is obtained by inserting a needle through the wall of rectum into the prostate (Figure 10.39). A positive screening test does not indicate where in the prostate a lesion might exist, so multiple biopsies are performed. Typically at least ten core biopsies are obtained to sufficiently sample the prostate tissue; the procedure is performed with local anesthetic. The precise positioning of the needle is guided by ultrasound imaging. Small fragments of the prostate are

then removed from the needle, processed, and examined under a microscope [71]. The cost of obtaining and processing a prostate biopsy is approximately $700 [70].

If prostate cancer is detected when it is still localized to the prostate, physicians generally recommend one of two courses of action. The first is radical prostatectomy, a surgical procedure to remove the prostate. While this procedure is usually curative, because it removes the cancerous cells, it has some very serious side effects. Because important nerves which control

Table 10.12. *The ten year survival rates for three grades of prostate cancer following either surgery or conservative treatment. The earlier the cancer is detected, the less difference between survival rates for the two treatments.*

Cancer grade	Surgery ten year survival	Conservative ten year survival
Grade I	94%	93%
Grade II	87%	77%
Grade III	67%	45%

Source: Lu-Yao, GL and Yao, SL. Population-based study of long-term survival of patients with clinically localised prostate cancer. *Lancet* 1997; **349**; 906–910.

bladder function and sexual function are located in the same area as the prostate, they can be damaged during this surgical procedure [64]. Following radical prostatectomy, between 2% and 10% of men experience incontinence, and between 30% and 90% of men experience impotence [11]. Because of the seriousness of these side effects, other physicians recommend more conservative management of prostate cancer, choosing to watch the patient until symptoms develop, and then offering treatment [64].

Because prostate cancer is a relatively slow growing cancer, there is some controversy over whether detection of very early disease makes a difference in patient outcomes. Localized prostate cancer is classified into three grades based on the severity of the disease. A study to examine the ten year survival rates for localized prostate cancer found that the survival rates for surgery and conservative therapy were nearly the same for grade I disease, but were substantially higher when grade II or grade III disease were treated surgically (Table 10.12) [72].

This illustrates one of the challenges of screening for prostate cancer. Prostate cancer is a slow growing cancer; the average patient does not show symptoms for an average of ten years following the initial development of prostate cancer. Because prostate cancer occurs later in life, most men with prostate cancer actually die of other causes. For example, a 50 year old man has a 42% chance of developing microscopic prostate cancer sometime in his life, a 10% chance of having this cancer diagnosed, but only a 3% chance of dying of it [73]. As many as 20–50% of men who have died with no symptoms of prostate cancer have been found to have prostate cancer at autopsy [69]. Since the treatment of prostate cancer has significant side effects, patients and physicians are faced with difficult decisions about whether to treat the disease or watch the disease.

Thus, the question of whether to screen the general male population for prostate cancer has a complicated answer. Localized prostate cancer is curable, and advanced prostate cancer is fatal, indicating the benefits of detecting disease early through screening. While screening clearly has potential benefits, it also has potential risks. A positive screening results leads to a prostate biopsy, an expensive, reasonably invasive and uncomfortable procedure. For those patients whose screening test is falsely positive, this biopsy is unnecessary. Furthermore, because prostate cancer is such a slow growing cancer found in older men, screening may lead to over-detection of latent cancers. If we screen, we may detect many cancers that would never have produced symptoms before the patients died of other causes.

Let's examine some of this clinical evidence regarding the efficacy of screening. In Tyrol, Austria, the mortality from prostate cancer was constant from 1970 to 1993, prior to the introduction of mass screening. In 1993, mass screening for prostate cancer using digital rectal examination and serum PSA began [74]. Between 1993 and 2000, the mortality associated with prostate cancer decreased 44% in Tyrol [75]. While this study was not designed with a control group, cancer mortality remained constant in other parts of Austria where screening was not performed.

Other, more carefully designed and controlled studies have shown contrasting results. One completed randomized clinical trial of digital rectal examination and PSA to screen found no difference in the number of prostate cancer deaths between groups randomized to screening and usual care [76]. One prospective clinical trial in Canada suggested that screening with PSA could reduce the mortality due to prostate cancer by 67%, although the study was widely criticized for design and analysis flaws. Two large randomized clinical trials of screening are underway – the European Randomized

Costs of screening for prostate cancer

We can examine the predictive value of the PSA screening test and the cost to find prostate cancer with this test. Let us assume that we test one million men between the ages of 50 and 59 for prostate cancer using a serum PSA test with a sensitivity of 44% and a specificity of 91%. The expected prevalence of prostate cancer is 10% in this population. The cost to screen is $50/patient, and a high serum PSA results in a follow up biopsy which costs $700. What are the positive and negative predictive values in this situation? What is the cost to screen the entire population? What is the cost to biopsy all men with positive tests? What is the cost/cancer found? To answer these questions, we fill in the 2×2 table shown here.

	Test Positive	Test Negative	
Disease Present	44,000	56,000	# with Disease = 100,000
Disease Absent	81,000	819,000	# without Disease = 900,000
	# Test Pos = 125,000	# Test Neg = 875,000	Total Tested = 1,000,000

The PPV and NPV are then PPV = 44,000/125,000 = 35% and NPV = 819,000/875,000 = 94%. Thus, a man with a negative PSA test has a 94% chance of not having prostate cancer. However, a man with a positive PSA test only has a 35% chance of having prostate cancer. The cost to screen the entire population is $50 million dollars. In addition 125,000 men will have a positive PSA test and require a biopsy. Note that 81,000 of these biopsies are unnecessary! The cost to biopsy this group is 81,000*$700 = $56,700,000. Using this strategy we will find 44,000 cancers at a cost per cancer found of $56,700,000/44,000 = $1288 [24, 69, 70].

Are the costs of screening with the PSA test a good use of healthcare resources? In Chapter 11, we will examine how to calculate the cost effectiveness of different health interventions.

study of Screening for Prostate Cancer (ERSPC) involving 200,000 men, and the Prostate, Lung, Colorectal, Ovarian cancer (PLCO) study involving 74,000 men in the USA – but results are not expected to be available for years [77].

The fact that prostate cancer is such a slowly growing cancer makes it difficult to perform a controlled experiment and test whether an intervention truly reduces mortality. Figure 10.40 shows the natural history of prostate cancer vs. time. Once a microscopic cancer develops, it typically takes ten years before symptoms develop which would lead to a diagnosis even without the use of any screening tests. In this scenario, a typical patient survives 15 years beyond the initial diagnosis [69]. How does this sequence of events change if we screen for early disease? By screening asymptomatic patients, we detect disease earlier, as much as ten years before symptoms develop. If our ability to detect prostate cancer early does not change the natural history of the disease, these screened patients do not live to be any older than patients who have not been

screened. However, screened patients do survive for a longer period following diagnosis of their cancer, only because their cancer was detected before it produced clinical symptoms. This apparent increase in survival time following diagnosis is called "lead time bias," indicating that the new intervention simply lead to earlier diagnosis without truly changing the outcome. Thus, randomized clinical trials must be carefully designed to minimize lead time bias [78].

Given the limited clinical evidence currently available, different countries approach prostate cancer screening in different ways. In the United States, there are conflicting recommendations regarding screening (Table 10.13). The American Cancer Society recommends men aged 50 or older with more than a ten year life expectancy should be screened with DRE and PSA. The American College of Preventive Medicine recommends that men aged 50 or older with greater than ten year life expectancy should be informed of the potential benefits and risks of screening and make their own decision [69]. The US Public Service Task Force in

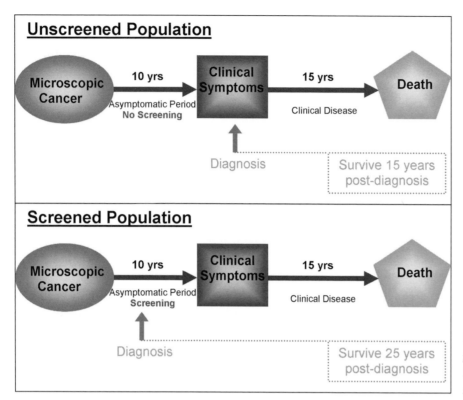

Figure 10.40. *The natural history of prostate cancer versus time. The apparent increase in survival associated with screening is called lead time bias.*

their Guide to Clinical Preventive Services recommends against screening using DRE or serum PSA [24]. While they find good evidence that PSA screening can detect early stage prostate cancer, they conclude that there is mixed and inconclusive evidence that early detection improves health outcomes. They note that screening is associated with important harms, including frequent false positive results and unnecessary anxiety, biopsies, and potential complications of treatment of some cancers that may never have affected a patient's health. They conclude that the available evidence is insufficient to determine whether the benefits outweigh the harms for a screened population. In Europe, screening is not currently recommended, because it is not believed that there is sufficient evidence to indicate that screening reduces mortality [69].

How has the incidence and mortality of prostate cancer changed in the USA following the introduction of screening? Routine screening has led to a dramatic increase in the number of cases of prostate cancer detected (Figure 10.41). The increase in incidence was accompanied by a shift in the stage at which prostate cancer is detected to earlier clinical stages. Over

the period 1950–1996, the incidence of prostate cancer increased by 190% in the USA (Table 10.14). Over this same period, the five year survival rate increased from 43% to 93%. However, the increase in five year survival rate may simply reflect the lead time bias associated with earlier detection. Over this same period, the mortality of prostate cancer in the USA has actually increased by 10%. In contrast, during this same period cervical cancer screening led to a 79% decrease in the incidence of cervical cancer and a 76% reduction in the mortality of cervical cancer.

The next type of cancer we will consider is ovarian cancer; in contrast to cervical cancer and prostate cancer where screening tests are available, there is currently no good screening test for ovarian cancer. Table 10.14 shows that the incidence and mortality of ovarian cancer have not changed appreciably from 1950–1996 [79].

Ovarian cancer

We have considered two cancers where screening tests are available: cervical cancer and prostate cancer. In our final example, we turn to a cancer where there is no

Table 10.13. *There is no consensus on how, if, or at what age men should begin being screened for prostate cancer in the USA. Modified from http://www.mayoclinic.com/health/prostate_cancer/HQ01273. With permission from Mayo Foundation for Medical Education and Research. All rights reserved.*

Organization	Recommendation
American Urological Association	Men over 50 should consider testing. Men at high risk should begin testing at age 45.
American Cancer Society	Offer the PSA and DRE tests annually beginning at age 50 to men who have a ten year life expectancy and to younger men at higher risk
CDC	Routine screening is not recommended because there is not consensus on whether screening and early treatment reduces mortality.
US Preventive Services Task Force	Evidence is insufficient to determine whether the benefits of screening outweigh the harms.
American Academy of Family Medicine	Physicians should counsel men between ages of 50 and 65 about known risks and uncertain benefits of screening so they may make an informed choice.
American College of Physicians	Physicians should describe potential benefits and known harms of screening, diagnosis and treatment, listen to patients' concerns and individualize the decision of whether to screen

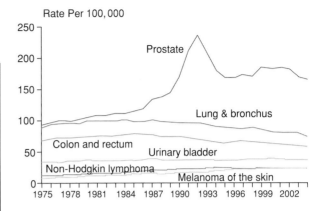

Cancer Incidence Rates* Among Men, US, 1975–2004

*Age-adjusted to the 2000 US standard population and adjusted for delays in reporting.
Source: Surveillance, Epidemiology, and End Results Program, Delay-adjusted Incidence database:
SEER Incidence Delay-adjusted Rates. 9 Registries, 1975–2004, National Cancer Institute, 2007.

Figure 10.41. *Cancer incidence rates for men in the US vs. time. A large increase in prostate cancer incidence was reported shortly after initiation of PSA based screening in the late 1980s to early 1990s [21]. Reprinted with permission from the American Cancer Society, Inc. from www.cancer.org. All rights reserved.*

22,430 new cases of ovarian cancer, representing 3.3% of all cancers in women.

It was estimated that 15,280 women would die as a result of ovarian cancer in 2007 in the USA [1]. Worldwide there were 190,000 new cases of ovarian cancer and 114,000 deaths in this same year. The highest rates of ovarian cancer occur in Scandinavia, Eastern Europe, USA, and Canada (Figure 10.43) [11].

The treatment for ovarian cancer involves surgery, and for advanced disease involves radiation therapy and chemotherapy. The five year survival rate for all stages of ovarian cancer is 45% [1]. There are four stages of ovarian cancer; when detected early, the five year

adequate screening test – ovarian cancer. The ovaries are part of the female reproductive system (Figure 10.42a,b) and are located adjacent to the fallopian tubes. In the USA in 2007, there will have been an estimated

Table 10.14. *Changes in survival rates and incidence for several cancer types since 1950 [79].*

	Five year survival, %			% Change (1950–1996)	
	1950–1954	1989–1995	Absolute increase in five year survival, %	Mortality	Incidence
Prostate	43%	93%	50%	+10%	+190%
Cervix	59%	71%	12%	−76%	−79%
Ovary	30%	50%	20%	−2%	+3%

*Source: JAMA 2000, **283**: 2975. Copyright 2000 American Medical Association.*

(a)

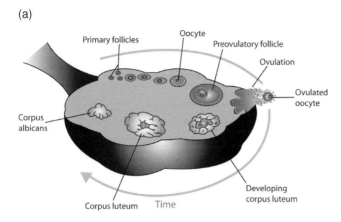

(b)

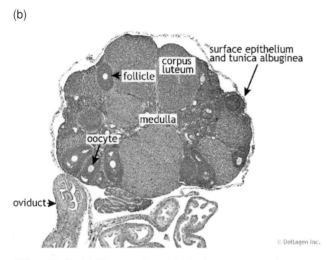

Figure 10.42 *(a) Diagram of ovary indicating stages of ovulation. (b) Histologic photograph of ovary.*

survival rates are much higher. Some 90% of women diagnosed with stage I ovarian cancer, when the disease is localized to the ovaries, survive five years beyond their initial diagnosis. However, the five year survival rate for metastatic, stage III–IV ovarian cancer is only 25–37% [80]. Unfortunately, because of the lack of good screening tests and the fact that early ovarian cancer produces relatively few symptoms, more than 70% of women diagnosed with ovarian cancer are diagnosed at stages III and IV [1]. Table 10.15 compares the ratio of mortality rate to the incidence rate for the ten most common cancers in women; ovarian cancer has one of the highest mortality to incidence ratios, second only to pancreatic cancer and lung cancer [81]. On average, women who die of ovarian cancer lose 18 years of life to the disease [82].

Ovarian cancer is said to "whisper" because the symptoms are so vague. Symptoms can include unexplained changes in bowel and/or bladder habits; gastrointestinal upset; unexplained weight loss or weight gain; pelvic and/or abdominal pain, discomfort, bloating or swelling; a constant feeling of fullness; fatigue; abnormal or postmenopausal bleeding and pain during intercourse. Frequently, women (and their physicians) will attribute these symptoms to those normally experienced with aging.

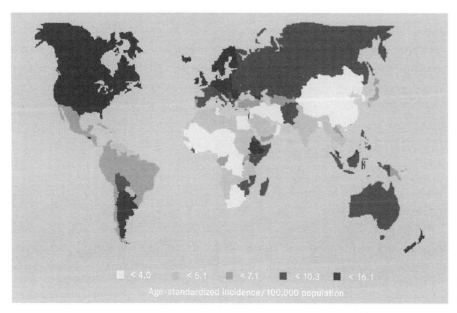

Figure 10.43. *Global incidence rates of ovarian cancer.*

Table 10.15. *Mortality/incidence ratio for ten most common solid cancers in women in the USA.*

Site	Incidence*	Mortality*	Mortality:Incidence Ratio
Breast	127.8	25.5	0.20
Lung & Bronchus	52.3	41.1	0.79
Colon & Rectum	44.6	16.4	0.37
Uterine Corpus	23.2	4.1	0.18
Non-Hodgkin Lymphoma	16.3	6.2	0.38
Melanoma of the Skin	14.9	1.7	0.11
Ovary	13.5	8.9	0.66
Thyroid	12.5	0.5	0.04
Pancreas	10.1	9.2	0.91
Urinary Bladder	9.4	2.3	0.24

* Age-adjusted rate per 100,000 women per year; based on US cases diagnosed in 2000-2004

From Rosenthal et al., *Clinical Obstet. and Gynecol.*, **49**(3)433–447, 2006.

There are a number of factors that put a woman at higher risk for developing ovarian cancer. The most important risk factors are a personal or family history of breast, ovarian, endometrial, prostate or colon cancer, particularly having one or more first-degree relatives (mother, sister, daughter) who have ovarian cancer. Ovarian cancer is sometimes associated with a mutation in the BRCA1 or BRCA2 gene. Hereditary ovarian cancer accounts for about 10% of cases [80]. In addition, the risk of ovarian cancer increases with the more lifetime cycles of ovulation that a woman has undergone. Thus, women who have undergone hormonal treatment for infertility, never used birth control pills, and who never became pregnant are at higher risk for ovarian cancer. In addition, the use of high dose estrogen for long periods without progesterone may also increase the risk of developing ovarian cancer.

The ovary is an almond-shaped organ that contains all the eggs that will be released over a woman's reproductive lifetime (Figure 10.42a,b). The ovary is lined by a single layer of epithelial cells. Beneath the epithelium, the ovary contains spherical follicles, each containing a single oocyte (egg), in a region known as the ovarian cortex. At the very center of the ovary, blood vessels bring in oxygenated blood and nutrients in a region known as the ovarian medulla. Each month, one or more follicles undergoes a transformation in preparation for ovulation. A primordial follicle enlarges and develops into a primary follicle. The follicle continues to enlarge and move toward the surface of the ovary. The secondary follicle then merges with the ovarian surface, ruptures and releases the oocyte. The defect in the ovarian surface must then repair itself. The scar left behind is known as a corpus albicans. Thus, the surface of the ovarian epithelium is constantly undergoing damage and repair. During this process, epithelial cells can become transformed and lead to ovarian cancer. As the frequency of this repair process increases, so do the chances that an ovarian epithelial cell will become transformed leading to an ovarian cancer. This probably explains why the use of oral contraceptives, pregnancy and breast feeding reduce the risk of ovarian cancer development.

Because ovarian cancer does not generally produce symptoms until very advanced stages, there has been substantial research to develop good early detection tools. Three are available, but all suffer from significant limitations; these techniques include: (1) pelvic and rectal examinations, (2) the CA-125 blood test, and (3) transvaginal ultrasound.

Pelvic and rectal examinations are normally conducted when a woman has a Pap smear. In this procedure, a physician manipulates the abdomen to feel the uterus and ovaries to find abnormality in shape or size. While this procedure can detect large changes associated with advanced ovarian cancer, it is unlikely to detect early stage ovarian cancer.

The CA-125 blood test is similar to the use of PSA to screen for prostate cancer. Ovarian cancer cells produce a protein called CA-125 which is released into the bloodstream; 80% of women with advanced ovarian cancer have elevated CA-125 levels [83]. In fact, physicians routinely used blood levels of CA-125 to monitor women following treatment for ovarian cancer – it is a sensitive indicator of persistent or recurrent disease [84]. Unfortunately, CA-125 levels are very unreliable for detecting early cancer, particularly in pre-menopausal women. The reasons are two fold: first, CA-125 levels are often not elevated in early ovarian cancer. Second, CA-125 levels can be elevated by conditions such as pregnancy, endometriosis, uterine fibroids, liver disease, and benign ovarian cysts [85]. Thus, in a pre-menopausal woman, an elevated CA-125 level is much more likely due to a benign cause than due to ovarian cancer [81]. The sensitivity and specificity of serum CA-125 levels in one large Norwegian study were an overall sensitivity of 30–35%, with a specificity of 95.4% [86, 87]. In general, the sensitivity is lower for detecting early stage disease.

Improved CA-125 tests

Recent attempts to improve the performance of screening using CA-125 have focused on using an algorithm that incorporates patient age, absolute levels of CA-125 and the rate of change of CA-125. Using this approach, a sensitivity of 83% and a specificity of 99.7% have been achieved [81].

Finally, ultrasound imaging can be used to visualize the ovaries. It is difficult to use ultrasound to visualize the ovaries through the abdominal wall. In order to view the small ovaries, an ultrasound probe is inserted into the vagina, and placed close to the ovaries. Using high frequency sound a picture of the ovaries is created (Figure 10.44a,b,c). Transvaginal ultrasound can detect

(a)

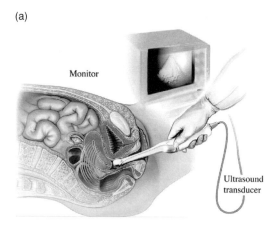

(b)

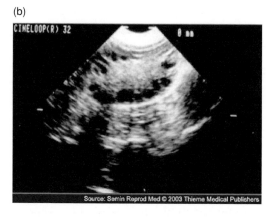

(c)

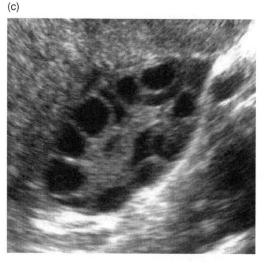

Figure 10.44 *Transvaginal ultrasound can be used to image the ovary. (a) Illustration Copyright © 2008 Nucleus Medical Art, All rights reserved. www.nucleusinc.com. (b) From Richard S. Legro, MD. Diagnostic criteria in polycystic ovary syndrome, Semin. Reprod. Med. **21**(3): 267–275, 2003. Reprinted by permission. (c) Reprinted with permission from Samuel Marcus MD, Medical Director. www.ivf-infertility.com.*

ovarian malignancies in asymptomatic women, based on the increase in ovarian volume, and the presence of complex cysts within the ovary [88]. However, it has poor accuracy in detecting early stage disease. A recent large study of transvaginal ultrasound to screen 14,469 asymptomatic women achieved a sensitivity of 81% and a specificity of 98.9% for the detection of ovarian cancer [89].

The only way to confirm a positive screening test for ovarian cancer is to perform a biopsy of the ovary. Because the ovaries are located in the abdominal cavity, this procedure involves surgical exploration of the abdomen to visualize and potentially biopsy the ovaries. Typically, this surgery is performed through a laparoscope (Figure 10.45). As we will see in detail in Chapter 14, in this procedure a small trochar is punched through the abdominal wall and the abdomen is inflated with CO_2 gas. Then, fiber optic laparoscopes are inserted through the abdomen to view the ovaries; small biopsy forceps can also be inserted to sample the tissue; a diagnosis of ovarian cancer can be definitely made by examining the biopsy in the same way that a cervical biopsy is examined. Approximately 1% of women undergoing laparoscopy will have a complication that will require an open surgical procedure [90].

Let's consider what happens when we screen a group of women for ovarian cancer using the available screening and diagnostic tests. If we screen 1,000,000 women in a setting with a 0.03% prevalence of undiagnosed ovarian cancer, there are a total of 300 cases that we can possibly detect [91]. Let's assume we use the CA-125 blood test to screen our patients, and recommend that those women with an elevated CA-125 have a laparoscopy. The sensitivity of CA-125 is 35% and the specificity is 95.4% [86]. The test costs about $60 to perform [92]. In this scenario, we will spend $60 million to screen our population; the screening test will identify 105 true positives, and a staggering 45,986 false positives, all of whom will undergo laparoscopy and biopsy, which is our gold standard. The cost of laparoscopy is approximately $1500 and 1% of women undergoing laparoscopy will suffer a serious complication requiring open surgery [90, 93]. In this scenario, we will spend $1,229,871 for each cancer that we find.

(a)

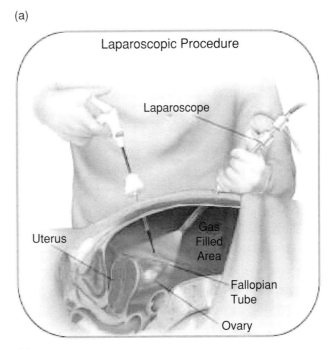

(b)

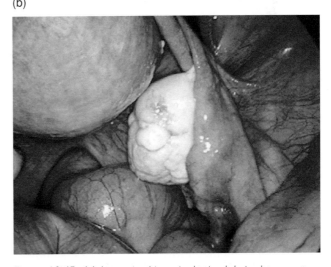

Figure 10.45. (a) An ovarian biopsy is obtained during laparoscopy. Courtesy of Allon Health Center, Center for Women's Medicine. (b) A fiber optic catheter is used to visualize the ovaries and guide biopsy direction. Reprinted with permission. Courtesy of John P.A. George, M.D., Gynecologic Endoscopy, Washington Hospital Center, Washington, DC.

Although we find only 105 cancers, 195 cancers will go undetected and 45,986 women will undergo an unnecessary laparoscopy and 460 women will suffer a complication as a result. In this scenario, the number of patients who suffer a serious complication caused by screening (460 women) exceeds the number of women correctly diagnosed with ovarian cancer (105 women).

The PPV of this screening strategy is only 0.23%, the NPV is 99.98%

If we use transvaginal ultrasound to screen our same population, the outcomes improve somewhat. The sensitivity of transvaginal ultrasound is 81% and its specificity is 98% [91]. The cost to perform this imaging procedure is approximately $200 [92]. In this scenario, we identify 243 of the 300 ovarian cancers, but 19,994 false positives lead to unnecessary laparoscopies, resulting in 202 complications. While the cost to detect a case of ovarian cancer is reduced to $947,965 in this strategy, the associated PPV is still dismally low at 1.2%; the NPV is 99.99%.

We can examine the use of transvaginal ultrasound in a population with a higher prevalence of ovarian cancer. If we screen post-menopausal women over the age of 45, the prevalence of undiagnosed ovarian cancer rises to approximately 0.2% [94]. In our cohort of 1,000,000 women, there are 2000 cases of ovarian cancer. In this population, transvaginal ultrasound correctly identifies 1620 women with ovarian cancer. The approach results in 19,960 false positives, and 216 serious complications. The cost to detect a single case of ovarian cancer is reduced to $143,438, and the PPV is 7.51%.

In this population, how high does the specificity of our screening test need to be in order to achieve a PPV of 10%? A simple calculation shows that the test specificity must reach 99.9% in order to achieve even a modest PPV, where one in every ten follow up laparoscopies will identify an ovarian cancer. This illustrates the difficulty of screening for a rare disease – in general, unless the specificity of the test is extremely high, the number of false positive results will far exceed the number of true positive results. If the follow up test carries any risk, then screening for a rare disease can actually cause greater harm than good.

Let's examine what has happened in an actual clinical trial of these technologies to screen for ovarian cancer. The most successful results have been obtained using a combination of approaches to screen for ovarian cancer. A randomized clinical trial of 22,000 women compared no screening to a combination of screening with CA-125 and transvaginal ultrasound [95]. In the group of women who were screened, CA-125 blood tests were performed annually for three years. If the CA-125 levels

exceeded a threshold level, transvaginal ultrasound was performed. In the screening group of 10,958 women, 468 women underwent 781 ultrasound exams because their CA-125 levels were elevated. Twenty-nine women underwent biopsy to detect six cancers. Thus, the overall positive predictive value of multi-modal screening was $6/29 = 20.7\%$. While the predictive value of this approach is higher, there are concerns that the sensitivity of this approach is not high enough. Despite the screening provided in this study, an additional ten women in the screening group developed ovarian cancer during a follow up of eight years. Five of the 16 cancers discovered in the screening group were stage I or II, whereas only two of the 20 cancers discovered in the control group were stage I or II.

Because ovarian cancer is such a devastating disease, there are a number of ongoing trials testing new screening approaches [81]. In the UK, a trial of 200,000 post-menopausal women is underway, comparing annual screening with CA-125 or transvaginal ultrasound to no screening [96]. Results are expected in 2012. In the USA, a trial of 78,000 is underway comparing the ability of annual serum CA-125 and transvaginal ultrasound to no screening [97]. Results of these clinical trials will help determine future screening recommendations throughout the world.

New screening tests for ovarian cancer

Because of the limitations of current screening tests, researchers are searching for additional markers that might be useful for ovarian cancer screening. Most current cancer screening tests look for a single protein in the serum (e.g. CA-125, PSA). However, serum contains many proteins; it may be possible to identify complex patterns of serum proteins which are predictive of cancer. This field is called proteomics. In this approach, researchers use techniques to analyze the patterns made by all proteins in the blood, without even knowing what they are.

The technique used to measure the pattern of serum proteins is known as mass spectrometry. In this technique, serum proteins are extracted, and bombarded with an electron beam. The electron beam has sufficient energy to fragment the proteins. This process produces

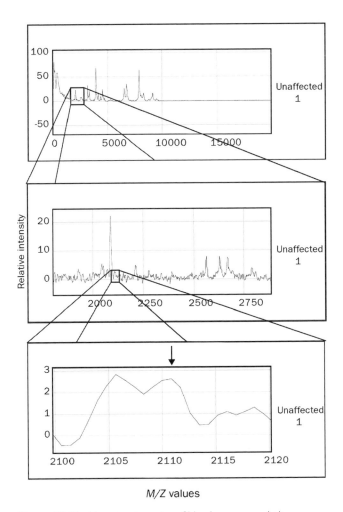

M/Z values

Figure 10.46. *Mass spectrometry of blood serum may help differentiate healthy individuals from those with cancer [83]. Reprinted with permission from Elsevier (The Lancet, 2002, Vol. **359** No. 9306, pp. 572–577).*

while the strength of the signal plotted on the y-axis is proportional to the amount of protein fragment with that mass in the sample. If one does mass spectrometry using a chemically pure sample, the mass of each fragment of the molecule enables one to determine the chemical structure of the sample, by working backwards to generate the original molecule. This technique is frequently used by chemists to identify the structure of an unknown chemical compound. However, serum contains a mixture of many proteins, with widely varying concentrations. In this case, instead of a series of a few sharp peaks, the resulting mass spectrum contains many peaks, of varying height. While one cannot use these data to work backwards and reconstruct the structure of each protein, it can be used to identify patterns of proteins that differ between healthy and diseased patients.

Recently a new blood test based on this technique to screen for ovarian cancer received widespread media attention. The test was first described in the medical literature in 2002 [83]. In this test, a blood sample is obtained from a patient. Serum proteins are isolated and the sample is analyzed using mass spectrometry. Scientists obtained blood from 50 women known to have ovarian cancer, 50 women known to be normal and 16 women with benign ovarian disease. They analyzed the resulting mass spectrometry data to search for protein peaks which differed in these two groups of patients. They examined thousands of proteins and identified a few which appeared to be different in the two groups. Using these differences they were able to define a diagnostic algorithm which correctly identified 50 out of 50 patients with ovarian cancer (sensitivity = 100%), and correctly identified 63 out of 66 women as normal (specificity = 95%). You can easily show that, in this setting, the positive predictive value of this test is 94%, significantly higher than what we calculated for CA-125 or transvaginal ultrasound.

charged fragments, most of which have a unit positive charge. These tiny charged fragments are then sprayed out of a nozzle through a magnetic field into a vacuum chamber. The positively charged fragments are accelerated in the vacuum chamber through a strong magnetic field. The time required for each fragment to travel down this chamber is dependent on the ratio of its mass to charge. The mass spectrometer produces a graph that shows distribution of masses in the sample. A computer program is then used to analyze patterns and distinguish blood from patients with cancer and from those without.

Figure 10.46 shows a typical mass spectrograph. The protein fragment mass is indicated on the x-axis,

> Do you think it is fair to compare the PPV of this test in this setting to our PPV calculations for CA-125 and transvaginal ultrasound? Why or why not?

Phase I: pattern discovery

Phase II: validation

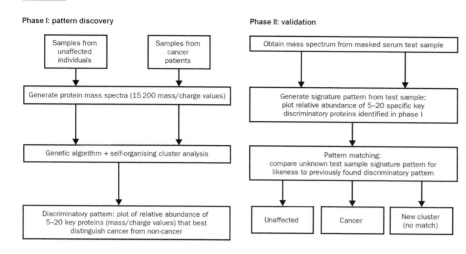

Figure 10.47. *The initial phase in developing the blood test to screen for ovarian cancer (left) and the subsequent phase, validating the pattern found in the first phase (right). Reprinted with permission from Elsevier (The Lancet, 2002, Vol. **359** No. 9306, pp. 572–577).*

Let's examine the development of this test in more detail. In the initial phase of the study, called pattern discovery (Figure 10.47, left), blood samples were obtained from patients known to have cancer and patients known to be normal. Protein mass spectra were obtained from each of these samples, and investigators examined the spectra. Each one contained the strength of the signal at 15,200 different mass/charge ratios. Different types of data analysis were applied to identify a small group of 5–20 key proteins which differed between the two groups of patients. The proteins were characterized by their mass/charge ratio and their relative abundance. This phase of the study is sometimes called the training phase, because it focuses on narrowing down a large number of data points, to identify a small group which provide diagnostically useful information. However, one limitation of this approach is that the number of proteins measured usually greatly exceeds the numbers of patients participating in the trial. Under these conditions, it is possible that differences in protein abundance between patients with cancer and patients without cancer arise due to simple chance fluctuations, and have nothing to do with the disease process at all [83].

From the thousands of peaks measured, the abundance of protein at only five different mass to charge ratios was found to vary between patients with and without cancer. Figure 10.48 shows data from four patients – two with cancer and two without. There is a very different peak intensity at a mass to charge ratio of 2111 in the spectra from cancer patients compared

Density plot

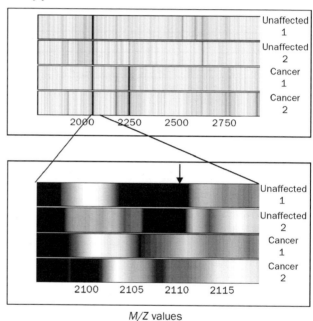

M/Z values

Figure 10.48. *Spectra illustrating that a difference in mass to charge ratio may differentiate patients with ovarian cancer from those without it [83]. Reprinted with permission from Elsevier (The Lancet, 2002, Vol. **359** No. 9306, pp. 572–577).*

to the spectra from unaffected individuals. To guard against the possibility that these fluctuations are due to chance, most clinical trials to test new diagnostic tests use a training phase to optimize the algorithm. Then, a second group of patients is recruited and the diagnostic algorithm is applied to data collected from this group; a phase generally referred to as validation (Figure 10.47, right). The performance of the diagnostic algorithm when applied to data in this validation group

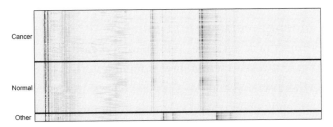

Figure 10.49. *A heat map representation of all the ovarian cancer screening study specimens; the shifted spectra at the bottom raise questions about the experimental protocol of the study [101].*

gives the best estimate of how well the algorithm performs [83].

Accounts describing the exciting promise of this new diagnostic test were widely reported in the media (see short article from the February 18, 2002 issue of *Newsweek* above). At the time, the lead author on the study, Lance Liotta, said, "The most important next goal is validating the promise of these results in large, multi-institutional trials [98]." While the general media responded with enthusiasm to the possibility of a new test which could improve the early detection of cancer, response from the scientific community was much more skeptical. Dr. Eleftherios P. Diamandis, head of clinical biochemistry at Mount Sinai Hospital in Toronto, expressed the concern that, "If you don't know what you're measuring, it's a dangerous black-box technology . . . They are rushing into something and it could be a disaster" [99]. Dr. Nicole Urban, head of gynecologic cancer research at the Fred Hutchinson Cancer Research Center in Seattle warned patients, "Certainly there's no published work that would make me tell a woman she should get this test" [99].

The datasets used to generate the ovarian cancer screening algorithm were made publicly available. When others tried to reproduce the results reported in the literature, several problems were identified. Most importantly, it appeared that there was a change in the experimental protocol for the measurements made from benign specimens that caused a systematic change in the data. Figure 10.49 shows a heat map representation of the 216 spectra from the pattern discovery and validation phases of the data. The m/z ratio runs along the x-axis, and the samples are grouped by diagnosis. There is a clear difference in the pattern of the benign speci-

mens shown at the bottom of the figure possibly due to a change in protocol for these specimens [101].

The use of proteomics technology at present can be thought of as a "black box technology." Serum samples are sent into the black box, and a diagnosis comes out. Because the approach does not rely on biological explanations, it is crucial that the approach be reliable and reproducible in any location. Further studies with additional samples are required to demonstrate the potential of this new technology.

Summary

In this chapter, we have seen the benefits and possible harms associated with cancer screening. Screening should be undertaken only when the following conditions have been met: (1) the effectiveness of the screening test has been demonstrated, (2) there are sufficient economic resources to screen all patients in the target group, (3) there are tools to confirm disease in patients with a positive screening test, (4) there are existing procedures to treat the disease, (5) and when disease prevalence is high enough to justify effort and costs of screening.

One of the challenges of screening is that it may reveal disease that might never be detected or cause problems otherwise. This is certainly true for screening for cervical cancer with the Pap test. Most abnormalities found on the Pap smear never become invasive cancer. However, there are relatively low cost, minimally invasive tools to follow an abnormal Pap smear, and treatment of high grade pre-cancer can prevent future development of cervical cancer. The use of screening has dramatically reduced both the incidence and the mortality of cervical cancer throughout the developed world. Likewise, screening for prostate cancer likely identifies many cases of prostate cancer which would otherwise have never produced any symptoms. Unlike the case of cervical cancer, screening with the PSA test has dramatically increased the apparent incidence of prostate cancer, while the mortality has largely remained unchanged. Ovarian cancer presents one of the most difficult challenges in cancer screening; because it is a relatively rare disease, any potential screening test must have a very high specificity to yield a reasonable predictive value. The relative inaccessibility of the ovaries makes it difficult and invasive to follow up an abnormal screening test. As a result, we do not currently screen for ovarian cancer although it is the most deadly of the female reproductive cancers.

While screening can have important medical benefits, it requires resources, both to test people and to follow up abnormal screening results. How do we decide if screening represents a good investment of healthcare resources? In the next chapter, we will examine the use of cost effectiveness analysis to make these decisions.

Bioengineering and Global Health Project
Project task 6: Gather information regarding current research and development efforts
What research and development efforts are currently underway to solve the health problem that you have identified? Write a one-page summary of this research, summarizing what is known about the effectiveness or limitations of these current procedures.

Homework

1. In the USA, what is the most prevalent cancer in (a) men and (b) women? Worldwide, what is the most prevalent cancer in (c) men and (d) women?

2. Cancer screening.
   a. What four types of cancer are routinely screened for in the United States? For each, describe the screening test that is used.
   b. Do most people in the USA adhere to screening recommendations? What factors cause people not to be screened?
   c. Discuss whether these screening tests are used throughout the rest of the world.

3. Lung cancer is the leading cause of cancer death for both men and women in the United States. More people die as a result of lung cancer than of colon, breast, and prostate cancers combined. Lung cancer is rare in people under the age of 40. The average age of people diagnosed with lung cancer is 60. In 2004 there are expected to be about 173,770 new cases of lung cancer in the United States [102]. About 160,440 people will die of this disease. The population of the United States in 2004 is 292,287,454.
   a. Calculate the annual incidence rate of lung cancer in the USA in 2004.
   b. Calculate the mortality rate of lung cancer in the USA in 2004.
   c. Why is the mortality rate of lung cancer so high?

4. Describe in your own words, **WITHOUT** using equations or other mathematical expressions or the

words "true", "false", "positive", or "negative" the following terms with regard to a screening test for ovarian cancer.

True Positive

False Positive

False Negative

True Negative

PPV

NPV

5. A diagnostic test is 92% sensitive and 94% specific. A test group is comprised of 500 people known to have the disease and 500 people known to be free of the disease. How many of the known positives would actually test positive? How many of the known negatives would actually test negative?

6. A screening test for a particular disease has a sensitivity of 96% and a specificity of 92%. You plan to screen a population in which the prevalence of the disease is 0.2%. How many false positives will be found by this screening procedure for each true positive that is found?

7. A clinical trial of a new automated mammography system was carried out in 50,000 women known to have breast cancer. If 37,500 women received a positive test result, what would the specificity of the new test be?

8. Based on all the information currently available, you estimate that the patient in your office has a one in four chance of having a serious disease. You order a diagnostic test with sensitivity of 95% and a specificity of 90%. The result comes back positive. Based on all the information available, what would be the chance your patient really has the disease? Source: [103].

9. A test with 99.9% sensitivity and 99% specificity is used to screen a population for a disease with 1% prevalence. What would be the proportion of test positives in the screen who actually have the disease? Source: [103].

10. The American Disease X Foundation reports that 6% of the population over 50 years of age has Disease X. You inquire as to the source of their information, and they cite disease population screening data in the literature which reports that 6% of that population was positive when screened. Referring to the literature, you discover that the screening test used had sensitivity of 95% and specificity of 98%. What proportion of the population over 50 years of age do you think really has the disease? Source: [103].

11. A recent study examined the expression of p53 (a protein found in many transformed cell lines derived from tumors) as a marker for ovarian cancer. The sensitivity and specificity of p53 as a marker for the diagnosis of ovarian cancer in this study were 82% and 93% respectively. Forty-seven patients with no family history of breast or ovarian cancer were included in the study. Fourteen of the 17 patients with ovarian cancer had p53 overexpression. Fifteen of the 47 patients had never given birth.

a. If p53 overexpression was used as a test for ovarian cancer, how many patients in this study received a false positive test result?

b. If p53 overexpression was used as a test for ovarian cancer, how many patients in this study received a false negative test result?

c. How much better are these results for a screening test than CA-125?

12. You are a physician for Mr. Jones, a 65 year old African American man who presents to you with complaints of difficulty urinating. Specifically, he has trouble starting urine flow and has an intermittent stream. He says he noticed this problem some time ago, and that it has slowly been getting worse. Mr. Jones says he has always been healthy and has not seen a doctor in thirty years. He was adopted and does not know his family history.

a. What disease discussed in Chapter 10 might explain Mr. Jones' symptoms?

b. What three risk factors for this disease does Mr. Jones have?

c. What initial tests are available that might aid your diagnosis of Mr. Jones?

d. If the initial tests are positive, what would be the next step in diagnosis?

e. Mr. Jones does indeed have the disease you suspected, and you recommend surgical intervention. Any surgical procedure has the risks of pain, bleeding, and infection. What are two specific risks associated with this particular surgery?

f. List two reasons why the tests listed in part c are controversial for use as screening tools.

g. In American males, prostate cancer is the most common, non-skin cancer (accounting for 33% of all new cancers), but is less deadly than might be expected, ranking behind both lung and colon cancer as the third leading cause of cancer death (9%). By contrast, ovarian cancer is the eighth most common new non-skin cancer in American women (3% of new diagnoses), but accounts for a surprising number of deaths; it ranks as the fifth leading cause of cancer death in this population (6%). Give three reasons for the discrepancy between the incidence and death rates for these two cancer types.

13. A patient comes to your office complaining of abdominal fullness and a change in bowel habits. She reports a family history of breast cancer and ovarian cancer. You suspect she may have ovarian cancer and order a serum CA-125 test. The sensitivity of this test is 35% and the specificity is 98.5%. The incidence of ovarian cancer in this population is 0.1%. The test comes back positive.

a. If you gave this test to 1,000,000 women, how many patients would have a true positive (TP) result, a false positive (FP) result, a true negative (TN) result and a false negative (FN) result?

b. Given her positive test result, what is the likelihood that your patient really has ovarian cancer?

c. What test would you recommend that your patient undergo next?

14. A company called BioCurex recently announced results of a clinical trial for a new test to detect lung cancer (see story below).

RANCHO SANTA MARGARITA, Calif.–(BUSINESS WIRE)–April 5, 2004–BioCurex Inc. announces results for lung cancer detection using its proprietary Serum-RECAF(TM) blood test. The results confirm 90% sensitivity with 95% specificity. The findings further substantiate the use of RECAF(TM) as a universal cancer marker with a potential market size of $2 billion per year for all cancers. The study included 32 lung cancer patients and 103 normal donors with statistical verification.

© 2003 Business Wire.
[http://www.biospace.com/ccis/
news_story.cfm?StoryID=15650520&full=1]

a. Calculate the number of patients with true negative (TN), true positive (TP), false positive (FP) and false negative (FN) test results in this trial.

b. What is the positive predictive value in this trial?

c. Do you think the PPV you calculated in part b is an accurate estimate of what to expect if the test is used to screen the general population for lung cancer? Why or why not?

15. Suppose we have two new screening tests for ovarian cancer – Test A and Test B. When tested in a large population, we find the sensitivity and specificity values for the two tests listed in the table below. Your mother is worried about her risk of ovarian cancer because both her mother and sister died of ovarian cancer at a young age. She asks your advice about which screening test to undergo. Which test would you recommend that she take? Why?

Test	Se	Sp
Test A	60%	95%
Test B	95%	60%

16. Consider the development of a new proteomics based screening test for ovarian cancer described in this chapter. Apply the five steps of technology assessment to this new technology. Does this assessment support the use of the technology?

References

[1] American Cancer Society. *Cancer Facts and Figures 2007*. Atlanta: American Cancer Society; 2007.

[2] Miniño AM, Heron MP, Smith BL. *Deaths: Preliminary Data for 2004*. National Vital Statistics Report. 2006 June 28;**54**(19).

[3] American Cancer Society. *Cancer Facts and Figures 2005*. Atlanta: American Cancer Society; 2005.

[4] Edwards BK, Brown ML, Wingo PA, Howe HL, Ward E, Ries LAG, *et al.* Annual Report to the Nation on the Status of Cancer, 1975–2002, Featuring Population-Based Trends in Cancer Treatment. *Journal of the National Cancer Institute*. 2005 October 5; **97**(19).

[5] WHO. *Mortality: Revised Global Burden of Disease (2002) Estimates*. Geneva: World Health Organization; 2002.

[6] Parkin DM, Bray F, Ferlay J, Pisani P. Global cancer statistics, 2002. *CA: a Cancer Journal for Clinicians*. 2005 Mar–Apr; **55**(2): 74–108.

[7] International Union Against Cancer. *Global Action Against Cancer*. Geneva: WHO/UICC; 2005.

[8] World Health Organization. *Global Cancer Rates Could Increase by 50% to 15 Million by 2020*. Geneva: World Health Organization; 2003 April 3.

[9] American Cancer Society. *The Case for Global Action*. 2007 [cited 2007 May 9]; Available from: http://www.cancer.org/docroot/AA/content/ AA_2_5_1x_Case_for_Global_Action.asp?sitearea=AA

[10] American Cancer Society. *Lifetime Probability of Developing or Dying from Cancer*. 2006 February 22 [cited 2007 May 9]; Available from: http://www.cancer.org/docroot/CRI/content/CRI_2_ 6x_Lifetime_Probability_of_Developing_or_Dying_ From_Cancer.asp?sitearea=

[11] Stewart BW, Kleihues P, eds. *World Cancer Report*. Lyon: IARC Press; 2003.

[12] American Cancer Society. *Cancer Statistics 2005* [PowerPoint]. Atlanta: American Cancer Society; 2005.

[13] MacKay J, Eriksen M. *The Tobacco Atlas*. Geneva: The World Health Organization; 2002.

[14] American Cancer Society. *Cancer Prevention and Early Detection Facts and Figures 2007*. Atlanta: American Cancer Society; 2007.

[15] Hanahan D, Weinberg RA. The hallmarks of cancer. *Cell*. 2000 Jan 7; **100**(1): 57–70.

[16] Dolan DNA Learning Center – Cold Spring Harbor Laboratory. *Inside Cancer*. [Multimedia Guide] 2004 [cited 2007 May 10]; Available from: www.insidecancer.org

[17] Cotran RS, Kumar V, Robbins SL. *Robbins Pathologic Basis of Disease*. 5th edn. W.B. Saunders Company; 1994.

[18] National Cancer Institute. *Milestone (1971): President Nixon declares war on cancer*. 2007 [cited 2007 May 10]; Available from: http://dtp.nci.nih.gov/timeline/noflash/milestones/ M4_Nixon.htm

[19] Niederhuber JE. *The Nation's Investment in Cancer Research*. Frederick, Maryland: National Cancer Institute; 2006 October.

[20] National Cancer Institute. *The NCI Strategic Plan for Leading the Nation*. Frederick, Maryland: National Cancer Institute; 2006 January.

[21] National Center for Health Statistics. *Health, United States, 2006 With Chartbook on Trends in the Health of Americans*. Hyattsville, MD; 2006.

[22] Carr DB, Loeser JD, Morris DB. *Narrative, Pain and Suffering (Progress in Pain Research and Management)*. Seattle, WA: IASP Press; 2005.

[23] Wang XS, Li TD, Yu SY, Gu WP, Xu GW. China: status of pain and palliative care. *Journal of Pain and Symptom Management*. 2002 Aug; **24**(2): 177–9.

[24] US Preventive Services Task Force. *Guide to Clinical Preventive Services*. 2nd edn. Alexandria, VA: International Medical Publishing; 1996.

[25] Elmore JG, Armstrong K, Lehman CD, Fletcher SW. Screening for breast cancer. *Jama*. 2005 Mar 9; **293**(10): 1245–56.

[26] Warner E, Plewes DB, Hill KA, Causer PA, Zubovits JT, Jong RA, *et al.* Surveillance of BRCA1 and BRCA2 mutation carriers with magnetic resonance imaging, ultrasound, mammography, and clinical breast examination. *Jama*. 2004 Sep 15; **292**(11): 1317–25.

[27] Thomas DB, Gao DL, Ray RM, Wang WW, Allison CJ, Chen FL, *et al.* Randomized trial of breast self-examination in Shanghai: final results. *Journal of the National Cancer Institute*. 2002 Oct 2; **94**(19): 1445–57.

[28] American Cancer Society. *Cancer Facts and Figures 2006*. Atlanta: American Cancer Society; 2006.

[29] Humphrey LL, Helfand M, Chan BK, Woolf SH. Breast cancer screening: a summary of the evidence for the

US Preventive Services Task Force. *Annals of Internal Medicine*. 2002 Sep 3; **137**(5 Part 1): 347–60.

[30] Parkin DM, Bray F. Chapter 2: The burden of HPV-related cancers. *Vaccine*. 2006 Aug 21; **24** Suppl. 3: S11–25.

[31] Nijhuis ER, Reesink-Peters N, Wisman GB, Nijman HW, van Zanden J, Volders H, *et al.* An overview of innovative techniques to improve cervical cancer screening. *Cell Oncology*. 2006; **28**(5–6): 233–46.

[32] Burchell AN, Richardson H, Mahmud SM, Trottier H, Tellier PP, Hanley J, *et al.* Modeling the sexual transmissibility of human papillomavirus infection using stochastic computer simulation and empirical data from a cohort study of young women in Montreal, Canada. *American Journal of Epidemiology*. 2006 Mar 15; **163**(6): 534–43.

[33] American Cancer Society. *FDA Approves New Cervical Cancer Screening Test*. 2003 March 31 [cited 2007 May 10]; Available from: http://www.cancer.org/docroot/ NWS/content/NWS_1_1x_FDA_Approves_New_ Cervical_Cancer_Screening_Test.asp

[34] Cohen J. Public health. High hopes and dilemmas for a cervical cancer vaccine. *Science (New York, NY)*. 2005 Apr 29; **308**(5722): 618–21.

[35] Alberts B. *Molecular Biology of the Cell*. 3rd edn. New York: Garland Publishing; 1994.

[36] Renshaw AA. Measuring sensitivity in gynecologic cytology: a review. *Cancer*. 2002 Aug 25; **96**(4): 210–17.

[37] Centers for Disease Control. *Regulations for Implementing the Clinical Laboratory Improvement Amendments of 1988: a Summary*. MMWR. 1992 (No. RR-2).

[38] Austin RM. Human papillomavirus reporting: minimizing patient and laboratory risk. *Archives of Pathology & Laboratory Medicine*. 2003 Aug; **127**(8): 973–7.

[39] Bogdanich W. Lax Laboratories: The Pap Test misses much cervical cancer through lab's errors. *The Wall Street Journal*. 1987 November 2.

[40] Nanda K, McCrory DC, Myers ER, Bastian LA, Hasselblad V, Hickey JD, *et al.* Accuracy of the Papanicolaou test in screening for and follow-up of cervical cytologic abnormalities: a systematic review. *Annals of Internal Medicine*. 2000 May 16; **132**(10): 810–19.

[41] Fahey MT, Irwig L, Macaskill P. Meta-analysis of Pap test accuracy. *American Journal of Epidemiology*. 1995 Apr 1; **141**(7): 680–9.

[42] Mitchell MF, Schottenfeld D, Tortolero-Luna G, Cantor SB, Richards-Kortum R. Colposcopy for the diagnosis of squamous intraepithelial lesions: a meta-analysis. *Obstetrics and Gynecology*. 1998 Apr; **91**(4): 626–31.

[43] Welch HG. *Should I be Tested for Cancer?* Berkely, CA: University of California Press; 2004.

[44] Ferenczy A, Franco E. Cervical-cancer screening beyond the year 2000. *The Lancet Oncology*. 2001 Jan; **2**(1): 27–32.

[45] Hutchinson ML. Assessing the costs and benefits of alternative rescreening strategies. *Acta Cytologica*. 1996 Jan–Feb; **40**(1): 4–8.

[46] Kurman RJ, Henson DE, Herbst AL, Noller KL, Schiffman MH. Interim guidelines for management of abnormal cervical cytology. The 1992 National Cancer Institute Workshop. *Jama*. 1994 Jun 15; **271**(23): 1866–9.

[47] Jeronimo J, Morales O, Horna J, Pariona J, Manrique J, Rubinos J, *et al.* Visual inspection with acetic acid for cervical cancer screening outside of low-resource settings. *Revista Panamericana de Salud Publica (Pan American Journal of Public Health)*. 2005 Jan; **17**(1): 1–5.

[48] Birdsong GG. Automated screening of cervical cytology specimens. *Human Pathology*. 1996 May; **27**(5): 468–81.

[49] Groopman J. Contagion: Papilloma Virus. *The New Yorker*. 1999 September 13.

[50] Saslow D, Runowicz CD, Solomon D, Moscicki AB, Smith RA, Eyre HJ, *et al.* American Cancer Society guideline for the early detection of cervical neoplasia and cancer. *CA: a Cancer Journal for Clinicians*. 2002 Nov–Dec; **52**(6): 342–62.

[51] Russell J, Crothers BA, Kaplan KJ, Zahn CM. Current cervical screening technology considerations: liquid-based cytology and automated screening. *Clinical Obstetrics and Gynecology*. 2005 Mar; **48**(1): 108–19.

[52] Health Technologies Advisory Committee – Minnesota. New technologies for cervical cancer screening: strategies to lower the rate of cervical cancer. *Health Services/Technology Assessment Text (HSTAT)*. National Library of Medicine; 1999.

[53] FDA. *FDA Approves Expanded Use of HPV Test*. 2003 March 31 [cited 2007 May 20]; Available from: http://www.fda.gov/bbs/topics/NEWS/2003/ NEW00890.html

[54] Schiffman M, Herrero R, Hildesheim A, Sherman ME, Bratti M, Wacholder S, *et al.* HPV DNA testing in

cervical cancer screening: results from women in a high-risk province of Costa Rica. *Jama*. 2000 Jan 5; **283**(1): 87–93.

[55] Arbyn M, Sasieni P, Meijer CJ, Clavel C, Koliopoulos G, Dillner J. Chapter 9: Clinical applications of HPV testing: a summary of meta-analyses. *Vaccine*. 2006 Aug 21; **24** Suppl. 3: S78–89.

[56] Digene Corporation. *Hybrid Capture® 2 High-Risk HPV DNA Test*™ [cited 2007 May 20]; Product Insert]. Available from: http://www.digene.com/pdf/L2290-P.I,%20hc2%20HPV%20DNA%20Test%20US.pdf

[57] Germar MJ, Marialdi M. Visual inspection with acetic acid as a cervical cancer screening tool for developing countries. *Review for the 12th Postgraduate Training Course in Reproductive Health/Chronic Disease 2003* [cited 2007 May 28]; Available from: http://www. gfmer.ch/Endo/Course2003/ Visual_inspection_cervical_ cancer.htm

[58] Coalition NCC. *NCCC Section on the HPV Vaccine*. 2007 [cited 2007 March 2]; Available from: http://www.nccc-online.org/hpv-vaccine.php

[59] Cohen J. High hopes and dilemmas for a cervical cancer vaccine. *Science (New York, NY)*. 2003; **308**: 618–21.

[60] Kane MA, Sherris J, Coursaget P, Aguado T, Cutts F. Chapter 15: HPV vaccine use in the developing world. *Vaccine*. 2006 Aug 21; **24** Suppl. 3: S132–9.

[61] Sankaranarayanan R, Shastri SS, Basu P, Mahe C, Mandal R, Amin G, *et al.* The role of low-level magnification in visual inspection with acetic acid for the early detection of cervical neoplasia. *Cancer Detection and Prevention*. 2004; **28**(5): 345–51.

[62] Pan American Health Organization. *Visual Inspection of the Uterine Cervix with Acetic Acid (VIA): A Critical Review and Selected Articles*. Washington, D.C.; 2003.

[63] Park SY, Follen M, Milbourne A, Rhodes H, Malpica A, MacKinnon M, *et al.* Automated image analysis of digital colposcopy for the detection of cervical neoplasia. Manuscript submitted for publication, 2007.

[64] Kasper D, Braunwald E, Fauci A, Longo D, Hauser S, Jameson JL, eds. *Harrison's Principles of Internal Medicine*. 16th edn. New York City: McGraw-Hill; 2005.

[65] Hutchinson K, Wener M. *National Health and Nutrition Examination Survey Lab Methods 2003–2004: Total Prostate-Specific Antigen (PSA)*. University of Washington Medical Center; 2007 May 4.

[66] Aureon Biosystems. *LumiPhos Plus*. [cited 2007 March 10]; Available from: http://www.aureonbio.com/ProductList.htm

[67] Lumigen. *Lumigen PPD*. 2006 [cited 2007 March 10]; Available from: http://www.lumigen.com/documents/ LumigenPPD.shtml

[68] Hernandez J, Thompson IM. Prostate-specific antigen: a review of the validation of the most commonly used cancer biomarker. *Cancer*. 2004 Sep 1; **101**(5): 894–904.

[69] Albertsen PC. Is screening for prostate cancer with prostate specific antigen an appropriate public health measure? *Acta Oncologica (Stockholm, Sweden)*. 2005; **44**(3): 255–64.

[70] Gottlieb RH, Mooney C, Mushlin AI, Rubens DJ, Fultz PJ. The prostate: decreasing cost-effectiveness of biopsy with advancing age. *Investigative Radiology*. 1996 Feb; **31**(2): 84–90.

[71] American Cancer Society. *How Is Prostate Cancer Diagnosed?* 2006 July 26 [cited 2007 May 20]; Available from: http://www.cancer.org/docroot/CRI/content/ CRI_2_4_3X_How_is_prostate_cancer_diagnosed_36.asp

[72] Lu-Yao GL, Yao SL. Population-based study of long-term survival in patients with clinically localised prostate cancer. *The Lancet*. 1997 Mar 29; **349**(9056): 906–10.

[73] Neal DE, Leung HY, Powell PH, Hamdy FC, Donovan JL. Unanswered questions in screening for prostate cancer. *European Journal of Cancer*. 2000 Jun; **36**(10): 1316–21.

[74] Bartsch G, Horninger W, Klocker H, Reissigl A, Oberaigner W, Schonitzer D, *et al.* Prostate cancer mortality after introduction of prostate-specific antigen mass screening in the Federal State of Tyrol, Austria. *Urology*. 2001 Sep; **58**(3): 417–24.

[75] Horninger W, Berger A, Pelzer A, Klocker H, Oberaigner W, Schonitzer D, *et al.* Screening for prostate cancer: updated experience from the Tyrol study. *The Canadian Journal of Urology*. 2005 Feb; **12** Suppl. 1: 7–13; discussion 92–3.

[76] Ilic D, O'Connor D, Green S, Wilt T. Screening for prostate cancer: a Cochrane systematic review. *Cancer Causes Control*. 2007 Apr; **18**(3): 279–85.

[77] Amling CL. Prostate-specific antigen and detection of prostate cancer: what have we learned and what should we recommend for screening? *Current Treatment Options in Oncology*. 2006 Sep; **7**(5): 337–45.

[78] Hennekens CH, Buring JE. *Epidemiology In Medicine.* Philadelphia: Lippincott Williams & Wilkins; 1987.

[79] Welch HG, Schwartz LM, Woloshin S. Are increasing 5-year survival rates evidence of success against cancer? *Jama.* 2000 Jun 14; **283**(22): 2975–8.

[80] Reynolds EA, Moller KA. A review and an update on the screening of epithelial ovarian cancer. *Current Problems in Cancer.* 2006 Sep–Oct; **30**(5): 203–32.

[81] Rosenthal AN, Menon U, Jacobs IJ. Screening for ovarian cancer. *Clinical Obstetrics and Gynecology.* 2006 Sep; **49**(3): 433–47.

[82] BC Cancer Agency. *Years of Life Lost.* 2005 July 25 [cited 2007 May 22]; Available from: http://www.bccancer.bc.ca/HPI/CancerStatistics/FF/LifeLost.htm

[83] Petricoin EF, Ardekani AM, Hitt BA, Levine PJ, Fusaro VA, Steinberg SM, *et al.* Use of proteomic patterns in serum to identify ovarian cancer. *The Lancet.* 2002 Feb 16; **359**(9306): 572–7.

[84] Urban N. Specific keynote: ovarian cancer risk assessment and the potential for early detection. *Gynecologic Oncology.* 2003 Jan; **88**(1 Pt 2): S75–9; discussion S80–3.

[85] Nahhas WA. Ovarian cancer. Current outlook on this deadly disease. *Postgraduate Medicine.* 1997 Sep; **102**(3): 112–20.

[86] Zurawski VR, Jr., Sjovall K, Schoenfeld DA, Broderick SF, Hall P, Bast RC, Jr., *et al.* Prospective evaluation of serum CA 125 levels in a normal population, phase I: the specificities of single and serial determinations in testing for ovarian cancer. *Gynecologic Oncology.* 1990 Mar; **36**(3): 299–305.

[87] Gladstone CQ. Screening for ovarian cancer. In: Canadian Task Force on Periodic Health Examination, ed. *Canadian Guide to Clinical Preventive Health Care.* Ottawa: Health Canada; 1994: 870–81.

[88] Menon U, Jacobs IJ. Ovarian cancer screening in the general population. *Current Opinion in Obstetrics & Gynecology.* 2001 Feb; **13**(1): 61–4.

[89] van Nagell JR, Jr., DePriest PD, Reedy MB, Gallion HH, Ueland FR, Pavlik EJ, *et al.* The efficacy of transvaginal sonographic screening in asymptomatic women at risk for ovarian cancer. *Gynecologic Oncology.* 2000 Jun; **77**(3): 350–6.

[90] Lok IH, Sahota DS, Rogers MS, Yuen PM. Complications of laparoscopic surgery for benign ovarian cysts. *The Journal of the American Association of Gynecologic Laparoscopists.* 2000 Nov; **7**(4): 529–34.

[91] Hensley ML, Spriggs DR. Cancer screening: how good is good enough? *Journal of Clinical Oncology.* 2004 Oct 15; **22**(20): 4037–9.

[92] Johnson D, Sandmire D, Klein D. Ovarian Cancer. In *Medical Tests That Can Save Your Life: 21 Tests Your Doctor Won't Order Unless You Know to Ask.* Emmaus, PA Rodale; 2004.

[93] *Fees Involved in Infertility Treatments.* 2005 [cited 2007 June 1]; Available from: http://www.sharedjourney.com/articles/eggfees.html

[94] Jacobs IJ, Skates S, Davies AP, Woolas RP, Jeyerajah A, Weidemann P, *et al.* Risk of diagnosis of ovarian cancer after raised serum CA 125 concentration: a prospective cohort study. *BMJ (Clinical Research Ed.)* 1996 Nov 30; **313**(7069): 1355–8.

[95] Jacobs IJ, Skates SJ, MacDonald N, Menon U, Rosenthal AN, Davies AP, *et al.* Screening for ovarian cancer: a pilot randomised controlled trial. *The Lancet.* 1999 Apr 10; **353**(9160): 1207–10.

[96] *UK Collaborative Trial of Ovarian Cancer Screening.* 2006 July 11 [cited 2007 May 20]; Available from: http://www.ukctocs.org.uk/index.html

[97] Buys SS, Partridge E, Greene MH, Prorok PC, Reding D, Riley TL, *et al.* Ovarian cancer screening in the Prostate, Lung, Colorectal and Ovarian (PLCO) cancer screening trial: findings from the initial screen of a randomized trial. *American Journal of Obstetrics and Gynecology.* 2005 Nov; **193**(5): 1630–9.

[98] FDA. *Protein Patterns May Identify Ovarian Cancer.* 2002 February 7 [cited 2007 May 20]; Available from: http://www.fda.gov/bbs/topics/NEWS/2002/NEW00797.html

[99] Pollack A. New cancer test stirs hope and concern. *The New York Times.* 2004 February 3.

[100] Underwood A. Testing: ovarian cancer. *Newsweek.* 2002 February 18: 12.

[101] Baggerly KA, Morris JS, Coombes KR. Reproducibility of SELDI-TOF protein patterns in serum: comparing datasets from different experiments. *Bioinformatics (Oxford, England).* 2004 Mar 22; **20**(5): 777–85.

[102] American Cancer Society, How Many People Get Lung Cancer, 2004; Available from: http://www.cancer.org/docroot/CRI/content/CRI_2_2_IX_How_many_people_get_lung_cancer_26.asp?sitearea=

[102] www.med.uiuc.edu/m2/epidemiology/ReviewQuestions/SensitivitySpecificity_Questions.htm.

Cost effectiveness of screening for disease

We began this book by considering the case study of a new treatment for advanced breast cancer – high dose chemotherapy and bone marrow transplant. Motivated by the disappointing performance of this treatment when it was offered prematurely to thousands of American women, we considered how to systematically evaluate new technologies in Chapter 2. Recall that our process of technology assessment consists of asking five questions about a new medical technology [1].

Biologic plausibility. Does our current understanding of the biology of the disease in question support the use of the technology?

Technical feasibility. Can we safely and reliably deliver the new technology to the target patients?

Clinical trials. Do the results of randomized clinical trials comparing the new technology to current standards of care show a benefit?

Patient outcomes. Are patients better off for having used the new technology?

Societal outcomes. What are the costs and ethical implications of the technology?

Thus far, we have examined the biological plausibility and technical feasibility of new technologies to prevent infectious disease and to detect cancers at an early stage. We have seen the crucial role that clinical trials play and the importance of adhering to ethical guidelines

in the development of new technologies. However, we have not yet developed the tools to determine whether a new technology is a cost effective use of healthcare resources. In Chapter 10, we considered the pros and cons of screening for early cancer. We found that while screening for cervical cancer can reduce both cancer incidence and mortality, there is considerable debate about whether to screen for prostate cancer. In both cases, we must make decisions about how to use limited healthcare resources. Potentially, screening can identify disease at an earlier stage, when it is less expensive to treat. If effective screening tests are available, we must decide who should be screened, how frequently screening should occur, and what test (or test combination) provides the most effective use of resources. In the present chapter, we will learn how to use cost effectiveness analysis to help make recommendations about whether new technologies should be adopted and how they can be used most effectively.

Types of economic evaluation of health technology

How do we approach the economic evaluation of a new health technology? In general, we are interested in comparing both the benefits and costs of a new health technology relative to some standard of care. In order to make this comparison meaningful, we must calculate

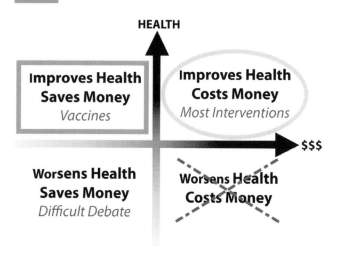

Figure 11.1. *A representation of health-policy space, comparing the costs and health benefits of interventions.*

the costs and benefits of the technologies in a way that can be compared. Figure 11.1 shows one way of thinking about this comparison. This figure represents what is sometimes called "health-policy space." One axis of the graph represents health, while the other represents cost. All health interventions fall in one of the four quadrants of this graph. In the lower right quadrant, we have interventions which both cost money and worsen health. Obviously we do not want to adopt such interventions. Technologies which save money but worsen health fall in the lower left quadrant; frequently there is difficult public debate associated with whether to adopt new interventions which may slightly worsen health but have potential for enormous cost savings. In the upper left quadrant, we have interventions which both save money and improve health. We obviously want to adopt interventions which fall in this quadrant; unfortunately, there are not many interventions which fall here. Many childhood vaccines are one of the few interventions which both improve health and save money. Most interventions fall in the top right quadrant, improving health but consuming resources. How do we determine which quadrant of the graph a new intervention falls in? And how can we compare two interventions within a quadrant? There are several approaches to evaluate the cost effectiveness of an intervention; here we will consider two in detail.

The first method is to compare clinical strategies in terms of clinical outcomes as measured in physical units (e.g., blood pressure, number of strokes, cases of cancer found, life years gained, etc.). Cost effectiveness ratios can be computed and compared, and may be expressed for example as dollars spent per case of cancer detected, or dollars spent per life year gained. The advantage of this approach is that it is generally simpler to calculate this type of outcome. The challenge of this approach is that it is difficult to compare the cost effectiveness of two very different types of clinical strategies (e.g. vaccination vs. end stage cancer therapy).

An alternative approach is to evaluate clinical benefits in terms of a common set of units. Quality adjusted life years are usually considered as the "yardstick" for this approach, since they are comparable across interventions. Recall that the quality adjusted life year is the number of years of full health that are considered equivalent to a given number of years in a reduced state of health. The advantage of this approach is that expressing cost effectiveness in a common set of units is extremely useful for making policy planning decisions. The disadvantage is that they are more difficult to calculate than physical units.

Components of an economic evaluation for health technology assessment

Just like the method of engineering design, we can follow a step-wise process to carry out an economic evaluation of a health technology. We first consider the steps in this process, and then turn to several examples. You will see that the economic evaluation includes many components in addition to the "economics"; in other words, to appropriately perform an economic evaluation, one must also perform an evaluation of the effectiveness of a technology as well.

Define the problem

This is the starting point for an economic evaluation. A problem should be well defined and posed in an answerable form [2]. Typically a trade-off is usually obvious and should be made explicit, e.g. if improved survival rates have been observed for an expensive intervention, one would like to determine if the improvement in clinical outcomes is worth the additional cost from the intervention. Similarly, if there are trade-offs between quality

**Healthcare outside the clinic:
July 5, 2007
Tessa Swaziland**

After two days of WFP work, I was able to escape the clinic and see a bit more of the healthcare facilities outside of the Baylor world. Wednesday, I went with Carrie and Julia to Vuvulane, a rural community about two hours from the COE. The clinic was run by nurses, and the only doctors ever out there were the Baylor doctors once a month. We met up with Good Shepherd (hospital in Manzini) nurses who have been going to these outreach sites before Baylor got involved. They provided the connections and the ARVs. We provided the medical expertise. I spent the day with one of the nurses. I was "helping" with pill-counting, but really I was just watching. The nurse was super nice and seemed to think I was just great, so we had quite a bit of fun while we worked.

As I watched her work, I kept noticing things she was doing incorrectly (as she filled out the adherence sheet). The whole point of pill-counting is to see how well a patient is adhering to their drug regimen. It is very important because if a patient has poor adherence, then their virus will build resistance to the drug, and the patient's viral load will increase. (Basically, they will get sicker, and they will be much harder to treat in the future.) So, I started asking her questions about why and how she did what she did. Like, when she wrote down 56 for the expected number of pills, I asked why. She said that the patient needed two a day, and since there are 28 days between visits, they should've taken 56. Looking at the records, I could see that there were 30 days between this visit and the last. When I asked her about that, she said, "We always do 28. That's just what we do." And that was that. There were quite a few other major problems I observed with the system, but ultimately, it was all meaningless, since they never calculated their adherence percent. And even if the person clearly had terrible adherence (one man had 50 extra pills!), they never counseled them or took them off the ART (anti-retroviral treatment). Drug resistance has implications for not only the individual who isn't taking his meds correctly, but also anyone he passes the disease to, who will suffer from drug-resistant HIV.

On the way to and from Vuvulane, we drove by Hlane National Park, where Carrie was lucky enough to see a lion. I was rummaging through my bag and missed it completely!

On the way back, we stopped at the private Mbabane hospital, which was much nicer than the government hospital but still had much to improve upon before it would ever meet American standards. We were checking up on a premature infant who was born at 28 weeks. (I could be off a few weeks there) I had never seen a preemie before and wasn't quite prepared for it. The mother even asked me if I was afraid of the baby because I stayed by the door and stayed there throughout the visit. The baby was miniscule – the diaper engulfed its entire body, and its leg was the thickness of my thumb. What I *was* afraid of was giving the baby some germ from the outside world that would do its premature immune system in. Carrie told me that a neo-natal unit in the states would be 10 times quieter and darker to simulate the conditions in the womb. This poor baby had a bright beam of sunlight pouring over him, in addition to a symphony of construction sounds reverberating from somewhere nearby in the hospital. Julia said that if the baby had been in the government hospital, there is no question that it would be dead by now.

On Thursday, I shadowed Dr. Eileen at the government hospital. Although I've never been interested in clinical medicine, and I never will be, I found rounds to be quite interesting. She and Dr. D (doctor at that hospital) went from bed to bed examining patients, asking the mothers questions, and looking at X-rays. One of the biggest challenges was deciding whether they had a weakened immune system due to AIDS or TB or both. Sometimes the x ray clearly indicated TB, but with some patients, it was more difficult to tell.

One of the most exciting things about our visit was the fact that, when we arrived, Dr. D and some nurses had sat down all of the patient's mothers (this was the pediatric ward) and were discussing the importance of washing hands and boiling water. This may seem insignificant, but in fact, if they can effectively communicate these messages to the mothers, they could more than halve the number of patients that needed to be there. They plan to do a lot more of this sort of doctor-patient interaction,

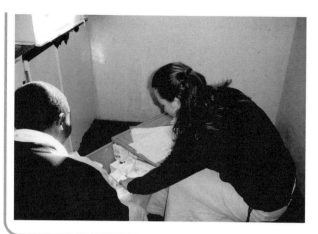

and they were also trying to make signs to accompany the campaign. In that respect, I was able to help. Making those signs is one of the smaller projects I've been working on.

On Friday, I presented an updated plan for WFP. Thursday, we had a volunteer come in, and Dave started training her. The plan for the following week was to transition from the old system (where Dave and I are very involved) to the new system (where Dave and I won't be here to do anything). A lot of the doctors weren't there, so that kind of caused problems, but nothing too big.

Steps in a cost effectiveness assessment

Define the problem
Identify the perspective
Identify the alternatives
Analyze the effectiveness
Analyze the costs
Perform discounting
Perform sensitivity analysis
Address ethical issues
Interpret the results

Identify the alternatives

Identifying the alternative technologies is a critical component for an economic evaluation of a new technology [3]. Sometimes the baseline strategy is the strategy of doing nothing. Other times, we will use the current standard of care as the baseline strategy.

It is also important to note that a "do nothing" strategy may or may not be an appropriate alternative to consider. Doing nothing has certain ethical implications, both from clinical and policy perspectives. "Doing nothing" is likely to have both economic and clinical consequences. In particular, it is rare that a "do nothing" strategy has no economic costs. Certainly, if "doing nothing" means that a patient will die, there are likely to be costs associated with the dying process, such as palliative care. Less dramatically, if "doing nothing" means "no screening", then associated costs of undetected disease must be incorporated in the analysis.

Analyze the effectiveness

The effectiveness is the clinical outcome to be evaluated in an economic evaluation. Ideally, the effectiveness is evaluated in terms of quality adjusted life years. Quality adjusted life years represent a standardized measure of health outcome that incorporates both length and quality of life into consideration. The number of quality adjusted life years is the number of years in perfect health that are valued comparably to the number of life years experienced in a less desirable health state. Similarly, the number of quality adjusted life years is the

and quantity of life, these should be noted explicitly as well.

Identify the perspective

In carrying out an economic analysis, one must consider whose costs to include; this is called the perspective of the analysis. The perspective is of critical importance when performing an economic evaluation. The ideal perspective is that of the "societal perspective," which incorporates all aspects of costs and benefits. In technology assessment, the perspective must clearly be identified, for it is possible to have different results of an analysis when different perspectives are taken. Potential perspectives for an analysis include: patient, health maintenance organization, payer, or society. Guidelines from the Panel on Cost-Effectiveness in Health and Medicine recommend that the "reference case" be done from the societal perspective.

number of years of life weighted by a measure of the quality of life experienced. The quality of life component is based on a 0–1 utility scale, where 0 represents death and 1 represents perfect health [3, 4].

Analyze the costs

Costs can be classified in several categories [5]. Direct healthcare costs include physician services, pharmaceuticals, tests, inpatient care, outpatient care, and administration regarding clinical facilities, including medical records, food, nursing, and supplies. Non-healthcare costs are costs that are borne as a result of seeking medical care. These may include costs of babysitting or travel during the time having medical care. Finally, the cost of time seeking medical care and lost productivity due to illness are other opportunity costs of healthcare interventions that do not result in money exchanging hands, though it truly is a cost of healthcare.

Perform discounting

Discounting is the procedure that calibrates outcomes that occur over time. For economic evaluation of healthcare programs, both costs and effectiveness must be appropriately discounted. The Panel on Cost-Effectiveness in Health and Medicine has recommended that a discount rate of 3% should be used. For example, at a discount rate of 3%, a cost of $1 spent next year is equivalent to the cost of 97 cents spent today. If the costs of an intervention total $1000 each year over a three year period, then the total cost attributed to that program in present dollars should be calculated to be $1000 + $1000/1.03 + $1000/(1.03^2)$. Over long periods of time, discounting can have a major impact. The cost of $1 spent in ten years' time is equivalent to spending only 74 cents today.

Perform sensitivity analysis

Sensitivity analysis is the process of varying parameters in an analysis to determine if the selected optimal alternative would remain optimal. It is crucial to examine whether the output of model-based cost effectiveness analysis is extremely sensitive to small changes in one or more input parameters. Most frequently, sensitivity analysis is performed by varying a single parameter of the model and determining how the bottom line result would change; for example, what would be the changes in the incremental cost-effectiveness ratio if the prevalence of disease would change.

Address ethical issues

Ethical issues need to be addressed when performing technology assessment studies. Many situations arise in cost effectiveness analysis in which the best decision for a single patient is not the same decision that would be made when taking into account society as a whole. Specifically, a clinical intervention may be found to be clinically effective, but it might not be cost effective. Thus, what might be best for a single individual might not be best for society as a whole. This happens when an expensive clinical intervention yields minimal benefits. Such a situation is frequent in the oncology literature, where expensive interventions yield marginal health benefits at a substantial cost.

Interpret the results

The interpretation of results of an economic evaluation is an important component of the summary of the analysis. Results are often presented as a ratio, where the numerator includes all the costs and the denominator includes all the outcomes. Many times, results are presented as an incremental analysis, comparing the costs and benefits of a new strategy to the next most effective strategy. For example, to compare the effectiveness of the use of an HPV vaccine and screening to that of screening alone one would calculate the following ratio:

$$\frac{[\text{costs of vaccination \& screening}] - [\text{costs of screening}]}{[\text{outcomes of vaccination \& screening}] - [\text{outcomes of screening}]}.$$
(11.1)

The implications of the results should be stated, with a particular emphasis on how the results may affect clinical practice or public policy. One important issue to address is the issue of magnitude. The economic analysis may yield an incremental cost effectiveness ratio and the important question to answer is whether this ratio is a "big number" or a "little number." In other words, is

Table 11.1. *A league table ranks interventions in terms of their relative cost effectiveness [15].*

Intervention	Cost effectiveness ratio
Pneumococcal vaccine for adults over 65 years old	Cost saving
Tobacco cessation counseling	Cost saving to $2,000/QALY saved
Chlamydia screening of women 15–24 years old	$2,500/QALY saved
Colorectal cancer screening for people more than 50 years old	$13,000/QALY saved

the bottom line result significant enough to make an impact or a difference? Should the result imply a change in clinical practice or public policy? Some investigators have chosen to use a "league table" to make comparisons between the results of the cost effectiveness analysis under investigation and the results of previously performed cost effectiveness analyses that have been previously published in the literature (Table 11.1). These league tables (named after the European soccer leagues, which publish the standings or rankings of teams in the newspapers) provide one measure of comparison. The reasoning goes that if the incremental cost effectiveness ratio for a given analysis is similar to the incremental cost effectiveness ratio for a previously performed analysis and the previously analyzed intervention is considered an acceptable clinical practice that is also "cost effective," then the intervention under investigation should also be considered a cost effective use of resources.

Many articles in the clinical evaluation literature have cited $50 000 per quality adjusted life year (QALY) as an appropriate threshold for determining the cost effectiveness of a healthcare intervention. This threshold was determined from earlier studies that have been considered acceptable and a good value (e.g. treatment of moderate hypertension) or acceptable though a controversial value (e.g. dialysis for early stage renal disease). However, there is no specific reason or benchmark that presently applies to indicate that $50,000 per QALY is the threshold for optimal decision making regarding resource allocation.

A lack of resources prevents the application of this rule in many developing countries. In this setting, an alternative that has been suggested is to consider interventions to be very cost effective if the amount that must be spent to gain one QALY is less than the per capita GDP, and cost effective if it is less than three times the per capita GDP [6].

Examples of cost effectiveness analysis

With this overview, we next consider several examples of cost effectiveness analysis in detail. In this chapter, we will focus on the problem of screening for cervical cancer. More than 50 million Pap tests are performed annually in the USA, and the costs of screening exceed several billion dollars annually [6]. With such a large investment, it is imperative to ensure that we continue to invest public health resources wisely. With so many new screening technologies available, we will find that cost effectiveness analysis can help us decide which test to use, who should be screened and how frequently they should be screened. Our goal is to maximize clinical benefits for women in the face of competing health problems and limited resources. We will see that the solutions to this trade-off vary substantially as we move from the developed to the developing world.

We begin our analysis by going back to 1988 in the USA. At that time, Medicare did not provide coverage for cervical cancer screening, even though the elderly accounted for 40% of cervical cancer deaths. Researchers wanted to determine whether cervical cancer screening would be cost effective in this population [7]. To answer this question, a study was carried out to examine the costs and benefits of screening elderly low income women; researchers adopted a societal perspective for all costs and benefits. The new technology to be assessed was the Pap smear and the alternative considered was no screening. To calculate the effectiveness and the costs of the new technology, researchers combined data from a real clinical trial with projections of future costs and benefits from an economic model. All cost and benefit calculations used a 5% discount rate. The results indicated that the use of Pap smear screening would be cost saving in this group of women;

Projecting costs and benefits into the future: Markov models

Cost effectiveness analysis can be carried out using real data from clinical trials or by using mathematical models to project costs and benefits. Since we are usually interested in assessing costs and benefits over a patient's lifetime, most studies use the model-based approach because it provides a rapid, inexpensive way to compare the impact of many different strategies. Markov modeling is a mathematical technique that allows us to follow a cohort of patients over a defined period of time. In a Markov model we simulate the health of a group of patients over a defined period of time. At short time intervals, we use a probabilistic approach to predict each patient's health state, and we keep track of the costs and benefits associated with a medical intervention. If we want to compare the costs and benefits of several interventions, we can use a Markov model to simulate the effects of different interventions on several groups of patients, comparing the costs and benefits of each intervention. Markov modeling is a useful tool because it allows us to quickly and economically compare many different strategies; we can use the results of Markov modeling to help guide the design of clinical trials to test interventions that appear to be most cost-effective based on modeling studies.

The first step in building a Markov model is to define a cohort of patients, and to define all the different health states that the patients can experience. Let's take the example of cervical cancer screening. As the starting point for analysis, we can take a cohort of 18-year-old women who initially present for screening of cervical pre-cancer. Using a Markov model, we wish to follow this cohort of women through their lifetimes to determine the expected economic costs and health benefits of using different strategies for the screening, diagnosis and management of cervical pre-cancer. The figure above show the different health states (relevant to cervical cancer) that our patients can experience. During the course of the initial model cycle, women can either develop HPV or not, or die from causes other than cervical cancer. In the simulation, women move through the model, either progressing, persisting, or regressing from health state to health state. The health states include "NORMAL", "HPV", "LGSIL", "HGSIL", "EICC" (Early-Invasive Cervical Cancer), "LICC" (Late-Invasive Cervical Cancer), and "DEATH". This Markov process models the natural history of cervical pre-cancer, and in order for our results to be meaningful, we must have reliable information about the probabilities of disease regression and progression for each health state. In order to implement a Markov model, we must choose a time interval to update each patient's health within the simulation. For analyses of cervical cancer screening, often a cycle length of six months is selected because six months is the length of time that occurs between follow-up visits should a woman either require follow-up after treatment, or require follow-up after a false positive screening test. To compare the cost effectiveness of different interventions using this approach, we can either let women progress through the model with no screening, or we can superimpose screening, diagnosis, and treatment on the model, to reflect the alternatives for evaluation.

sensitivity analysis indicated that this result was true over a broad range of model parameters. The researchers recommended that screening should be made available to previously unscreened elderly women, which raised the ethical dilemma of whether Medicare should restrict a benefit to only those women who had never before been screened.

Let's examine the details of this analysis. In the study, a group of women aged 65 years and older who were seeking care in a municipal hospital outpatient clinic were invited to participate in a cervical cancer screening program; 816 women agreed to participate in the study, and 25% of these women had never had a Pap smear. The researchers monitored the number of cervical cancers and pre-cancers in the screened group; they found that 11 women had abnormal Pap smears. Follow up testing confirmed two patients with invasive cancer, two with high grade pre-cancer and three with low grade pre-cancers. They then added up the actual costs of screening and treatment in this group of women which amounted to $59,733.

To estimate the benefits of screening, they used a **Markov model** to predict the number of years of life gained by early detection. They assumed that patients with pre-cancer have a 100% survival rate; using known values of progression rates of cervical pre-cancer and mortality rates of cervical cancer, they calculated the life expectancy of women if screened and if not screened. The difference in these projections yielded the number of years of life gained through early detection. They found that screening saved a total of 30.33 years of life in this group of 816 women. Adjustments for the decrease in quality of life due to cervical cancer were also taken into account, in order to estimate the number of quality adjusted life years gained by the intervention. Because the quality of life was better with early detection, the intervention added 36.77 quality adjusted years of life in this group.

To estimate the costs associated with the alternative of no screening, researchers used the results of their Markov model to estimate the costs associated with treating the cancers that would have developed in the absence of screening in this same group of women. They estimated that it would have cost $107,936 to treat the same group of women if their cancers had not been identified until they sought treatment because of symptoms.

Comparing the cost of the new technology ($59,733) to the alternative of no screening ($107,936), shows that this cervical cancer screening program is COST SAVING! It saves medical costs by early detection of precancer and improves life expectancy. For every 100 Pap tests performed, the program saved 3.72 years of life and generated $5907 in savings by averting the need for future expensive cancer treatments [7].

Summary of cost effectiveness analysis of cervical cancer screening for low income, elderly women [7]

New technology:	Pap smear screening in low income elderly women
Alternative:	No screening
Number of Pap tests performed:	816
Costs of technology:	$59,733
Benefits of technology:	30.33 life years gained 36.77 QALYs gained
Costs of alternative:	$107,936 to treat invasive cancers that develop in the absence of screening
Net costs of intervention:	$59,733 − $107,936 = −$48,203 (intervention saves money)
Cost effectiveness:	SAVE $1311/QALY

In part, as a result of this study, Medicare benefits were extended to cover triennial screening with Pap smears in 1990 for all women with no upper age limit [8]. Is this a cost effective use of resources? The study we just considered examined only a one-time screen in a population with limited prior access to screening. Should these results be generalized to an entire population of elderly women? Should there be any upper age limit on screening? These questions were considered in a follow-up study by the same group that showed that programs targeting women who have never been screened will be cost saving and that triennial screening is the most cost effective approach [9]. For a population representative of the USA, where most women have had some screening, triennial screening will cost $2254 per year of life saved, whereas annual screening will cost $7345 per year of life saved. For women who have regularly been screened, costs of annual screening increase to $33,752 per year of life saved. Based on these results, researchers recommend that women over the age of 65 can stop screening if they have a history of negative smears.

In Chapter 10, we saw a number of new approaches to improve cervical cancer screening, including liquid cytology and HPV DNA testing. Cost effectiveness analysis can also help us compare the net benefits and costs of adopting these new technologies compared to screening with the conventional Pap test.

To compare the cost effectiveness of new interventions for cervical cancer screening, an economic analysis was recently reported. Strategies considered included: (1) the conventional Pap test, (2) liquid based cytology followed by HPV DNA testing in positive cases, (3) HPV DNA testing and conventional Pap in women over the age of 30, and (4) HPV DNA testing and liquid based cytology in women over the age of 30. In all cases, the frequency of screening was varied and the alternative was no screening. The costs and performance assumed for each technology are summarized in Table 11.2. A societal perspective was adopted [11].

Results of the analysis are shown in Figure 11.2. For each strategy, the figure shows the average life expectancy as a function of the per person total costs. Both costs and life years were discounted at an annual rate of 3%. Because life years have been discounted at this rate, the total discounted life expectancy in the

Table 11.2. *Performance and cost of new technologies for cervical cancer screening [11].*

Technology	Sensitivity	Specificity	Cost per Test
Liquid Cytology	84%	88%	$71
Pap	69%	97%	$58
HPV	88%	95%	$49
HPV + cytology	94%	93%	$120

absence of screening is only 28.7 years. With screening this rises to 28.78 years. Results are coded according to the frequency of screening: blue and red correspond to more frequent screening (annual or biennial) and black and yellow represent less frequent screening (triennial or every four years). Compared to no screening, screening every four years yields a large gain in life expectancy at a nominal cost. The cost per year of life saved using the conventional Pap smear every four years is $9400 (Point 1).

We see that as the frequency of screening increases, there are small gains in life expectancy, but large gains in cost. What is the marginal cost of these additional gains? To assess the incremental cost effectiveness, we divide the increase in life expectancy by the increase in cost. For example, as we go from screening with the

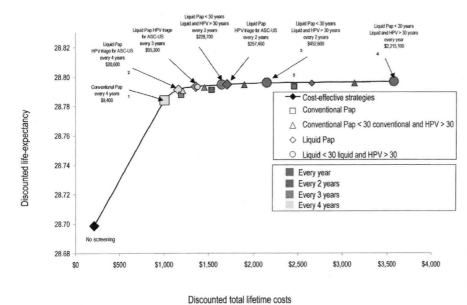

Figure 11.2. *Discounted life expectancy versus per person costs of cervical cancer screening for several different technologies and screening intervals [11].*

Impact of cost effectiveness study of cervical cancer screening for low income, elderly women

Before 1989, Medicare did not provide coverage for routine Pap tests. Representative Cardis Collins of Illinois introduced a bill to add coverage for Pap smears to Medicare every year for 15 years before this benefit was added in 1989. The economic analysis carried out by Jeanne Mandelblatt and colleagues was very influential in convincing legislators to vote for the bill. Dr. Mandelblatt recalls:

Courtesy of Jeanne Mandleblatt, MD, MPH, Georgetown University Medical Center.

"I previously worked in the Harlem community and other New York City neighborhoods that were very poor in resources: housing, healthcare, and other resources. The issue I wanted to address was whether we should screen older women for cervical cancer. The reason I, as opposed to someone else, did this is that I was the only person in the primary care clinic who knew how to do gynecologic examinations, and I was the first person in 10 years to observe that the examination tables had stirrups! This was the beginning of my life's work.

In the first few years of our screening program, the nurse practitioner and I screened more than 800 women. They were on average 74 years old and had largely been unscreened previously. As a result, we found that screening these women actually saved lives as well as healthcare costs (3.72 lives and $5,907 saved for every 100 Pap smears done)-an ideal program.

But then serendipity came into play. We were doing this work at a time when there was an explosion in the growth of the older population and members of Congress were receiving a lot of pressure from their older constituents to include preventive services.

Along I came with my Pap smear analysis and showed that if we were to screen the average elderly population at that point Pap smear screening would be a good buy. It would cost about $2,200 per year of life saved. Of great importance was that we could save money if we targeted screening to women who had not been screened previously, but the cost-effectiveness would worsen by more than 10-fold if screening were applied to women who had already been regularly screened.

What were our responsibilities and what were the issues that came out of this work? When we presented this work to the Office of Technology Assessment, we proposed considering cervical cancer screening as a targeted benefit and perhaps even including benefits to do outreach to women who have never been screened. The OTA said that under Medicare, benefits must be included for all (or no) women, so our recommendation could not be implemented.... The actual cost-effectiveness for Medicare might not be as favorable as it could have been if targeted to the highest-risk women" [10].

conventional Pap every four years (Point 1) to screening with liquid cytology followed by HPV DNA testing (Point 2), we must spend $20,600 for every additional year of life gained. This incremental cost effectiveness is simply the inverse of the slope of the line connecting two adjacent points on the graph. As the slope approaches zero (horizontal line), the incremental cost effectiveness decreases. For example, if we compare screening women over the age of 30 with a combination of liquid cytology and HPV DNA testing every two years (Point 3) with annual testing (Point 4), we see that we must spend an additional $2,215,100 for every year of life gained – not a cost effective use of resources.

Let's use this graph to compare the cost effectiveness of an annual conventional Pap smear (Point 5) with less frequent use of liquid cytology and HPV DNA testing every 2 years (Point 3). We see that the model predicts we will save more lives and spend less money with this new strategy. This type of analysis can be used by professional societies to help make policy recommendations about screening strategies.

We can use a similar strategy to make recommendations about screening in the developing world. Researchers compared the cost effectiveness of different strategies for cervical cancer screening in five developing countries: India, Kenya, Peru, South Africa, and Thailand [12]. The strategies considered included direct visual inspection with acetic acid (DVI – also called VIA), conventional Pap testing, and HPV DNA testing, compared to the alternative of no screening (Table 11.3). The number of clinic visits required was also varied including strategies that required separate visits for screening, colposcopic diagnosis, and treatment (three visit strategies), and separate visits for screening and treatment (two visit strategies). In the case of DVI, results are available immediately, so that a one visit strategy with screening and treatment on the same day was also considered. In each case, the frequency of screening was varied, including screening once, twice, or three times within a woman's life, and screening every five years.

Table 11.4 compares the demographic and economic characteristics of the five countries. The costs of clinical care were based on local labor costs. Despite wide variations in the incidence of cervical cancer and the availability of resources, results indicated that one or two

Table 11.3. *Performance of cervical cancer screening strategies considered in cost effectiveness analysis in developing countries [12].*

Intervention	Sensitivity	Specificity
VIA	76%	81%
Pap	63%	94%
HPV DNA	88%	93%

Discounted life expectancy

According to the CDC, the life expectancy at birth for women in the United States in 2004 is 80 years. If we discount the value of future life by 3% per year, what is the equivalent discounted life expectancy at birth?

To make this calculation, we simply add up the present value of all future years of life, discounting the value of future life years by 3% per year.

$$\text{Discounted life expectancy} = 1 + 1/1.03 + 1/(1.03)^2 + \cdots + 1/(1.03)^{79}$$

You can easily use a spreadsheet to show that the discounted life expectancy is only 31.1 years.

visit screenings with DVI or HPV DNA testing once per lifetime reduced the lifetime risk of cancer by approximately 25–36% at a cost of less than $500 per year of

Table 11.4. *Demographic and economic characteristics of five developing countries considered in a study of cost effectiveness of cervical cancer screening [12]. © 2005 Massachussetts Medical Society. All rights reserved.*

Demographic and Economic Characteristics of Five Developing Countries	India	Kenya	Peru	South Africa	Thailand
Total population (millions)	1016	30	26	44	61
Rural population (% of total)	72.34	64.11	27.23	44.51	68.86
Population density (# of persons/km²)	341.69	52.87	20.26	36.03	118.87
Women 35–39 yr of age (% of total pop.)	3.28	2.18	3.21	3.35	4.1
Literacy rate among women ≥15 yr of age (%)	45.39	76.02	85.24	84.56	90.52
Women employed in informal sector (% employed women)	86	83	58	58	54
Average hourly wage rate (2000 international dollars)	0.48	1.94	2.26	9.9	2.59
Female life expectancy at birth (yrs)	63.56	47.37	71.69	48.97	71.06
Cervical cancer incidence (age-standardized inc./100,000)§	186.5	200.1	238.3	174.8	129.6
HIV prevalence among adults (% of total pop.)	0.7	14	0.4	19.9	2.2
Per capita GDP (2000 international dollars)	2430	1005	4747	9486	6373

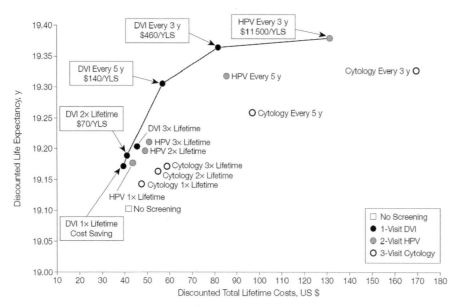

Figure 11.3. *Discounted life expectancy versus discounted total lifetime costs of screening for cervical cancer in South Africa from [13]. Copyright © 2001, American Medical Association.*

life saved, a figure that is less than the per person GDP in each of the five countries [12].

Let's examine the results from South Africa in more detail. Figure 11.3 shows the results of the analysis, graphing discounted life expectancy versus discounted total lifetime costs per patient. Note that the base discounted life expectancy in South Africa is only 19 years. Single visit strategies based on DVI are shown as black circles, two visit strategies based on HPV testing are shown as gray circles and three visit strategies based on cytology are shown as open circles. In general, single visit strategies save more lives and cost less than multi-visit strategies. As the frequency of screening is increased, more lives are saved but at higher costs. The incremental increase in cost effectiveness jumps from $140 per year of life saved for screening with DVI every five years to $460 per year of life saved when DVI is used every three years [13].

Summary

What can cost effectiveness analysis tell us about the best use of resources for cervical cancer screening? In developed countries, cost effectiveness decreases as the frequency of screening increases to more than every two or three years. Strategies that improve sensitivity (liqud based cytology, HPV testing) offer little benefit at drastically increased costs, unless the screening interval is increased. When the screening interval is increased to three to four years with these tests, results can be extremely cost effective. Small changes in specificity have a large influence on the cost effectiveness of strategies that involve frequent screening. Screening after age 65 is not cost effective for women who have had consistently negative screening results. For older women who have had no prior screening, initiating screening is very cost effective [6].

In the developing world, screening efforts should target women age 35 or older and should screen all women at least once in their lives. If a large fraction of women can be screened, then screening two to three times per lifetime could reduce lifetime cancer risk by 25–40%. If three lifetime screens are offered, efforts should target women aged 30–50, with a screen about every five years [6].

It is interesting to directly compare the cost effectiveness of screening in the developing and developed world. Figure 11.4 shows the reduction in lifetime risk of developing cervical cancer as a function of the lifetime costs per person for cervical cancer screening in South Africa and the United States. In the lower left region of the graph, increasing the frequency of screening in South Africa from once to twice to three times in a woman's lifetime shows a region of rapidly escalating benefits at low incremental cost relative to doing nothing. Three strategies from the USA (triennial,

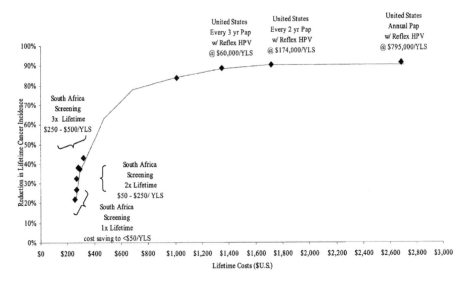

Figure 11.4. *Reduction in lifetime risk of developing cervical cancer versus total lifetime costs of screening. The graph compares strategies in a developing country (South Africa) to that of a developed country (United States). Used with permission from [14]. This article was published in* Virus Research, *Vol.* **89**, *S. J. Goldie, Health economics and cervical cancer prevention, pp. 301–9, copyright Elsevier (2002).*

Bioengineering and Global Health Project

Project task 7: Develop an alternative solution that meets the constraints you specified

This could be either an improvement to an existing device or a completely new strategy. For example, if your project is to develop alternative TB therapies, you might propose an alternate drug delivery mechanism (e.g. patches, controlled-release injections, implantable devices, etc.) that could be used to deliver six to nine months of TB therapy without requiring frequent physician visits or daily drugs. Remember, improvements are not limited only to changes in device design, but may also include increasing availability of the device to healthcare workers in the field or decreasing the cost to manufacture and distribute a health technology. Turn in a one-page summary of your new solution. The summary should include the following: (1) overview of your design, (2) scientific principles of the design, (3) expected benefits of the design, (4) potential risks associated with the design, (5) a plan to determine whether the design meets the constraints specified in Task 6.

biennial and annual screening) are on a region of the curve where the slope is quite flat and there are very small incremental benefits at high incremental costs associated with more frequent screening. This graph illustrates the very different challenges facing developed and developing countries. The challenge for the USA is to ensure that the price we pay for achieving small gains in life expectancy does not increase disparities in access to care. In contrast, the challenge for developing countries is to simply get on the curve. The extreme disparity in cost effectiveness can be illustrated by comparing how much additional life one can buy for an investment of $50,000. In the USA, an additional investment of $50,000 toward cervical cancer screening can buy 15 weeks of life. In South Africa, this same investment can buy more than 1000 years of additional life [14]!

Thus, the process of health technology assessment is critical for decision making in all countries. It can be used to help answer both clinical and policy questions and should be performed at all stages of the technology development process, to ensure appropriate and efficient allocation of scarce healthcare dollars.

Homework

1. What is a QALY? How much is our society willing to spend to gain one QALY?
2. Draw and label a graphical representation of health policy space. In which quadrant of the graph would the following interventions be located?
   a. Measles vaccinations for children.
   b. Anti-retroviral drug therapies for HIV infected patients.

c. Screening all women for ovarian cancer using serum CA-125 levels.

3. Explain why it is important to consider "additive procedures" when analyzing the cost effectiveness of new medical technologies. Give a specific example in which additive procedures influenced cost effectiveness.

4. Cardiovascular disease is the leading cause of death in the United States.

   a. What are four major treatments for coronary artery disease?

   b. Currently, coronary artery disease is diagnosed using coronary angiography, a painful and expensive technique. You have just developed a new, painless technique to diagnose coronary artery disease. The technique is also substantially less expensive than coronary angiography. Somewhat surprisingly, some health economists predict that the introduction of your new technique will actually cause healthcare expenditures to grow. Why might they make this prediction?

5. When detected early, ovarian cancer is curable for 95% of women. Unfortunately, in the majority of cases, ovarian cancer is not detected until widespread metastasis has occurred. In this circumstance, ovarian cancer is fatal approximately 63% of the time. The American Cancer Society estimates that there will be about 25,580 new cases of ovarian cancer in the United States in 2004. About 16,090 American women will die of the disease in 2004. On average, 22 years of life are lost when a woman dies of ovarian cancer. You have developed a new blood test that can detect ovarian cancer in the earliest possible stages. Each blood test costs $200 to perform. There are 292 million Americans, approximately 70 million of whom are women over age 40.

   a. How much money would we spend annually if all women over age 40 were screened with this new test?

   b. Calculate the annual mortality rate of ovarian cancer without the use of the new test. Compare this to the expected annual mortality rate of ovarian cancer with the use of the new test.

   c. If the new test was used, how many years of life would be gained?

   d. If this test was administered annually to all women over age 40, how many $ would we spend per year of life gained?

   e. Based on your answer to part d, do you think this test would be adopted in the developed world? In the developing world? Explain your reasoning.

6. In the USA, HIV infection is often discovered at an advanced state, when patients seek treatment for complications due to AIDS. When detected early, HIV infection can be controlled using HAART. On average, early detection of HIV infection extends the life of HIV patients by 1.8 years. Recent articles in the February 10, 2005, issue of the *New England Journal of Medicine* argue that we should consider voluntary screening of all patients, particularly those in higher risk groups. In 2006, the CDC recommended that people aged 13–64 should be screened for HIV regardless of risk factors. In this problem, you will be asked to do some calculations to determine whether you agree with their conclusions. The prevalence of HIV infection in the general US population is 0.33%. There are 292 million Americans, 140 million of whom are over age 40. The cost of screening for HIV is approximately $2.

   a. How many Americans are infected with HIV?

   b. If the test was administered to all Americans, how much would we spend in testing?

   c. If the new test was used, how many years of life would be gained?

   d. If this test was administered to all Americans, how many $ would be spent per year of life gained?

   e. Based on your answer to part d, do you think this test would be adopted in the developed world? In the developing world? Explain your reasoning.

   f. Approximately one person per 200,000 individuals tested will have a false positive test result. In other words, their HIV test will be positive even though they do not have HIV. If we test all Americans, how many people will receive a false positive result?

7. You have developed a new technology that could detect pre-cancerous cells in the sputum. This technology can enable much earlier detection of lung cancer, reducing the fraction of lung cancer patients that die of their disease from 90% to 15%. Your test costs $100 to perform. Assume that on average, 18 years of life are lost when a person dies of lung cancer. There are 292 million Americans, 140 million of whom are over age 40. There are 173,770 new cases of lung cancer in the United States each year.

   a. How much money would we spend annually if all adults over age 40 were screened with this new test?

   b. Calculate the mortality rate of lung cancer without the use of the new test. Compare this to the expected mortality rate of lung cancer with the use of the new test.

   c. If the new test was used, how many years of life would be gained?

   d. If this test was administered annually to all adults over age 40, how many $ would we spend per year of life gained?

   e. Based on your answer to part d, do you think this test would be adopted in the developed world? In the developing world? Explain your reasoning.

References

[1] Littenberg B. Technology assessment in medicine. *Academic Medicine*. 1992 Jul; **67**(7): 424–8.

[2] Drummond MF. *Methods for the Economic Evaluation of Health Care Programmes*. 2nd edn. Oxford: Oxford University Press; 1997.

[3] Cantor SB, Ganiats TG. Incremental cost-effectiveness analysis: the optimal strategy depends on the strategy set. *Journal of Clinical Epidemiology*. 1999 Jun; **52**(6): 517–22.

[4] Bergus GR, Cantor SB, Ebell MH, Ganiats TG, Glasziou PP, Hagen MD, *et al.* A glossary of medical decision-making terms. *Primary Care*. 1995 Jun; **22**(2): 385–93.

[5] Cantor SB. Pharmacoeconomics of coxib therapy. *Journal of Pain and Symptom Management*. 2002 Jul; **24**(1 Suppl.): S28–37.

[6] Goldie S. A public health approach to cervical cancer control: considerations of screening and vaccination strategies. *International Journal of Gynaecology and Obstetrics: the official organ of the International Federation of Gynaecology and Obstetrics*. 2006 Nov; **94** Suppl. 1: S95–105.

[7] Mandelblatt JS, Fahs MC. The cost-effectiveness of cervical cancer screening for low-income elderly women. *Jama*. 1988 Apr 22–29; **259**(16): 2409–13.

[8] Power EJ. From the Congressional Office of Technology Assessment. *Jama*. 1990 Jun 13; **263**(22): 2996.

[9] Fahs MC, Mandelblatt J, Schechter C, Muller C. Cost effectiveness of cervical cancer screening for the elderly. *Annals of Internal Medicine*. 1992 Sep 15; **117**(6): 520–7.

[10] Hagen MD, Garber AM, Goldie SJ, Lafata JE, Mandelblatt J, Meltzer D, *et al.* Does cost-effectiveness analysis make a difference? Lessons from Pap smears. Symposium. *Medical Decision Making*. 2001 Jul–Aug; **21**(4): 307–23.

[11] Goldie SJ, Kim JJ, Wright TC. Cost-effectiveness of human papillomavirus DNA testing for cervical cancer screening in women aged 30 years or more. *Obstetrics and Gynecology*. 2004 Apr; **103**(4): 619–31.

[12] Goldie SJ, Gaffikin L, Goldhaber-Fiebert JD, Gordillo-Tobar A, Levin C, Mahe C, *et al.* Cost-effectiveness of cervical-cancer screening in five developing countries. *The New England Journal of Medicine*. 2005 Nov 17; **353**(20): 2158–68.

[13] Goldie SJ, Kuhn L, Denny L, Pollack A, Wright TC. Policy analysis of cervical cancer screening strategies in low-resource settings: clinical benefits and cost-effectiveness. *Jama*. 2001 Jun 27; **285**(24): 3107–15.

[14] Goldie SJ. Health economics and cervical cancer prevention: a global perspective. *Virus Research*. 2002 Nov; **89**(2): 301–9.

[15] Salinsky, E. "Clinical Preventive Services: When is the Juice Worth the Squeeze?" *National Health Policy Forum Issue Brief*. 2005; 806.

Technologies for treatment of heart disease

We have examined how new technologies can be used to prevent infectious disease. When prevention fails, we need effective methods to both detect and treat disease. In Chapter 10, we showed that new imaging technologies can play a critical role in improving the early detection of cancer. By identifying cancers and pre-cancers at an early stage, we provide an opportunity to intervene when the disease is most responsive to therapy. We now examine the use of technology to treat disease. Our focus is on cardiovascular disease, where the development of medical therapies, new surgical procedures and implantable cardiac devices (Figure 12.1) has led to dramatic reductions in cardiovascular mortality in developed countries over the past 50 years. Despite these advances, we will see that cardiovascular disease is still the leading cause of death in the United States. Moreover, the mortality of cardiovascular disease is rapidly increasing in developing countries. We will see that future advances in cardiovascular devices such as drug eluting stents, surgical robots, and implantable artificial hearts have the promise to further reduce cardiovascular mortality. However, these technologies are expensive and require infrastructure beyond that which is currently available in most developing countries. More cost effective treatments, together with a greater emphasis on the prevention of cardiovascular disease, are needed to address this growing burden of disease.

The global burden of cardiovascular disease

Almost 25% of the USA population – 61 million Americans – suffers from cardiovascular disease. Today, heart disease accounts for more than 40% of all deaths in the USA, with 950,000 Americans dying of heart disease every year. The costs of cardiovascular disease in the USA exceed $350 billion per year. More than $200 billion is spent annually on direct healthcare expenditures and $140 billion in productivity is lost each year due to premature death and disability.

Globally, cardiovascular disease was responsible for 16.6 million deaths in 2001, representing nearly 1/3 of global mortality. More than 80% of deaths due to cardiovascular disease occur in low and middle income countries, where access to interventional therapies is least available. By 2010, cardiovascular disease is predicted to be the leading cause of death in developing countries. Heart disease is also an important cause of global morbidity. Every year, more than 20 million people survive heart attacks and strokes, and the cost of caring for these patients is high [1].

The changing patterns of incidence of cardiovascular disease present a unique global challenge (Figure 12.2). At the beginning of the twentieth century, cardiovascular disease accounted for less than 10% of all deaths worldwide; today it accounts for nearly half of mortality in the developed world and one quarter of mortality

(a)

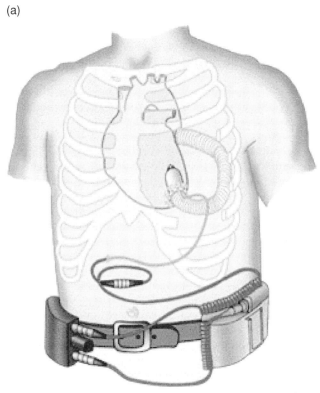

Copyright 2000, Texas Heart Institute

(b)

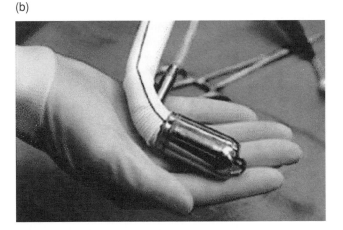

Figure 12.1. *(a) The use of small implantable cardiac devices which can be temporarily or permanently placed in the chest cavity can assist patients suffering from end stage heart failure. (b) The Jarvik 2000, is an example of this new type of device which assists the left ventricle in pumping oxygenated blood to the rest of the body. Texas Heart® Institute.*

32 million heart attacks and strokes per year ...only the tip of the iceberg

Undetected billions are at high cardiovascular risk... due to hypertension, diabetes, high lipids, tobacco use, physical inactivity and unhealthy diet

Figure 12.2. *Unhealthy lifestyles put billions of people at risk for cardiovascular disease and sudden death. World Health Organization,* Integrated Management of Cardiovascular Risk: Report of a WHO Meeting, 2002.

in the developing world [2]. The increasing incidence and mortality associated with cardiovascular disease is partly due to improvements in public health which have reduced the incidence and mortality of infectious disease; as people live longer lives, they are more likely

to suffer from the chronic diseases of middle and old age. However, the increasing global mortality of cardiovascular disease also partly reflects changes in lifestyle. The risk factors for developing cardiovascular disease include tobacco use, low levels of physical activity, inappropriate diet, high blood pressure, and high serum cholesterol levels. The rate of increase in the number of persons at risk for cardiovascular disease is most pronounced in developing countries.

In Chapter 4, we considered the epidemiologic transition that is occurring in many developing countries. We saw that, as societies develop and life expectancy increases, the causes of death shift from primarily infectious disease to chronic diseases [2]. This shift is driven by a combination of factors, including changes in diet, patterns of physical activity, and tobacco consumption [1]. Food market globalization and the increased availability of cheap, higher fat foods lead to increased caloric intake, while mechanization and urbanization lead to decreased caloric expenditure [2]. Resultant increases in the underlying risk factors of

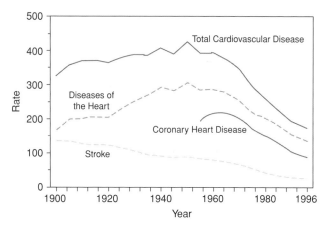

Figure 12.3. *Changing mortality rates of heart disease per 100,000 population throughout the epidemiologic transition in the United States [4].*

cardiovascular disease, such as obesity, serum cholesterol level, hypertension and diabetes, drive increases in the incidence of cardiovascular disease. In the USA, this epidemiologic transition took place in the first half of the twentieth century and the incidence of heart disease increased during this period (Figure 12.3). In the latter half of the twentieth century, advances in treatment of heart disease led to declines in the mortality of cardiovascular disease. From 1965 to 1990, mortality due to heart disease fell by 50% in Australia, Canada, France, and the USA, and fell by 60% in Japan [3].

Today, the epidemiologic shift that occurred in the USA in the early 1900s is being played out in many developing countries. Because of the global availability of processed foods, developing countries are entering a period of increased incidence of cardiovascular disease sooner in the epidemiologic transition than occurred in the USA. For example, in China the number of upper

income people consuming a high fat diet rose from 23% to 67% from 1989 to 1993 [3]. As a result, the increase in cardiovascular disease incidence is outpacing the availability of effective treatments that normally accompany the economic growth of the epidemiologic shift. Left unchecked, cardiovascular disease has the potential to severely reduce economic growth. Because cardiovascular disease strikes people in their mid-life years the economic impact is especially severe [1]. If the head of a household dies at an early age due to cardiovascular disease, there is a devastating impact on the entire family [2]. For example, in Bangladesh, a dependent child who loses an adult parent or guardian is more than 12 times likely to die themselves [2]. Of great concern is that fact that people in developing countries tend to die at an earlier age as a result of cardiovascular disease than do people in developed countries [3]. In developed countries, only one quarter of deaths due to cardiovascular disease occur in people under the age of 70. In contrast, nearly half of deaths due to cardiovascular disease in developing countries occur in people under the age of 70 years.

Types of CVD

Recall from Chapter 4 that there are two main forms of cardiovascular disease (CVD): ischemic heart disease and cerebrovascular disease (stroke). Ischemic heart disease can cause heart attack, and is the leading cause of death in the USA. It is also the leading cause of premature, permanent disability among working adults. Globally, ischemic heart disease is the second leading cause of death in adults aged 15–59 and the leading cause of death among adults over the age of 60 (Table 12.1).

Table 12.1. *Top ten causes of global mortality among adults aged 15–59 years and 60 + [5].*

Mortality – adults aged 15–59			Mortality – adults aged 60+		
Rank	Cause	Deaths (000)	Rank	Cause	Deaths (000)
1	HIV/AIDS	2279	1	Ischaemic heart disease	5825
2	Ischaemic heart disease	1332	2	Cerebrovascular disease	4689
3	Tuberculosis	1036	3	Chronic obstructive pulmonary disease	2399
4	Road traffic injuries	814	4	Lower respiratory infections	1396
5	Cerebrovascular disease	783	5	Trachea, bronchus, lung cancers	928
6	Self-inflicted injuries	672	6	Diabetes mellitus	754
7	Violence	473	7	Hypertensive heart disease	735
8	Cirrhosis of the liver	382	8	Stomach cancer	605
9	Lower respiratory infections	352	9	Tuberculosis	495
10	Chronic obstructive pulmonary disease	343	10	Colon and rectum cancers	477

Stroke is the third leading cause of death in the USA, and the second leading cause of death worldwide in adults over the age of 60.

Frequently, cardiovascular disease results in sudden death: 1.3 million Americans experienced a heart attack in 2006, and 40% of those who have a heart attack will die from their disease, and half of these deaths occur within one hour of symptom onset, before people reach the hospital [6]. Because so many of these deaths occur without warning, it is important to focus on opportunities to either prevent cardiovascular disease or to diagnose and treat ischemic heart disease before a heart attack occurs.

Opportunities to prevent CVD

Modifiable risk factors for developing cardiovascular disease (smoking, unhealthy diet, physical inactivity) lead to symptoms of hypertension, diabetes, obesity, and high serum cholesterol [7]. More than 50% of death and disability due to cardiovascular disease could theoretically be prevented through efforts to reduce high blood pressure, high cholesterol, obesity and smoking [1]. In particular the importance of smoking cessation programs could be enormous. In 1995, there were 1.1 billion smokers in the world, and tobacco contributed to three million deaths. In 2001, tobacco contributed to five million deaths, and this is expected to rise to ten million by 2020.

We can screen for those at risk for developing cardiovascular disease by measuring blood pressure annually: 15–37% of the world's population suffers from hypertension [7]. Hypertension is considered to be a blood pressure above 140/90 mm Hg, while pre-hypertension to be a blood pressure that stays between 120–139/80–89 mm Hg, and normal blood pressure should be <120/80 mm Hg for an adult. The higher (systolic) number is the pressure while the ventricles contract. The lower (diastolic) number is the pressure when the ventricles fill.

A 12–13 point reduction in blood pressure can reduce the risk of heart attack by 21%. Reducing systolic blood pressure by 11 mm Hg can be achieved by diet changes. We detect 32–64% of patients with hypertension in high income countries, but rates of detection are much lower in developing countries. However, although we detect most cases of high blood pressure in developed countries, it is frequently not treated adequately. Of those with high blood pressure in the USA, over 70% have uncontrolled hypertension (Figure 12.4). For those on treatment, only 13–29% have their hypertension adequately controlled. In African countries, control rates are only 2% [7].

Another screening tool is to monitor serum cholesterol levels every five years. Table 12.2 shows total cholesterol levels considered desirable, borderline and abnormal. A 10% drop in total cholesterol can reduce the risk of heart attack by 30%. There are two types of lipoproteins which carry cholesterol in the blood: LDL and HDL. Cholesterol carried by LDL, or low density lipoprotein, contributes to the development of atherosclerosis. Cholesterol carried by HDL, or high density lipoprotein, seems to protect against heart attack, possibly by transporting to the liver to be safely excreted. Unfortunately, 80% of patients with high cholesterol levels do not have them adequately under control.

The cardiovascular system

Figure 12.5 shows the anatomy of the heart and a diagram of blood flow in the cardiovascular system. As we have seen in Chapter 4, the heart is a four chambered pump. Oxygenated blood from the lungs travels to the left atrium via the pulmonary veins. It is pumped through the bicuspid valve to the left atrium, where it exits through the aortic valve to the ascending aorta and to the tissues of the body. Deoxygenated blood returning from the body enters the right atrium via the inferior and superior vena cava. Blood flows from the right atrium through the tricuspid valve to the right ventricle. The right ventricle delivers blood via the pulmonary valve to the pulmonary artery and the lungs. The atria and ventricles contract in succession. The ventricles relax and fill during ventricular diastole; during ventricular systole, the myocardium contracts, ejecting blood out of the ventricles.

Blood pressure

We measure blood pressure using a device called a sphygmomanometer. A cuff is placed on the upper arm and the cuff pressure is increased until it is higher than the systolic pressure; this stops blood flow into the arm. We gradually release the cuff pressure. At the moment when the cuff pressure is equal to the systolic pressure, blood will begin to flow again in the arm. The turbulent flow through the artery produces a sound called the Korotkoff sound. When the cuff pressure is equal to the diastolic pressure, the artery is no longer compressed and the Korotkoff sounds are no longer audible. Because results are somewhat subjective, people must be trained to use conventional sphygmomanometers. There are more than 500 automatic devices to measure blood pressure on the market, but only about 15 have been validated [7]. Developing an accurate, affordable device to monitor blood pressure in low resource settings could have a large impact on reducing mortality due to cardiovascular disease. The high capital cost, maintenance cost, and cost of replacing batteries can be problematic in low resource settings [7, 17].

The figures in this box are copyright © 2004 by Pearson Education, Inc. Reprinted with permission.

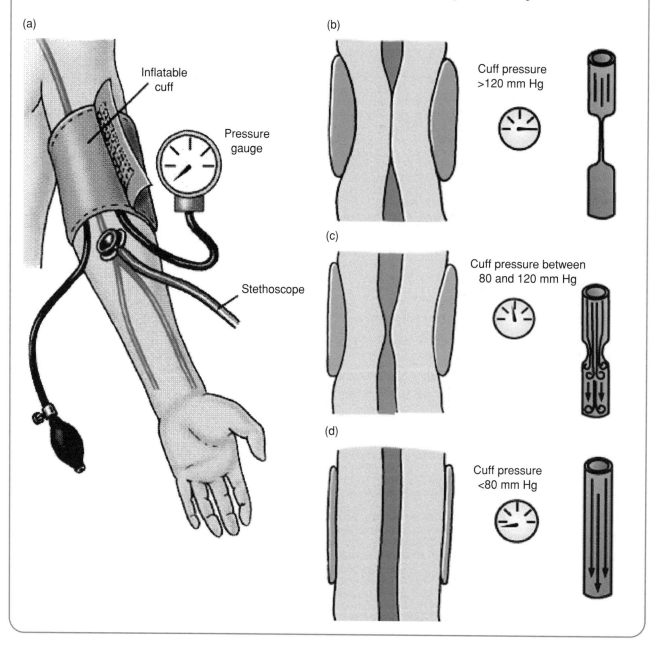

(a) Inflatable cuff — Pressure gauge — Stethoscope

(b) Cuff pressure >120 mm Hg

(c) Cuff pressure between 80 and 120 mm Hg

(d) Cuff pressure <80 mm Hg

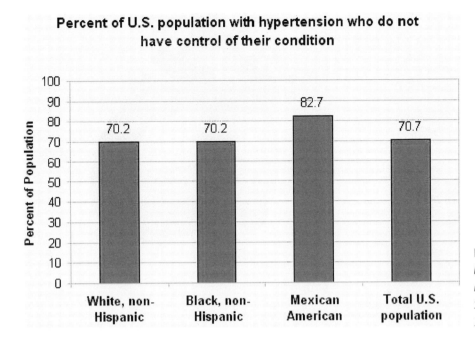

Percent of U.S. population with hypertension who do not have control of their condition

Figure 12.4. *Of those Americans with high blood pressure, more than 70% do not have it under adequate control. Source: MMWR, Vol. 54, No. 1, Jan 14, 2005.*

Table 12.2. *Diagnostic guidelines for levels of serum cholesterol.*

Interpretation of Serum Cholesterol			
	Total Cholesterol	**LDL**	**HDL**
Optimal		Under 100	Above 60
Desirable	Under 200	Under 130	
Borderline	200-239	130-159	
Abnormal	Over 240	Over 160	Below 40

American Heart Association, AHA recommendations-cholesterol levels.

Cardiovascular disease compromises the performance of the cardiovascular system. In order to monitor the severity and progression of cardiovascular disease and assess the impact of treatment, physicians typically examine several quantitative measures of heart performance.

The heart rate (HR) is the number of heart beats per minute; a normal heart rate is 60–90 bpm at rest. The stroke volume (SV) is defined as the amount of blood pumped by ventricle with each heart beat; a normal stroke volume at rest is 60–80 ml. The cardiac output

(CO) is the total volume of blood pumped by the ventricle per minute. We can calculate the cardiac output in terms of the heart rate and stroke volume as:

$$CO = HR \times SV. \tag{12.1}$$

A normal cardiac output at rest is 4–8 L/min. The total volume of blood in the circulatory system is normally about 5 L. Thus, each minute, the entire blood volume is pumped through the heart! In young adults during periods of maximal exercise, the heart rate can increase up to 180 bpm and the CO may increase to more than 25 L/min [8].

The ejection fraction (EF) is the fraction of blood pumped out of ventricle relative to total volume (at end diastole). If the total volume at the end of diastole is denoted as EDV, then the ejection fraction can be calculated as:

$$EF = \frac{SV}{EDV}. \tag{12.2}$$

A normal ejection fraction is approximately 60% or greater. We can measure the ejection fraction using a technique called echocardiography. In this procedure, ultrasound imaging is used to image the heart

(a)

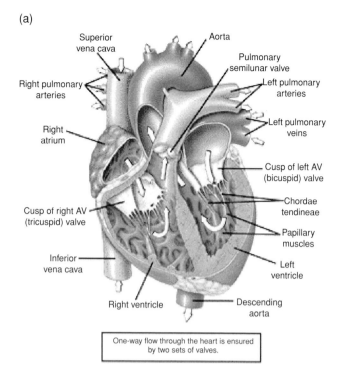

One-way flow through the heart is ensured
by two sets of valves.

(b)

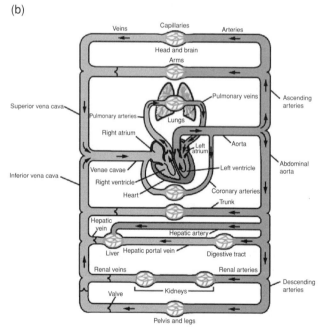

Figure 12.5. *The anatomy of the heart (a) and the circulatory
system (b). An animation of the cardiac cycle can be found at:
http://www.pbs.org/wgbh/nova/eheart/human.html. Copyright © 2004
by Pearson Education Inc. Reprinted with permission.*

throughout a cycle of contraction. We can use echocar-
diographic images to estimate the volume of the ventricle during systole and diastole; these values are then
used to calculate the ejection fraction.

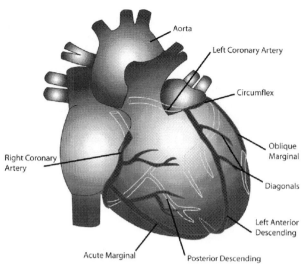

Figure 12.6. *The coronary arteries supply oxygenated blood to the
myocardium.*

Echocardiography

The decreased ejection fraction associated with
heart failure can be seen using echocardiography.
A normal echocardiogram can be found at:
http://www.kumc.edu/kumcpeds/cardiology/
movies/nllongecholabeled.html

An echocardiogram of a dilated heart in end
stage heart failure shows a much lower ejection
fraction and can be found at: http://www.kumc.
edu/kumcpeds/cardiology/movies/sssmovies/
dilcardiomyopsss.html

How do heart attacks happen?

Of the 32 million heart attacks and strokes that occur
each year, about 12.5 million are fatal. Some 40–75% of
heart attack victims die before reaching the hospital; this
is true in both developed and developing countries [7].
As we saw in Chapter 4, heart attacks result from coronary artery atherosclerosis. The heart muscle requires a
constant supply of oxygenated blood. This is provided
via the coronary arteries, which take their origin from
the aorta and branch to deliver oxygenated blood to
the entire myocardium (Figure 12.6). At their widest
point, the coronary arteries are approximately 3 mm in
diameter.

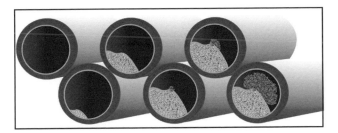

Figure 12.7. *When an atherosclerotic plaque ruptures, the thrombogenic necrotic core is exposed to the blood. A clot can rapidly form and occlude the vessel leading to a heart attack.*

(a)

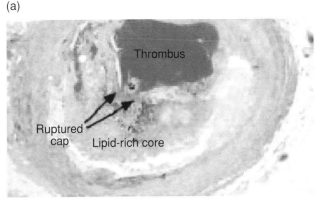

(b)

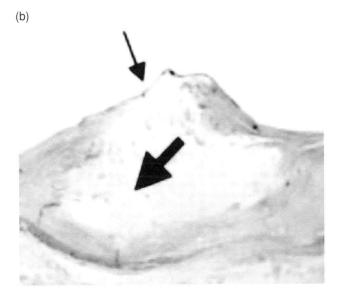

Figure 12.8. *(a) Histology of coronary artery from patient who died of MI. (b) Histology of vulnerable plaque. Small arrow indicates thin fibrous cap and large arrow indicates a large lipid pool and necrotic core [9].*

In response to injuries caused by hypertension, elevated serum cholesterol, or serum nicotine levels, atherosclerotic blockages can arise in the coronary arteries. These blockages can impinge on the lumen of the coronary arteries, reducing the supply of oxygenated blood to the heart. When the myocardium receives insufficient oxygen, angina (chest pain) can result. As more and more of the vessel lumen is compromised by an atherosclerotic narrowing, the frequency of angina increases, often occurring at rest in severe cases. As we saw in Chapter 4, if the fibrous cap covering the atherosclerotic plaque breaks, the contents of the necrotic core are then exposed to the blood. These contents are highly thrombogenic and result in the formation of a clot at the site of the lesion (Figure 12.7). The clot results in the complete cessation of flow of oxygenated blood to the myocardium served by that coronary artery, and the result is a heart attack – the myocardial tissue which is served by the occluded artery suffers oxygen deprivation and dies unless treatment restores oxygenated blood flow. The consequences of a heart attack depend on the amount of myocardial tissue damaged and the severity of the damage.

Not all patients with coronary atherosclerosis experience angina. In many patients, sudden coronary death is the first sign of cardiovascular disease. These vulnerable patients have vulnerable plaques (Figure 12.8) that are not sufficiently large to produce angina but are at high risk of breaking open and leading to thrombosis. The most common type of vulnerable plaque is one with a thin fibrous cap and the presence of inflammatory cells [9].

What symptoms does a typical patient undergoing a heart attack experience? Many heart attacks start slowly; symptoms may come and go (see following case study). Most heart attacks involve discomfort in the center of the chest which can be intermittent. The discomfort can resemble an uncomfortable pressure, squeezing, fullness, or pain. Discomfort in other areas of the upper body can also be present, including pain or discomfort in one or both arms, the back, neck, jaw, or stomach. Patients frequently experience shortness of breath, along with or prior to chest discomfort. Other symptoms can include breaking out in a cold sweat, nausea, or light-headedness.

Case study of myocardial infarction

Three months following his first visit to your office, Mr. Solomon presents to the ER in the early morning, with chest pain of one hour duration. Mr. Solomon describes the pain as being severe and "like someone was sitting on his chest." The pain, located "in the lower part of my breast bone," awakened him from his sleep. Although he tried to relieve the pain by changing positions in bed, sitting up and drinking water, it remained unchanged. He did not sleep well because "I had an upset stomach an acid-burning feeling." He attributed these symptoms to over eating and drinking at a Christmas party. He has no pain or discomfort in his arms but says he has an "acheness" in his left jaw which he attributes to "bad teeth." Physical examination reveals the patient to be anxious, pale, diaphoretic and in obvious discomfort. He is unshaven and accompanied by his wife. He tries to relieve his pain by belching. He coughs occasionally. Mr. Solomon says "the flu has been going around the office, and I've had a little cough and fever all week." Source: [10]

How do we treat acute myocardial infarction?

Because heart attacks result from the formation of a clot at the site of an atherosclerotic narrowing, one treatment is to immediately infuse a thrombolytic (agent that can dissolve blood clots). Tissue plasminogen activator (tPA) is one such agent and it is approved for use in certain patients having heart attack or stroke. A number of clinical studies have shown that tPA and similar clot-dissolving agents can reduce damage to heart muscle and reduce mortality. To be effective, they must be administered within a few hours after symptoms begin. They are typically administered through an intravenous (IV) line in the arm. Patients treated within 90 minutes after onset of chest pain are one-seventh as likely to die compared to patients who receive therapy at later times. There are risks associated with using thrombolyt-

ics, principally the risk of intra-cranial hemorrhage; this risk is highest in patients over the age of 70 years.

Act in Time: a video profile of three heart attack survivors illustrates the importance of seeking medical care quickly during a heart attack and can be found at: http://www.nhlbi.nih.gov/actintime/video.htm

How do we treat coronary atherosclerosis?

There are several procedures available to treat atherosclerotic lesions prior to plaque rupture and heart attack. Here we will consider three: coronary artery bypass grafting (CABG), percutaneous transluminal angioplasty (PTCA) and stents.

Open heart surgery

One method to treat patients with obstructions in the coronary arteries is to graft a new conduit for blood flow which bypasses the blockage in the coronary artery (Figure 12.9). This procedure is known as coronary artery bypass grafting (CABG) and generally requires open heart surgery. In the procedure, a new vessel is first harvested from a different location in the patient. For example, a section of the saphenous vein can be taken from the leg or a segment of the radial artery can be taken from the arm and used to create a bypass.

In the CABG procedure, the patient is prepared for surgery and placed under general anesthesia. Access to the cavity is gained by sawing through the breastbone (sternum) and inserting a chest retractor (Figure 12.10). The graft vessel is retrieved. The pericardial sac enclosing the heart is cut open to expose the heart. The beating heart must be stopped in order to carry out the procedure, so the patient must be placed on cardiopulmonary bypass. Blood is diverted through the **heart–lung machine**, the heart is stopped, and the graft is inserted. Circulation is then returned to the heart, and the incision is closed.

In 2001, 516,000 CABG procedures were performed in the USA. The procedure typically takes four to six hours and patients remain in the hospital for five to

How do we detect atherosclerosis?

We typically suspect coronary atherosclerosis when a patient experiences symptoms of angina. When patients have symptoms, we often perform a procedure called angiography to visualize the coronary arteries and determine whether atherosclerotic blockages are present. When detected early, patients can be treated with interventions to prevent a future heart attack. In coronary angiography, a radio-opaque dye is injected into the coronary arteries and X-ray movies of the heart are acquired to image the presence of the dye traveling through the coronary arteries.

The photos in this box show: (a) an angiography suite used to obtain coronary angiograms; (b) patient lying awake during the procedure; and (c) angiogram with focal obstruction.

(a)

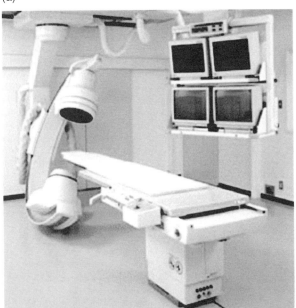

(b)

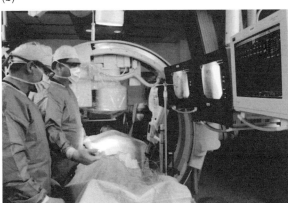

(c)

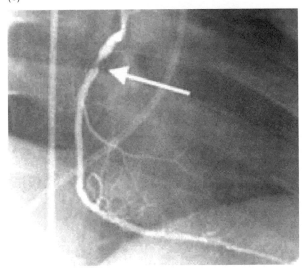

(a) Interventional room at Oregon Health and Science University Hospital with Siemens Asion Artis System. Photo courtesy of the Dotter International Institute. (b) Photo courtesy of Ozarks Medical Center. (c) Texas Heart ® Institute.

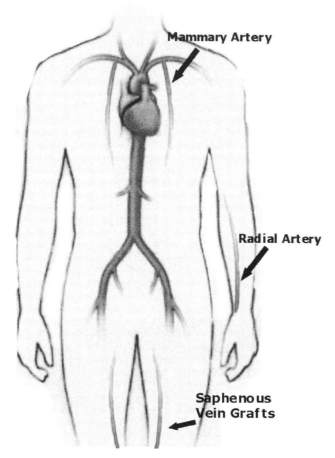

Figure 12.9. *Vessels frequently harvested for use in CABG. The MetroHealth System, Cleveland Ohio.*

(a)

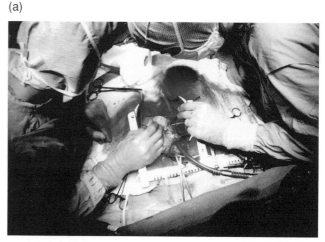

(b)

Coronary Artery Bypass

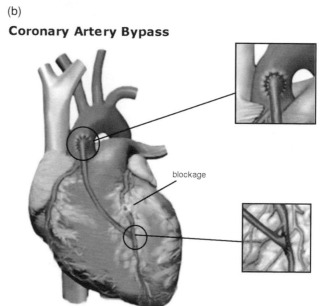

blockage

Figure 12.10. *(a) A chest retractor top is used to gain access to the cardiothoracic cavity during open heart surgery. Courtesy of NIH. (b) Detail of coronary artery bypass; a blood vessel graft is used to bypass the blockage. The MetroHealth System, Cleveland Ohio.*

seven days. The long term success of the procedure is excellent; most grafts remain open and functioning for 10–20 years [11]. There are risks associated with the procedure. About 5% of patients suffer a heart attack as a result of the procedure. About 5% experience stroke; the risk is greatest in those over 70 years old. The mortality rate of the procedure is 1–2%. A small portion of patients (1–4%) develop an infection in the sternal wound. More common side effects of the procedure include a "post-pericardiotomy syndrome," which occurs in about 30% of patients undergoing CABG, who report fever and severe chest pain which occurs a few days to six months after surgery. In addition, some people report memory loss and loss of mental clarity or "fuzzy thinking" following CABG. This is sometimes called "pump head," and is thought to be a side of effect of the cardiopulmonary bypass.

Surgical techniques to perform CABG continue to advance – even though it is generally being performed in a population that is increasingly older and sicker, outcomes continue to improve [11]. Recent surgical advances have made it possible to reduce the invasiveness of CABG. One important advance is the ability to perform CABG on the beating heart. A special stabilizer is used to hold portions of the beating heart sufficiently still to carry out the surgery. Because the procedure obviates the need for the heart–lung machine, it has the potential to reduce the incidence of pump head.

Heart–lung machine

(a)

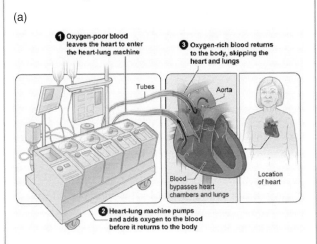

① Oxygen-poor blood leaves the heart to enter the heart-lung machine

③ Oxygen-rich blood returns to the body, skipping the heart and lungs

Tubes

Aorta

Location of heart

Blood bypasses heart chambers and lungs

② Heart-lung machine pumps and adds oxygen to the blood before it returns to the body

(b)

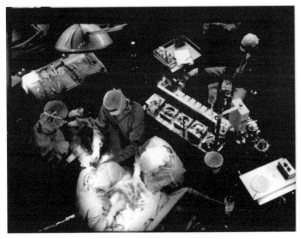

During CABG, the heart–lung machine is used to take over the function of the cardiac and respiratory systems. The heart–lung machine consists of a chamber that receives the blood from the body. Blood is pumped by machine through an oxygenator. The oxygenator removes CO_2 and adds oxygen. The pump then pumps this newly oxygenated blood back to the body. The machine is connected to the patient by a series of tubes that the surgical team places.

(Figure source: National Heart, Lung, and Blood Institute as a part of the NIH and the US Department of Health and Human Services.)

Off-pump CABG

A video showing an off-pump CABG procedure can be found at:

http://www.surgery.usc.edu/divisions/ct/videos-mpeg-offpumpcoronaryarterybypassgrafting.html

About 20% of CABGs can be performed **off-pump** [12]. There is some debate about whether off-pump CABG is a better alternative than CABG with bypass. Currently, researchers believe that patients can achieve a good outcome with either type of procedure [13]. Patients treated with off-pump CABG are likely to have less blood loss, less need for transfusion following surgery, less early neuro-cognitive dysfunction following the procedure and less renal insufficiency than patients treated with traditional CABG. However, it is harder to treat multi-vessel disease with off-pump CABG. The length of hospital stay, mortality rate and long term neurologic function and cardiac outcome appear similar in patients treated with or without bypass. Advances in robotic surgery have even made it possible to perform endoscopic CABG through the closed chest.

Angioplasty

While CABG is effective, it is a major surgical procedure. A less invasive alternative is percutaneous transluminal coronary angioplasty (also called balloon angioplasty). In this procedure, a balloon tipped catheter is advanced under angiographic guidance to the site of the blockage in the coronary artery (Figure 12.11). The balloon is then inflated, crushing the blockage and restoring patency to the artery. Balloon angioplasty cannot always be performed successfully. If the patient has diffuse disease rather than a local obstruction, it is difficult to treat with balloon angioplasty, which is best suited for treating focal lesions. If the patient has a near total occlusion, the balloon cannot be advanced across the lesion. If the lesion contains calcified disease, the force of the balloon may not be strong enough to restore patency. Yet, the biggest problem with balloon angioplasty is the problem

da Vinci Robotically Assisted Coronary Surgery System

One of the major disadvantages of coronary artery bypass graft surgery is the invasive nature of the procedure. Advances in robotic surgery have recently made it possible to perform endoscopic cardiac surgery. In 1998, the first CABG procedure was performed without opening the chest [14]. The da Vinci Surgical System is a new device which enables robotically assisted cardiac surgery.

The device consists of a **master console** which allows for remote control of micro-surgical instruments contained on a surgical cart placed beside the patient. The **surgical cart** contains robotic instruments which are mounted on three arms. A middle arm holds a stereo-endoscope that allows the surgeon sitting at the master console to view the tissue. Two parallel endoscopes separated slightly produce a stereo image pair which is displayed in 3D at the master console. The video image is magnified by ten times and presented to the surgeon. Left and right arms hold microsurgical instruments and are manipulated by the surgeon sitting at the master console. An additional video cart holds a light source, CO_2 insufflator, image processing system and computer monitor which allows the rest of the surgical team to view the procedure in 2D. The cameras on the middle arm are inserted into the patient's chest through a 12 mm hole. The instruments at the tips of the left and right arms are inserted into the chest cavity through two 10 mm ports. The left and right arms are articulated like the human wrist [15] and hold instruments which move with six degrees of freedom, following a scaled translation of the surgeon's movements. The computer controlling the system applies a filter to remove any tremor associated with the surgeon's movements. Thus, the surgeon can perform the procedure robotically without the need to open the chest. One of the most significant challenges is the lack of tactile

(a)

(b)

(c)

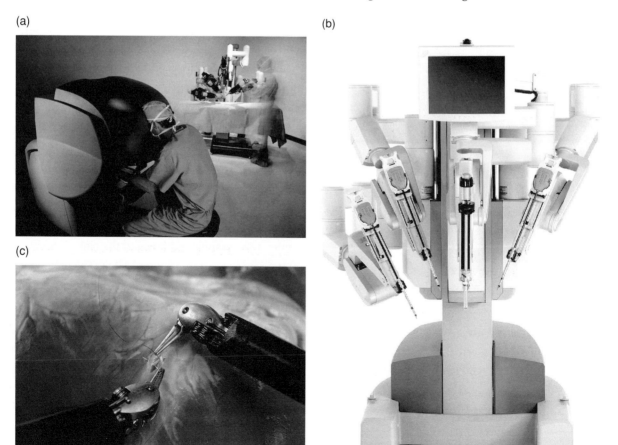

feedback which surgeon's normally use to guide their motion. In robotically assisted surgery, surgeons must learn to "feel with their eyes."

Endoscopic bypass can also be performed on the beating heart, without the need for cardiopulmonary bypass. The procedure was first done successfully in 2001. In endoscopic off-pump bypass, a fourth arm is used to introduce a stabilizer that is placed in contact with the tissue to be anastamosed. The patient is placed on anesthesia. The left lung is deflated and the cameras are inserted in between the ribs at the fifth intercostal space. The thoracic cavity is insufflated with CO_2 gas to create space to manipulate the surgical instruments. The two arms containing the robotic instruments are inserted between the third and seventh intercostal spaces. The most difficult part of the off-pump endoscopy bypass procedure is the **anastamosis of the graft vessel** to the diseased coronary artery. The lack of tactile feedback, combined with motion of the heart complicates this delicate portion of the procedure. Operating times are long for the anastamosis – about 20 minutes, which is three to four times longer than conventional CABG.

Recently, a series of 13 patients was reported in which 11 patients underwent off-pump endoscopic cardiac bypass to treat single vessel disease [15]. The mean stay in the ICU was 1.2 days, and the mean hospital stay was 4.5 days. All lesions were patent three months post-procedure.

The figures in this box show (a) the master console, (b) the surgical cart, and (c) the anastomosis of the graft vessel [14, 15] in the da Vinci Surgical System CAGB procedure. The images are courtesy of Intuitive Surgical.

A video showing a closed chest CABG procedure can be found at:
http://www.hsforum.com/stories/storyReader$1537

of restenosis, or the recurrence of an atherosclerotic blockage at the site of treatment. Between 25% and 54% of patients treated with balloon angioplasty will develop restenosis, usually within six months of the procedure. Because of the high rate of restenosis, balloon angioplasty is usually accompanied by the placement of a stent.

Stents

In a stent procedure, an expandable metal grid is placed atop the balloon tipped catheter and advanced to the site of the blockage. When the balloon is inflated, the stent expands permanently, and is left behind at the site of treatment. The use of stents (Figure 12.12) reduces the rate of restenosis associated with balloon angioplasty. However, it does not eliminate the problem, and the main limitation of stenting remains restenosis [11].

Two recent clinical trials indicate that restenosis rates are even lower when drug eluting stents are used. These stents are designed to slowly release drugs which reduce the proliferation of vascular cells and are FDA approved to treat discrete new atherosclerotic lesions in vessels from 2.5–3.5 mm in diameter based on these trials.

Despite promising results, there is some concern that these early trials may have overstated the advantages of drug eluting stents because they compared the rate of restenosis of drug eluting stents to that of bare metal stents with thick struts, and these stents have much higher restenosis rates in small diameter lesions [16]. However, drug eluting stents are being widely adopted, fueled in part by patient demands due to wide media coverage of lower restenosis rates [11]. In 2006, more than 85% of coronary interventions in the USA were performed with drug eluting stents [16]. Drug eluting stents cost three to four times more than bare metal stents. Also, there is concern that they may lead to a rare but fatal side effect known as in-stent thrombosis. Stent thrombosis occurs when a blood clot forms in the stent before it has been re-endothelialized [9]. Since drug eluting stents delay the healing process, stent thrombosis may be more likely than with bare metal stents. There is some concern that the initial clinical trials of drug eluting stents may have been too small to see this complication which occurs in 0.4–0.6% of cases with bare metal stents. There is growing concern that the enthusiasm for drug eluting stents may exceed

(a)

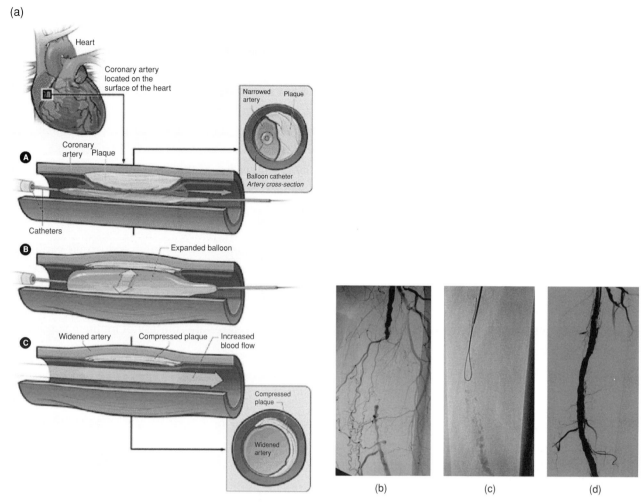

(b)   (c)   (d)

Figure 12.11. *(a) In a balloon angioplasty procedure a balloon tipped catheter is inflated at the site of an atherosclerotic blockage. The angiogram shows the site of the blockage before (b), during (c), and after (d) the procedure. Source: (a) National Heart, Lung, and Blood Institute as a part of the NIH and the US Department of Health and Human Services; (b), (c), and (d) Scott et al. 2007. Abstracts: Subintimal Angioplasty for the Treatment of Claudication and Critical Limit Ischemia: 3-year results. Southern Association of Vascular Surgery.*

the strength of scientific evidence supporting their advantages [16].

Comparison of treatment methods for atherosclerosis

The available treatments for atherosclerosis vary substantially in the degree of invasiveness, the long term success rate and cost. CABG is the most invasive procedure, and patients are typically hospitalized for four to seven days following the procedure. In contrast, angioplasty or stent placement require only a one to two day hospital stay. The rate of restenosis is lowest following CABG – generally, only 5–6% of patients treated with CABG experience restenosis, and it usually occurs after five or more years following the procedure. Rates of restenosis are highest following angioplasty (25–45%, usually within six months), while 15–20% of patients experience restenosis following stent placement, usually within six months.

Because angioplasty and stents are less invasive than CABG, the number of CABG procedures has fallen by 1/3 in the last decade – about 365,000 CABG procedures were performed in 2006. At the same time, the number of stent procedures has risen dramatically, to nearly a million in the same year [12]. Is the trend toward performing an increasing number of less invasive

(a)

(b)

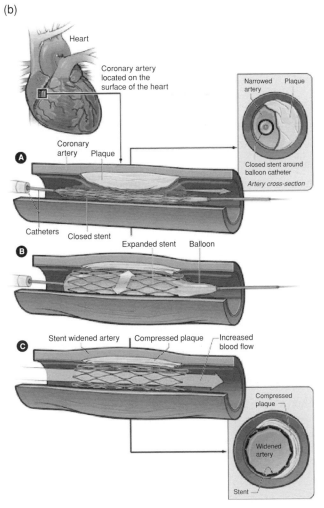

Figure 12.12. *The placement of a stent can reduce the restenosis rate. Image (a) courtesy of HeartHealthywomen.org. Image (b) courtesy of National Heart, Lung and Blood Institute as a part of the National Institutes of Health and the U.S. Department of Health and Human Services.*

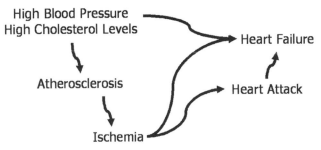

Figure 12.13. *Many forms of heart disease culminate in a process called heart failure.*

procedures cost effective? Angioplasty (approximate average cost $17,000) and stent placement (average cost $19,000) are less expensive than CABG, which has an approximate average cost $30,000. However, patients treated with angioplasty or stent more frequently require a second and even third procedure due to the high restenosis rates. For example, within five years, 20–40% of patients treated with balloon angioplasty have a second PTCA, and 25% have CABG due to restenosis. The costs of these additive procedures offset some of the initial savings. Immediately after treatment, the per patient costs of PTCA are only 30–50% those of CABG, but after one year this rises to 50–60%, after three years is 60–80%, and after five years is greater than 80% of the costs of treating with CABG.

Because of the additional costs associated with drug eluting stents, the price of CABG and stent procedures are now comparable for patients who have multiple blockages, costing about $30,000 [12].

What is heart failure?

Although there are many types of cardiovascular disease, they frequently culminate in a process known as heart failure (Figure 12.13). Heart failure occurs when the left or right ventricle loses the ability to keep up with amount of blood flow [8]. As a result, the heart cannot maintain sufficient cardiac output to meet the energy demands of the body. We have seen that cardiac output is a function of both heart rate and stroke volume; as the energy needs of the body increase, cardiac output rises as a result of increases in both the heart rate and the stroke volume. In order to increase the stroke volume, the contractility of the cardiac muscle must be increased. Myocardial contractility is a function of both

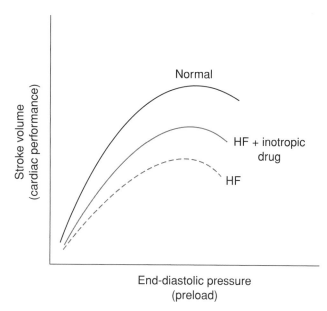

Figure 12.14. *Increases in the filling pressure during diastole give rise to an increase in cardiac contractility and stroke volume. This relationship is known as Starling's Law and provides for cardiac reserve. In heart failure, the curve shifts down and there is less cardiac reserve. Heart failure can be treated with inotropic drugs which increase myocardial contractility, but this typically does not completely restore cardiac function. From* The Merck Manual of Diagnosis and Therapy, *Edition 18, edited by Mark H. Beers. Copyright 2006 by Merck & Co., Inc., Whitehouse Station, NJ. Available at: http://www.merck.com/mmpe. Accessed June 7, 2007.*

the filling conditions of the heart (sometimes called the preload) as well as the resistance to blood flow out of the ventricle (sometimes called the afterload). During diastole, the ventricles fill with blood, a portion of which is pumped out during systole. When the filling pressure of diastole (preload) is increased, the end diastolic volume increases; as the myocardial cells are stretched, their contractility increases, so that the stroke volume rises. This relationship between filling pressure and stroke volume is known as Starling's Law (Figure 12.14). The increase in contractility that accompanies the increased filling pressure provides for substantial cardiac reserve to meet the increased energy demands associated with exercise. In heart failure, damage to the heart muscle results in a downward shift of this curve and a reduced cardiac performance at rest and decrease in cardiac reserve.

Heart failure is common. Five million Americans have heart failure, representing 1–2% of the adult population and it affects 6–10% of those over the age of 65 [17, 18].

Around 700,000 Americans die each year from heart failure [18]. There are more than 500,000 newly diagnosed heart failure patients each year in the USA [19]. It is the single most common cause of hospital admission for those over the age of 65. Heart failure is associated with significant morbidity. It reduces qualify of life more than any other chronic illness: 30–40% of patients with heart failure die within one year of diagnosis, and 60–70% die within five years of diagnosis. People aged 40 have a 20% chance of developing heart failure sometime in their lives, and a one in three chance of dying within a year of diagnosis [17].

The economic costs of heart failure are also substantial. The total costs of caring for patients with heart failure is about $60 billion per year; of this, $25 billion per year is spent on in patient hospital care for these patients [19, 20]. The costs to treat patients with heart failure consumes more than 2.5% of the healthcare budget in the UK [21].

In left ventricular heart failure, the left ventricle loses ability to contract and can't push enough blood into circulation. As the ability to pump decreases, blood coming into the left chamber from lungs "backs up," causing an increase in the pressure in the pulmonary veins. This increase in pressure causes fluid to leak out of the pulmonary capillaries and into the lungs (edema). Fluid in the lungs interferes with the ability to properly oxygenate blood, and leads to shortness of breath and persistent coughing or wheezing (Table 12.3). As left ventricular heart failure progresses, shortness of breath develops with less and less exertion, and patients sometimes experience shortness of breath even at rest. When patients lay down at night, shortness of breath increases.

Blood which has pooled in the lower extremities during the day is displaced to the thoracic cavity. The failing left ventricle cannot cope with the additional preload, and additional fluid extravasates into the lungs. As heart failure progresses, it becomes more and more difficult for patients to sleep due to the shortness of breath and coughing that occur in the recumbent position. The progression of heart failure is frequently tracked by monitoring the number of pillows that a patient must sleep with, and some patients find that they must sleep sitting up [20].

Table 12.3. *A summary of some common symptoms of heart failure and why they occur. Reprinted with permission © 2009. American Heart Association, Inc.*

Symptom:	Why it Happens:	People May Experience:
Shortness of breath (also called dyspnea)	Blood "backs up" in pulmonary veins (the vessels that return blood from the lungs to the heart) because the heart can't keep up with the supply. Causes fluid to leak into lungs	Breathlessness during activity, at rest, or while sleeping, which may come on suddenly and wake them up. Often have difficulty breathing while lying flat; may need to prop up upper body and head on pillows
Persistent coughing or wheezing	Fluid builds up in lungs	Coughing that produces white or pink blood-tinged phlegm.
Buildup of excess fluid in body tissues (edema)	As flow out of heart slows, blood returning to heart through veins backs up, causing fluid build up in tissues.	Swelling in feet, ankles, legs or abdomen or weight gain. May find that shoes feel tight
Increased heart rate	To "make up for" loss in pumping capacity, heart beats faster	Heart palpitations, which feel like the heart is racing or throbbing.
Confusion, impaired thinking	Changing levels of blood substances, such as sodium, can cause confusion	Memory loss and feelings of disorientation.
Lack of appetite, nausea	Digestive system receives less blood, causing problems with digestion	Feeling of being full or sick to their stomach.
Tiredness, fatigue	Heart can't pump enough blood to meet needs of tissues. Body diverts blood away from less vital organs (limb muscles) and sends it to heart & brain.	Tired feeling all the time and difficulty with everyday activities, such as shopping, climbing stairs, carrying groceries or walking.

(Symptoms of Heart Failure)

Table 12.4. *The four stages of heart failure. Reprinted with permission © 2007. American Heart Association, Inc.*

NYHA Class	Functional Capacity	Objective Assessment
Class I	Patients with cardiac disease but without resulting limitation of physical activity. Ordinary physical activity does not cause undue fatigue, palpitation, dyspnea or anginal pain.	No objective evidence of cardiovascular disease
Class II	Patients with cardiac disease resulting in slight limitation of physical activity. They are comfortable at rest. Ordinary physical activity results in fatigue, palpitation, dyspnea or anginal pain.	Objective evidence of minimal cardiovascular disease
Class III	Patients with cardiac disease resulting in marked limitation of physical activity. They are comfortable at rest. Less than ordinary activity causes fatigue, palpitation, dyspnea or anginal pain.	Objective evidence of moderately severe cardiovascular disease
Class IV	Patients with cardiac disease resulting in inability to carry on any physical activity without discomfort. Symptoms of heart failure or the anginal syndrome may be present even at rest. If any physical activity is undertaken, discomfort increases.	Objective evidence of severe cardiovascular disease

In right ventricular failure, the right ventricle loses the ability to contract. The increase in systemic venous pressure causes fluid extravasation out of the peripheral capillaries (peripheral edema). Swelling in the feet, ankles, and legs are common with right sided heart failure. (Table 12.3).

As ventricular function becomes increasingly impaired, a higher preload is required to maintain cardiac output at reasonable levels. The ventricles attempt to compensate by becoming more dilated, and larger (hypertrophy). Initially these changes help to compensate for decreased cardiac performance, but ultimately they increase ventricular wall stiffness and compromise cardiac performance. The severity of heart failure is usu-

ally classified using the New York Heart Association (NYHA) system (Table 12.4). There are four stages of heart failure; the mortality of heart failure increases with increasing stage. The one year mortality is 5–15% for patients with class II disease, 20–50% for patients with class III disease, and greater than 50% for those with class IV heart failure [21].

How do we treat heart failure?

While early stage heart failure can be treated with drugs to help restore ventricular contractility (see Figure 12.14) and diuretics to reduce fluid retention, late stage heart failure is much more difficult to treat. About half of patients with heart failure die suddenly due to

(a)   (b)

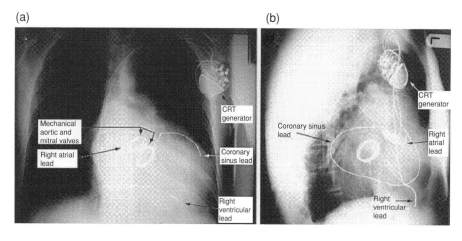

Figure 12.15. *(a) Chest X-ray of patient with a cardiac resynchronization therapy (CRT) device implanted and replacement mitral and aortic valves. (b) A CRT generator takes over the function of the intrinsic conduction system in the heart. From [21]. Reproduced with permission from the BMJ Publishing Group.*

cardiac arrhythmia [17]. Unfortunately, this is not treated well with anti-arrhythmic drugs. Instead, an implantable cardiac defibrillator can be used to automatically detect arrhythmias and treat them instantly. In chronic heart failure, conduction disturbances can also lead to dyssynchronous ventricular contractions. This can be treated by implanting a cardiac resynchronization device to act as a pacemaker (Figure 12.15). The cardiac resynchronization device mimics the intrinsic conduction system and ensures that atrial and ventricular contraction are synchronized [21]. The use of implantable defibrillators reduces risk of death associated with heart failure by about 25% [17].

Heart transplant

The mortality associated with heart failure is so high because there are a limited number of ways to treat end stage heart failure. Most are invasive, expensive and have major side effects. The treatment of choice for stage IV heart failure is **heart transplant** [18].

> ### Heart Transplant
> An interactive tutorial which guides you through **performing a heart transplant** can be found at: http://www.pbs.org/wgbh/nova/eheart/transplantwave.html [22].

Transplantation of the human heart was first introduced by Christiaan Barnard in the 1960s. In this procedure, a donor heart is implanted into the chest cavity of the recipient. After the patient has been prepared for surgery, the physician uses a scalpel to make an incision in the chest. A Stryker saw is then used to saw through the sternum. A chest retractor which spans the incision in the sternum is inserted and expanded to gain access to the thoracic cavity. A scalpel is used to cut open the pericardium and expose the recipient's heart. The patient is placed on cardiopulmonary bypass and circulation is diverted to the heart–lung machine. Once the patient is on bypass, the lower part of the patient's heart is removed, leaving behind the back walls of the right and left atria. The back of the donor heart is cut away so that it can be sutured in place to these atrial walls. Similarly, the patient's pulmonary artery and aorta are sutured in place to the donor heart. The patient is removed from bypass and the donor heart takes over circulation. The sternum and overlying skin are sutured closed.

The initial efforts to perform heart transplants were disappointing, primarily due to the patient's immune response to the transplanted organ which led to high rejection rates [18]. In the early 1980s, immunosuppressive drugs became available and survival improved dramatically. In order to minimize the likelihood of rejection, physicians attempt to match the MHC molecules on the surface of donor and recipient cells. Recall that T cells inspect MHC proteins and use this as a signal to identify infected cells. There are two types of MHC molecules: class I MHC molecules are found on all nucleated cells; class II MHC molecules are found on antigen presenting immune cells. Self-tolerance of the immune system is enabled because T cells which recognize class I MHC-self-antigens are destroyed early in development.

How to become an organ donor

1. Speak with your family about your decision to donate. Make sure they know about your wish to be an organ donor
2. Sign a Uniform Donor Card, and have two family members sign the card as witnesses
3. Carry the card in your wallet at all times. Download Uniform Organ Donor Card: http://www.organdonor.gov/get_involved/materials.htm#a2

If you haven't told your family you're an organ and tissue donor – you're not! Sharing your decision with your family is more important than signing a donor card. In the event of your death, health professionals will ask your family members for their consent to donate your organs and tissues. This is a very difficult time for any family, and knowing your wishes will help make this decision easier for them. They will be much more likely to follow your wishes if you have discussed the issue with them. Remember – signing an organ donor card is NOT enough. Discuss your decision with your family [23].

http://www.organdonor.gov
http://www.tdh.state.tx.us/agep/become.htm
http://www.lifegift.org/default.html
http://www.lifegift.org/UD_Organ_Donation.html
http://www.shareyourlife.org/

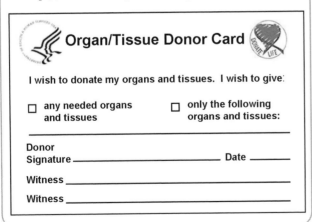

When certain types of organ transplants are performed, there is an attempt to match the MHC receptors between the donor and host. The greater the difference in peptide sequences of MHC receptors between donor and recipient, the stronger the immune response and the greater the chance of organ rejection. Donor matching involves comparing the MHC receptors, which are unique to each individual. Matching histocompatibility antigens of the donor and the recipient as well as possible can minimize the likelihood of rejection; however, this is not currently done for heart transplants due to time limitations in preserving the donor heart.

Cyclosporine, azathioprine and low-dose steroids are given to patients undergoing heart transplantation to reduce T-cell activation. Because patients taking these drugs are in an immuno-compromised state, they are susceptible to virus-related diseases, such as B-cell lymphomas (Epstein–Barr virus), squamous cell carcinomas (human papillomavirus), Kaposi's sarcoma (a herpes virus), and viral infections (cytomegalovirus). In addition, these patients can develop graft-versus-host disease. This is caused by alloreactive T-cells left within the donor tissue that can cause tissue damage in the recipient. Because of the serious risk of rejection, heart transplant patients often must undergo routine heart biopsies to monitor for this complication.

The risk of rejection is highest right after surgery. In one study, in the first year after transplant, 37% of patients did not experience any rejection episodes, 40% had one rejection episode, and 23% had more than one episode. To reduce the likelihood of early rejection, transplant patients are often given induction therapy, which is the use of drugs to heavily suppress the immune system right after transplant surgery. Patients must keep taking some anti-rejection drugs for the rest of their life.

Despite the availability of immunosuppressive drugs, heart transplantation has serious complications. Coronary atherosclerosis frequently develops in the transplanted heart [18]: 40% of transplant patients develop CAD requiring retransplantation within six years [24]. Drugs to reduce the likelihood of rejection can lead to frequent infections. Today, about 80% of heart transplant recipients are alive two years after the operation, and 50% survive five years.

Table 12.5. *Some important milestones in the history of developing a totally implantable artificial heart [19].*

History of the total artificial heart	
1813	LeGallois discovers importance of perfusion for organ survival
1937	Demikov sustains animal on artificial pump
1953	Gibbon develops heart–lung machine
1958	Atsumi creates hydraulic and roller pump models
1958	Liotta designs first total artificial heart prototype
1964	President Johnson starts the US Artificial Heart Program
1965	Kolff and Akutsu sustain calf for 44 hours on an artificial pump
1969	Cooley and Liotta implant first total artificial heart
1970	Kolff and Jarvik create Jarvik-7 prototype
1982	Jarvik-7 first implanted
2001	Abiomed begins AbioCor implants
2004	FDA approves Cardiowest (renamed from Jarvik-7) total artificial heart for bridge to transplant

Reprinted from [19] with permission from Elsevier.

Today, there are about 2000 heart transplants performed each year in the USA. The number of patients awaiting heart transplant is many times higher (about 50,000–70,000) [18]. Only about one in 24 patients who needs a transplant actually receives a donor heart, primarily due to a lack of supply of donor hearts [25].

Totally implantable artificial heart

Since the supply of donor hearts is limited, there is substantial interest in developing an artificial heart. In fact, the concept of a total artificial heart was first suggested in the early 1800s (Table 12.5). However, work did not begin in earnest until the 1950s after Gibbon demonstrated with the heart–lung machine that an artificial device could mimic function of the heart [18].

(a)

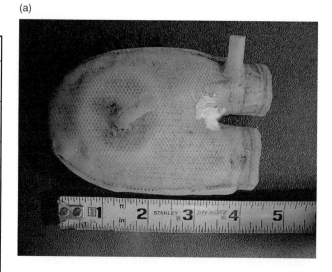

(b) 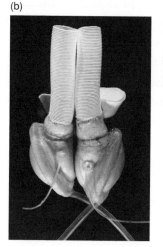 (c)

Figure 12.16. *History of the artificial heart. (a) 1958: this heart, designed by Drs. Willem Kolff and Tetsuzo Akutsu, was made of polyvinyl chloride. It was used to sustain a dog for 90 minutes. Courtesy Duare Ausherman, Department of Artificial Organs, Cleavelend Clinic. (b) 1969: Dr. Domingo Liotta designed the first artificial heart to be implanted in a human as a bridge to transplant. The patient survived for three days with the artificial heart and 36 hours more with a transplanted heart. © Bettmann CORBIS. (c) 1982: Drs. Willem Kolff, Donald Olsen, and Robert Jarvik designed the Jarvik-7 heart. This device was the first to be implanted in a human as destination therapy. Courtesy of the University of Utah and the National Heart, Lung, and Blood Institute as a part of the National Institutes of Health.*

From an engineering point of view, developing a simple pump to propel blood through the body seems like a minor task, but it has taken almost 100 years to accomplish (Figure 12.16). Why has it been so difficult to achieve this goal? Let's consider the engineering design requirements in detail. An artificial heart must be capable of providing an average flow rate of 6 L/min, at a mean arterial pressure of 120 mm Hg, at a beat rate of not more than 120 beats per minute, with a filling pressure of 20 mm Hg. It must operate continuously without stopping for a minimum of 10 years, beating 40 million times per year. Owing to lack of space in the thoracic cavity, it is difficult to design in redundant, failsafe features. The artificial heart must not damage red blood cells or platelets. It should respond to biologic demand and alter output according to the patient's needs. It must not produce blood clots that can lead to stroke. It must operate in people of a variety of sizes from children to adults. It must operate in a warm saline environment without corrosion. It must efficiently use power and minimize the amount of heat generated because surrounding tissues are damaged at temperatures above 41°C [26].

In 1964, the NHLBI set a goal of designing an artificial heart by 1970 [18]. It was initially estimated that this effort would require a total budget of $300 million, but only $100 million was allocated [19]. The design goals

---

Controversy surrounding the first artificial heart implanted in a patient

On April 4th, 1969, Haskell Karp, a patient suffering from class IV heart failure, became the first human to have an artificial heart implanted. Mr. Karp consented for surgery and possible use of an artificial heart if he could not be weaned from bypass [19, 34]. On April 4, he underwent surgery. When he could not be removed from bypass, an artificial heart was implanted.

Surgeon Denton Cooley performed the operation.

Following the procedure, Dr. Cooley reported that, "Mr. Karp has regained organ function indicating that the mechanical heart is feasible."

Karp survived several days with the artificial heart. A donor heart was located, and a human heart transplant was performed; however, Karp died 14 hours later.

His family later expressed their concerns about the media attention. Karp's wife, Shirley Karp, said "He could not say anything. I don't think he was really conscious. One day they removed the tube from his throat, they put a sheet over all the apparatuses in back of him and had the media take their pictures. Immediately after this was done they put back the tube and opened up everything that had closed up."

The heart that Karp received was developed by a team of scientists led by Dr. Michael Debakey. At that time, Debakey's group was testing the artificial heart in animals. Dr. Liotta was the principal scientist developing the artificial heart, and had made a prototype that worked in calves. Two air driven diaphragms served as the pump. The device was made of Dacron and was driven by an external power source. Dr. Liotta proposed that the heart be implanted into a patient, despite poor results in animal trials (at the time 4 of 7 calves receiving the heart died after implant). Debakey rejected this proposal. In frustration, Liotta secretly went to Dr. Cooley who agreed to implant the heart into a patient. The IRB was not informed.

Dr. Cooley justified his actions, "Dr. Debakey seemed to show little interest in ever using it. Dr. Liotta thought he was just wasting his years in a laboratory. The time had come to really give it a test and the only real test would be to apply it to a dying patient. In those days I didn't feel like we needed permission. I needed the patient's consent. I think if I had sought permission from the hospital, I think I probably would have been denied and we would have lost a golden opportunity."

Dr. Debakey later said, "I was in Washington when I read in the morning papers about the use of this artificial heart. I was shocked. I didn't know he had taken it from the laboratory."

The procedure sparked a major political and legal controversy which was widely covered in the media. The Liotta heart was never used again and it was 12 years before the next human artificial heart implant was performed [19, 34].

of this effort were to create a reliable artificial heart with a non-thrombogenic blood contacting surface, that was small enough to fit in the chest cavity, and give variable output using a pumping action that avoids blood trauma.

Figure 12.16 illustrates some of the early designs of artificial hearts, some of which were tested in animal models and in patients. In 1969, Dr. Denton Cooley implanted the first artificial heart in a patient undergoing resection of a left ventricular aneurysm [18]. The device was developed by Dr. Domingo Liotta. The Liotta heart supported the recipient for three days until a donor heart could be found. A donor heart was implanted successfully but the patient died shortly afterward.

The Jarvik-7 artificial heart was developed and implanted into patients first in 1982. This artificial heart was powered by a large external console [18]. Five implants were performed from 1982 to 1985. The longest implant sustained a patient for 620 days. Barney Clark, a dentist, was the first patient to receive the Jarvik-7 artificial heart; he survived for 112 days [19]. He became a symbol of medical technology and provided hope for patients with end stage heart disease. However, he suffered renal failure, seizures, device failure, and pneumonia before dying [24]. In general, patients receiving the Jarvik-7 were hemodynamically stable, but suffered many complications, including hemorrhage, stroke, and sepsis. Patients were tied to a large power source and could not leave the hospital. This heart used pulses of air to create the pumping action of the ventricles; as these air pulses were delivered, patients experienced a physical jolting that was uncomfortable. The electrical leads running from the heart to the external power source were spots for infection. To avoid this complication, early engineers thought they could design implantable artificial hearts with a radioactive source of power (Plutonium-238 fuel capsule). This approach was tested in vitro and in vivo in animal models, but was later abandoned due to public fears over use of radioactive fuel [20, 26].

In 1998, the NIH started a program to fund a totally implantable artificial heart [19]. Two groups received funding under this program, and the efforts led to a commercial device called the AbioCor artificial heart

> ### Debate over Jarvik-7 artificial heart
>
> "Is the artificial heart a valuable new device that can save the lives of tens of thousands of people in the US each year? Or is it instead something that will prolong dying and drain precious medical resources? [22]." These questions were the focus of a debate between artificial heart inventor Robert Jarvik and Daniel Callahan of the Hastings Center, a biomedical ethics research institute, shortly after the first clinical implantation of the Jarvik-7 artificial heart in 1985.
>
> Callahan estimated that the use of the artificial heart would add between $2.5–$5B to healthcare costs in the USA and that this money could be better spent on health education to prevent heart disease. Jarvik countered that $3B is spent each year on video games and that the use of the artificial heart could allow people to remain productive members of society [27].

(Figure 12.17). AbioCor is the first totally implantable artificial heart [18]. The unit, which is implanted into the chest cavity, weighs about 1000 g and is the size of a grapefruit. It includes two artificial ventricles, their corresponding valves, and a motor driven hydraulic pumping system. The heart also includes implantable control electronics. It has two batteries – an internal battery and an external, belt-worn battery. The internal battery is constantly recharged by the external battery using transcutaneous energy transmission. The internal battery can provide power to the device for 30 minutes, while the external battery provides power for many hours. Using this device, cardiac output can reach 8 L/min with a stroke volume of 60 ml [19].

The procedure to implant the AbioCor artificial heart is very similar to a heart transplant. Surgeons implant an energy-transfer coil in the abdomen. The chest is opened and a heart–lung machine is used to take over the patient's circulation. To implant the artificial heart, the surgeon removes most of patient's diseased heart, leaving behind the atria. The device is attached to the atrial chambers and to the aorta [18]. Because there is

Transcutaneous energy transmission

To recharge the internal battery of the totally implantable artificial heart, a new technology had to be developed to enable the transmission of electrical energy across the skin. This technology, known as transcutaneous energy transmission, relies on two electrical coils, one located outside the body, and the other located just underneath the skin. An external power source drives an oscillator which causes the external coil to radiate electrical energy. This energy is transferred through the skin and picked up by the implanted coil. The induced voltage in the implanted coil is converted from an alternating current to a direct current and used to charge the internal battery. The efficiency of power transfer is about 70% for coil separations ranging from 3 mm to 10 mm. Output currents range from 1.5 A to 3.6 A at voltages of about 10–25 V [14, 18]. Courtesy of Medical Science Monitor.

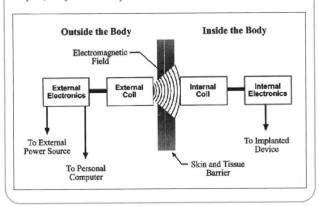

no connection to an outside power source, infection is greatly reduced.

In 2001, the AbioCor device was first implanted into a human as part of a clinical trial (Figure 12.18). The goals of this initial clinical trial were to determine whether AbioCor™ could extend life with acceptable quality for patients with less than 30 days to live and no other therapeutic alternative, and to learn what is needed to develop the next generation of AbioCor in order to treat a broader patient population. The criteria that a patient must meet to be eligible for the trial (inclusion criteria) included: bi-ventricular heart failure, greater than eighteen years old, high likelihood of dying within the next

I've been following the case of a ten-year-old boy all week. It's been very interesting and very sad.

He's ten. And HIV+. But he didn't get it vertically. (His mother is negative, meaning she can't have passed it to him.) His father and twin brother are also negative. This leaves us with very strongly suspecting sexual abuse. (There is a possibility as well that he got it from poor biohazard practices of a traditional healer. His mother swears up and down it isn't sexual abuse but that he also hasn't been to a traditional healer before he started getting sick a lot.)

He's also in severe physical pain from three arthritic joints. His right wrist, left ankle and his left knee are very swollen and warm. He can't move them because of the pain. He's completely non-ambulatory. He cries when you move him around.

We did a lot of tests – X-rays, blood tests, and a joint tap – but haven't gotten any useful results. Well, the X-rays show that the joints are not septic (because they aren't degrading). His white count isn't elevated (maybe. It's hard to know because we don't know what his baseline is . . . as that's affected by the HIV.) The lab didn't do any of the tests on the joint fluid because they don't have reagents. (They probably just chucked it but didn't tell anyone so the poor family waited around all day for lab results.)

Really, we don't know what's causing this and can't get helpful information. It's very sad.

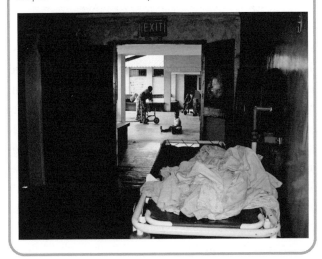

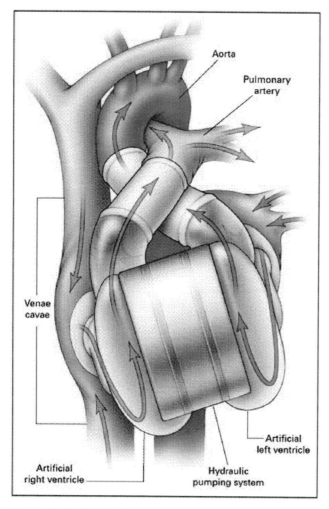

Aorta

Pulmonary artery

Venae cavae

Artificial right ventricle

Hydraulic pumping system

Artificial left ventricle

Figure 12.17. *The AbioCor totally implantable artificial heart. Jessup (2001). Mechanical Cardiac Support Devices – Dreams and Devilish Details,* **345**(20): 1490–93. Copyright © 2001. Massachusetts Medical Society. All rights reserved.

30 days, unresponsive to maximum existing therapies, ineligible for cardiac transplantation, and successful AbioFit™ analysis. In addition, patients were ineligible to participate in the study if they had any of the following criteria (exclusion criteria): heart failure with significant potential for reversibility, life expectancy >30 days, serious non-cardiac disease, pregnancy, psychiatric illness (including drug or alcohol abuse), and an inadequate social support system. The endpoints of the clinical trial were: all-cause mortality through 60 days, and quality of life assessments at 30-day intervals until death. The number of patients initially approved was five, with the possibility to expand to 15 patients in increments of five if the 60-day experience was satisfactory to the FDA.

As of 2006, implants had been performed in 14 patients. For the first seven patients, the 30-day survival was 71% compared with a predicted survival rate of 13%; 43% were still alive at 60 days. Five deaths occurred, one due to intra-operative bleeding at the time of implantation, two due to stroke, one due to pulmonary embolism and one due to multi-organ failure. No serious device complications were observed in the first seven patients. In June, 2005, AbioCor's request for a humanitarian device exemption was denied by the FDA due to concerns regarding the need for anti-coagulation and quality of life [19].

Despite the impressive engineering achievements of the AbioCor device, it does have some serious limitations. Chief among these are its large size. Because it is so large, the device fits in only 50% of men and 20% of women and does not fit in any children [19]. AbioCor costs about $75,000 now, excluding cost of surgery and follow up [18]. There is debate as to whether this expense is justified. Currently, the last year of life for many patients with congestive heart failure is expensive; hospitalization runs about $90,000/month. Surgical and hospital costs for heart transplant are at least $500,000 [18]. It is predicted that the AbioCor price will drop to about $25,000. If this device can give people five years of life, most experts agree that it will likely be adopted.

Left ventricular assist devices

Most patients with heart failure have failure of the left ventricle. Rather than replacing the entire heart, an alternative approach is to implant a pump to assist the weakened left ventricle. Such a device is called a left ventricular assist device (LVAD), and it consists of a small pump which is placed inside the left ventricle. The output of the pump is connected to the ascending aorta (Figure 12.19a,b). The LVAD acts to partially unload the left ventricle, and can slow the progression of heart failure and may even allow for some myocardial recovery. The LVAD can be implanted without the need to remove the patient's heart. In 1994, the FDA approved the use of the LVAD as a bridge to transplantation for those patients with end stage heart failure who were awaiting a donor heart. Self-contained LVADs were approved in 1998 [18]. Short term use of an LVAD can improve the

(a)

(b)

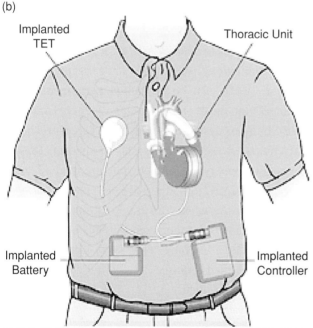

Implanted TET

Thoracic Unit

Implanted Battery

Implanted Controller

The AbioCor System has four main parts
that are implanted inside the body.

Figure 12.18. *(a) Photograph of the AbioCor totally implantable artificial heart. Reprinted with permission from Abiomed, Inc. (b) Model of the AbioCor device implanted in the chest cavity. Reprinted with permission from Abiomed, Inc. (c) Robert Tools, the first recipient of the AbioCor totally implantable artificial heart. Courtesy of John Lair, Jewish Hospital, University of Louisville Health Sciences.*

(c)

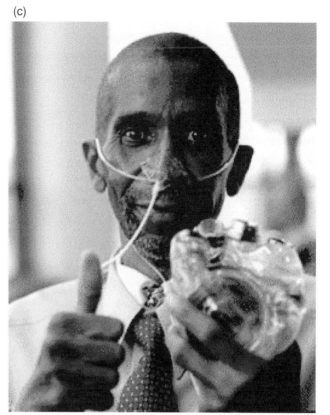

Figure 12.18. *(cont.)*

health of patients awaiting transplantation. In addition, it enables patients with end stage heart failure to remain in their homes with skilled nursing care (at a cost of $50/day) rather than needing to remain in the ICU at a cost of $5000/day [24].

There are two main types of LVADs: positive displacement pulsatile pumps and rotary continuous flow pumps (Table 12.6) [25]. The LVADS initially approved for clinical use were pulsatile pumps that mimic the cycle of diastole and systole in the heart. These LVADS contain mechanical parts that can wear out, such as ball bearings, diaphragms or valves [28]. The next generation of LVADs were based on pumps that deliver continuous flow (Figure 12.20). The advantage of this approach is that continuous flow pumps have fewer mechanical parts, are smaller in size, are quieter to operate, require less power to operate, and are less expensive compared to pulsatile pumps [28, 29]. Despite these advantages, there is debate about whether pulsatile flow is critical to maintain organ function. Patients who receive continuous flow LVADs do not

(a)

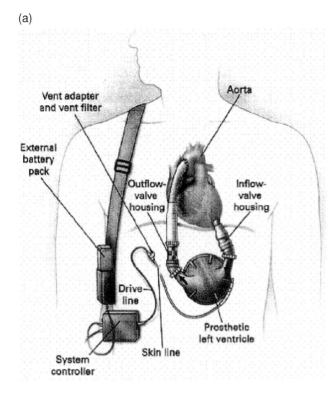

(b)

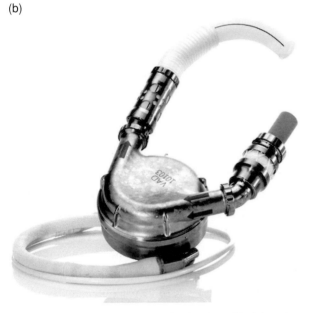

Figure 12.19. *Schematic diagram (a) and photo (b) of the HeartMate LVAD. (a) Rose et al. (2001). Long-term use of a ventricular assist device for end stage heart failure. NEJM. **345**(20): 1435–43. Copyright © 2001 Massachusetts Medical Society. All rights reserved. (b) Reprinted with permission from Thoratec Corporation.*

Table 12.6. *The two main categories of LVADs and some examples of each [29].*

Pulsatile flow pumps	Continuous flow pumps (rotary pumps)	
Positive displacement pumps	**Axial pumps**	**Centrifugal (radial) pumps**
• HeartMate	• DeBakey	• VentrAssist
• Novacor	• Jarvik 2000	• DuraHeart
• LionHeart	• HeartMate II	• HeartQuest
• SynCardia TAH	• Incor	• HeartMateIII
• AbioCor TAH		• Gyro Pump

From [29]. Reprinted with permission from the *Journal of Artificial Organs*.

have a pulse and you can't measure their blood pressure. On the other hand, pulsatile pumps fail more frequently and they are larger than axial flow pumps [25, 27].

Given the success of LVADs as a temporary bridge to transplantation, researchers have begun to study whether LVADs can be implanted permanently as a treatment for heart failure. This use of LVADs is referred to as destination therapy. A recent clinical trial compared the use of LVADs to medical therapy in patients who were not eligible for cardiac transplant. The REMATCH study (Randomized Evaluation of Mechanical Assistance for the Treatment of Congestive Heart Failure) randomized patients to receive an LVAD or medical therapy. This study used the Heart-Mate LVAD which required an external connection for power (shown in Figure 12.19). Patients receiving an LVAD had both improved survival (Figure 12.21) and quality of life. At one year, 48% of LVAD patients were still alive, compared to only 26% of those receiving medical therapy.

At two years, the survival rate in the LVAD group was 26%, compared to 8% for medical management [19]. However, patients receiving the LVAD had higher complication rates, with 28% of patients developing infection at the device site by three months and 42% of patients experiencing bleeding by six months, and a 35% device failure rate at two years [18]: 41% of

(a)

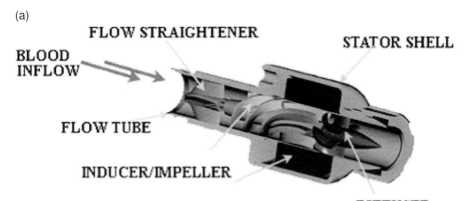

FLOW STRAIGHTENER   STATOR SHELL

BLOOD
INFLOW

FLOW TUBE

INDUCER/IMPELLER

DIFFUSER

(b)

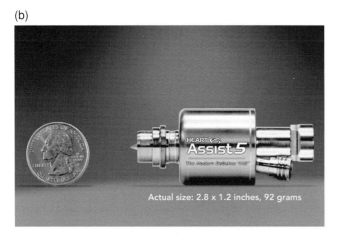

Actual size: 2.8 x 1.2 inches, 92 grams

Figure 12.20. *There are currently two types of continuous flow pumps in development: centrifugal and axial flow pumps. The Micromed DeBakaey VAD shown here is an example of an axial flow pump [29]. Courtesy of micromedcv.com. This device was developed jointly between NASA, Baylor and MicroMed Technology Inc. The LVAD is only 25 mm in diameter and 75 mm in length and requires less than 10 W of input power. The axial flow pump generates a rotational speed of 10,000 rpm and can deliver flow rates of up to 10 l/min. From [25].*

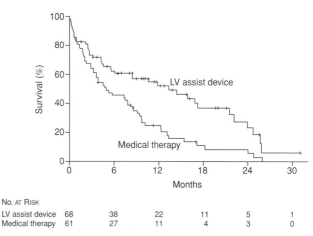

No. AT RISK

LV assist device	68	38	22	11	5	1
Medical therapy	61	27	11	4	3	0

Figure 12.21. *Results of REMATCH clinical trial. Rose et al. (2001). Long-term use of a ventricular assist device for end stage heart failure. NEJM. **345**(20): 1435–43.* © 2001 Massachusetts Medical Society. All rights reserved.

deaths in the LVAD group were due to sepsis, and 17% of deaths in the LVAD group were due to device failure indicating the challenges associated with mechanical circulatory assist devices [29].

Currently, there are three ways of thinking about the future clinical uses for LVADS. (1) They can be used as a bridge to recovery in patients who have heart failure. Temporary implantation of an LVAD can provide time for the myocardium to remodel and heal [30]. (2) They can be used as a bridge to transplant for those patients in end stage heart failure, sustaining them until a donor heart can be found or an artificial heart is available. (3) They can be used as destination therapy for patients

with end stage heart failure. As of 2005, the FDA had approved devices for the last two indications [19]. Based on the available clinical data, it appears that patients do better if LVADs are implanted early, before multi-organ failure begins. As a result of these trials, LVADs have evolved more towards being permanent devices. One challenge for the future is that as more people awaiting transplantation are sustained for longer periods of time on LVADs, the backlog of patients awaiting heart transplants will grow. In order to be successfully used as destination therapy, future LVADS will need to be substantially more durable [31].

Prevention of cardiovascular disease

We have seen the economic, technical and clinical challenges of treating end stage heart disease. Yet, the risk factors associated with heart disease are well known

HeartMate III

(a)

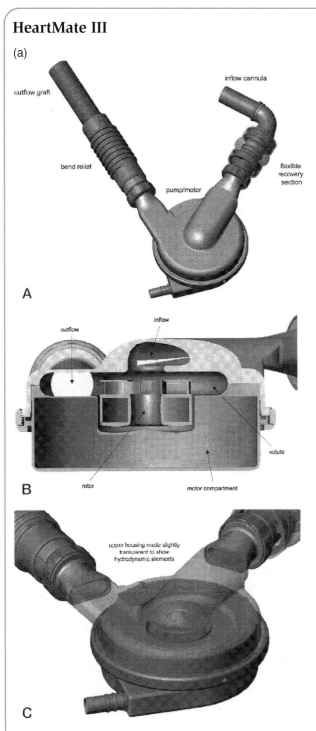

A

B

C

One continuous flow LVAD under development is the HeartMate III. The pump has no mechanical bearings, but instead uses a magnetically suspended rotor which eliminates contact between moving parts and extends wear life [28]. The device has an upper housing with inflow and outflow channels and a lower housing that contains the motor and rotor. The device diameter is 69 mm, and the height is 30 mm. It uses transcutaneous energy transmission to provide power. The device is made of titanium except for polyester grafts which connect to the patient's heart and aorta. The titanium surfaces are textured to promote the development of a layer of biological material at the surface upon contact with blood. This pseudo-intima reduces the need for anticoagulation and reduces risk of thrombus development. Blood flows through the device in a circular pattern [28].

The figures in this box are reproduced from Bourque, Kevin, Gernes, David B., HeartMate III: pump design for a centrifugal LVAD with a magnetically levitated rotor. *ASAIO Journal.* Volume **47**(4), July 2001, pp. 401–5.

(b)

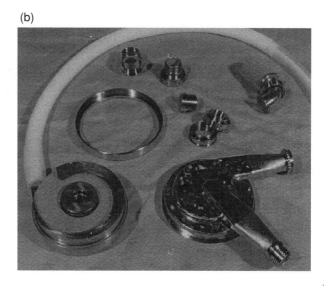

and many can be avoided or reduced through changes in lifestyle. There is growing awareness of the benefits of an increased emphasis on prevention of cardiovascular disease, particularly in developing countries facing a future epidemic of heart disease.

There are successful examples of governmental policies to combat heart disease. For example, the Polish government cut subsidies for and imposed taxes to raise the price of animal fats. In response, people cut their consumption of animal fats and used more vegetable oil. The result was a decrease in cardiovascular disease mortality in the 1990s [32].

Is prevention cost effective? We spend more than $100 B in the USA on coronary heart disease. Studies suggest that educational programs dedicated to smoking reduction cost less than $500 per year of life saved, and may even be cost saving [33]. Exercise programs probably cost $10,000 per year of life saved and less than $5000 per QALY saved. For people who actually enjoy exercise, then the health benefits of exercise measured in QALYs may be cost saving! Screening for hypertension and treating with drugs costs about $20,000 per year of life saved.

Despite the potential impact of prevention policies, it is frequently more difficult to get people galvanized around prevention efforts in developing countries. In developed countries, public health efforts commenced when the epidemic of cardiovascular disease was at its peak and there were few treatments available [3]. As a result, the public was receptive to these activities. In contrast, developing countries are grappling with the double burden of infectious and chronic diseases and community awareness of the dangers of cardiovascular disease is not high.

When implemented, prevention too often focuses on a single risk factor. However, it is much more effective to focus on integrated management of risk: for example, by simultaneously addressing hypertension, smoking cessation and diabetes. A barrier to implementation of such approaches in low and middle income countries is the lack of infrastructure. For example, the lack of basic equipment such as blood pressure measuring devices can make implementing programs to control hypertension impossible. In Nigeria, only 10% of facilities could measure blood lipid levels and 11% of healthcare facilities did not have devices to measure blood pressure. A recent study of hypertension patients showed that 98% of patients have to pay out of pocket for prescriptions. This cost burden is mainly borne by patients. The inability to pay for drugs was cited as a major reason for not taking prescriptions. In Pakistan, the cost of a monthly prescription for hypertension medications ranges from 50% to 200% of monthly per capita income. In moving forward to combat heart disease, in many ways, the challenge is simply to "use what we know [7]."

Bioengineering and Global Health Project
Project task 8: Construct a prototype for an in-class demonstration and design review
The audience for your presentation will be a scientific review panel from the World Health Organization. The panel will review the proposed designs and decide which efforts will receive funding to move into the development phase. Therefore, you must convince the panel of the severity of the problem as well as the efficacy of the proposed design. Your prototype does not need to actually function, but it should illustrate the scientific principles of the device as well as the ways in which it satisfies the design requirements. In constructing your prototype, we suggest that you use common household materials (plastic wrap, aluminum foil, cardboard, etc.) to illustrate how the device would work. You are limited to a total materials cost of $10. You should be prepared for questions from the review panel following your presentation.

Homework

1. When a sphygmomanometer is used to measure blood pressure, what two values are measured? Describe how the sphygmomanometer is used to perform these measurements.

2. If a person's cardiac output is 5.5 l/minute at a heart rate of 75 beats/minute, what is the stroke volume in ml? How many seconds will it take for this person's entire 4.9 l volume of blood to be pumped through the heart?

3. An avid runner is out for her morning jog before classes.
   a. Before she starts her jog she measures her pulse for 15 seconds. She counts about 17 beats in the 15 seconds. What is her heart rate in beats per minute?

b. This same jogger buys a heart rate monitor because she wants to see her heart rate through different parts of her run. When she is in the middle of her workout the monitor reads her heart rate as 119 bpm. If she were to feel her pulse at this same time for 15 seconds about how many beats would she feel?

c. Assuming that her heart stroke volume is about 70 ml, what would her cardiac output be in both parts a and b?

d. If her ejection fraction (EF) is 77%, using the stroke volume from above, what is her end-diastolic volume (EDV)?

e. On average, the normal stroke volume is 70 ml. The HR can drop to as low as 20 bpm when sleeping, while it is normally 70 bpm when awake. Compare the cardiac output when a person is at rest to when they are asleep.

f. After a heart attack, a patient's ejection fraction drops to only 20%. His EDV is 135 ml and heart rate is 90 bpm. Calculate SV and CO and compare these values to the normal values.

4. In our unit on heart failure, we considered what happens to the ejection fraction as disease progresses.

What is the ejection fraction?

What is a normal value for the ejection fraction?

How does the ejection fraction change as heart failure develops?

How do we measure the ejection fraction?

5. Cardiovascular diseases were responsible for 871,500 of the 2,428,033 American deaths in 2004. In 2004, death rates from CVD were 335.7 for white males, 448.9 for black males, 239.3 for white females, and 331.6 for black females. (Death rates are given per 100,000 population). From 1994 to 2004, death rates due to CVD declined 25 percent. This decline has been attributed to a combination of smoking cessation, weight loss, and blood-pressure control.

a. What percent of all deaths were due to cardiovascular diseases?

b. Calculate the number of deaths for each demographic group due to cardiovascular diseases in 2004: white males, black males, white females and black females.

c. How do smoking cessation, weight loss, and blood-pressure control each contribute to reducing cardiovascular mortality?

6. Cardiovascular disease is the leading cause of death in the USA.

a. What are the two major forms of cardiovascular disease (CVD) and which form is more common in the USA?

b. Name three factors that increase the risk of CVD.

c. Name a major screening tool for CVD commonly performed during a routine physical. Describe the procedure for the test and specify levels indicative of CVD progression.

7. What are the early warning signs of a heart attack? Why is it important to seek treatment quickly in the event of a heart attack? How do these signs differ from heart failure?

8. A patient complains of mild chest pain lasting several minutes at a time and shortness of breath. He says that in just the past week the frequency of these symptoms has increased notably. You order a coronary angiogram.

a. Describe the procedure used to obtain this photograph and note any abnormalities.

b. Assume that the best films for angiography are acquired if the contrast agent is well mixed in the blood supply. Assume that this occurs after blood re-circulates ten times through the body. The perfusionist tells the X-ray technician that the patient is sedated and has a heart rate of 20 beats per minute. Estimate total blood volume, assume a slightly lower than normal stroke volume of 50 ml, and show a calculation to determine how long the technician should wait before taking the chest X-ray.

c. The physician tells the patient that his condition will require treatment and explain that he has three options: coronary artery bypass grafting (CABG), percutaneous transluminal coronary angioplasty (PTCA), or stent implantation. The patient has several questions; please provide answers to each.

   i. "If autologous tissue is used for the bypass, where do the vessels come from?"

   ii. "Which procedure will likely require the longest hospital stay?"

   iii. "Which procedure is least likely to result in restenosis one year after the surgery?"

   iv. "Which procedure has the highest initial cost?"

9. A 55-year-old man presents to the emergency department complaining of shortness of breath, swelling in his ankles, tiredness and chest pain. The physician is worried that the patient is suffering from dilated cardiomyopathy, which is a disease that causes the heart to become enlarged. An echocardiogram of the patient's heart is immediately performed. The physician reviews the results of the echocardiogram and concludes that the maximum volume of blood in the ventricle during a cardiac cycle is 140 ml and the amount of blood pumped out of the ventricle each cycle is 60 ml.

   a. Calculate the end-systolic volume (ESV) and the ejection fraction of the patient. Is the ejection fraction normal or abnormal? Explain your reasoning.

   b. Briefly define systolic failure, diastolic failure and pulmonary edema.

   c. Which side of the heart is more likely to be affected first in heart failure?

   d. Which type of failure do you think that the patient is experiencing? Explain your reasoning. (Hint: take a look at part c.)

   e. Explain why he is experiencing shortness of breath, tiredness and edema (ankle swelling).

10. A person dies of cardiovascular disease-related causes every 33 seconds. While the overall rate of heart disease in the USA has shown a promising decline in recent years, one type of heart disease, heart failure, is on the increase. Patients with milder forms can be treated with a number of medications. However, patients with more advanced disease have few options.

   a. List three treatment options that can be used in patients with advanced heart failure.

   b. What are the main advantages and disadvantages of each approach?

11. About 4000 people in the United States await heart transplants each year. The first problem all heart transplant recipients face if they survive surgery is donor organ rejection.

   a. Explain how the immune system rejects the donor heart by considering it a foreign invader.

   b. Describe the process of organ donor-matching.

   c. What are the requirements to become an organ donor? Which is the most important step?

12. Heart disease typically develops over a period of decades.

   a. Describe two approaches to prevent the development of heart disease.

   b. Has our society focused on prevention of disease or treatment of disease? Should this change?

13. Read the article below and answer the following questions.

March 21, 2004
New Studies Question Value of Opening Arteries
By GINA KOLATA
A new and emerging understanding of how heart attacks occur indicates that increasingly popular aggressive treatments may be doing little or nothing to prevent them. The artery-opening methods, like bypass surgery and stents, the widely used wire cages that hold plaque against an artery wall, can alleviate crushing chest pain. Stents can also rescue someone in the midst of a heart attack by destroying an obstruction and holding the closed artery open. But the new model of heart disease shows that the vast majority of heart attacks do not originate with obstructions that narrow arteries.

Instead, recent and continuing studies show that a more powerful way to prevent heart attacks in patients at high risk is to adhere rigorously to what can seem like boring old advice – giving up smoking, for example, and taking drugs to get blood pressure under control, drive cholesterol levels down and prevent blood clotting. Researchers estimate that just one of those tactics, lowering cholesterol to what guidelines suggest, can reduce the risk of heart attack by a third but is followed by only 20 percent of heart patients.

"It's amazing and it's completely backwards in terms of prioritization," said Dr. David Brown, an interventional cardiologist at Beth Israel Medical Center in New York. Heart experts say they understand why the disconnect occurred: they, too, at first found it hard to believe what research was telling them. For years, they were wedded to the wrong model of heart disease.

"There has been a culture in cardiology that the narrowings were the problem and that if you fix them the patient does better," said Dr. David Waters, a cardiologist at the University of California at San Francisco.

The old idea was this: Coronary disease is akin to sludge building up in a pipe. Plaque accumulates slowly, over decades, and once it is there it is pretty much there for good. Every year, the narrowing grows more severe until one day no blood can get through and the patient has a heart attack. Bypass surgery or angioplasty – opening arteries by pushing plaque back with a tiny balloon and then, often, holding it there with a stent – can open up a narrowed artery before it closes completely. And so, it was assumed, heart attacks could be averted.

But, researchers say, most heart attacks do not occur because an artery is narrowed by plaque. Instead, they say, heart attacks occur when an area of plaque bursts, a clot forms over the area and blood flow is abruptly blocked. In 75 to 80 percent of cases, the plaque that erupts was not obstructing an artery and would not be stented or bypassed. The dangerous plaque is soft and fragile, produces no symptoms and would not be seen as an obstruction to blood flow. That is why, heart experts say, so many heart attacks are unexpected – a person will be out jogging one day, feeling fine, and struck with a heart attack the next. If a narrowed artery were the culprit, exercise would have caused severe chest pain.

Heart patients may have hundreds of vulnerable plaques, so preventing heart attacks means going after all their arteries, not one narrowed section, by attacking the disease itself. That is what happens when patients take drugs to aggressively lower their cholesterol levels, to get their blood pressure under control and to prevent blood clots.

Yet, researchers say, old notions persist. "There is just this embedded belief that fixing an artery is a good thing," said Dr. Eric Topol, an interventional cardiologist at the Cleveland Clinic in Ohio. In particular, Dr. Topol said, more and more people with no symptoms are now getting stents. According to an analysis by Merrill Lynch, based on sales figures, there will be more than a million stent operations this year, nearly double the number performed five years ago. Some doctors still adhere to the old model. Others say that they know it no longer holds but that they sometimes end up opening blocked arteries anyway, even when patients have no symptoms.

Dr. David Hillis, an interventional cardiologist at the University of Texas Southwestern Medical Center in Dallas, explained: "If you're an invasive cardiologist and Joe Smith, the local internist, is sending you patients, and if you tell them they don't need the procedure, pretty soon Joe Smith doesn't send patients anymore. Sometimes you can talk yourself into doing it even though in your heart of hearts you don't think it's right."

Dr. Topol said a patient typically goes to a cardiologist with a vague complaint like indigestion or shortness of breath, or because a scan of the heart indicated calcium deposits – a sign of atherosclerosis, or buildup of plaque. The cardiologist puts the patient in the cardiac catheterization room, examining the arteries with an angiogram. Since most people who are middle-aged and older have atherosclerosis, the angiogram will more often than not show a narrowing. Inevitably, the patient gets a stent. "It's this train where you can't get off at any station along the way," Dr. Topol said. "Once you get on the train, you're getting the stents. Once you get in the cath lab, it's pretty likely that something will get done."

One reason for the enthusiastic opening of blocked arteries is that it feels like the right thing to do, Dr. Hillis said. "I think it is ingrained in the American psyche that the worth of medical care is directly related to how aggressive it is," he said. "Americans want a full-court press." Dr. Hillis said he tried to explain the evidence to patients, to little avail. "You end up reaching a level of frustration," he said. "I think they have talked to someone along the line who convinced them that this procedure will save their life. They are told if you don't have it done you are, quote, a walking time bomb."

Researchers are also finding that plaque, and heart attack risk, can change very quickly – within a month, according to a recent study – by something as simple as intense cholesterol lowering. "The results are now snowballing," said Dr. Peter Libby of Harvard Medical School. "The disease is more mutable than we had thought." The changing picture of what works to prevent heart attacks, and why, emerged only after years of research that was initially met with disbelief.

Early attempts to show that opening a narrowed artery saves lives or prevents heart attacks were unsuccessful. The only exception was bypass surgery, which was found to extend the lives of some patients with severe illness but not to prevent heart attacks. It is unclear why those patients lived longer; some think the treatment prevented their heart rhythms from going awry, while others say that the detour created by a bypass might be giving blood an alternate route when a clot formed somewhere else in the artery.

Some early studies indicated what was really happening, but were widely dismissed. As long ago as 1986, Dr. Greg Brown of the University of Washington at Seattle published a paper showing that heart attacks occurred in areas of coronary arteries where there was too little plaque to be stented or bypassed. Many cardiologists derided him.

Around the same time, Dr. Steven Nissen of the Cleveland Clinic started looking directly at patients' coronary arteries with a miniature ultrasound camera that he threaded into blood vessels. He found that the arteries were riddled with plaque, but almost none of it was obstructing blood vessels. Soon he began proposing that the problem was not the plaque that produced narrowings but the hundreds of other areas that were ready to burst. Cardiologists were skeptical.

In 1999, Dr. Waters of the University of California got a similar reaction to his study of patients who had been referred for angioplasty, although they did not have severe symptoms like chest pain. The patients were randomly assigned to angioplasty followed by a doctor's usual care, or to aggressive cholesterol-lowering drugs but no angioplasty. The patients whose cholesterol was aggressively lowered had fewer heart attacks and fewer hospitalizations for sudden onset of chest pain. The study "caused an uproar," Dr. Waters said. "We were saying that atherosclerosis is a systemic disease. It occurs throughout all the coronary arteries. If you fix one segment, a year later it will be another segment that pops and gives you a heart attack, so systemic therapy, with statins or antiplatelet drugs, has the potential to do a lot more." But, he added, "there is a tradition in cardiology that doesn't want to hear that."

Even more disquieting, Dr. Topol said, is that stenting can actually cause minor heart attacks in about 4 percent of patients. That can add up to a lot of people suffering heart damage from a procedure meant to prevent it. "It has not been a welcome thought," Dr. Topol said.

Stent makers say they do not mislead doctors or patients. Their new stents, coated with drugs to prevent scar tissue from growing back in the immediate area, are increasingly popular among cardiologists, and sales are exploding. But there is not yet any evidence that they change the course of heart disease.

"It's really not about preventing heart attacks per se," said Paul LaViolette, a senior vice president at Boston Scientific, a stent manufacturer. "The obvious purpose of the procedure is palliation and symptom relief. It's a quality-of-life gain."

a. Discuss the advantages of placing more emphasis on prevention of heart disease.

b. Discuss the challenges of testing the efficacy of prevention efforts.

c. Why do you think our society places so much more emphasis on treatment of end stage heart disease rather than preventive measures?

14. The heart-lung machine was developed in the 1950s. It enabled many types of cardiac surgery which were previously impossible.

a. Describe the function of a heart–lung machine.

b. What is the major disadvantage of using the heart–lung machine?

c. What are two new technologies which eliminate the need for use of the heart–lung machine?

15. List the major surgical steps involved in implanting an AbioCor total artificial heart.

References

[1] World Health Organization. *Cardiovascular Disease: Prevention and Control* [cited 2007 June 7]; Available from: http://www.who.int/dietphysicalactivity/publications/facts/cvd/en/

[2] Levenson JW, Skerrett PJ, Gaziano JM. Reducing the global burden of cardiovascular disease: the role of risk factors. *Preventive Cardiology*. 2002 Fall; **5**(4): 188–99.

[3] Reddy KS, Yusuf S. Emerging epidemic of cardiovascular disease in developing countries. *Circulation*. 1998 Feb 17; **97**(6): 596–601.

[4] Centers for Disease Control and Prevention. Achievements in Public Health, 1900–1999: Decline in deaths from heart disease and stroke – United States, 1900–1999. *MMWR*. 1999; **48**(30): 649–56.

[5] Beaglehole R, Irwin A, Prentice T. *The World Health Report 2003: Shaping the Future*. Geneva: World Health Organization; 2003.

[6] Borden WB, Faxon DP. Facilitated percutaneous coronary intervention. *Journal of the American College of Cardiology*. 2006 Sep 19; **48**(6): 1120–8.

[7] World Health Organization. *Integrated Management of Cardiovascular Risk: Report of a WHO Meeting*. Integrated Management of Cardiovascular Risk; 2002 9–12 July. Geneva: WHO; 2002.

[8] Merck. *Heart Failure*. 2005 November [cited 2007 June 7]; Available from: http://www.merck.com/mmpe/sec07/ch074/ch074b.html

[9] Waxman S, Ishibashi F, Muller JE. Detection and treatment of vulnerable plaques and vulnerable patients: novel approaches to prevention of coronary events. *Circulation*. 2006 Nov 28; **114**(22): 2390–411.

[10] Chandrasekhar A. *Patient with Chest Pain*. Loyal University Medical Education Network [cited 2007 June 7]; Available from: http://www.meddean.luc.edu/lumen/meded/mech/cases/case2/Case_f.htm

[11] Reul RM. Will drug-eluting stents replace coronary artery bypass surgery? *Texas Heart Institute Journal/from the Texas Heart Institute of St.* 2005; **32**(3): 323–30.

[12] Feder BJ. In the stent era, heart bypasses get a new look. *New York Times*. 2007 February 25.

[13] Sellke FW, DiMaio JM, Caplan LR, Ferguson TB, Gardner TJ, Hiratzka LF, *et al.* Comparing on-pump and off-pump coronary artery bypass grafting: numerous studies but few conclusions: a scientific statement from the American Heart Association council on cardiovascular surgery and anesthesia in collaboration with the interdisciplinary working group on quality of care and outcomes research. *Circulation*. 2005 May 31; **111**(21): 2858–64.

[14] Wimmer-Greinecker G, Deschka H, Aybek T, Mierdl S, Moritz A, Dogan S. Current status of robotically assisted coronary revascularization. *American Journal of Surgery*. 2004 Oct; **188**(4A Suppl.): 76S–82S.

[15] Mishra YK, Wasir H, Sharma KK, Mehta Y, Trehan N. Totally endoscopic coronary artery bypass surgery. *Asian Cardiovascular & Thoracic Annals*. 2006 Dec; **14**(6): 447–51.

[16] Tung R, Kaul S, Diamond GA, Shah PK. Narrative review: drug-eluting stents for the management of

restenosis: a critical appraisal of the evidence. *Annals of Internal Medicine.* 2006 Jun 20; **144**(12): 913–19.

[17] McMurray JJ, Pfeffer MA. Heart failure. *The Lancet.* 2005 May 28–Jun 3; **365**(9474): 1877–89.

[18] Zareba KM. The artificial heart – past, present, and future. *Medical Science Monitor.* 2002 Mar; **8**(3): RA72–7.

[19] Gray NA, Jr., Selzman CH. Current status of the total artificial heart. *American Heart Journal.* 2006 Jul; **152**(1): 4–10.

[20] Zevitz ME. Heart failure. *eMedicine.* 2006 June 15; www.emedicine.com/med/topic3552.htm.

[21] Patwala AY, Wright DJ. Device based treatment of heart failure. *Postgraduate Medical Journal.* 2005 May; **81**(955): 286–91.

[22] Groleau R. *Operation: Heart Transplant or How to Transplant a Heart in Nineteen Easy Steps.* NOVA Online 2000 November [cited 2007 June 7]; Available from: http://www.pbs.org/wgbh/nova/eheart/transplant.html

[23] US Department of Health and Human Services. *OrganDonor.gov. 2007* [cited 2007 June 7]; Available from: http://www.organdonor.gov/

[24] Westaby S. The need for artificial hearts. *Heart* (*British Cardiac Society*). 1996 Sep; **76**(3): 200–6.

[25] Song X, Throckmorton AL, Untaroiu A, Patel S, Allaire PE, Wood HG, *et al.* Axial flow blood pumps. *Asaio J.* 2003 Jul–Aug; **49**(4): 355–64.

[26] Poirier VL. The LVAD: A case study. *The Bridge.* 1997 Winter; **24**(4).

[27] Artificial heart: the debate goes on. *Science News.* 1986 February 22; **129**: 122.

[28] Bourque K, Gernes DB, Loree HM, Richardson JS, Poirier VL, Barletta N, *et al.* HeartMate III: pump design for a centrifugal LVAD with a magnetically levitated rotor. *Asaio Journal.* 2001; **47**: 401–5.

[29] Fukamachi K. New technologies for mechanical circulatory support: current status and future prospects of CorAide and MagScrew technologies. *Journal of Artificial Organs.* 2004; **7**(2): 45–57.

[30] Stevenson LW, Rose EA. Left ventricular assist devices: bridges to transplantation, recovery, and destination for whom? *Circulation.* 2003 Dec 23; **108**(25): 3059–63.

[31] Chen JM, Naka Y, Rose EA. The future of left ventricular assist device therapy in adults. *Nature Clinical Practice.* 2006 Jul; **3**(7): 346–7.

[32] Mitka M. Heart disease a global health threat. *Jama.* 2004 Jun 2; **291**(21): 2533.

[33] Brown AI, Garber AM. A concise review of the cost-effectiveness of coronary heart disease prevention. *The Medical Clinics of North America.* 2000 Jan; **84**(1): 279–97, xi.

[34] Nova Transcripts, *"Electric Heart"*, PBS Airdate December 21, 1999, www.pbs.org/wgbh/nova/transcripts/2617eheart.html.

13

Clinical trial design and sample size calculation

In Chapter 12, we considered the many limitations of the currently available treatments for heart disease; because these invasive procedures are expensive and have serious side effects, there is an important global need to develop more cost effective ways to prevent heart disease. In the early 1990s, a series of small, highly publicized studies offered hope that a simple intervention – taking high doses of vitamin E (Figure 13.1) – might reduce the risk of developing cardiovascular disease by as much as 40% [1, 2]. A subsequent randomized clinical trial compared the rates of myocardial infarction and cardiovascular death in a group of 1035 patients taking vitamin E to those in a group of 967 patients taking a placebo; results were reported in 1996 and also showed that vitamin E provided a protective effect [3]. Yet, a pivotal clinical trial involving 9541 patients in 2000 indicated that there might be no reduction in the risk of cardiovascular disease for those who take vitamin E supplements and when these patients were followed for more than seven years, it appeared that the use of vitamin E supplements may actually increase the risk of developing heart failure [4, 5].

This series of studies provides a good illustration of both the process and challenges of clinical research. Generally, early studies involving small numbers of patients provide data which allow us to generate

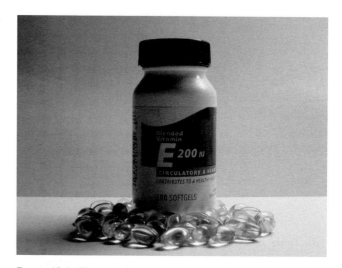

Figure 13.1. *There has been considerable disagreement among different epidemiologic and clinical studies designed to determine whether taking vitamin E supplements can reduce the risk of developing cardiovascular disease.*

hypotheses; these hypotheses must then be tested rigorously in larger clinical studies. Because of inherent biological variability, it is not uncommon for results of early studies to be contradicted by larger, more rigorously designed studies. In fact, a recent study compared conclusions presented in highly cited articles to those of subsequent studies with larger sample size or better controlled design [6]. Results showed that nearly 1/3 of highly cited studies were later contradicted and that this was most likely for studies where patients were not randomly assigned to a treatment or control group!

In this chapter, we will examine how clinical studies and clinical trials are designed. Our goal is to develop the tools to properly interpret and use the results of clinical research. We begin by examining how to characterize biological data and its associated variability. Next, we provide an overview of the different types of clinical studies and clinical trials used to generate and test hypotheses. Finally, we examine how to calculate whether a study to test a hypothesis includes sufficient numbers of patients to ensure that any resulting conclusions are based on real differences in the data and not just statistical variability.

As we consider these important issues, we will find that the process of designing clinical research studies and clinical trials has many similarities to the scientific method that we learned about in Chapter 7. In order to help form hypotheses, we begin by carrying out observational and epidemiologic studies. Ultimately, we must carry out carefully controlled, randomized clinical trials designed to isolate the effect of just one factor and determine whether it has the predicted impact. The challenge in this process is that because of biological variability, we must always ensure that studies include a large enough group of patients to be sure that any differences we see are real and not just due to chance.

Descriptive statistics

We can use descriptive statistics to analyze the average core body temperature for normal adults. The following dataset contains the core body temperature from a group of 65 men and 65 women. We can calculate parameters which measure the central tendency of the data as well as the dispersion in the data [7].

Using the definitions in the text, we find the following values.

96.3	97.6	98.2	98.7	97.2	98.1	98.6	99.0
96.7	97.7	98.2	98.7	97.2	98.2	98.6	99.0
96.9	97.8	98.2	98.8	97.4	98.2	98.6	99.1
97.0	97.8	98.3	98.8	97.6	98.2	98.7	99.1
97.1	97.8	98.3	98.8	97.7	98.2	98.7	99.2
97.1	97.8	98.4	98.9	97.7	98.2	98.7	99.2
97.1	97.9	98.4	99.0	97.8	98.2	98.7	99.3
97.2	97.9	98.4	99.0	97.8	98.3	98.7	99.4
97.3	98.0	98.4	99.0	97.8	98.3	98.7	99.9
97.4	98.0	98.5	99.1	97.9	98.3	98.8	100.0
97.4	98.0	98.5	99.2	97.9	98.4	98.8	100.8
97.4	98.0	98.6	99.3	97.9	98.4	98.8	
97.4	98.0	98.6	99.4	98.0	98.4	98.8	
97.5	98.0	98.6	99.5	98.0	98.4	98.8	
97.5	98.1	98.6	96.4	98.0	98.4	98.8	
97.6	98.1	98.6	96.7	98.0	98.5	98.8	
97.6	98.2	98.6	96.8	98.0	98.6	98.9	

Mean core temperature: 98.25 °F
Median core temperature: 98.3 °F
Mode core temperature: 98 °F
Standard deviation: 0.73 °F

Using a bin size of 0.4 °F, starting at a minimum value of 96.3 and ending at a maximum value of 100.4, we find the frequency histogram shown below (*Journal of Statistics Education*, Volume **4**, Number 2 (July 1996)).

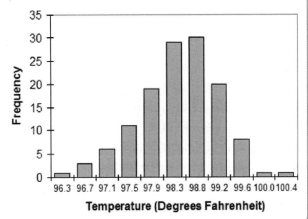

Descriptive statistics

In clinical research, we are constantly concerned with determining whether differences observed between groups of patients are due to an intervention given only to one group or are simply due to biological variability. Generally speaking, we carry out clinical trials in groups of patients that we believe are representative of the greater population. How do we characterize these groups? An important first step is to use **descriptive statistics** to assess demographic variables (e.g. gender, age) for the patients in our sample and to assess clinical parameters for these patients (e.g. blood pressure, serum nicotine levels).

When characterizing data like this, we generally need to assess three main factors: the central tendency of the data (e.g. mean, mode), the spread in the data (e.g. variance, standard deviation), and how the data are distributed across the range of possible values (e.g. normally distributed) [8].

There are several methods to characterize central tendency of the data. If we are interested in a single, continuous variable, x, we define the mean to be the average of measured values for our population [8]. The median is that value which occupies the middle rank when the values are rank ordered from least to greatest. The mode is the most commonly observed value.

Similarly, there are several measures of the dispersion of the data. Two of the most common are the variance and the standard deviation. The variance is the sum of the squared deviations from the mean.

$$\text{Variance} = \sum_{i=1}^{N} (X_i - \bar{X})^2 \qquad (13.1)$$

The standard deviation is the square root of the variance.

In order to examine the distribution of our data, we can display the data graphically in a histogram, in which we divide the data into bins and count the number of values that fall into each bin. A relative frequency histogram is a plot of the fraction of the observations that fall within that bin. The area under a relative frequency histogram is unity.

The distributions of data often tend to follow similar curves. One of the most commonly encountered dis-

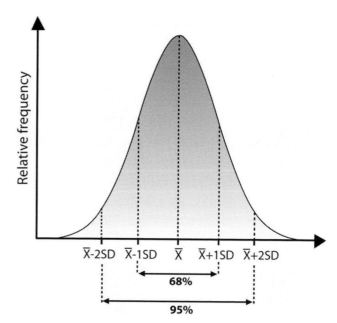

Figure 13.2. *Normal distribution. Gore and Altman, BMA London. Copyright BMJ Publishing Group, with permission.*

tributions in biological data is the normal distribution (Figure 13.2). In normally distributed data, the relative frequency histogram is determined completely by the mean and the standard deviation, and the mean, mode and median are identical. One standard deviation on either side of the mean contains 68% of the area under the relative frequency histogram for a normal distribution: 95% of the data are contained within $+/-1.96$ SD of the mean.

Types of clinical studies

There are two major types of clinical studies: those conducted with the goal of hypothesis generation and those designed to test a specific hypothesis. Hypothesis generating studies are often referred to as epidemiologic studies, while hypothesis testing studies are called clinical trials.

Hypothesis generation

There are several types of studies which can be used to generate clinical hypotheses. For example, in a case study, we identify one patient and attempt to understand the factors related to their illness, either by reviewing their history or through physical examination or clinical testing. Similarly, in a case series, we identify a group

Infant Diagnosis:
June 13, 2007
Kim Malawi

The project that I'm going to be spending the rest of my summer working on is a Malawi Ministry of Health pilot project that is supported by a variety of NGOs, including BIPAI (Baylor International Pediatric AIDS Initiative), UNICEF, the Clinton Foundation, etc. It is a part of their "HIV-Free Generation" goal.

This project aims to identify HIV exposure or infection as early as possible to begin providing care as early as possible. Ideally, all pregnant women will be routinely screened for HIV at their prenatal care visits (ANC). Those that are pregnant will get CD4 counts, be clinically staged and given a single dose of Nevirapine (NVP) to take at the onset of labor. If they qualify (by CD4 count or WHO stage) they can be referred to an ARV clinic and started on ARVs. The single dose of NVP isn't needed for women on full ARV therapy, but does reduce mother to child transmission by about 50% if the women aren't otherwise on therapy. (The baby is also supposed to receive a dose when it is born for this to be most effective.)

Since this is obviously not ideal, right now the program is also involved in identifying infants below the age of 18 months, which is an age group that cannot be identified by the currently available rapid HIV tests. Those tests check for anti-HIV antibodies in the blood, but babies can have their mother's antibodies (passed through the placenta) for up to 18 months. But, we are now able to do DNA PCR testing on dry blood spots. These tests identify viral DNA in the blood, which you only have if you are infected. It can identify babies as early as six weeks old. The only problem is that while a positive is a definite positive, a negative is not a definite negative in infants that age. Continued breastfeeding is continued exposure, so infants have to be retested six weeks after they stop breastfeeding.

And breastfeeding is a huge problem. It would be best to not do it at all, but babies would almost all starve to death. Formula is far too expensive. So most women breastfeed. In that situation, it would be best to breastfeed exclusively for six months and then rapidly wean to solid foods/juice. But they also give other foods (mixed feeding). The problem with mixed feeding is that it actually increases the chances of transmission through a couple mechanisms. First, anything other than breast milk acts as an irritant and makes micro perforations in the baby's intestines, essentially breaking a barrier between the baby's body and maternal HIV. Secondly, the baby's body recognizes anything but mother's milk as foreign, which initiates an immune response, drawing CD4 cells to the intestinal tract, and thereby bringing the HIV's host cells into close proximity to the virus. Many mothers start mixed feeding as young as two months old because they are culturally pressured to do so (if you don't give your baby other food as young as possible, you clearly don't love them enough) and continue with the mixed feeding until nearly two years old.

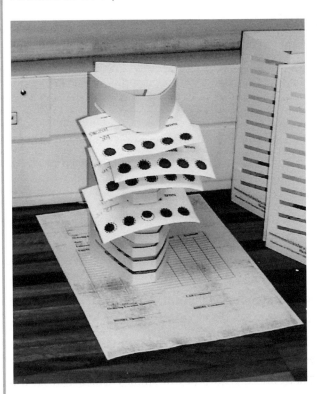

Right now our job is to train health care providers at these seven pilot sites about the new testing algorithm and procedures. We then provide a week to two weeks of onsite support and coaching to make sure they get it right. We are also attempting to identify potential HIV+ kids in their pediatric wards, under five clinics and outpatient clinics.

So, that's what I'm up to. It's very interesting and very exhausting.

of patients with a similar illness and try to ascertain similar factors that may be responsible for their disease. We have already seen a good example of a case series. Recall that, in 1981, a series of four previously healthy homosexual men were identified who developed *Pneumocystis carinii* pneumonia and mucosal candidiasis – the first description of AIDS and its association with behavioral risk factors.

Hypothesis testing

Once we have formulated a hypothesis, there are a number of different types of clinical studies which can be used to test the hypothesis. Hypothesis testing studies are divided into two major groups: observational and experimental. In an observational study, we identify a group of patients with disease and a similar group of patients without disease. We collect data from these two groups of patients and compare them to test our hypothesis. Observational studies have the advantage of being relatively simple and inexpensive to carry out, but they can suffer from important problems. One of the most common is bias – suppose we identify a group of patients with lung cancer and a matched group of healthy controls. We notice that the lung cancer patients report an increased frequency of consuming alcohol, and we conclude that alcohol consumption may be related to the development of lung cancer. It is also possible that alcohol consumption is strongly correlated with cigarette smoking and that smoking is actually the factor which is responsible for the increase in lung cancer. In an observational study, we can't control the administration of the intervention to decisively show cause and effect – we can only retrospectively analyze differences in our two groups of patients [9].

To avoid this and other sources of error, we can design experimental approaches to test our hypothesis. A clinical trial is a research study to evaluate the effect of a new diagnostic or therapeutic in a group of patients [11]. Clinical trials help us to determine whether new interventions are safe and effective. Clinical trials may involve a single group of patients or multiple groups of patients, but in general, their goal is to isolate all but a single variable and measure the effect of that variable. Clinical trials are done in a prospective way: they are

How to find a clinical trial

The National Institutes of Health maintains a publicly accessible database of clinical trials to test experimental therapies for serious or life-threatening diseases conducted with federal and private support.

http://clinicaltrials.gov

The database provides a summary of the purpose of the study, patient recruiting status, criteria for patient participation, the location of the trial, the research study design, the phase of the trial, and the disease or condition and drug or therapy under study. Currently, information can be found for more than 36,000 ongoing trials [10].

first planned, then data are collected; this enables the investigators to manipulate the intervention in a controlled setting to show cause and effect [9].

A single arm study involves a single group of subjects; subjects receive an intervention and then we monitor them to see if their condition improves [12]. Usually, we compare the change in their condition before and after the intervention (change from baseline) so that each patient serves as their own control. One problem with a single arm study is that improvements following an intervention may be partially due to the placebo effect – where patients or their physicians think they are getting better simply because they believe they are receiving an intervention – and this can change their perception of symptoms.

To avoid this problem, clinical trials can be carried out using two groups of patients, one of which receives the intervention and the other which receives a placebo. Randomized clinical trials, in which different subjects are randomly assigned to different treatment and control groups, are generally considered to be the strongest type of clinical evidence [12]. In a double-blinded, randomized clinical trial, participants are randomized in a way so that neither the participants nor their physician knows which group they are part of. This type of study design eliminates conscious bias. Subjects must be randomized in a way so that the treatment and control groups are similar on average; in this way we can attribute any differences in the outcomes between the groups to the intervention and not any other factors that may have differed between the groups.

Clinical trials: drugs

Clinical trials of new drugs are carried out in different phases to assess the safety and efficacy of the new drug.

In a phase I trial, researchers assess the safety of the drug in a series of 20–80 normal volunteers.

In a phase II trial, the experimental drug is given to a larger number of volunteers (100–300) to assess both its safety and efficacy.

In a phase III trial, the new drug is given to a large number of patients in order to compare its effectiveness to standard therapies and to monitor for side effects. Typically, phase III trials are randomized, placebo controlled, double-blinded studies. It is especially important to carry out a rigorous calculation of the sample size required for phase III trials.

Sample size

The sample size is the number of subjects in each arm of a study needed to detect a predetermined difference [13]. Determining the necessary sample size for a clinical trial is a careful balancing act – as the sample size becomes larger, we reduce the risk that differences between the two groups arise due to chance, but the cost and complexity of doing the study increase. Our goal in setting a sample size for a randomized clinical trial is to appropriately balance these factors.

In general, setting a sample size is a complex statistical calculation. However, for data which can be described by the normal distribution, there is a relatively straightforward process to estimate the number of patients required. Let's examine the process to determine the sample size for a two-arm, randomized clinical trial. In this type of trial, we have both a treatment group (receives the intervention) and a control group (receives a placebo). We will monitor the clinical outcomes

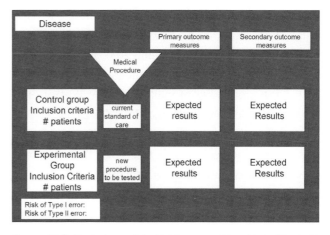

Figure 13.3. *To design a clinical trial, you must be able to fill out all the information in the template shown.*

in each group. We must select a primary outcome – this is the outcome we are most interested in comparing between the two groups. In selecting our sample size, we want to ensure that any differences between the treatment and control group are real, knowing that there will be some statistical uncertainty associated with the primary outcome. We will choose our sample size so that this uncertainty is sufficiently lower than the difference in primary outcome we wish to detect between the control and treatment groups. Figure 13.3 provides a template which illustrates all of the information needed to calculate the sample size for such a trial.

What happens if we choose our sample size to be too small? Essentially, there are two types of mistakes that can result. The first is that we mistakenly conclude that there is a difference between the two groups, when in reality there is no difference. This is called a Type I error, and can result in adopting an intervention, even when it is not truly effective. The second type of error that we can make is to mistakenly conclude that there is not a difference between the two, when in reality there is a difference. This is called a Type II error and can result in us discarding a potentially effective intervention. Our sample size will be dictated by the risk that we are willing to accept for making each of these types of errors.

The risk of making a Type I error is called the significance level of the study. Typically, studies are designed so that the risk of making a Type 1 error (significance level) is between 1% and 5%. The risk of making a

Type II error is usually denoted by the variable β. Typically, studies are designed so that the risk of making a Type II error is between 10% and 20%. Often, we refer to the power of a study, where the power is defined to be 1 – the probability of making a Type II error. The power of a study usually ranges from 0.8 to 0.9.

With these definitions, we can now outline the process of calculating a sample size [14]. First we must select a primary outcome. Our primary outcome can be either a binary measure (did the patient have a heart attack?) or a continuous variable (what is the patient's systolic blood pressure?). If the outcome variable is binary, we must estimate the expected rate of the primary outcome in the treatment group and the control group. If the primary outcome variable is continuous, we must estimate the size of the difference in the average outcome for the two groups and the expected standard deviation associated with this variable. Next we set acceptable levels of Type I and II errors. This determines the p-value and the power of the study. With this information, we can use a graphical tool called Altman's nomogram to estimate the sample size required in each group.

To use Altman's nomogram, we must first calculate the standardized difference we expect to find in the primary outcome variable. For a binary outcome variable, the standardized difference is calculated as:

$$\text{standardized difference} = \frac{P_1 - P_2}{\sqrt{\bar{p}(1 - \bar{p})}} \qquad (13.2)$$

$$\bar{p} = \frac{P_1 + P_2}{2}, \qquad (13.3)$$

where P_1 is the fraction of patients in the treatment group who experience the primary outcome and P_2 is the fraction of patients in the control group who experience the primary outcome.

For a continuous outcome variable, the standardized difference is calculated as:

$$\text{standardized difference} = \frac{\text{effect size}}{\text{standard deviation}}, \qquad (13.4)$$

where the effect size is the difference in the average value of the primary outcome we expect between the treatment and control groups and the standard deviation is that expected for the primary outcome variable.

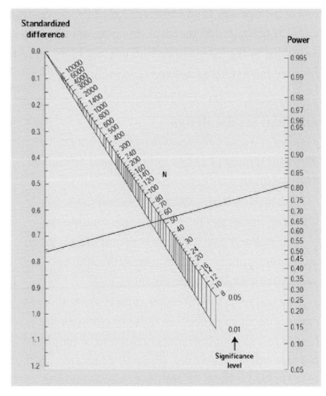

Figure 13.4. *Altman's nomogram, showing sample size calculation for a standard difference of .85 and a power of .8. Altman's nomogram is used to calculate the sample size required in each arm of a study, given the expected standardized difference, desired power and significance level. Note that for the results obtained using Altman's nomogram to be valid, we assume that the continuous variable follows the normal approximation to the binomial distribution and the two groups being compared are of equal size. Altman (1982). How large a sample? In* Statistics in Practice. *Eds S. M. Gore and D. G. Altman. BMA London. Copyright BMJ Publishing Group, with permission.*

We use Altman's nomogram (Figure 13.4) to determine the number of subjects required in each group by drawing a straight line to connect the standardized difference we wish to detect and the desired power of the study. The point at which this line intersects the line drawn at the desired significance level gives the sample size required for each group.

Suppose we wish to carry out a randomized clinical trial to compare the effectiveness of a new drug eluting stent to that of the standard uncoated stent. We design a study where patients requiring treatment for coronary artery disease are randomized to receive either the standard stent or the new drug eluting stent. We decide that our primary outcome will be to follow the patients and determine whether or not they experience restenosis during the first six months following the procedure. This is an example of a binary outcome variable, since each patient either does or does not experience restenosis. A pilot clinical trial with the new drug eluting stent suggests that only 10% of patients experience restenosis following treatment with this device. A review of the literature shows that, on average, the number of patients who experience restenosis within six months following treatment with a standard stent is approximately 45%. We wish to design a trial with a 5% significance level and a power of 0.8, meaning that we are willing to take a 5% chance of making a Type I error and a 20% chance of making a Type II error.

We begin by calculating the standardized difference:

$$\text{standardized difference} = \frac{P_1 - P_2}{\sqrt{\bar{p}(1 - \bar{p})}}$$
$$= \frac{0.45 - 0.1}{\sqrt{0.275(1 - 0.275)}} = 0.78 \tag{13.5}$$

$$p = \frac{P_1 + P_2}{2} = \frac{0.45 + 0.1}{2} = 0.275. \tag{13.6}$$

With this, we turn to Altman's nomogram, and draw a line connecting a power of 0.8 to a standardized difference of 0.78. We look to see where this line intersects that representing a 0.05 significance level and see that 54 patients will be required in each arm of the study. In the case of a continuous outcome variable, we follow an identical procedure, except when calculating the standardized difference.

In summary, when designing and analyzing clinical trials there are four major parameters of interest.

1. The significance level of the study.
2. The power of the study.
3. The effect size one can detect.
4. The sample size.

For a fixed significance level and power, as we increase the sample size we can detect smaller and smaller differences in the outcomes between the treatment and control groups. It is important to place the effect size we wish to determine in the proper clinical context. There is an important difference between the statistically significant difference we can detect given the

parameters above and what may be clinically meaningful. The smallest difference that is clinically meaningful is sometimes called the minimum clinically important difference (MCD). It is important not to waste time and money enrolling patients in order to detect differences that are substantially less than the MCD, since these differences are of no clinical importance.

We have seen that in order to calculate a sample size, we need to estimate the difference that might exist between the treatment and control groups and the expected variance in the data. In practice, the difference may be larger than predicted or the spread in the data may be smaller than expected. In this case, the number of patients required to reach statistical significance will be smaller than initially predicted [13]. However, it is also possible that the differences between the treatment and control groups will be smaller than predicted or the spread in the data will be larger; in order to reach significance, more patients will be required than originally estimated [13].

Frequently, special committees called Data Safety and Monitoring Boards (DSMB) are convened in order to monitor interim data in clinical trials to ensure the safety of participating subjects [15]. The federal government requires that a DSMB be used to monitor all phase III clinical trials. Typically, a DSMB is composed of scientists and clinicians who are knowledgeable about the field of interest, but are not otherwise involved in the study. The role of the DSMB is to analyze all adverse events reported in a trial and to perform interim analysis of clinical outcomes to determine whether a trial should be stopped early.

If an interim analysis indicates that a new treatment is substantially better than the standard of care, then the study may be stopped early and the new treatment offered to both the study group and the control group [11]. Conversely, if an interim analysis reveals substantially increased risk associated with a new treatment, then a study may also be stopped early to prevent additional harm. Making decisions to stop a trial early are frequently difficult, because they force researchers to make judgments about when interim results cross the boundary from suggestive to conclusive. Making decisions to stop a trial or proceed with a trial following an interim analysis can also raise serious ethical questions.

Weekend in Mochudi and watching the Zebras (soccer team): June 18, 2007
Rachel Botswana

After a PAC doctor, Heather's suggestion and Liz's approval, we went ahead and started screening patients in hopes of casting a larger net to catch the patients with less then perfect adherence (although for the most part, all the patients in Gabarone have good adherence, especially when compared to Francistown, a clinic where the pharmacist constructed a list of 31 reasons why he would not do pill counts, and is described as eccentric . . .) which brings me to an idea for another project at Rice which would not be life changing here, but would be a pleasant convenience – a pill counter!

I am thinking something akin to those change counters where you dump the change on top and it sorts it and counts it. The most important part would be cost, simplicity and ease of use so that at the Francistown clinic, there would be no excuse

for not pill counting, a practice which reminds me of DOTS (directly observed therapy) for TB. While not being exactly direct, DOTS is helpful for making sure people continue to take their pills and be counseled when they miss doses.

After talking to the nurse head, Mamapula (which means Mother of Rain) I found out that we probably should've been communicating better. She said she was a little in the dark as to what exactly we were doing, but after about 5 minutes explaining, she was good to go and helped show us how the system works at reception – where to pick up the binders of the patients to be seen – and how to do the pill counts – where to find the pharmacy sheet of how many pills the patient left with. I guess we hoped to expedite the process in the exam rooms with the doctors, but we weren't able to "catch" all the patients that we had hoped to, only ending up with a couple. On the other hand, considering that the patients are taking their pills so well, this is great news for the clinic!

Tomorrow we will go back to the data extraction project in an effort to see the trends in the resistance tests (RT) that patients who were failing on line 1 or 2 of the ARVs. RTs are expensive and usually only done in extreme circumstances but the results of collecting the mutations of the HIV are pivotal in letting the doctor know which drugs will be ineffective and which ones will work.

http://hivdb.stanford.edu/pages/algs/HIVdb.html

For example, a recent trial was carried out to assess the efficacy of a new treatment for severe sepsis (blood-stream infection) [15]. Researchers designed a trial to compare a new drug, tifacogin, to that of a placebo; the primary outcome was mortality. The trial was designed with a sample size of approximately 1500 subjects. An interim analysis conducted after 722 patients had been accrued showed a mortality rate of 38.9% in the placebo group and 29.1% in the tifacogin group. This difference was significant at the 0.006 level. The DSMB was forced to decide whether to stop the trial early, given the apparently lower mortality rate associated with the new drug. Ultimately, the board decided that the results were not sufficiently compelling and the trial was continued. Neither the researchers nor the prospective trial participants were informed of the interim results.

Clinic: dosing guide:
June 22, 2007
Lindsay Botswana

As we still attempt to fashion some sort of trial for our dosing guide, we are working on another project Dr. Lowenthal presented to us. For our "trial," we just don't have enough patients with poor adherence to create any kind of pool big enough to make our results somewhat reliable. For now, we're for the most part just giving everybody dosing guides and not creating a control group (those w/o guides). As unscientific as this is, the doctors and nurses are referring patients to us whom they think really need the guide, at a rate of one to two per day. Given the current situation, it makes more sense to just allow the people who need guides to get them – we won't be able to bring together a "scientific" study. There are many confounding factors in creating a "scientific" study, mainly that all patients referred to us will also have to go through adherence class again, which itself may raise adherence rates. I would rather rely on the patients' comments when they come back to the clinic for their next appointment to see if the guide is helpful. I think this may end up being a lot more subjective than we all expected, but I'm okay with that.

We had an 11-year-old patient who came in the other day with her mother, who had recently taken custody of her due

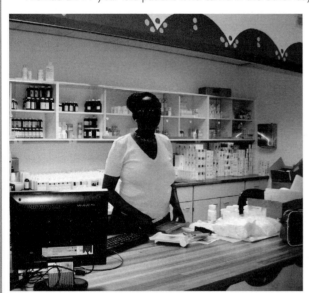

to the death first of a nanny and then of an aunt, who had been caring for the girl very well and making sure she took her meds. After I created a dosing guide for the mother (the girl wasn't with us), she went back to the waiting room as usual. As I went back to the nurses to make sure the guide was correct, they told me about the family's situation – the mother was quite irresponsible, and the 11-yr-old girl was taking the ARVs on her own! She had been doing a very good job despite the complexity of her regimen, but she still needed help, which she wouldn't get from her mother. Her eyes lit up as I explained to her the dosing guide and how she could use it, as the mom's eyes wandered around the room in boredom. I think we are better off helping the five or six children needing help who come in each week, rather than spending our time creating a trial. A doctor at clinic suggested that we try to set up a visit at one of the SOS Villages (orphanages) to see if we would be able to give all the "mothers" there dosing guides for all the kids who need them.

The importance of the minimum clinically important difference

"Samples which are too small can prove nothing; samples which are large enough can prove anything" [14].

Was this decision scientifically sound? Was it ethical? The interim analysis, although not definitive, suggested that the new drug reduced mortality by about 25% compared to placebo. If the researchers had been informed of the interim results, it might have made them more hesitant to enroll their patients in the trial and to fol-

low the study protocol. In the end, recruitment was continued to the final sample size. When all the data were analyzed, it was determined that mortality in the tifacogin group was 34.2% compared to 33.9% in the placebo group – a difference that was neither clinically nor statistically significant.

To address these ethical concerns regarding decisions made by DSMBs, it has been recommended that informed consent documents be modified to indicate if a trial will be monitored by a DSMB. If so, during informed consent, the role of the DSMB should be explained, noting that the DSMB may make recommendations to continue a trial even in the face of evidence suggesting effectiveness of one of the treatments. The informed consent process should make clear that, in the interests of maintaining the scientific integrity of the study, interim results will not necessarily be made available to patients enrolled in the study [15].

Bioengineering and Global Health Project
Project task 9: Design a clinical trial to test your new technology

In this task, you will design a clinical trial to test the new technology that you have developed. As part of this design, you must choose the subjects to be tested including the control group (receives standard of care) and the experimental group (receives new technology). You must specify the primary and secondary outcomes that you will monitor in the trial. You must calculate a statistically justified sample size for the trial. Complete the table below and provide a one page summary of the trial design and sample size justification.

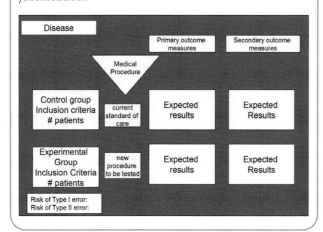

Homework

1. Complete the Human Participant Protections Education for Research Teams course which can be found at: http://cme.cancer.gov/clinicaltrials/ learning/humanparticipant-protections.asp. This course presents information about the rights and welfare of human participants in research. The tutorial is designed for those conducting research involving human participants. Complete the exercises at the end of each of the six content areas.

Print a certificate of completion upon completing the course.

2. You are a researcher working on a vaccine for malaria, which is caused by a parasitic protist. You are interested in performing clinical trials for your vaccine.

   a. During phase I testing, you must recruit about 100 healthy volunteers. What are you trying to determine through this testing?

   b. During phase II testing, what should your malaria vaccine be tested against?

   c. Phase III testing involves a double-blind study. What does this mean and why is it important?

3. You are designing a study to test a new implantable artificial kidney for patients with end-stage renal disease. You divide your patients into two groups: one will receive an implanted artificial kidney, the other will receive tri-weekly hemodialysis (standard of care in end stage renal disease). Your primary endpoint is mortality for all causes at one year, a secondary endpoint is patient quality of life at one year, which will be assessed via questionnaire.

   a. What is a Type I error? What are the possible consequences of making a Type I error in this study?

   b. What is a Type II error? What are the possible consequences of making a Type II error in this study?

   c. Define the p-value.

   d. Define power (if you use Greek letters in your definition, you must define these as well).

   e. Why would blinding be difficult in this study?

   f. Which of your endpoints is more likely to be affected by a lack of blinding?

   g. Assuming you expect 30% mortality at one year for the control group and 20% for the treatment group, what sample size would be required to achieve 80% power? What p-value should you use? Justify this value.

4. You are designing a clinical trial to compare the performance of a new thrombolytic agent to dissolve blood clots associated with acute myocardial infarction. You design a trial to compare the new agent to the standard of care, streptokinase. You choose the 30 day mortality as your primary

endpoint. There will be some statistical uncertainty associated with the measured mortality rate in the treatment and control groups. Your goal in selecting the sample size for the trial is that this uncertainty be significantly less than the difference in the mortality rate between the control and treatment groups. We must set acceptable levels for the risks of Type I and II error.

a. Define Type I error and Type II error.

b. Suppose you expect a mortality rate in the group treated with the new drug stent of 5%, while the expected restenosis rate in the group treated with the current stent is 7%. You calculate a standardized difference of 0.19. If you can tolerate a 20% risk of Type II error and a 5% risk of Type I error, how many patients are needed in the trial? Use Altman's nomogram to indicate how you calculated your answer.

c. If the mortality rate for the new drug was expected to be 1%, would the required sample size increase or decrease?

d. List one secondary outcome you would want to monitor in this trial.

5. Consider the Acute Respiratory Distress Syndrome Network trial of low versus traditional tidal volume ventilation in patients with acute lung injury and acute respiratory distress syndrome published in 1996. Mortality rates in the low and traditional volume groups were 31% and 40%, respectively, corresponding to a reduction of 9% in the low volume group. What sample size would be required to detect this difference with 90% power using a cut-off for statistical significance of 0.05? Source: [16].

References

[1] Rimm EB, Stampfer MJ, Ascherio A, Giovannucci E, Colditz GA, Willett WC. Vitamin E consumption and the risk of coronary heart disease in men. *The New England Journal of Medicine*. 1993 May 20; **328**(20): 1450–6.

[2] Stampfer MJ, Hennekens CH, Manson JE, Colditz GA, Rosner B, Willett WC. Vitamin E consumption and the risk of coronary disease in women. *The New England Journal of Medicine*. 1993 May 20; **328**(20): 1444–9.

[3] Stephens NG, Parsons A, Schofield PM, Kelly F, Cheeseman K, Mitchinson MJ. Randomised controlled trial of vitamin E in patients with coronary disease: Cambridge Heart Antioxidant Study (CHAOS). *The Lancet*. 1996 Mar 23; **347**(9004): 781–6.

[4] Yusuf S, Dagenais G, Pogue J, Bosch J, Sleight P. Vitamin E supplementation and cardiovascular events in high-risk patients. The Heart Outcomes Prevention Evaluation Study Investigators. *The New England Journal of Medicine*. 2000 Jan 20; **342**(3): 154–60.

[5] Lonn E, Bosch J, Yusuf S, Sheridan P, Pogue J, Arnold JM, *et al.* Effects of long-term vitamin E supplementation on cardiovascular events and cancer: a randomized controlled trial. *Jama*. 2005 Mar 16; **293**(11): 1338–47.

[6] Ioannidis JP. Contradicted and initially stronger effects in highly cited clinical research. *Jama*. 2005 Jul 13; **294**(2): 218–28.

[7] Shoemaker AL, College C. What's normal? – temperature, gender, and heart rate. *Journal of Statistics Education*. 1996; **4**(2).

[8] Neely JG, Stewart MG, Hartman JM, Forsen JW, Jr., Wallace MS. Tutorials in clinical research, part VI: descriptive statistics. *The Laryngoscope*. 2002 Jul; **112**(7 Pt 1): 1249–55.

[9] Bernson M. Tutorial: an introduction to clinical trials Part I of II: purposes and phases. *The Next Generation*. 2005; **1**(4).

[10] US National Institutes of Health. ClinicalTrials.gov. 2007 [cited 2007 June 7]; Available from: http://www.clinicaltrials.gov/

[11] The Patient Education Institute. *Clinical Trials*. MedlinePlus 2005 January 25 [cited 2007 June 7]; Available from: http://www.nlm.nih.gov/medlineplus/tutorial.html

[12] Bernson M. Tutorial: An introduction to clinical trials part II of II: statistics and experimental design. *The Next Generation*. 2005; **1**(5).

[13] Neely JG, Karni RJ, Engel SH, Fraley PL, Nussenbaum B, Paniello RC. Practical guides to understanding sample size and minimal clinically important difference (MCID). *Otolaryngol Head Neck Surgery*. 2007 Jan; **136**(1): 14–18.

[14] Ogundipe L. Sample size determination in clinical research: 2. *Hospital Medicine*. 2000 Nov; **61**(11): 797–8.

[15] Slutsky AS, Lavery JV. Data safety and monitoring boards. *The New England Journal of Medicine*. 2004 Mar 11; **350**(11): 1143–7.

[16] Whitley E, Ball J. Statistics Review 4: Sample size calculations. *Critical Care*. 2002 May 10; **6**: 335–341.

Technology diffusion

We have considered the development of new technologies to improve the prevention of infectious disease, the early detection of cancer and the treatment of cardiovascular disease. Through this journey, we have seen the often integrated processes of scientific research, engineering design and clinical research in action. Through this coordinated process, we develop new healthcare technologies which are safe and effective. How do we ensure that the results of these efforts actually reach the patients who need them? We will now consider this important question, as we examine how new health technologies are managed. In the present chapter we will examine the diffusion of new innovations – what drives the adoption of technologies that are newly proven to be beneficial? How does the diffusion of new technologies differ throughout the world? How can we speed the adoption of effective new technologies?

The diffusion of new interventions has been historically slow. Consider the example of the use of vitamin C to treat scurvy [1]. Scurvy was a major cause of mortality for early sailors. In 1497, Vasco Da Gama lost 100 out of 160 crew members to scurvy sailing around the Cape of Good Hope. A dietary connection was suspected, and in 1601, British Navy Captain James Lancaster was in command of four ships traveling from England to India when he performed a clinical trial to test whether lemon juice could reduce the incidence of scurvy. He required

Figure 14.1. *Lemon juice, a simple intervention to prevent scurvy, was not adopted until more than 200 years after first proven effective.*

sailors to take three teaspoons of lemon juice daily on one ship, while sailors on the other three ships served as the control group. The results were astonishing; while 110 of 278 (40%) sailors died in the control group, there were no deaths in the experimental group. Despite this impressive result, the intervention was not widely adopted. In fact, in 1747, British Navy Physician James Lind repeated the study with similar results. It was not until 1865 that the British Navy finally adopted the innovation, 264 years after Captain Lancaster's successful clinical trial!

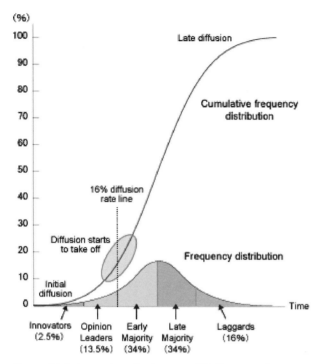

Figure 14.2. *Innovation adoption curve [3]. Courtesy of Sustain.co.uk.*

Has the situation changed today? Remarkably, a recent study found that it now takes an average of 17 years for new knowledge generated in randomized clinical trials in the United States to be implemented in clinical practice, and even then, implementation is highly non-uniform [2]. Technology diffusion is even slower in developing countries, which frequently do not have the capacity to carry out medical research, and must instead import new technology developed in other settings [3].

Let's examine the pace at which adoption of new technologies proceeds. If we graph the fraction of a target population who has adopted a new technology over time, we find that the curve frequently has an S-shape (Figure 14.2). Initially, diffusion is slow. As more of the population adopts the innovation, the speed of diffusion increases. Often, we find that there is a tipping point, when between 15% and 20% of the population have adopted an innovation, its spread becomes difficult to stop. At some point, the market is nearly saturated and diffusion again slows. The shape of this curve has been found to be remarkably similar for many different types of innovations.

It is interesting to think about how the people responsible for innovations and their adoption interact during this process. The group of people who most rapidly adopt new technologies are the innovators who develop these new technologies. Innovators are willing to take risks to develop new technologies; they are often perceived as risk takers or mavericks, who may be overly invested in the success of their own innovation. As a result, they may be viewed as incautious and are frequently socially disconnected. Next to adopt a new technology are the early adopters; this group of people is usually involved in testing and comparing the efficacy of several new interventions in order to determine which is most effective. Early adopters are viewed as well connected, social opinion leaders; their actions are watched by the communities in which they practice. The early majority generally follows the lead of early adopters to accept new innovations. They rely on the conclusions of people they know before deciding to test a new change. The late majority wait to see local proof that a new intervention is effective before adopting it. Finally, the most conservative group, sometimes called laggards, wait to adopt a new innovation until it has become the status quo or standard of care.

Let's examine a case study of the diffusion of a relatively new technology – that of laparoscopic cholecystectomy – or removal of the gall bladder using fiber optic guided surgery. This procedure was one of the first minimally invasive surgical procedures to be developed and the story of its adoption illustrates many of the concepts we have just introduced.

The gall bladder (Figure 14.3) acts to store bile made by the liver. After you eat a meal, the gall bladder contracts, and secretes bile into a duct which empties into the small intestine. Bile provides an important aid in digestion. In some cases, liquid bile may precipitate into solid stones called gall stones. This is a relatively common occurrence, with nearly 1/5 of North Americans and 1/4 of Europeans developing gall stones at some point in their lives. If gall stones block the bile duct, patients can experience severe abdominal discomfort and pain, heartburn, and indigestion.

Prior to 1990, gall stones were treated in an open surgical procedure to remove the gall bladder. At this time,

I have to admit that I found myself the night before last feeling a bit nervous about rolling out the activities and interacting on an educational level with the students. Although I have worked with youth before and have had similar experiences with refugees from various African countries, I was still unsure of how the students would respond to our teaching styles and activities. I now know what some teachers describe as the butterflies in their stomach the night before their very first day of teaching.

After setting up our projector in the community hall, students began trickling in, one by one. We introduced ourselves, and again I was embarrassed by my lack of ability to pronounce their names. The initial ice was broken when they asked if I had music on my computer. They requested hip-hop or R&B and knew a few Usher songs and a Will Smith song I played.

We kicked off the first day with an ice breaker game in which the children split into two groups and lined up behind a blanket. When the blanket dropped, the students had to shout the name of their classmate on the other side of the blanket as fast as possible. It was so funny to watch their reactions as the blanket dropped and to keep them from peeking underneath the blanket.

I had my moment as the strict teacher and did my share of nagging. It turns out the concept of a pre-test is not common around here and even when we told them they could leave their names off the test, we had the toughest time convincing them that the pre-test we were giving out was only for our own knowledge – that it would never be graded or shown to their teachers. They were whispering to each other and trying to share answers left and right. I definitely did my part to stand over the mischievous ones, pace back and forth to scare them a bit, and laughed inside as their little eyes turned quickly away from their neighbors' papers as they looked up and saw me coming by.

I am really not sure how much they liked the powerpoint style of presentation, but I think it was something different and kept their attention fairly well. They were happy to get up in between slides and play the snowball game. We had them write down a question they were curious about relating to HIV, science, health, life in general and keep it to themselves. They then split up into two teams on opposite sides of the room and we told them to crumple up the papers and begin throwing them at the opposing team. It was absolute mayhem and turned out even better than we imagined.

There was a question asking what love is. This age group was 12 years old and above, and this question puzzled me at first. Remembering that many of these children are orphans and know only the love of a house mother who takes care of nine others, I took back a bit of my confusion and began thinking of the millions of orphaned children who most likely have never known unconditional love. Nothing pains me more than thinking of the plight of orphaned children, and every day that I walk through SOS, I am so thankful for the care and opportunities available to these children. I only wish this experience could be multiplied for so many others.

We prepared this morning for our afternoon at SOS again by finding appropriate images to describe the male and female reproductive systems since we found ourselves motioning around our own bodies most of the time yesterday. It's always such a pleasure to walk into the SOS Village because there are always children walking around who are so excited to see us and greet us with the warmest smiles. I want to get to know every child I pass, and I wish I had time to spend with each one of them.

Today's lesson and activities also went well, and the children are gaining so much knowledge that I think the questions I asked during some of the games were too easy. No one really got a kick out of the Twister board we brought, though, and they were all too shy to be the ones to play on the board. Overall, we found the boys to be more open and curious than the girls. The girls had questions and would whisper them to their most vocal friend, but the questions were few and I would really like the opportunity to sit in a smaller group setting with the girls. They are all sweet and welcoming, and seem to want speak with us on a more intimate level.

After the presentation and games, we got to run around with the younger ones a bit and catch them on the playground. Again, they loved posing for pictures and seeing themselves when the picture came up on the digital camera. I cannot explain the way the children look at me, but maybe a few of the shots can capture their spirit, personality, and love of life.

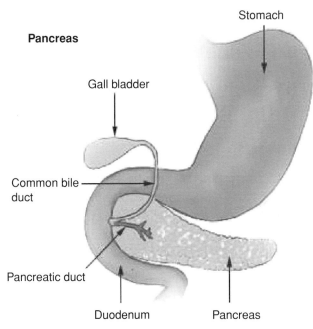

Figure 14.3. *The gall bladder is located just beneath the liver. It stores bile made in the liver. Upon contraction, bile is secreted into the small intestine to aid in digestion. When gall stones form in the gall bladder, they can block the outflow of bile, leading to acute inflammation which must be treated surgically. SEER Training Modules, Pancreatic & Biliary Cancer, US National Institutes of Health, National Cancer Institute 2009.*

http://training.seer.cancer.gov/.

surgical removal of the gall bladder was the most common non-obstetric surgical procedure in many countries. While effective and relatively safe (mortality rate 0.3–1.5%), this procedure required that patients stay in the hospital for about a week, and most lost 30 days of time from work.

Laparoscopic cholecystectomy was a new procedure developed in the 1980s which allowed for shorter hospitalization, more rapid recovery, an earlier return to work, and significant financial savings. In this procedure (Figure 14.4), the patient receives anesthesia, but only a small incision is made at the navel enabling a thin fiber-optic video camera to be inserted. The surgeon then fills the abdomen with carbon dioxide gas. Two needle-like instruments are inserted into the space created by filling the abdomen with gas. These instruments can be seen on the video monitor and serve as tiny hands, used to pick up the gall bladder and to manipulate the intestines. Using such instruments, the surgeon clips the gall bladder artery and bile duct, and safely dissects and removes the gall bladder and gall stones. The gall bladder is then removed

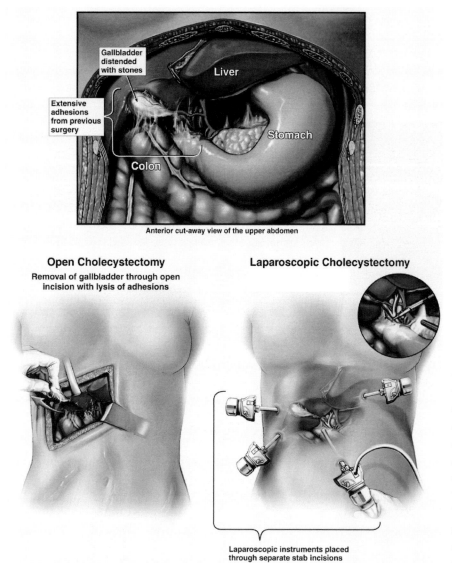

Gallbladder distended with stones

Liver

Extensive adhesions from previous surgery

Stomach

Colon

Anterior cut-away view of the upper abdomen

Open Cholecystectomy
Removal of gallbladder through open incision with lysis of adhesions

Laparoscopic Cholecystectomy

Laparoscopic instruments placed through separate stab incisions

Figure 14.4. *In laparoscopic cholecystectomy, the abdomen is filled with CO_2 gas to create a space for surgery. A camera inserted through the navel is used to visualize internal organs and small surgical instruments are used to manipulate tissues. Arteries and ducts leading to the gall bladder are clipped so that the gall bladder can be removed. When resected, the gall bladder is removed through the incision in the navel. Illustration © 2008 Nucleus Medical Art, all rights reserved. www.nucleusinc.com.*

through the navel incision. The entire procedure takes 30–60 minutes. It requires only three puncture wounds which do not need sutures. The procedure does not result in a scar; the three puncture wounds leave very slight blemishes and the navel incision is barely visible.

Laparoscopic cholecystectomy has been called the most significant major surgical advance of the 1980s. This procedure was the forerunner of a new era of minimally invasive surgery. The benefits of the less invasive procedure include the ease of recovery, since there is no incision pain as occurs with standard abdominal surgery. Following laparoscopic cholecystectomy, about 90% of patients can go home the same day. Within sev-

eral days, they can resume normal activities. The rate of complications for the laparoscopic procedure is about the same as for standard gall bladder surgery. Complications include nausea and vomiting which may occur after the surgery. In addition, injury to the bile ducts, blood vessels, or intestine can occur, requiring corrective surgery. In about 5 to 10% of cases, the gall bladder cannot be safely removed by laparoscopy. In these cases, standard open abdominal surgery is then immediately performed.

Let's examine the rate of diffusion for this new surgery. In fact, no technique in modern times has become so popular as rapidly as laparoscopic cholecystectomy. Its diffusion in healthcare is unprecedented.

The tools of laparoscopic surgery

The figures in this box show (a) an insufflator, (b) a trocar, (c) a camera, (d) laparoscopic cholecystectomy (surgery to remove the gall bladder), and (e) laparoscope with illuminator and camera system. Copyright 2009 by the University of Georgia, College of Veterinary Medicine. All rights reserved.

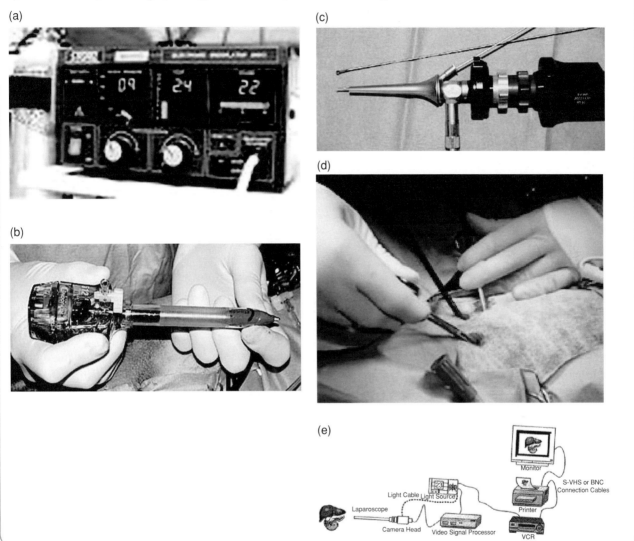

The technique was introduced in 1989; Figure 14.5 shows the percentage of gall bladder removal surgeries which were performed laparoscopically in the subsequent years. By 1992, 50% of all cholecystectomies in Medicare populations and 75% to 80% of all cholecystectomies in younger populations were performed through the laparoscope. Today, the laparoscopic procedure is the most widely used treatment for gall stone disease. Laparoscopic cholecystectomy resulted in a 22% decrease in the operative mortality rate for cholecystectomy [4]. In fact, the introduction of this new, minimally invasive procedure led to a substantial increase in the overall rate of cholecystectomy, as patients who were previously not good candidates for an open surgery could tolerate the risks of the minimally invasive procedure.

Given the success and rapid diffusion of this procedure, how was its innovator viewed? The man largely responsible for the innovations that enabled **laparoscopic surgery** is Kurt Semm. Semm, a gynecologist,

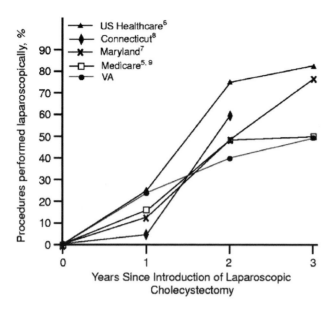

Figure 14.5. *Rate of diffusion of laparoscopic cholecystectomy [4].*

contributed more than 80 medical device inventions during his life. They included the electronic insufflator, the use of thermocoagulation, the loop ligator, and methods to suture structures during laparoscopy [5].

Semm's brother and father owned a medical instrument company which rapidly produced instruments for him that he could test. His ability to rapidly prototype and test new instruments enabled Semm to attempt to perform increasingly complex procedures endoscopically. Semm developed minimally invasive approaches to perform surgical procedures in both the areas of gynecology and general surgery.

In 1985, Semm's techniques were used to perform the world's first laparoscopic appendectomy. At the time, the laparoscopic approach was said to reduce the problem of adhesions which frequently formed following open surgeries. How did Semm's peers react to these advances? Semm was said to have gone "absolutely crazy." He was asked to undergo a brain scan by his colleagues. His lectures were initially greeted with laughter and derision. His techniques were initially viewed as too expensive and too dangerous. Further, colleagues said that Semm exaggerated the problems of adhesions associated with open surgeries. At the time, most surgeons saw no reason to change a well established working method into a complex technical manner. Because he

was a gynecologist, he was accused of having "surgeon envy," and was accused of trying to enter into general surgery to bolster his "operation ego."

Later, reflecting back Semm said, "Both surgeons and gynecologists were angry with me. All my initial attempts to publish on laparoscopic appendectomy were refused with the comment that such nonsense does not and will never belong to general surgery." As we know today, Semm displayed an ability to push his ideas through despite skepticism and suspicion. In fact, without Semm, the laparoscopic revolution may have been postponed by many years.

Can we use our growing knowledge about the determinants of technology diffusion, to speed the adoption of new health technologies which are proven to be safe and effective? A number of strategies have been suggested [1], including the following.

1. Find sound innovations. In the next chapter, we will consider how medical research is funded, and how this impacts the development and diffusion of new innovations.
2. Find and support innovators. As the story of Kurt Semm indicates, innovators who lead major changes often come from outside the field, and their contributions are not always initially appreciated.
3. Invest in early adopters. Finding leaders within the field to champion a new change can decrease the resistance to that change. The speed of diffusion can be increased by providing avenues to connect innovators and early adopters.
4. Make the activity of early adopters more visible. In making decisions about whether to adopt new innovations, the early majority looks to the activity of early adopters. Providing formal opportunities for these groups to interact can help increase the speed of diffusion.
5. Trust and enable reinventions. Often early adopters and the early majority adopt new innovations only in part, adapting them to make them most effective in their local setting. Innovators are sometimes resistant to these changes, believing that they are an indication of resistance. Supporting effective local reinvention can speed diffusion.

6. Create room for change. Adoption of new innovations requires energy; if people are not given time and resources to support this, new innovations cannot diffuse.

7. Lead by example. Adoption of new innovations requires change at all levels in a system; leaders themselves must be open to change if diffusion is to occur.

New innovations arise and are proven effective through a combination of basic science research, engineering design and clinical trials. In the next chapter, we will consider how health-related research is funded and regulated. We will see that the ways in which research is financed and regulated have an enormous impact on both innovation and the diffusion of innovation. This provides a unique opportunity to modify funding and regulatory policies to implement and reinforce the seven lessons considered above.

Homework

1. Contrast the rate of diffusion of the following two innovations: vitamin C to prevent scurvy in sailors, and laparoscopic cholecystectomy.

2. How are innovators, such as Kurt Semm, often viewed by others when introducing new technologies for health problems that replace treatments considered by leaders in the field as already successful?

3. Why does it take so long for a promising new technology to reach the market in the United States? What hurdles must researchers overcome to deliver an innovation that is marketable?

References

[1] Berwick DM. Disseminating innovations in health care. *Jama.* 2003 Apr 16; **289**(15): 1969–75.

[2] Balas E, Boren S. Managing clinical knowledge for health care improvement. In: Bemmel J, McCray A, eds. *Yearbook of Medical Informatics*. Stuttgart, Germany: Schattauer; 2000.

[3] Papageorgiou C, Savvides A, Zachariadis M. International Medical Technology Diffusion. *Journal of International Economics.* 2007; **72**(2): 409–27.

[4] Ferreira MR, Bennett RL, Gilman SC, Mathewson S, Bennett CL. Diffusion of laparoscopic cholecystectomy in the Veterans Affairs health care system, 1991–1995. *Effective Clinical Practice.* 1999 Mar–Apr; **2**(2): 49–55.

[5] Litynski GS. Kurt Semm and the fight against skepticism: Endoscopic haemosis, laparoscopic appendectomy and the Semm's impact on the 'laparoscopic revolution.' *JSLS.* 1998; **2**: 309–13.

Regulation of healthcare technologies

As technologies diffuse from the research laboratory into clinical practice, we frequently encounter a tension between the goal of ensuring that products are fully tested before they are made available to the general public and the desire to make potentially life-saving technologies rapidly available to patients in need. In the United States, the Food and Drug Administration (FDA) is responsible for ensuring that new health technologies made available to patients are safe and effective. The FDA regulates the manufacture, testing and sales of chemical and biological agents and medical devices; its jurisdiction encompasses a range of products that together account for one-fourth of all consumer spending in the USA [1].

In this chapter we will consider the process that the FDA follows in considering whether to approve new drugs and medical devices. We will see that this process has evolved over time – over the past hundred years, the role of the FDA has shifted from a system in which a company could market a drug unless the FDA could prove that it was unsafe to one where drug manufacturers must first obtain permission from the FDA at nearly every step in the testing, production and marketing process [1]. This increasing burden of regulation has occurred largely in response to tragedies which could have been prevented. Yet, we will see that, even today,

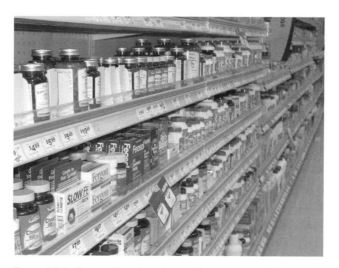

Figure 15.1. *Six out of ten Americans take one or more dietary supplements a day (multi-vitamins, amino acids, weight-loss cures, and herbal tonics) [3].*

FDA regulation of dietary supplements is substantially less stringent than that of drugs, sometimes making it difficult for the FDA to act and prevent harm before it occurs.

The distinction is important because more than 50% of Americans use some form of dietary supplements, including vitamins, minerals, amino acids, enzymes, herbs and other botanicals (Figure 15.1) [2]. In 2004, Americans spent more than $19 billion on dietary supplements [3]. We have already seen the important health benefits of dietary supplements. At the middle of the eighteenth century, scurvy killed more British sailors than war; simply supplementing the sailors' diet with vitamin C prevented scurvy [3]. Adequate dietary folic acid early in pregnancy can prevent neural tube defects in the fetus. Calcium builds strong bones and can prevent osteoporosis [2, 3]; low levels of vitamin B_{12} may lead to dementia [3]. Yet, unlike drugs and medical devices, the FDA does not review the safety and effectiveness of dietary supplements before they are marketed [4]. The FDA can only act to prohibit sale of a dietary supplement if it can prove that it presents a risk of injury [4].

The dangers of this approach are illustrated by the recent ban on sales of the dietary supplement ephedra (Figure 15.2), following the death of Steve Bechler, the 23 year old pitcher for the Baltimore Orioles [2, 3]. Bechler died in February, 2003 of heatstroke after taking an over-the-counter product containing ephedra. When it was sold, ephedra was the most popular supplement in USA, generating sales of more than \$1B/year, and accounting for approximately 10% of the annual sales of the US supplement industry [3]. Nearly ten years before Bechler's death, the FDA began to hear reports of the risks associated with ephedra. Yet, in 1999, it was estimated that more than 12 million people in the USA were using ephedra [2]. The risks of ephedra use (particularly when used with caffeine), include an increased risk of heart attack, stroke, palpitations, anxiety, psychosis, and death [3]. There is no evidence that ephedra containing supplements are effective except for short term weight loss [5]. Although ephedra products accounted for less than 1% of all dietary supplement sales, a study in 2003 found that they accounted for 64% of all adverse events associated with dietary supplements [4]. The rate of strokes was found to be statistically higher in users taking doses of ephedra above 32 mg a day; many ephedra containing supplement labels recommended that patients take more than 100 mg of ephedra per day [4]. In February of 2003, the FDA sent warning letters to manufacturers of ephedra challenging them to remove unproven claims on their product labels [2]. In February of 2004, the FDA moved to ban sales of ephedra because it posed an unreasonable

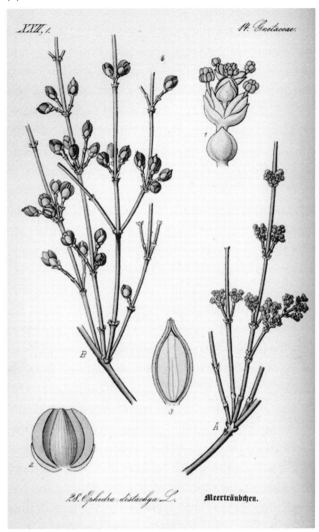

(a)

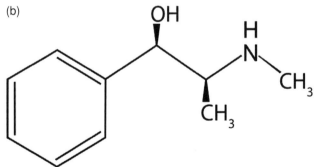

(b)

Figure 15.2. *(a) The dietary supplement ephedra is a naturally occurring substance derived from plants [4]. Wilhelm Thomé. 1885. Flora von Deutschland, Österreich und der Schweiz. Gera, Gemany. (b) The primary active ingredient in ephedra is ephedrine. Synthetic ephedrine is found in many over-the-counter remedies and some prescription drugs for treating colds, asthma, and nasal congestion [5], including Sudafed. Although the two products contain the same active ingredient (albeit at very different dosages), they are subject to a very different regulatory process in the USA.*

risk to patient safety. The ban took effect in April of 2004 [5].

In contrast to dietary supplements, the FDA requires that manufacturers provide scientific evidence to indicate that new drugs are both *safe* and *effective* before they can be marketed. Generally, data from both pre-clinical and clinical trials are required to gain approval to market a new drug. Pre-clinical testing in cells and animals must be completed to assess potential toxicity before FDA approval for human studies is granted. In order to test a new drug in people, a manufacturer must obtain an Investigational New Drug (IND) approval from the FDA. With an IND, human clinical trials are then allowed. In phase I trials, the safety of a compound is assessed by administering low doses of the new drug to a small group of healthy volunteers (20–100 volunteers). If successful, then phase II trials to assess the effectiveness of a compound can commence, usually in 100–300 patients who suffer from the condition targeted by the drug. The final step before seeking FDA approval to market a new drug is to carry out a randomized phase III clinical trial. After conducting phase I, II, and III clinical trials, a manufacturer can file an NDA (New Drug Application) for permission to market the new drug. If the NDA is approved, the manufacturer must still carry out studies to monitor for unanticipated complications of the drug and to study the longer term effects of drug exposure. This is called post-market surveillance. During this period, any adverse effects observed must be reported to the FDA [1].

Figure 15.3 illustrates the current drug development and approval process in the USA. The process is expensive, time consuming and complex. Not many drugs make it through this entire process. For every 5000–10,000 drugs that enter pre-clinical testing, on average only ONE makes it to market. It has been estimated that the cost of developing one new drug today ranges from $0.8 billion–$1.7 billion [6].

Do the current regulations governing approval for new drugs effectively balance protection from risk associated with potentially unsafe new drugs against gaining access to potentially effective new drugs? Are the fundamentally different approaches to the current regulation of drugs and dietary supplements appropriate? To help answer these important questions, it is helpful to examine the history of how the FDA regulates drugs, dietary supplements and medical devices.

In contrast to the current complex regulatory process for drugs, in the early 1900s, the manufacture and marketing of drugs was much less closely regulated. In the early 1900s, the patent medicine business accounted for more newspaper ads (Figure 15.4) than any other kind of product, a situation that is eerily similar to today's widespread Internet advertising for dietary supplements [3]. At the time, many supplements contained substantial amounts of alcohol; others were even laced with cocaine, caffeine, opium and morphine [3]. Manufacturers were not required to disclose the contents of patent medicines on the product label.

Largely in reaction to the sickening conditions of the meat-packing industry described in *The Jungle* by Upton Sinclair (Figure 15.5), Congress passed the Pure Food and Drug Act in 1906 [7]. This law permitted the newly formed Bureau of Chemistry, precursor to the FDA, to ensure that labels on foods and drugs contained no false or misleading advertising; in addition, labels were required to contain accurate levels of 11 dangerous ingredients including alcohol, heroin and cocaine [7].

While the Food and Drug Act provided some protection for consumers, it did not require that companies obtain approval before marketing new drugs. The FDA could act only after harm had occurred. In the early 1930s, sulfanilamide was used as an antibiotic to treat streptococcal infections. It had been used safely as a pill for years; to help make the drug easier to use for children who have difficulty swallowing pills,

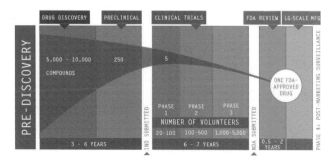

Figure 15.3. *Current drug development and approval process in the USA. Pharmaceutical Research and Manufacturers of America.*

Day 2 at SOS: Sex-ed and answering Snowball questions: July 3, 2007
Sophie Lesotho

On our second day at SOS, we incorporated a sex-ed lesson into our presentation. We explained a bit about the female and male reproductive systems using anatomy cartoon visuals that we found online.

The kids were very curious and asked many many questions. We inevitably had to deal with much giggling during our session, which I guess is normal. I remember having sex-education taught to me in 5th grade and thinking it was the funniest thing in the world.

We also focused on answering questions that came up the day before during our Snowball Activity. At the beginning of the activity we asked everyone to write down any question they had about health, HIV/AIDS, sex, etc., that they would normally be embarrassed to ask their teachers or ask in front of their peers. The purpose of this activity was to decrease the stigma associated with HIV . . . and just to answer any questions they had in a friendly and non-judgmental environment. After everyone was done writing their questions, we split them up into two teams and had them crumple up the pieces of paper they had written on into "snowballs." At the count of three, we had a huge snowball fight! This part was fun – the kids got really into this and were screaming and laughing. We felt that it was important to answer the questions that the kids had on this day since we came across many reoccurring questions on HIV/AIDS such as:

— What is the difference between HIV and AIDS?
— How does HIV attack the body?
— What are ARVs?
— What is a white blood cell?
— Can HIV be cured?

And we came across other serious questions such as:

— If a girl is living with her brother and gets raped by him, what will happen?
— What is love?
— We were told at school that condoms are not 100% safe. Why are they not and what are we supposed to do?

Later in the session, we played a version of Jeopardy asking the kids different questions covering the topics of immunity, Global AIDS, Prevention/Transmission, and Treatment. The kids participated well in this game and seemed to enjoy it very much. We found quickly that the kids knew too much for the questions we had made up for our game! They were answering almost all of them correctly.

Massengill, a Tennessee company found they could dissolve the drug in diethylene glycol (antifreeze) [7, 8]. They tested their formulation for flavor, appearance, and fragrance, but not for toxicity [3]. At the time, safety tests were not required by law [8]. The new product was called Elixir Sulfanilamide; it was shipped all over the country, and within weeks, more than 100 people were dead, most children [7]. After taking Elixir Sulfanilamide, they experienced severe abdominal pain, nausea, vomiting, and convulsions [3]. One woman who lost a child to the tragedy wrote to President Franklin Delano Roosevelt: "Even the memory of her is mixed with sorrow for we can see her little body tossing to and fro and hear that little voice screaming with pain and it seems as if it would drive me insane" [3].

Figure 15.4. *Lydia Pinkham's Vegetable Compound was an example of a patent medicine advertised in the newspaper. Pinkham's compound was advertised as "a positive cure for all those painful complaints and weaknesses so common to our female population." In 1914, the American Medical Association analyzed Pinkham's compound and found it to contain 20% pure alcohol and 80% pure vegetable extracts [3]. Schlesinger Liberary, Radcliffe Institute, Harvard University.*

Figure 15.5. *Upton Sinclair on the cover of* Time *Magazine. Associated Press.*

In reaction to the tragedy, in 1938, Congress passed the Food, Drug and Cosmetic Act (FD&C Act) [3, 7, 8]. The FD&C Act gave the FDA the authority it needed to regulate such products before they came on the market. It required that companies test new drugs for safety before marketing. Companies were required to submit test results to the FDA in a New Drug Application (NDA).

At this point in time, companies could test drugs in patients without approval from the FDA. It took yet another tragedy to change this provision. In the 1960s, the William Merrell company hoped to market the drug thalidomide in the USA. It submitted an NDA to the FDA to market the drug [8, 9]. However, an FDA medical officer, Frances Kelsey (Figure 15.6), refused to approve the NDA because of insufficient data regarding the safety of the drug. Even though the drug was never approved to be marketed in the USA, the company distributed over two million tablets in the USA for investigational use, which was not regulated by the FDA at that time. By 1962, it became clear that thalidomide caused severe birth defects, and the FDA quickly acted to seize the supply of thalidomide that had been distributed.

In response to this averted tragedy, President Kennedy signed the Drug Amendments Act into law in 1962. The new law required that FDA be provided with full details of planned clinical investigations of new drugs, requiring previous animal testing before initiating human clinical trials. In addition, it required that manufacturers prove new drugs to be both safe developing heart disease, diabetes, and cancer. In 1984, Kellogg's, together with the National Cancer Institute, launched a marketing campaign for All-Bran cereal that illustrated how a low-fat, high-fiber diet might reduce risk of certain cancers [3].

Research in alternative medicine

The National Center for Complementary and Alternative Medicine (NCCAM) is the branch of the NIH which oversees scientific research on complementary and alternative medicine (CAM).

The mission of NCCAM is to rigorously and scientifically explore complementary and alternative healing practices, to train complementary and alternative medicine researchers, and to disseminate accurate information to the public and professionals.

http://nccam.nih.gov/

Over this same period, the popularity of dietary supplements increased dramatically. Between 1990 and 1997, the use of herbal remedies increased almost four-fold in the USA [10]. In 1994, after intense lobbying from the dietary supplement industry, Congress passed the Dietary Supplement Health Education Act (DSHEA). DSHEA permitted supplement manufacturers to make statements about the role of their products in health without FDA approval, provided that they could substantiate the claims with scientific evidence and that they included a disclaimer that the statements had not been evaluated by the FDA [7]. Unlike drug manufacturers, companies selling supplements are not required to prove products are effective or safe before marketing them [3]. Table 15.1 outlines the types of claims which can be made regarding dietary supplements under DSHEA, and whether FDA review of supporting data and approval are required [10]. Manufacturers are not permitted to make claims linking their product to a specific disease without FDA review. For example, without FDA review, manufacturers CANNOT say that a product reduces cholesterol but they CAN say it maintains healthy cholesterol levels if they have

Figure 15.6. *Frances Kelsey received the President's Distinguished Federal Civilian Service Award in 1962, from President John F. Kennedy; she was the second woman ever to receive the award [8, 9]. Her actions at the FDA are credited with preventing thousands of birth defects in the United States. Courtesy of the National Library of Medicine, NIH.*

and effective before marketing them. The FDA was given control of all prescription drug advertising; previously this had been regulated by the Federal Trade Commission.

While the FDA is also charged with regulating food safety, up until the 1960s, the line between foods and drugs was fairly clear and it was usually straightforward to determine whether a new product should be regulated as a food or a drug. However, over the past few decades, the line dividing foods and drugs has become increasingly less clear [3]. For example, in the 1970s, several government commissions issued recommendations encouraging Americans to alter their diets if they wanted to have longer, healthier lives [3]. We have seen that changes in diet can reduce a patient's risk of

Challenges of health technology regulation in developing countries

Many developing countries struggle with the challenges of effectively regulating foods, drugs and medical devices. As commerce becomes increasingly global, these challenges are a concern for patients and consumers everywhere. The recent deaths of more than 100 people in Panama due to cold medicines and antihistamines containing diethylene glycol are eerily reminiscent of the 1938 Elixir Sulfanilamide tragedy in the United States.

The Panamanian tragedy resulted when the government of Panama manufactured cold medicines and antihistamine syrups with what it believed was pharmaceutical grade glycerin imported from Barcelona. On its way to Panama, the solvent passed through three trading companies on three continents, originating in China. Unfortunately, the barrels of solvent sent to Panama did not contain pharmaceutical grade glycerin, which costs about $1815/ton; instead the Chinese manufacturer had substituted the less expensive solvent diethylene glycol which is similar in appearance to glycerin, but costs only $725–845/ton. The company was not certified to make pharmaceutical grade ingredients, and falsely certified the purity of the solvent. After Panamanian children began to die as a result of taking the cough syrup, Panama asked the CDC to test the medicine and found that it contained the poison diethylene glycol. The figures in this box show the original source of the solvent in China, and tracks its shipment and labeling as it made its way to Panama (copyright © 2007 New York Times Graphics).

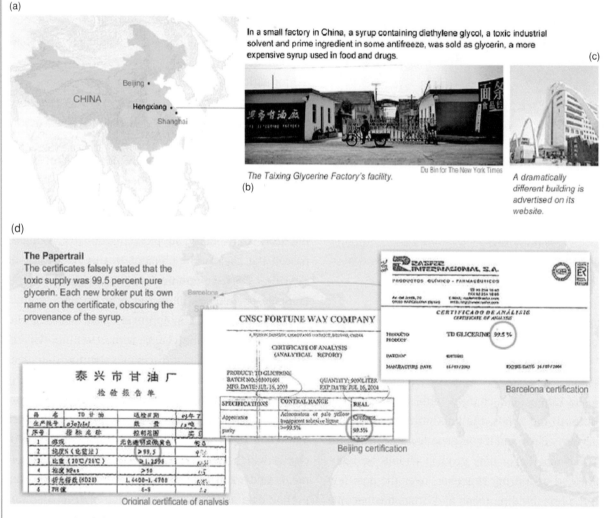

(a)

In a small factory in China, a syrup containing diethylene glycol, a toxic industrial solvent and prime ingredient in some antifreeze, was sold as glycerin, a more expensive syrup used in food and drugs.

CHINA

Beijing •

Hengxiang •

Shanghai

(c)

Du Bin for The New York Times

The Taixing Glycerine Factory's facility.

(b)

A dramatically different building is advertised on its website.

(d)

The Papertrail
The certificates falsely stated that the toxic supply was 99.5 percent pure glycerin. Each new broker put its own name on the certificate, obscuring the provenance of the syrup.

Original certificate of analysis

Beijing certification

Barcelona certification

As a result of the tragedy, the FDA has now recommended that all glycerin shipments be tested for diethylene glycol. More recently, in 2007 the FDA issued a consumer advisory warning people to discard all toothpaste made in China after federal officials discovered toothpaste containing diethylene glycol in Miami, Los Angeles, and Puerto Rico. Diethylene glycol was not listed as an ingredient on the label of the tainted toothpaste, but was discovered after FDA officials began to test Chinese made toothpaste following similar discoveries in toothpaste shipped from China to Latin America [19].

Table 15.1. *Claims that can be made about dietary supplements, according to the FDA. From [10].*

Table 1. Types of Dietary Supplement Claims Recognized and Permitted by the Food and Drug Administration

Type of Claim	Definition	FDA Preapproval Required	Acceptable Examples
Health	Approved claims of relationships between a nutrient and a disease or condition, provided certain other components to the claim are included	Yes	"A diet with enough calcium helps teen and young adult white and Asian women maintain good bone health and may reduce their high risk of osteoporosis later in life"[7] "Healthful diets with adequate folate may reduce a woman's risk of having a child with a brain or spinal cord defect"[8]
Nutrient content	Descriptions of the relative amounts of a nutrient in a product, as per specific FDA regulations	Yes	Low in sodium, fat free
Structure or function	Role of a nutrient intended to affect the structure or function in humans or that characterize the documented mechanism by which a nutrient acts to maintain such structure, provided that such statements are not disease claims	No	Saw palmetto supports prostate function Echinacea helps to maintain immune function

> **Echinacea** is one of the most commonly used cold remedies in USA. A recent clinical was carried out to determine whether echinacea could reduce the duration of the common cold. In a placebo controlled trial involving 400 children with common colds over four months, researchers found that the placebo worked just as well, and that children taking echinacea were more likely to develop a rash than those receiving placebo [3].

evidence to support this. They CANNOT say that **echinacea** cures disease, but with evidence CAN say it has natural antibiotic activities and is considered an excellent herb for infections of all kinds [3].

Do the majority of supplement retailers follow these rules? A recent study published in the *Journal of the American Medical Association* reviewed information presented about the eight most widely used herbal supplements found using the five most popular Internet search engines [10]. More than 80% of retail sites contained statements that made health-related claims (Table 15.2); more than half claimed to treat, prevent, diagnose or cure specific diseases even though these disease-related claims had not been reviewed by the FDA as required [10]. In addition, half of the sites with health-related claims omitted the required disclaimer that the statements had not been evaluated by the FDA [10].

Under DSHEA, there are almost no standards that regulate how dietary supplement pills are made, and they are not required to be tested once they are made [3]. DHSEA also places the burden of proving that a supplement is mislabeled on the FDA [7]. In response to the ban on ephedra, Congress has considered, but not passed, several bills that would modify 1994 law so that many unregulated botanical supplements would be treated more like drugs than like foods.

How does the regulation of medical devices compare with that of drugs and dietary supplements? The FDA did not regulate medical devices before 1938. Similar to drugs, after the Food, Drug and Cosmetic Act was passed in 1938, the FDA could only challenge the sale of medical devices it believed were unsafe and could only remove them from the market after patient injuries had occurred as a result of their use.

We have seen the rapid innovations in medical technology that occurred throughout the 1960s; in the field of cardiology alone, the heart – lung machine, prosthetic heart valves and many other technologies first came into use. The FDA initially tried to regulate many medical devices as drugs. After a number of catastrophic failures of life sustaining medical devices in the 1970s, including some types of heart valves and pacemakers, there was broad recognition that different rules were needed to regulate devices [11].

In 1976, Congress passed the Device Amendments to FD&C Act. These amendments recognized that no single regulatory policy would work for all devices. The rules for regulating the manufacturing and sales of a

Table 15.2. *Illegal health-related claims about dietary supplements are commonly found on retail websites.*

Sample Disease-Related Claims from Dietary Supplement Web Sites: From [10].

Gingko Biloba
"Its effects in improving circulation also contribute to its use for impotency and peripheral vascular insufficiency...Gingko treats depression, headaches, memory loss and ringing in the ears (tinnitus). It is also recommended for Alzheimer's, asthma, eczema, heart and kidney disorders."

St. John's Wort
"St. John's Wort is effective in the treatment of mild to moderate depression...recent studies have shown that it could have a potent anti-viral effect against enveloped viruses."

Echinacea
"Because it has natural antibiotic actions, Echinacea is considered an excellent herb for infections of all kinds. In addition, it works to boost lymphatic cleansing of the blood, enhances the immune system and has cortisone like properties which contribute tot its anti0imflammatory action. It is recommend for stubborn viral infections, yeast infections and for arthritic conditions."

Ginseng
Q: "(I have) high blood pressure (170/90). Will American Ginseng lower blood pressure, and if so, how much should one take and how long before results show"?
A: "While American Ginseng will help, we have combination product that will do a much better job. Look at product #1960 American Ginseng/Garlic/Tien Chi. This is a great product."
"It is potentially beneficial for AIDS, radiotherapy, and chemotherapy patients, as it reduces the side effects of toxic drugs by increasing red and white blood cell counts. Dang Shen is given for breast cancer, asthma, diabetes, heart palpitations, memory or appetite loss and insomnia."

Saw Palmetto
"The lipophilic extract of the saw palmetto (ser repens) berries is the most widely used herbal preventive and therapeutic agent for benign prostatic hypertrophy (BPH)."

Kava Kava
"It is a valuable urinary antiseptic, helping to counter urinary infections and to settle an irritable bladder...Kava kava's analgesic and cleansing diuretic effect often makes it beneficial for treating rheumatic and arthritic problems such as gout."

Valerian Root
"The herb valerian is most effective in treating a wide range of stress conditions such as irritability, depression, fear, anxiety, nervous exhaustion, hysteria, delusions, and nervous tension...The herb is useful for treating shingles, sciatica, neuralgia, multiple sclerosis and epilepsy."

---

tongue depressor should be different from those which are applied to an artificial heart.

The 1976 Amendments recognized three classes of medical devices. Class I devices pose the least risk to patient; they are not life sustaining, and encompass about 30% of medical devices including devices such as X-ray film, tongue depressors, and stethoscopes. The FDA requires that manufacturers use a process called Good Manufacturing Practices when making some Class I devices (stethoscopes, but not tongue depressors); essentially this entails keeping extensive records about the source of suppliers for all parts of the device. Class II devices are not life sustaining, but must meet performance standards. Class II devices make up about 60% of medical devices and include things like blood pressure monitors, and catheter guide wires. Class III devices are for use in supporting or sustaining human life, and they pose the greatest risk to patient. About 10% of medical devices are considered to be class III devices, including stents, heart valves, and LVADs. These devices face the most stringent approval process [1, 11, 12].

Manufacturers must show that a new device is both safe and effective and receive approval from the FDA prior to marketing the device [1, 11, 12]. Because medical devices were not formally regulated until 1976, there are two paths to obtain approval to market a new device.

Pre-market Notification Process (510K)

The 510K is the approval path for new class I or II devices which are considered to be substantially equivalent to a device already on the market prior to 1976. The manufacturer notifies the FDA 90 days prior to when they plan to market the new device through a 510K application. In order to be considered substantially equivalent, a new device must have the same indications for use, and be no more risky and no less effective than a pre-1976 device [1].

Pre-market Approval Application (PMAA)

The PMAA is the approval path for all new class III devices, as well as the path for class I or II devices that

History of regulations

The history of regulating drugs in the USA is the repeating story of misfortune, disaster, and tragedy – leading to reforms in drug regulation

1906: Pure Food and Drug Act

Drug labels could not contain any statement regarding therapeutic effect which is false and/or fraudulent. The FDA could act only after drugs were marketed [6]. To prevent marketing of an ineffective drug, it was not enough to show that the product did not work; the government had to show that the seller knew the claims it made were false [6].

1938: Food, Drug and Cosmetic Act

Marked the beginning of era in which it is illegal to market a new drug without FDA approval. New drugs could not be marketed without first notifying the FDA and allowing agency time to assess safety. The seller's belief regarding the product's value was no longer relevant. The central issue became does the product really work [6]?

1962: Drug Amendments to FD&C Act

Converted pre-market notification system into pre-market approval system where the FDA must review evidence of drug safety and effectiveness. Required that evidence of safety and efficacy come from well-controlled investigations by qualified experts. The FDA has the authority to prevent harm before it occurs [6].

1976: Device Amendments to FD&C Act

Devices are assigned to one of three classes. Based on the class of the device, the FDA may require pre-market approval or simply provide oversight of the manufacturing process and device labeling [6].

1994: Dietary Supplement Health & Education Act

Deregulated the dietary supplement industry. Manufacturers are not required to notify the FDA before they market supplements and they are not required to test safety or efficacy. Manufacturers are allowed to make health-related claims without FDA review if: they are supported by scientific evidence, they do not mention a specific disease, and are accompanied by a disclaimer that indicates statements have not been evaluated by the FDA [9, 10].

have no pre-1976 equivalent. Manufacturers submit a PMAA and provide the FDA with full reports of studies to show the safety and efficacy of the new device. All investigational studies must have been carried out with prior FDA approval under an IDE [1].

Investigational Device Exemption (IDE)

An IDE is required for clinical trials using an investigational device which will be used in a way that could pose a significant risk to study participants [11, 12]. An IDE application includes a complete description of the device as well as the planned study.

In the device approval process, the FDA considers the device and its intended use together. After the manufacturer submits a request for marketing approval, the FDA convenes an advisory panel to act on the request. The advisory panel includes physicians and scientists with expertise in the field of use relevant to the device. In addition, the panel includes two non-voting members: one consumer representative and one industry representative. Although the FDA is not required to follow the recommendations of panel, it usually does [12].

Once a device is approved for use, a Medical Device Reporting System is used to detect device related

Humanitarian use exemption

In order to provide an incentive for companies to develop devices for rare conditions, the FDA provides a third path for device approval. If a device is designed to treat or diagnose a condition that affects fewer than 4000 patients/year, would not otherwise be available without exemption, and no comparable device is available, companies can apply for approval under a humanitarian use exemption. To obtain approval, the manufacturer must show that patients will not be exposed to unreasonable or significant risk of injury or illness by device.

Recently approved devices

Information about newly approved medical devices can be found on the FDA's website.
http://www.accessdata.fda.gov/scripts/cdrh/cfdocs/cfTopic/MDA/mda-list.cfm?list=1

For example, on October 24, 2003, the agency approved a new type of coronary artery stent , the NIRflex Stent System, to treat coronary artery atherosclerosis. A summary of the action and the evidence of the safety and efficacy of the device can be found at:
http://www.fda.gov/cdrh/mda/docs/p020040.html

The approval letter from the FDA specifies the conditions of the approval, and a copy of the letter can be found at:
http://www.fda.gov/cdrh/PDF2/P020040a.pdf

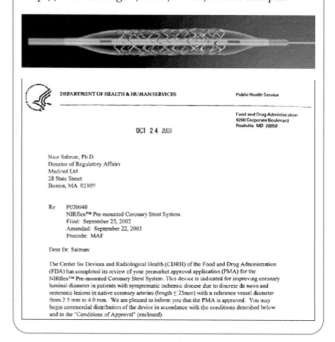

problems in a timely manner. Clinicians are required to report serious injuries or deaths that may have been caused by or related to a medical device within ten days.

The FDA requires post-marketing surveillance for both new devices and drugs. In 2004–2005, two new drugs were removed from the market due to side effects identified after they had been approved for use. The case of the selective non-steroidal anti-inflammatory drugs, Celebrex, Bextra and Vioxx, provides a good example of both the advantages and challenges of the regulatory process [13–16]. Vioxx was manufactured by Merck, and was introduced in 1999 as an alternative to the available non-selective, non-steroidal, anti-inflammatory drugs (NSAIDs) for treatment of osteoarthritis related pain [17]. At the time, the existing non-selective NSAIDs were effective in treating this pain, but these drugs produced gastrointestinal side effects such as ulcers. These existing non-specific NSAIDs inhibited two types of COX enzymes; it was thought that inhibition of COX 2 was responsible for the anti-inflammatory effect (and pain reduction), while inhibition of COX 1 caused the gastrointestinal side effects. Vioxx selectively inhibited COX 2, so it was thought that it would relieve pain without gastrointestinal side effects [18]. Vioxx received

approval from the FDA in 1999, and sales grew rapidly [18].

A 1996 study sponsored by Merck showed that COX 2 inhibition could also precipitate the formation of thrombus in the vascular endothelium [18], which suggested that the drug might have cardiovascular side effects. Despite the biological plausibility of this hypothesis, none of the clinical studies that were part of the 1998

Grassroots Soccer week: July 19, 2007
Sophie Lesotho

We have been collaborating with an organization called Grassroots Soccer, which mixes HIV education with sports activities. A coordinator for Grassroots Soccer in Lesotho, Refiloe, helped us all of last week to conduct the different activities with the children. One of the best games was "HIV Dodge." The activity clearly shows the role of the immune system, how HIV attacks it, and how ARVs attack HIV. It is a dodgeball-like game where all of the people standing around the human in a circle act as germs to attack the human in the middle with a dodgeball. The human cannot move, but the immune system acts as a defense to prevent the human from getting hit. The children then count how many times the human is hit. In the next round, HIV comes in and holds the hands of the immune system behind his/her back. The human is hit many more times than when HIV was absent. In the final round, ARVs come in and hold HIV's hands behind his/her back and the human is once again protected against the attack of germs.

All of the games were cleverly thought out and fun to participate in. What bothered me though was that the kids didn't seem to remember the messages as much as they did the games themselves. This somewhat surprised me because I had the notion the games would always work to drive in the educational messages. Unfortunately, this does not always seem to be the case. Overall however, the kids participated well in the activities and I hope that the activities helped to increase their understanding of HIV concepts.

At the end of the week, we gave the kids post-test evaluations to see how much they learned during the weeks we have spent with them (we gave them pre-test evaluations during our first session). I also interviewed some of the children on what they have learned about HIV/AIDS, what their favorite activities were, and also what their aspirations are for the future. We will continue working with the SOS children over the next two weeks that we are here, but we will be reiterating what we have already taught them – except this time, their role will be to teach others. We are having the kids perform and put together "Immunity Skits" that illustrate the role of the immune system, how HIV affects it, and how ARVs work (much like the dodgeball game except using acting). We will record these skits on the video camera and hopefully show this tape in the clinic waiting room and in schools back in Houston.

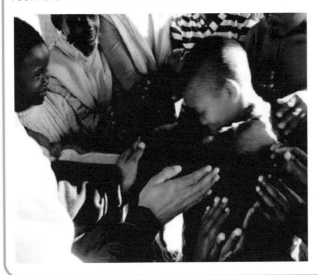 

NDA for Vioxx were designed to evaluate cardiovascular risk associated with the drug [17].

Subsequent to its initial approval, a large study of 8000 patients was carried out to expand the possible indications for use of Vioxx [17, 18]. In late 1999, the DSMB monitoring the study noted a 79% greater risk of cardiovascular death or serious cardiovascular event for those taking Vioxx [18]. The board recommended that the study continue, but that a plan to monitor adverse cardiac events be developed. A number of the members of the DSMB had consulting relationships with Merck, and it is unclear whether this potential conflict of interest influenced their recommendation [17]. When data from the study were finally published, the study authors used different stopping dates for the cardiovascular adverse events than they did for the primary study outcome. This highly irregular procedure was not described in the methods section of the paper. More importantly, it had the effect of underestimating the risk of cardiovascular adverse events [17]. In 2002, the FDA required that the Vioxx label be changed to indicate that physicians should exercise caution in prescribing Vioxx for patients with a history of ischemic heart disease [18].

Another large study in 2004 reported that Vioxx was associated with increased cardiovascular risk, but only after 18 months of use [17]. Subsequently, the study authors, who included five Merck employees and other authors who had received consulting fees from Merck, noted that their study contained flaws in the statistical analysis which understated the risk of cardiovascular side effects [17].

Not long afterwards, two other randomized studies noted the increased cardiovascular risk of Vioxx [18], causing the FDA to issue a cautionary note warning of the risk of cardiovascular events associated with the drug, and requiring a black box warning to be added to the drug label. Subsequently, Vioxx was voluntarily withdrawn from the market by Merck [17], but only after more than 80 million patients had taken the drug [18].

While the continued surveillance of Vioxx did ultimately identify the increased risk and lead to FDA action, the action may have been delayed as a result of conflicts of interest. Because Merck had billions of dollars at stake, and also paid consulting fees to some of the academic investigators who carried out the clinical studies, published the results, and served on the DSMB of the trials, ethical concerns were raised [17, 18]. In fact, in later reviewing the adverse cardiovascular event data from 18 different clinical trials of Vioxx, investigators showed that the only source of variation in whether Vioxx was judged to be associated with increased risk of myocardial infarction was whether

the events in each trial were examined by an independent, external committee or not [18]. As a result, many have recommended reforms to the drug development and regulatory process. For example, it has been suggested that academics engaged in industry-designed studies should have full access to clinical data and that these data should be made accessible to the public [17].

After Merck withdrew Vioxx from the market, the FDA began a systematic reexamination of the risks associated with both selective and non-selective NSAIDs. In 2005, FDA asked Pfizer to withdraw the selective NSAID Bextra from the market because the risk-benefit ratio was judged to be unfavorable [16]. The FDA advisory panel reviewing the data from Bextra voted 17 in favor, 13 against keeping Bextra on the market; but FDA judged that, given the closely split decision, regulatory action should be taken [13]. The advisory committee assessing the risk of Celebrex unanimously recommended that it remain on the market; FDA concurred and recommended labeling changes to highlight the increased cardiovascular risk associated with the drug. Finally, FDA also recommended labeling changes for non-prescription non-selective NSAIDs (e.g. Advil, Motrin) to highlight the potential cardiovascular and gastrointestinal risks associated with these products.

The story of Vioxx illustrates an important challenge in the development of new health technologies. The current development path for new medical products has become increasingly challenging and costly [6]. Just over the last decade, the costs of developing new drugs have escalated rapidly (Figure 15.7) to more than $1.7 billion. In part, late failures of new drugs due to unexpected adverse effects contribute to these increasing product development costs.

The increasing costs of development have been paralleled by increases in health-related research and development (R&D) funding in the United States. Figure 15.8 shows that, over the period 1993–2003, R&D funding increased dramatically, both in the private sector and from the federal government primarily via the National Institutes of Health. Despite this increase in R&D spending, there has been a dramatic decline in the number

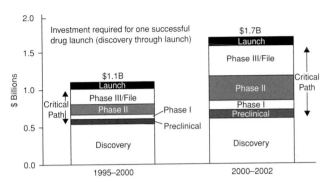

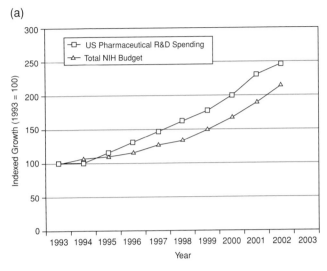

Figure 15.7. *The costs of developing one new drug have rapidly escalated [6]. Rebuilding Big Pharma's Business Model In Vivo. Nov. 2003 © 2008. Windhover Information Inc. This article cannot be reprinted without express permission of Windhover Information.*

of new applications submitted to the FDA for approval of new drugs and biological agents over the same time period (Figure 15.8). If the cost and complexity of developing new health technologies continues to increase in a way that outpaces R&D investments, health innovation will likely decline [6]. In the final part of this chapter, we examine how health-related research is funded in the USA and what actions are being taken to sustain and expand innovation.

Figure 15.9 provides a detailed picture of who carries out research in the USA and who pays for it. In this graph, research is divided into three types: basic research, applied research and development. The graph on the right illustrates that most basic research is carried out in universities and colleges. As we move to applied research and development, we see that most is carried

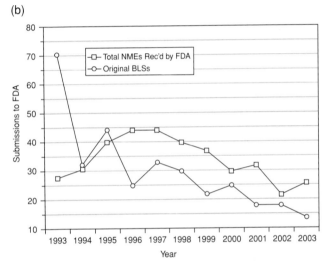

Figure 15.8a,b. *Despite an increase in R&D spending, the number of applications for new drugs and biologic agents has declined dramatically over the past decade [6].*

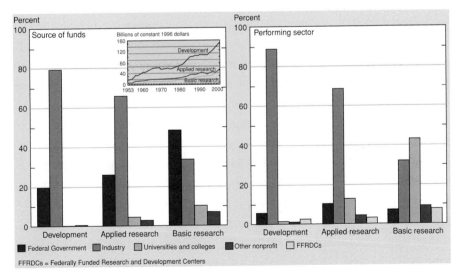

FFRDCs = Federally Funded Research and Development Centers

Figure 15.9. *Source of funds and performing sector for US research and development in 2000. Courtesy of FDA.*

out in private industry. The graph on the left shows who pays for R&D; basic research is largely supported by the federal government, whereas applied research and development is funded primarily by industry.

The federal government funds approximately 36% of all medical research in the USA. This is mostly funded through the **National Institutes of Health** which is organized into a number of different disease- and technology-related institutes. In 2006, the annual NIH budget was $28 billion. The NIH budget doubled from 1998 to 2003, and has been flat through 2007. While the focus of the NIH is on basic research, the contributions to development of new products has been substantial. A US Senate Report in May 2000 examined the role of public funding in the development of the 21 new drugs introduced between 1965 and 1992 considered by experts to have had highest therapeutic impact on society. The report found that public funding of research was key in developing 15 of the 21 drugs. Three of these drugs, captopril (Capoten), fluoxetine (Prozac), and acyclovir (Zovirax), had more than $1 billion in sales in 1994 and 1995. Others, including AZT, acyclovir, fluconazole (Diflucan), foscarnet (Foscavir), and ketoconazole (Nizoral), had NIH funding and research to help in clinical trials.

The NIH carries out internal research and also makes grants to academic and industrial researchers to carry out health related research. In general, the NIH sets research priorities by issuing requests for proposals through one of its institutes. Investigators write a proposal, describing their hypothesis, the significance of the proposed research, any preliminary results they have obtained, their planned research design and methods, and the plans to protect any animals or human subjects to be used. The NIH convenes a panel of experts in the field who critique the proposal and provide a numeric score to be used in determining whether the work should be funded. Following the review, the investigator receives both the score and the written comments. Each institute within the NIH then reviews the scores of all submitted proposals and decides which should receive funding. The scores range from a best of 100 and a worst of 500; typically proposals scored

National Institutes of Health

- National Cancer Institute
- National Eye Institute
- National Heart, Lung, & Blood Institute
- National Human Genome Research Institute
- National Institute on Aging
- National Institute on Alcohol Abuse & Alcoholism
- National Institute of Allergy & Infectious Diseases
- National Institute of Arthritis and Musculoskeletal and Skin Diseases
- National Institute of Biomedical Imaging & Bioengineering
- National Institute of Child Health & Human Development
- National Institute on Deafness & Other Communication Disorders
- National Institute of Dental & Craniofacial Research
- National Institute of Diabetes and Digestive and Kidney Diseases
- National Institute on Drug Abuse
- National Institute of Environmental Health Sciences
- National Institute of General Medical Sciences
- National Institute of Neurological Disorders & Stroke
- National Institute of Nursing Research
- National Library of Medicine

in the range of 100–170 are funded. Approximately 10–15% of submitted proposals are funded, although the fraction of funded proposals has been dropping steadily over the last four years because the NIH budget has not kept pace with the number of submitted proposals.

What does the future of biomedical research and technology diffusion hold? Today, many researchers and policy makers are concerned that the converging challenges of flat research budgets and the increasing complexity of new technologies and their regulation will limit our ability to achieve the potential health gains that could result from new basic science knowledge. Both the NIH and the FDA have introduced new programs to address these challenges. Recognizing the increasing time required for new technologies to diffuse from the bench to the bedside, in 2003 the NIH announced a series of new initiatives called the Roadmap to increase the emphasis on translational research. These initiatives are focused on creating new research approaches and teams that can work to speed the transition from the research laboratory to the clinical environment [6].

In parallel, to address the challenges associated with the increasingly complicated product development path, the FDA announced the Critical Path Initiative [6]. The goal of this initiative is to develop new scientific and technical methods to improve the predictability and efficiency on the path from laboratory prototype to commercial product [6]. For example, one pharmaceutical company has estimated that drugs which initially appeared promising but later failed due to liver toxicity observed in clinical trials have cost more than $2 billion in the past decade [6]. New predictive biological or computational models to predict liver failure early in the drug development process could have an enormous impact to speed and reduce the cost of drug development. These may include animal or computer methods, or biomarkers to predict risk of future toxicity, or new clinical evaluation techniques. Figure 15.10 illustrates the process of product development, from basic research to FDA approval, and illustrates where improvements in translational research and the critical path can help speed the successful development and diffusion of new health technologies.

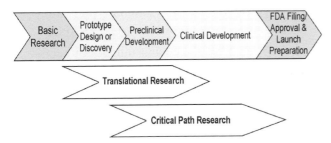

Figure 15.10. *The process of developing new health technologies and its relation to translational research and the critical path [6]. Adapted from:* Innovation or Stagnation – FDA paper.

Missing our SOS Kids Already: July 27, 2007
Sophie Lesotho

I am in love . . . with a little girl named Lisemelo. She is the most beautiful and sweetest little girl I have ever met in my life. Every time we walk by her home at the SOS orphanage, she is always waiting on the side of the road for us because she knows by now the times that we usually walk by. She just stands there and smiles at you so sweetly. Then, when you run up to her to give her a hug, she opens her arms so wide for you to pick her up (she is a tiny two-year-old). I wanted to cry today though . . . instead of walking back to the clinic from the SOS Village, Christina and I were picked up. As we drove by little Lisemelo's home, she was standing out on the street and she could see us in the car as we approached. I waved to her and her face looked so sad when she realized that there would be no hugs today. I looked out the back car window after we had passed and I could see her walking up closer to the street and looking after our car. I was so sad . . . oh goodness, what do I do? I am so in love with little Lisemelo! But! I love all of our SOS kids. They are so funny. I caught a recording of them doing a dance-off of sorts in front of the camera today. I was filming while Christina was trying to teach the kids a hip-hop dance that they are learning as part of their Immunology Skit which they will be performing next week. Their awesome personalities shined through so well in this recording. I wish that there was some way to post it up on the blog. As I filmed, all of the kids (from ages four all the way up to 16) were laughing and jumping in front of the camera showing off their best moves. They kept pushing each other aside to hog the camera view and were laughing as they tried to steal the spotlight from each other. It is one of the funniest and cutest things I have ever seen. I love our SOS kids . . . I don't know what I'm going to do without them.

Christina teaching the SOS kids a dance:

Homework

1. You have developed a new type gold nanoparticle to improve a physician's ability to detect cancer at the earliest possible stages. Many Internet based health food stores sell solutions of identical gold nanoparticles to "lift your body's performance, and fight off germs, viruses, bacteria, allergens, pathogens and pollution." Yet, it will be several years before you can begin clinical trials to determine whether this same nanoparticle can improve early cancer detection.
   http://alchemistsworkshop.com/_wsn/page8.html

   a. Describe the current differences in FDA regulation of this particle when it is used as a dietary supplement vs. as a drug.

   b. Briefly summarize the history of government regulation of drugs in the USA, noting the year of major changes in legislation and the primary changes in regulation associated with these laws.

2. Consider the differences in regulation of drugs and dietary supplements.

   a. If I wish to market a new dietary supplement which I claim will improve immune function, am I obligated to provide scientific data to the FDA indicating that it is safe and/or effective before I can sell it?

   b. If I wish to market a new drug to treat pancreatic cancer, am I obligated to provide scientific data to the FDA indicating that it is safe and/or effective before I can sell it?

   c. The following is an actual Internet ad for an herbal supplement. If you were employed by the FDA to monitor and investigate these ads please indicate statements you see that may not follow legal guidelines in the United States. Explain why you chose each of your selections. If you did not identify anything, explain why.

   There's a new nutritional supplement available for people suffering from Type 2 diabetes. It's called The Body Rejuvenator, marketed by Lafayette Miracle Solutions. It contains two key ingredients – green tea extract and cinnamon. The first thing to realize is that nutritional supplements can very successfully control blood sugar in diabetics. Both green tea and cinnamon are well-known to help control blood sugar so that you don't have such wild blood sugar swings (and potentially don't need as much insulin either). Also, there are many other benefits documented from taking both green tea and cinnamon. Green tea is noted for its anti-cancer effects, as well as its ability to aid in weight loss, which is something that diabetics are typically concerned with.

3. I wish to market a new dietary supplement. Indicate which of the statements in the list below I am legally allowed to put on the product label. If a statement would not be allowed, indicate why not.

   a. Acidophilus, Bifidus & Bulfaricus promote the health of the digestive tract.

   b. Black Currant Oil contains essential fatty acids that provide dietary support for normal healthy blood lipids and helps to support the cardiovascular system.

   c. SkinAnswer, a glycoalkaloid skin cream, as a treatment for skin cancer.

   d. Ephedra-free Total Lean^{tm} helps dieters increase their metabolism and boost their energy.

   e. MGN-3, a rice-bran extract, a treatment for HIV, the virus that causes AIDS.

   f. ZantrexTM-3 promises 546% more weight loss than the leading ephedra-based diet pill and that's a fact. Here's another fact: Zantrex-3 is way beyond ephedra, way beyond fat-burners, way beyond everything on the market today. Zantrex-3 is a new category of bifurcated weight loss compounds providing both rapid weight loss and incredible energy combined into a single power-packed Super Pill. New Zantrex-3 is so powerful you won't find it in some Wal-Mart next to some "Flintrock" vitamin for kiddies.

   g. BeneFin, which is produced from shark cartilage, is a treatment for cancer.

4. Over the past 100 years, the role of the FDA in regulating drugs has changed significantly. Briefly describe the history of these changes. Contrast changes over time in the history of FDA regulation of medical devices.

5. Read the following article. Explain how, after passing all the safeguards of pre-clinical testing,

phases I–III of clinical trials, and required scrutiny of an FDA panel, problems such as these could occur.

April 23, 2004

F.D.A. Seeks Reports of Stent Problems

By GINA KOLATA

The Food and Drug Administration is actively seeking reports of possible problems with a stent that came on the market last month, saying it has heard of serious medical complications in some cases.

But Dr. Daniel G. Schultz, director of the agency's office of device evaluation, said in a telephone interview on Tuesday that it was too soon to say whether there was a problem with the stent and, if so, what was causing it and what advice to give doctors and patients. The F.D.A. knows of 20 to 25 incidents, Dr. Schultz said, but the reports range from sketchy to highly detailed.

"We're fairly early in the process of assessing the reports", Dr. Schultz said. "At this stage, our main goal is to gather more information."

The device's maker, Boston Scientific, says that its stent is safe and is performing excellently and that any problems are extraordinarily rare.

Paul LaViolette, senior vice president at the company, said more than 70,000 of the stents had been used in the United States since the device went on sale in March.

"We have to conclude, and I will say this with a lot of experience, that this product is performing extremely well," Mr. LaViolette said.

But a few cardiologists reported in telephone interviews that they got into trouble after the stent, a small wire tube used to hold open arteries, was slipped into place.

Like all stents, Boston Scientific's stent, the Taxus Express2, comes packaged with a deflated balloon inside. A cardiologist threads the stent with its balloon into an artery. When the site of the blockage is reached, the doctor inflates the balloon, pressing the stent against the artery wall. Then the balloon is deflated and the catheter and balloon withdrawn, leaving the stent flush against the artery, holding the vessel open.

Some doctors said the balloon stuck on the stent when they were removing it. Some were able to free the balloon; some were not. Dr. William Campbell, director of the cardiac catheterization laboratory at Borgess Heart Institute in Kalamazoo, Mich., said a patient was rushed into emergency open heart surgery to remove the balloon and stent. Others, like Dr. Alejandro Prieto of Michigan State University, said that the balloon did not deflate and that he had to use a sharp wire to pop it. But then he also punctured the patient's artery.

"Those are serious problems," said Dr. Schultz, who said the F.D.A. had received similar reports.

Dr. Andrew Carter of Providence St. Vincent's Medical Center in Portland, Ore., said that while his medical center had used several hundred taxus stents without incident he was nonetheless worried.

"I have never had a balloon that did not deflate, or a device entrapment," he said, referring to balloons that got stuck on stents. "You would never expect to see it. Period."

6. In 1937, a drug manufacturer attempted to modify sulfanilimide, an antibiotic for streptococcal infections, so that it was easier for children to take. Sulfanilimide had been used safely as a pill for years; however, most children can't swallow pills. A company in Tennessee found they could dissolve the drug in ethylene glycol (antifreeze). The company tested their new solution for flavor, appearance, fragrance, but NOT for toxicity. They proceeded to ship it all over the country. Within weeks, many children had died.

   a. Was this legal at the time?

   b. How and when were federal laws reformed to prevent this from happening in the future?

7. Suppose we wish to track the progress of a promising new drug through all stages of development and testing.

   a. How long does it typically take for a promising new drug to go from the research laboratory to the market in the United States?

   b. Describe the phases of study researchers must go through to develop a new drug before it can be marketed. For those phases of study which involve giving the drug to patients, give the typical number of patients involved and the goals of the clinical trial.

   c. What fraction of promising drug candidates actually make it to the market?

   d. What is the cost of developing a new drug today in the USA?

   e. Recently, several drugs to treat arthritis were withdrawn from the market or given a black box warning. Why were the problems with these drugs not discovered until after the FDA had approved their sale?

8. Contrast the role of the NIH and industry in providing funding to support medical research in the United States.

9. Read the article below and answer the following questions.

Experiment: Closed-Heart Surgery
Associated Press 16:30 PM Apr, 01, 2006. Used with permission from the Associated Press. Copyright © 2009. All rights reserved.

   Dr. Samuel Lichtenstein cut a 2-inch hole between an elderly man's ribs. Peering inside, he poked a pencil-sized wire up into the chest, piercing the bottom of the man's heart. Within minutes, Bud Boyer would have a new heart valve – without having his chest cracked open. Call it closed-heart surgery. "I consider it some kind of magic," said Boyer, who left the Vancouver, British Columbia, hospital a day later and was almost fully recovered in just two weeks.

   In Michigan, Dr. William O'Neill slipped an artificial valve through an even tinier opening. He pushed the valve up a patient's leg artery until it lodged in just the right spot in the still-beating heart.

   The dramatic experiments, in a few hospitals in the United States, Canada and Europe, are designed to find easier ways to replace diseased heart valves that threaten the lives of tens of thousands of people every year. The experiments are starting with the aortic valve that is the heart's key doorway to the body.

   The need for a less invasive alternative is great and growing. Already, about 50,000 people in the U.S. have open-heart surgery every year to replace the aortic valve. Surgeons saw the breastbone in half, stop the heart, cut out the old valve and sew in a new one. Even the best patients spend a week in the hospital and require two months or three months to recuperate. Thousands more are turned away, deemed too ill to survive that operation and out of options. Demand is poised to skyrocket as the baby boomers gray; the aortic valve is particularly vulnerable to rusting shut with age. The new experiments are a radical departure from that proven, if arduous, surgery.

   The artificial valves do not even look like valves, squished inside metal cages until they are wedged into place. Barely 150 of any type have been implanted worldwide, most in the last year. It is unclear if they will work as well as traditional valve replacements, which last decades.

   For now, the only patients who qualify for these valves are too sick to be good candidates for regular valve replacement.

   Some deaths during the earliest attempts at implanting the devices forced doctors to come up with safer techniques. Clinical trials apparently are back on track, and even the most skeptical cardiologists and heart surgeons are watching how these pioneers fare. The hope is that one day, replacing a heart valve could become almost an overnight procedure.

   "There's lots of technical challenges that need to be overcome," said Dr. Robert Bonow, a valve specialist at Northwestern University, who is monitoring the research for the American Heart

Association. "Most of us do think this is the future," he said.

O'Neill's first successful patient in March celebrated the one-year anniversary of his through-the-leg implant. "I call it a new birthday," chuckled Fred Grande, 78, a Richmond, Michigan, car collector who took one of his beloved models for a fast spin less than a week after the procedure.

"That's the home run we want to hit with all the patients," said O'Neill, cardiology chief at William Beaumont Hospital in Royal Oak, Michigan.

"It's gratifying" to watch people once deemed beyond help bounce back, added Dr. Jeffrey Moses of New York-Presbyterian Hospital/Columbia University, who with O'Neill is leading the U.S. study. One of Moses' first patients is playing golf at age 92.

The heart has four valves – one-way swinging doors that open and close with each heartbeat to ensure blood flows in the right direction. More than 5 million Americans have moderate to severe valve disease, where at least one valve does not work properly, usually the aortic or mitral valves. Worldwide, roughly 225,000 valves are surgically replaced every year.

Topping that list is the aortic valve. It can become so narrowed and stiff that patients' hearts wear out trying harder and harder to push oxygen-rich blood out to the rest of the body. Calcium deposits accumulate on its tender leaflets. Touch one chipped out of a patient and it feels almost like a rock.

With minimally invasive valve replacement, doctors do not remove that diseased valve. Instead, they prop it open and wedge an artificial one into that rigid doorway.

"It's ironic. You use the disease process to actually help hold your valve in place," said Lichtenstein, of St. Paul's Hospital in Vancouver, who helped create the between-the-ribs method.

Edwards LifeSciences in Irvine, California, the biggest maker of artificial heart valves, and Paris-based CoreValve are testing versions of a collapsible valve made of animal tissue that is folded inside a stent, a mesh-like scaffolding similar to those used to help unclog heart arteries.

The difference is how doctors get the new valve to the right spot, pop open its metal casing and make it stick.

The U.S. studies thread the Edwards valve through a leg artery up to the heart, known as "percutaneous valve replacement." Unlike with open-heart surgery, doctors do not stop the patient's heart. So the trickiest part is keeping regular blood flow from washing away the new valve before it is implanted.

Once the device is almost in place, doctors speed the heartbeat until normal pumping pauses for mere seconds – and quickly push the new valve inside the old one. Inflating a balloon widens the metal stent to the size of a quarter, lodging it into place and unfolding the new valve inside, which immediately funnels the resuming blood flow.

So far, 19 Americans have been implanted this way, plus more than 80 other people worldwide, most of them in France by the procedure's inventor, Dr. Alain Cribier, and in Vancouver by Lichtenstein's colleague, Dr. John Webb.

Fourteen people in Canada, Germany and Austria have received the Edwards valve through the ribs. That is a more direct route to the heart for patients whose leg arteries are too clogged to try the other experiment. Doctors make a tiny hole in the bottom of the heart muscle so the new valve can enter. Then they use the same balloon technique to wedge it inside the old valve.

Talks have begun with the Food and Drug Administration about opening a similar U.S. study later this year.

CoreValve's slightly different valve is being tested in Europe and Canada. It, too, is threaded up the leg artery. But it is made of pig tissue instead of horse tissue and has a self-expanding stent that requires no balloon. Doctors remove a sheath covering it and the stent's metal alloy, warmed by the body, widens until it lodges tight against the old, rocky valve. More than 45 have been implanted; CoreValve hopes to begin a U.S. study next year. Lead researcher Dr. Eberhard Grube of The Heart Center in Siegburg, Germany, expects within months to begin testing a newer version small enough to thread through an artery at the collarbone, another more direct route to the heart.

The experiments come with some significant risks. Edwards temporarily halted the U.S. study last year after four of the first seven U.S. patients died. Initially, doctors threaded the valve up a leg vein, not an artery, a route that required tortuous turns inside the heart and sometimes damaged a second valve, O'Neill said. Twelve people have been implanted since the study restarted in December using the artery route considered easier and safer. All but one have survived and are faring well, researchers say.

O'Neill and Moses – plus doctors at a third hospital, the Cleveland Clinic – have government permission to implant eight additional patients in the U.S. pilot study, which will be expanded if it goes well.

CoreValve's first four patients died as doctors struggled to develop and learn the through-the-artery technique, Grube said. For doctors, pushing the large valve through tiny, twisting arteries – against regular blood flow and guided by X-rays – is laborious. Occasionally, they are not able to wedge it into position. Because they are squeezing a round valve into an irregular-shaped opening, there is a risk that the new valve will leak blood backward into the heart, also problematic.

But once researchers master how to get the valve into place safely, the question becomes how much recipients benefit. Do these very ill patients live longer than expected? If not, does quality of life improve enough to warrant the procedure anyway?

Three of French inventor Cribier's original patients have lived 2 1/2 years so far, with a "return to normal life and no sign of heart failure,"

he said. Eleven others have lived a year and counting. CoreValve reports five patients faring well a year later. Aside from those who did not survive the implantation, others have died from their advanced illnesses even though their new valve was working.

It is the cases of astounding successes – Grande and Boyer, for example – that have other heart specialists taking note, Northwestern's Bonow said.

"Patients have to know what they're getting into," he said. Many of the seriously ill are willing to chance the experimental procedure because "they're so debilitated and . . . there have been some good examples of patients who have gotten better." The bigger challenge, Bonow added, is whether to expand the studies to include less sick patients who could survive open-heart valve replacement but want to avoid its rigors. Already, there are such patients clamoring to be included. That is a difficult decision because even 80- and 90-year-olds successfully can have regular valve replacement. When performed by the most skilled surgeons, risk of death from the operation is about 2 percent – but in less experienced hands, it can reach 15 percent, Bonow said.

Just as using a balloon to unclog heart arteries is sometimes done on patients who would fare better with bypass surgery, researchers eventually will have to ask if patients would accept a less-than-perfect aortic valve if they could skip surgery's pain and risks, said Dr. Michael Mack of Medical City Hospital in Dallas. "There is a trade-off, and how you make that trade-off is a totally gray area," he said.

But Vancouver's Boyer, who had two previous open-heart surgeries for clogged arteries, said avoiding that kind of pain is not a trivial issue for patients. "They're doing something to the field of medicine that's going to make life a hell of a lot easier to people who've got that problem," said a grateful Boyer, describing how he could finally breathe easy after the through-the-ribs valve implant. "I think I'll have a bunch of other parts go bad before I have a problem with this."

    a. Discuss the factors which are likely to affect the diffusion of this technology. Do these factors always benefit the patient?

    b. Why do you think the sample sizes are so low for the studies reported here? Consider what we learned in Chapter 12 about the trials of the AbioCor artificial heart. What factors do you think the FDA considers in decisions regarding the clinical trials reported here?

10. Consider a new implantable device that does not have any reasonably similar products already in the market.

    a. What class of device would the FDA consider this product?

    b. Would this device need a 501K or PMA approval process?

    c. Based on your answer to part b, describe the remaining steps needed to carry this product through to the market.

References

[1] Merrill RA. Regulation of drugs and devices: an evolution. *Health Affairs*. 1994 May 1, 1994; **13**(3): 47–69.

[2] Administration USFaD. *Evidence on the Safety and Effectiveness of Ephedra: Implications for Regulation*. February 28, 2003.

[3] Specter M. Miracle in a bottle. *The New Yorker*. 2004 February, 2004: 64–75.

[4] Medicine NCfCaA. *Reducing ephedra-related risks*. 2003 [cited June 8, 2007]; Available from: http://nccam.nih. gov/health/alerts/ephedra/022803.htm

[5] Rados C. Ephedra ban: no shortage of reasons. *FDA Consumer*. 2004 March–April 2004.

[6] Administration USFaD. *Innovation or stagnation?: Challenge and opportunity on the critcal path to new medical products*. In: Services USDoHaH, ed. 2004.

[7] Administration USDoFaD. *History of the FDA*. [cited 2007 June 8, 2007]; Available from: http://www.fda.gov/oc/history/historyoffda/

[8] Administration USFaD. *A brief history of the Center for Drug Evaluation and Research*. November 1997 [cited 2007 June 8, 2007]; Available from: http://www.fda.gov/cder/about/history/Histext.htm.

[9] Administration USFaD. *Thalidomide: important patient information*. July 7, 2005 [cited 2007 June 9, 2007]; Available from: http://www.fda.goc/cder/news/thalidomide.htm

[10] Morris CA, Avorn J. Internet marketing of herbal products. *JAMA*. 2003 September 17, 2003; **290**(11): 1505–9.

[11] Wholey MH, Haller JD. An introduction to the Food and Drug Administration and how it evaluates new devices: establishing safety and efficacy. *Cardiovascular and Interventional Radiology*. 1995 Mar–Apr 1995; **18**(2): 72–6.

[12] Pritchard WF, Jr., Carey RF. U.S. Food and Drug Administration and regulation of medical devices in radiology. *Radiology*. 1997 October 1, 1997; **205**(1): 27–36.

[13] Administration USFaD. *FDA regulatory actions for the CoX-2 selective and non-selective non-steroidal anti-inflammatory drugs (NSAIDs): questions and answers*. April 7, 2005 [cited 2007 June 9, 2007]; Available from: http://www.fda.gov/cder/drug/ infopage/COX2/COX2qa.htm.

[14] Administration USFaD. *COX-2 selective (includes Bextra, Celebrex, and Vioxx) and non-selective non-steroidal anti-inflammatory drugs (NSAIDs)*. July 18, 2005 [cited 2007 June 9, 2007]; Available from: http://www.fda. gov/cder/drug/infopage/COX2/default.htm

[15] Administration USDoFaD. *FDA public health advisory: FDA announces important changes and additional warning for COX-2 and non-selective non-steroidal anti-inflammatory drugs (NSAIDs)*. April 7, 2005 [cited 2007 June 9, 2007]; Available from: http://www.fda.gov/cder/drug/advisory/COX2.htm

[16] Administration USFaD. *FDA News: FDA announces series of changes to the class of marketed non-steroidal anti-inflammatory drugs (NSAIDs)*. April 7, 2005 [cited June 9, 2007]; Available from: http://www.fda.gov/bbs/ topics/news/2005/NEW01171.html

[17] Krumholz HM, Ross JS, Presler AH, Egilman DS. What have we learnt from Vioxx? *BMJ*. 2007 January 20, 2007; **334**(7585): 120–3.

[18] Zarraga IGE, Schwarz ER. Coxibs and heart disease: what we have learned and what else we need to know. *Journal of the American College of Cardiology*. 2007 January 2, 2007; **49**(1): 1–14.

[19] Bogdanich W and Hooker J. From China to Panama, a trail of poisoned medicine. *The New York Times*, May 6, 2007.

Future of bioengineering and world health

As we conclude our study of bioengineering and world health, we stop to examine some of the health challenges that developing countries face and consider how efforts to develop new, appropriate technologies may help address these challenges.

Ten million children under the age of five die every year throughout the world; 98% of these deaths occur in developing countries. This is more than twice the number of children born each year in the USA and Canada combined. It has been estimated that 2/3 of childhood deaths could be prevented today with available technologies feasible for low income countries. Yet, current technologies do not reach millions of children in need – in many cases because these technologies are presently too expensive, require infrastructure that is unavailable (e.g. refrigeration for heat labile vaccines), or cannot be delivered because of a lack of effective healthcare systems [1].

Advances in the biosciences, bioengineering, and public health are responsible for the dramatic gains in life expectancy achieved over the last century. Throughout this book, we have seen that these advances are not equally available to people throughout the world (Figure 16.1). A recent global checkup of human health observes "In far too many countries health conditions remain unacceptably – and unnecessarily – poor" [2].

Figure 16.1. *In the pediatric ward at Kamuzu Central Hospital in Lilongwe, Malawi, there is one nurse on staff for every 80 pediatric patients. Owing to a shortage of staff, children cannot be admitted without a guardian to provide care for them. Because there are not enough beds, most patients must share beds, while their parents sleep on the floor beside them.*

Despite these inequities, 90% of the $70 billion spent each year on health research and development is devoted to diseases that predominantly affect industrialized nations.

It is clear that many of the technologies we have examined in this book – for example, left ventricular assist devices – are simply not available in many

(a)

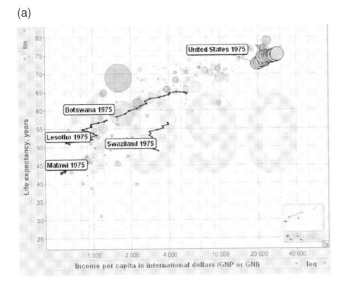

(b)

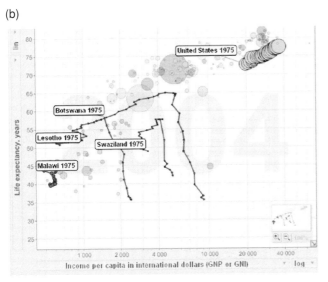

Figure 16.2. *(a) Graph showing changes in life expectancy and per capita income for different countries throughout the world from 1975 to 1990. Progress in four sub-Saharan African countries is tracked over this period. Data from the USA are shown for comparison. (b) Graph showing these changes from 1975 to 2004. Note the dramatic drop in life expectancy due to the AIDS pandemic beginning in 1990. The free software tool at Gapminder World provides an excellent way to explore global changes in income and health metrics over time. http://www.gapminder.org*

developing countries. At the same time, developing countries are facing a growing burden of cardiovascular disease. Furthermore, the impact of **HIV/AIDS** has slowed, and even reversed much of the progress to improve health in many African countries (Figure 16.2).

What stops life saving tools from reaching the world's poorest children? The Disease Control Priorities Project notes ". . . health inequities have arisen largely from uneven adoption and implementation of health interventions associated with technical progress" and cites as priorities developing effective low cost interventions for neglected diseases as well as developing effective strategies to get interventions to neglected people [1].

The country of Swaziland, located in southern Africa, illustrates some of the challenges of **HIV/AIDS** in sub-Saharan Africa. Swaziland has a population of about one million people. In 2005, the HIV prevalence rate among adults was estimated to be a staggering 34%, the highest in the world. Life expectancy at birth in 2006 was only 41 years. More than 70,000 Swazi children have been orphaned as a result of HIV/AIDS [3].

In short, biotechnology and bioengineering research continues to transform the future of healthcare in developed countries, but ensuring that the benefits of research are available to all world citizens requires a new way of thinking, which must incorporate technology development as well as public policy and management of healthcare delivery. Often, due to limited infrastructure, it is not simply enough to provide existing technologies designed for use in developed nations; in many cases a new kind of technology – one which is robust and does not require disposable supplies – is needed to function effectively in the developing world (Figure 16.3).

The growing importance of global health problems has received substantial attention lately. Concerns about worldwide spread of emerging infectious diseases, humanitarian desires for equity in access to healthcare, and the fragile balance of sustaining past advances in world health in the face of increasing conflict and nat-

ural disasters all highlight the importance of increased efforts to address global health disparities. Thanks in large part to the substantial investments and efforts of the Gates Foundation, a number of high-profile scientific planning activities have delineated the research challenges most likely to lead to long term, substantial benefits in global health. These include the need for improved, cost effective point-of-care diagnostic

Departing thoughts: August 3, 2007

Yesterday, a young boy at the clinic lay paralyzed while several of the doctors struggled to keep him alive. His tiny body was on the examining table while beeping noises filled the room. The young boy could no longer breathe on his own. There are no child ventilators in this country. Why? None of the people we spoke to really knew for sure, but most speculated that this was due to a lack of organization of those who are in charge of inventory of medical supplies. The physicians decided to take the child to Bloemfontain, which is the town with the nearest South African hospital. I just thought about how this boy would have surely died if he was not in the care of the doctors at our clinic. The staff at the government hospital seems too overstretched to have ever taken the care to send the child to Bloemfontain where a simple ventilator could be obtained to save this child's life.

There is such needless dying and suffering in this country. And when I hear that many are dying due to carelessness or disorganization – something inside me just burns with anger. The country always runs out of CD4 reagents . . . As a result, I have heard it announced repeatedly throughout my stay here that there are no CD4 counts for the week, which is crucial for monitoring HIV patients and starting them on antiretroviral treatment. In a country where 1/4 of the population has HIV, having CD4 counts available is a necessity. Again, there are speculations that the CD4 reagents run out not because of a lack of funding, but instead because of people who are disorganized.

Another thing that deeply angered and frustrated me today was seeing about 20 spotless Mercedes Benzs lined up at the airport along with several beautiful black Audis and about 20 shiny Toyota Camrys . . . The government here pays for these cars for their highest government officials, while the lower down government officials all get Camrys (but not to worry, starting next year all the Camrys get upgraded to Lexus cars). It is all so wasteful . . . Especially now, after the government has declared a state of emergency. This country suffered from its worst drought in 30 years – many were expected to starve to death or die from malnutrition this year. Why is the government paying for these fancy cars and not spending it on their people who are suffering and dying from a mere lack of food?

The 500,000 rand or so spent on each of these cars could also easily be invested to send more kids to school. There is a lack of access to higher education for too many children of this country because they cannot afford to pay for high school tuition.

Furthermore, while the local government hospital is in shambles, a beautiful new ministry building is being built right next to it. The conditions at the hospital are terrible – it is overcrowded, the staff is too few, the building is too old – there are even cockroaches crawling out of the children's beds during the summer. There is absolutely no excuse for not making the rebuilding of the hospital a priority. People are dying needlessly due to the dreadful conditions of the place. But yet, a beautiful new ministry building is being built . . . and the rebuilding of the hospital will continue to be put off while the government officials keep driving their sparkling cars. . . .

devices, the need for improved vaccines to prevent infectious disease, and the need for novel delivery methods for vaccines and medications.

Lack of educational opportunity also contributes to poor health in the developing world. Young people with little or no schooling are up to twice as likely to contract HIV as those who have completed primary education. Comprehensive solutions to global health challenges must include strategies to expand access both to appropriate technologies and to education.

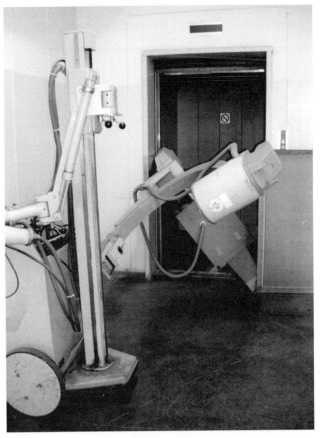

Figure 16.3. *Another challenge of providing appropriate health technology in developing countries is the frequent lack of infrastructure, technical supplies and the difficulties associated with maintaining and repairing instrumentation. The photo, taken at a hospital in Swaziland, illustrates that the availability of technology does not always translate into the ability to use that technology to address health needs. The photo shows an X-ray imaging system in front of the hospital elevator. Both the elevator and the X-ray machine are broken, and parts to repair them could not be obtained. The X-ray machine is now used to block the entrance to the elevator.*

Technologies for the developing world must adapt not only to inadequate resources, but also to unique economic, cultural, social, and environmental realities. Designing technologies like this is an exercise in extreme engineering, which can be approached through both novel high-tech and low-tech solutions. Efforts to design appropriate health technologies can improve care both in developing countries and in wealthy countries as well. New, simple health tools that work in hospitals in developing countries will also perform in rural, remote, and underprivileged healthcare settings in the industrialized world. We have seen that in many developed countries, healthcare costs continue to escalate, partly due to progressively more complex and expensive technologies that offer only incremental health benefit. New, cost effective technologies have the potential to not only reduce inequities in healthcare, but also to reduce the costs of care for all.

Low-tech solutions to health challenges in developing countries

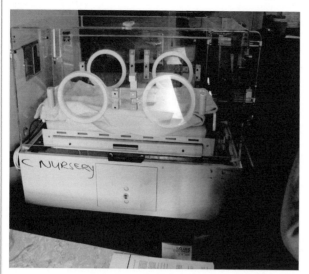

Queen Elizabeth Central Hospital is the main government hospital in Blantyre, Malawi. The neonatal intensive care unit at the hospital has only one commercially made neonatal incubator. Unfortunately, when the thermostat in the incubator broke, no spare parts were available to repair the incubator. Instead, Dr. Liz Molyneaux, chair of the Department of Pediatrics, and her colleagues invented their own incubator, which can be made for less than $100 using locally available materials. The Blantyre Hot Cot consists of a wooden crib, with a hinged Plexiglass cover. Four 60 W light bulbs which can be turned on independently are installed beneath the crib. The bulbs warm the air beneath the baby. Warm air rises up into the crib, and the temperature is controlled by adjusting the number of bulbs which are turned on. The neonatal intensive care unit contains 12 Hot Cot incubators; an excellent example of a low-tech solution which addresses the health challenge in a way that is affordable and can be operated in the current infrastructure.

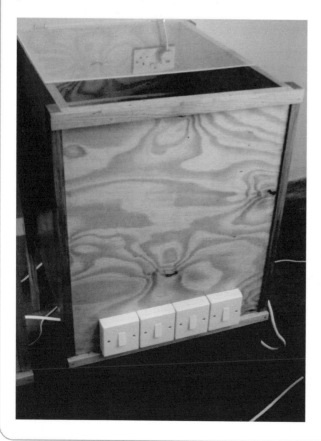

Things I have seen:
August 7, 2007

It has been difficult accepting the fact that my time is over here because there are few times when I have left a foreign place feeling at home. It is true that I have missed my family and close friends, but I am leaving feeling as if I have become acquainted with an incredible family at Baylor. The physicians, the staff, the patients, and the people I have come in contact with have been some of the most amazing individuals I have known, and I will never forget their hospitality and their welcoming spirits. I am forever grateful to the Beyond Traditional Borders program, BIPAI, our program directors and mentors who have created such an incredible opportunity and have allowed me to be part of such a wonderful mission.

I have been told by many who have worked in developing settings or in some sort of volunteer work that those who attempt to teach others or contribute to a problem in some way end up leaving having learned much more than they could have ever taught others or end up gaining much more than they could have been left behind as some sort of contribution. I have found this to be true, and I have learned so much about the people, culture, health, education, challenges, and opportunities of this country in such a short period of time. I have felt such a unique combination of emotions all packed into a series of encounters and experiences that seem to blend into one another like one of the beautiful tapestries woven in rural villages here. It is as if all of life's emotions can be packed into a single day's work – happiness, frustration, empathy, anger, desperation, fulfillment. . . . I could go on and on.

I am leaving with a refreshed and renewed perspective on global health and the complexities that exist when working on problems of such magnitude. I will miss the daily challenge of working on any aspect of HIV/AIDS and the tough questions I asked that ended up consuming my thoughts and conversations late into the nights. I have seen for myself the tragic truth that many speak of . . . of the needless deaths that occur daily and the completely preventable illnesses that young and innocent children die from. I have seen the "accidents of latitude" that Bono and Sachs speak of when they talk about the unthinkable disparities that exist among those who have been born in the developed world and those who have been born in areas such as sub-Saharan Africa. I have seen the struggling face of a baby who died of a simple case of diarrheal infection, and the face of her mother who thought her child was on her way to improvement. I have seen the determined faces of a medical team that went to great lengths to take a child to a South African hospital just to put a baby on a life-saving ventilator, a simple tool they lacked in this country. I have felt the pang of injustice, not injustice I have personally faced, but injustice that I have felt through my close encounters with children, mothers, grandmothers, health professionals, and people from all over the world working in this country. I have seen things and felt emotions that have left a lasting impression, and I only hope that I have been able to contribute a fraction of the impact I have felt myself and that I have been able to leave just one child with a fraction of the knowledge I have gained.

Homework

1. The table shows the global disease burden and the R&D funding for several diseases.
   a. Define DALY. Why is the DALY a better measure of disease burden than mortality rate?
   b. Based on this chart, what recommendations would you make for future funding of R&D funding?

Disease Burden and Funding Comparison

CONDITION	GLOBAL DISEASE BURDEN (million) DALYs*	R&D FUNDING ($ Millions)	R&D FUNDING per DALY*
Cardiovascular	148.190	9402	$63.45
HIV/AIDS	84.458	2049	$24.26
Malaria	46.486	288	$6.20
Tuberculosis	34.736	378	$10.88
Diabetes	16.194	1653	$102.07
Dengue	0.616	58	$94.16

2. Over the past decade, describe the epidemiologic shift in burden of disease that has occurred in developing countries. Given this epidemiologic shift, recommend two specific changes that should be made to R&D funding priorities for chronic diseases. Discuss one challenge associated with each recommendation.

3. You are working for a non-profit organization dedicated to implement new health technologies in the world's least developed countries. The board of directors wants to allocate their 2006 budget to the development of biotechnologies that will most improve health in developing countries. To encourage the successful application of these technologies to global health, they have requested that you conduct a study to determine which biotechnologies could have the largest positive impact on health in developing countries. Your final report must address the following questions.
   a. Mention at least three technologies that you would recommend for implementation.
   b. Explain the potential positive impact that each technology could have on health.
   c. Describe the criteria you used to select these technologies.

4. You are provided with $100 million dollars to make an investment in one of the three following areas.
   a. Launch of a public health program to educate people regarding the risks of lung cancer and the benefits of smoking cessation and which offers free smoking cessation programs.
   b. Development of a new screening test for early lung cancer which costs $100 and has a specificity of 95% and a sensitivity of 98%.
   c. Development of a novel pharmaceutical compound to treat lung cancer. The new treatment reduces side effects by improving the specificity of targeting and improves five year survival for lung cancer patients by 30%.
   Choose one of these options and explain your rationale for selecting this option. Be sure to justify why you feel the option that you selected is a better investment than the other two options. Please be as quantitative as possible in your justification.

5. Reflect on what we have learned about world health in relation to your own personal and career goals. How can YOU work to improve world health? Write down one personal goal illustrating how you will try to improve world health as a result of something you learned in this text.

References

[1] Jones D, Steketee R, Black R, Bhutta Z, Morris S, and The Bellagio Child Survival Study Group. How many child deaths can we prevent this year? *The Lancet.* **362** (9377): 65–71, 2003.

[2] Jamison DT, Breman JG, Measham AR, Alleyne G, Claeson M, Evans DB, Jha P, Mills A, Musgrove P, eds. *Priorities in Health*, The World Bank, Washington, DC, 2006.

[3] World Health Organization, *Key WHO Information*, www.who.int/countries/SWZ/en.

Index